AF213377

NATO ASI Series

Advanced Science Institutes Series

A series presenting the results of activities sponsored by the NATO Science Committee, which aims at the dissemination of advanced scientific and technological knowledge, with a view to strengthening links between scientific communities.

The Series is published by an international board of publishers in conjunction with the NATO Scientific Affairs Division

A Life Sciences	Plenum Publishing Corporation
B Physics	London and New York
C Mathematical and Physical Sciences	Kluwer Academic Publishers
D Behavioural and Social Sciences	Dordrecht, Boston and London
E Applied Sciences	
F Computer and Systems Sciences	Springer-Verlag
G Ecological Sciences	Berlin Heidelberg New York
H Cell Biology	London Paris Tokyo Hong Kong

Series H: Cell Biology Vol. 31

The ASI Series Books Published as a Result of Activities of the Special Programme on CELL TO CELL SIGNALS IN PLANTS AND ANIMALS

This book contains the proceedings of a NATO Advanced Research Workshop held within the activities of the NATO Special Programme on Cell to Cell Signals in Plants and Animals, running from 1984 to 1989 under the auspices of the NATO Science Committee.

The books published as a result of the activities of the Special Programme are:

Vol. 1: **Biology and Molecular Biology of Plant-Pathogen Interactions.** Edited by J. A. Bailey. 1986.

Vol. 2: **Glial-Neuronal Communication in Development and Regeneration.** Edited by H. H. Althaus and W. Seifert. 1987.

Vol. 3: **Nicotinic Acetylcholine Receptor: Structure and Function.** Edited by A. Maelicke. 1986.

Vol. 4: **Recognition in Microbe-Plant Symbiotic and Pathogenic Interactions.** Edited by B. Lugtenberg. 1986.

Vol. 5: **Mesenchymal-Epithelial Interactions in Neural Development.** Edited by J. R. Wolff, J. Sievers, and M. Berry. 1987.

Vol. 6: **Molecular Mechanisms of Desensitization to Signal Molecules.** Edited by T. M. Konijn, P. J. M. Van Haastert, H. Van der Starre, H. Van der Wel, and M. D. Houslay. 1987.

Vol. 7: **Gangliosides and Modulation of Neuronal Functions.** Edited by H. Rahmann. 1987.

Vol. 9: **Modification of Cell to Cell Signals During Normal and Pathological Aging.** Edited by S. Govoni and F. Battaini. 1987.

Vol. 10: **Plant Hormone Receptors.** Edited by D. Klämbt. 1987.

Vol. 11: **Host-Parasite Cellular and Molecular Interactions in Protozoal Infections.** Edited by K.-P. Chang and D. Snary. 1987.

Vol. 12: **The Cell Surface in Signal Transduction.** Edited by E. Wagner, H. Greppin, and B. Millet. 1987.

Vol. 19: **Modulation of Synaptic Transmission and Plasticity in Nervous Systems.** Edited by G. Hertting and H.-C. Spatz. 1988.

Vol. 20: **Amino Acid Availability and Brain Function in Health and Disease.** Edited by G. Huether. 1988.

Vol. 21: **Cellular and Molecular Basis of Synaptic Transmission.** Edited by H. Zimmermann. 1988.

Vol. 23: **The Semiotics of Cellular Communication in the Immune System.** Edited by E. E. Sercarz, F. Celada, N. A. Mitchison, and T. Tada. 1988.

Vol. 24: **Bacteria, Complement and the Phagocytic Cell.** Edited by F. C. Cabello and C. Pruzzo. 1988.

Vol. 25: **Nicotinic Acetylcholine Receptors in the Nervous System.** Edited by F. Clementi, C. Gotti, and E. Sher. 1988.

Vol. 26: **Cell to Cell Signals in Mammalian Development.** Edited by S. W. de Laat, J. G. Bluemink, and C. L. Mummery. 1989.

Vol. 27: **Phytotoxins and Plant Pathogenesis.** Edited by A. Graniti, R. D. Durbin, and A. Ballio. 1989.

Vol. 31: **Neurobiology of the Inner Retina.** Edited by R. Weiler and N. N. Osborne. 1989.

Neurobiology of the Inner Retina

Edited by

Reto Weiler

Department of Neurobiology, University of Oldenburg
P.O. Box 2503, 2900 Oldenburg, FRG

Neville N. Osborne

Laboratory of Ophthalmology, Oxford University
Walton Street, Oxford OX2 6AW, U.K.

Springer-Verlag Berlin Heidelberg New York
London Paris Tokyo Hong Kong
Published in cooperation with NATO Scientific Affairs Division

Proceedings of the NATO Advanced Research Workshop on the Neurobiology of the Inner Retina held in Oldenburg, FRG, from August 8–12, 1988

ISBN-13:978-3-642-74151-7 e-ISBN-13:978-3-642-74149-4
DOI: 10.1007/978-3-642-74149-4

Library of Congress Cataloging-in-Publication Data. NATO Advanced Research Workshop on the Neurobiology of the Inner Retina (1988 : Oldenburg, Germany) Neurobiology of the inner retina / edited by Reto Weiler, Neville N. Osborne. p. cm.—(NATO ASI series. Series H, Cell biology ; vol. 31) "Proceedings of the NATO Advanced Research Workshop on the Neurobiology of the Inner Retina held in Oldenburg, FRG, from August 8–12, 1988"—T.p. verso. "Published in cooperation with NATO Scientific Affairs Division." Includes index. ISBN-13:978-3-642-74151-7 (U.S.)
1. Retina—Physiology–Congresses. 2. Retina—Cytochemistry—Congresses. 3. Molecular neurobiology—Congresses. I. Weiler, Reto, 1947-. II. Osborne, Neville N. III. North Atlantic Treaty Organization. Scientific Affairs Division. IV. Title. V. Series. [DNLM: 1. Retina—cytology—congresses. 2. Retina—physiology—congresses. WW 270 N279n 1988] QP479.N36 1988 612.8'43—dc 20 DNLM/DLC for Library of Congress 89-11323

This work is subject to copyright. All rights are reserved, whether the whole or part of the material is concerned, specifically the rights of translation, reprinting, re-use of illustrations, recitation, broadcasting, reproduction on microfilms or in other ways, and storage in data banks. Duplication of this publication or parts thereof is only permitted under the provisions of the German Copyright Law of September 9, 1965, in its version of June 24, 1985, and a copyright fee must always be paid. Violations fall under the prosecution act of the German Copyright Law.

© Springer-Verlag Berlin Heidelberg 1989
Softcover reprint of the hardcover 1st edititon 1989

2131/3140-543210 – Printed on acid-free paper

PREFACE

The relatively simple, stratified nature of the retina and its specified use in the visual process has long made it an inviting tissue to study both for its own sake and as a model for the more complex processes of the brain. For these dual purposes, the retina can be thought of as basically consisting of two functional parts. First, the outer retina, comprised of the photoreceptor cells and attendant pigment epithelium, serves to capture the photic energy and convert it into a neurochemical response. Second, the inner layers of the retina, mainly bipolar, amacrine and ganglion cells (and their attendant Müller cells), function more clearly as a typical part of the CNS, transmitting the photic signals to the brain.

Between the 8th and 12th of August 1988 more than seventy scientists from all over the world gathered in Oldenburg (Federal Republic of Germany) for a meeting "The neurobiology of the inner retina" which was devoted entirely to the neural mechanism of the inner synaptic layer of the vertebrate retina. The meeting comprised twenty-three separate lectures and four specially arranged discussion groups. In addition, a number of posters were displayed and a period was allotted specifically for the discussion of these posters. The articles contained in this book will serve as a record of the papers delivered at the Oldenburg Meeting and illustrate the advances made in trying to understand the importance of the diversity of amacrine cell morphology and physiology in retinal function.

All who attended the meeting agreed that it was an intellectually stimulating and satisfying event. This was made possible by NATO who sponsored the advance workshop meeting, and also by the University of Oldenburg who provided the facilities. The meeting highlighted the advances made in our understanding of the functioning of the retina and allowed us to predict and focus on future trends. The personal contacts made will also be to the advantage of future research!

The editors express their sincere thanks to the many people in Oldenburg who gave their time and energy to the organisation of the meeting. We also acknowledge NATO who was the sole contributor of financial support. Finally, we thank the participants, not only for making the meeting successful, but also for producing scholarly manuscripts as manifested in the contents of this book.

Neville N. Osborne and Reto Weiler

TABLE OF CONTENT

POSTER PRESENTATIONS

CONTRIBUTORS

ADOLPH, A.R. — Eye Research Institute and Harvard Medical School, 20 Staniford St., Boston, MA 02114, USA

AMMERMÜLLER, J. — Zoolog. Institute, University of Munich, Luisenstr. 14, 8000 München 2, FRG

ARKIN, M. — Department of Physiology, University of Minnesota, 6-255 Millard Hall, Minneapolis, MN 55455, USA

BAI, Sun-Ho. — Department of Biophysical Sciences, School of Medicine, State University of New York, Buffalo, NY 14214, USA

BALL, A.K. — Department of Anatomy, McMaster University, Hamilton, Ontario, L8N 3Z5, CANADA

BARNETT, N.L. — Laboratory of Ophthalmology, Oxford University, Walton Street, Oxford OX2 6AW, UK

BEHRENS, U.D. — Department of Neurobiology, University of Oldenburg, P.O.Box 2503, 2900 Oldenburg, FRG

BERG, K. — Institute of Pathology, University of Tübingen, Liebermeisterstraße 8, 7400 Tübingen, FRG

BOELEN, M.K. — Neuroscience Research Unit, Department of Biology, University of Wollongong, Wollongong, NSW 2500, AUSTRALIA

BRECHA, N.C. — Department of Anatomy and Cell Biology and Department of Medicine, UCLA School of Medicine and Veterans Administration Medical Center, Los Angeles, CA 90073, USA

BRUNKEN, W.J. — Department of Biology, Boston College, Boston, MA 02167, USA

CALAS, A. — Université de Bordeaux I, Bordeaux, FRANCE

CAPP, A.J. — Imperial College (University of London), Department of Pure and Applied Biology, Neurobiology Group, London SW7 2BB, UK

CARRAS, P.L. — School of Optometry, Indiana University, Bloomington, IN 47405, USA

CASSIDY, C. — Departments of Physiology and Neurosurgery, Harvard Medical School, Boston, MA 02115, USA

CHUBB, I.W. — Neuroscience Research Unit, Department of Biology, University of Wollongong, Wollongong, NSW 2500, AUSTRALIA

CHUN, M.H. — Department of Neuroanatomy, Max-Planck-Institute for Brain Research, Deutschordenstr. 46, 6000 Frankfurt 71, FRG

COLEMAN, P. — Department of Physiology, University of Minnesota, 6-255 Millard Hall, Minneapolis, MN 55455, USA

COLLIN, S.P. National Vision Research Institute of Australia, Carlton, Vic. 3053, and Vision Touch & Hearing Research Centre, University of Queensland, St. Lucia, Old. 4067, AUSTRALIA

CRISWELL, M.H. Northeastern State University, Tahlequah, OK 74464, USA

CUNNINGHAM, J.R. Department of Pharmacology, United Medical and Dental Schools of Guy's and St. Thomas's Hospitals (UMDS), St. Thomas's Hospital, London SE1 7EH, UK

DATUM, K.H. Laboratories of the Max-Planck-Institute for Physiological & Clinical Research, Mathildenstr. 8, 8000 München 2, FRG

DAW, N.W. Department of Cell Biology and Physiology, Washington University, Medical School, St. Louis, MO 63110, USA

DEPLANO, S. Institute of Comparative Anatomy, University of Genoa, Viale Benedetto XV, 5, 16132 Genova, ITALY

DeVOE, R.D. School of Optometry, Indiana University, Bloomington, IN 47405, USA

van DIJK, B.W. The Netherlands Ophthalmic Research Institute, P.O.Box 12141, 1100 AC. Amsterdam-Zuidoost, NL

DJAMGOZ, M.B.A. Imperial College (University of London), Department of Pure and Applied Biology, Neurobiology Group, London SW7 2BB, UK

DOWNING, J.E.G. Imperial College (University of London), Department of Pure and Applied Biology, Neurobiology Group, London SW7 2BB, UK

DOWTON, M. Neuroscience Research Unit, Department of Biology, University of Wollongong, Wollongong, NSW 2500, AUSTRALIA

DAHLHEIM, P. Laboratories of the Max-Planck-Institute for Physiological and Clinical Research, Mathildenstr. 8, 8000 München 2, FRG

EHINGER, B. Department of Ophthalmology, University of Lund, 22101 Lund, SWEDEN

FAMIGLIETTI, E.V. Department of Anatomy and Lions' Sight Centre, University of Calgary, Faculty of Medicine, Calgary, Alberta, T2N 4N1, CANADA

FRIEDMAN, D.L. Department of Neurobiology and Anatomy, University of Texas Medical School, P.O.Box 20708, Houston, TX 77225, USA

GHAZI, H. Laboratory of Ophthalmology, Oxford University, Walton Street, Oxford OX2 6AW, UK

GLÄSENER, G. Institute of Zoology, TH Darmstadt, Schnittspahnstr. 3, 6100 Darmstadt, FRG

GUY, R.G.	Royal Melbourne Institute of Technology, Victoria, AUSTRALIA
IKEDA, H.	Vision Research Unit, The Rayne Institute, St. Thomas's Hospital, London SE1 7EH, UK
ISHIDA, A.T.	Department of Animal Physiology, University of California, Davis, CA 95616, USA
ISHIMOTO, I.	Department of Ophthalmology, Osaka University Medical School, 1-1-56 Fukushima, Fukushimaku, Osaka 553, JAPAN
JENSEN, R.J.	Southern College of Optometry, Memphis, TN 38104, USA
KAMERMANS, M.	The Netherlands Ophthalmic Research Institute, P.O.Box 12141, 1100 AC Amsterdam-Zuidoost, NL
KANEKO, A.	National Institute for Physiological Sciences, Okazaki 444, JAPAN
KATO, S.	Department of Neurophysiology, Neuroinformation Reserch Institute, University of Kanazawa, School of Medicine, 13-1 Takara-machi, Kanazawa 920, JAPAN
KAY, C.	Vision Research Unit, The Rayne Institute, St. Thomas's Hospital, London SE1 7EH, UK
LAM, D.M.K.	Alice McPherson Laboratory of Retina Research, The Center for Biotechnology, Baylor College of Medicine, 4000 Research Forest Drive, The Woodlands, TX 77381, USA
LEE, B.B.	Department of Neurobiology, Max-Planck-Institute for Biophysical Chemistry, 3400 Göttingen, FRG
LINKE, R.	University of Bremen, Department of Biology, P.O. Box 330 440, 2800 Bremen, FRG
LOW, J.C.	Imperial College (University of London), Department of Pure and Applied Biology, Neurobiology Group, London SW7 2BB, UK
LUKASIEWICZ, P.D.	Graduate Group in Neurobiology, University of California, Berkeley, CA 94720, USA
MAGUIRE, G.	Neurobiology Group, University of California, Berkeley, CA 94720, USA
MAITRE, M.	Centre de Neurochimie du CNRS and INSERM, Strasbourg, FRANCE
MANTEUFFEL, G.	University of Bremen, Department of Biology, P.O. Box 330 440, 2800 Bremen 33, FRG
MARC, R.E.	University of Texas Graduate School of Biomedical Sciences, Sensory Sciences Center, 6420 Lamar Fleming, Houston, TX 77030, USA

MARTIN-MARTINELLI, E.	Laboratoire de Neuro-Cytologie Oculaire, INSERM U-86, 15 rue de l'Ecole de Médecine, 75270 Paris 06, FRANCE
MASLAND, R.H.	Departments of Physiology and Neurosurgery, Harvard Medical School, Boston, MA 02115, USA
MASSEY, S.C.	Sensory Sciences Center, University of Texas, Graduate School of Biomedical Sciences, Houston, TX 77030, USA
McREYNOLDS, J.S.	Department of Physiology, The University of Michigan, Ann Arbor, MI 48109, USA
MILLAR, T.J.	Save Sight & Eye Health Institute, Department of Ophthalmology, Sydney Eye Hospital, Sir John Young Crescent, Woolloomooloo, NSW 2011, AUSTRALIA
MILLER, R.F.	Department of Physiology, University of Minnesota, 6-255 Millard Hall, Minneapolis, MN 55455, USA
MORGAN, I.G.	Centre for Visual Sciences, and Visual Sciences Group, Research School of Biological Sciences, Australian National University, Acton ACT 2601, AUSTRALIA
MÜLLER, F.	Department of Neuroanatomy, Max-Planck-Institute for Brain Research, Deutschordenstr. 46, 6000 Frankfurt 71, FRG
NEAL, M.J.	Department of Pharmacology, United Medical and Dental Schools of Guy's and St. Thomas's Hospitals (UMDS), St. Thomas's Hospital, London SE1 7EH, UK
NEGISHI, K.	Department of Neurophysiology, Neuroinformation Research Institute, University of Kanazawa School of Medicine, 13-1 Takara-machi, Kanazawa 920, JAPAN
NGUYEN-LEGROS, J.	Laboratoire de Neuro-Cytologie Oculaire, INSERM U-86, 15, rue de l'Ecole de Médecine, 75270 Paris 06, FRANCE
O'MALLEY, D.M.	Departments of Physiology and Neurosurgery, Harvard Medical School, Boston, MA 02115, USA
OSBORNE, N.N.	Laboratory of Ophthalmology, Oxford University, Walton Street, Oxford OX2 6AW, UK
PAN, Z.H.	Department of Biophysical Sciences, School of Medicine, State University of New York, Buffalo, NY 14214, USA
PINTO, L.H.	Department of Neurobiology and Physiology, Northwestern University, Evanston, IL 60201, USA
REDBURN, D.A.	Department of Neurobiology and Anatomy, University of Texas, Medical School, P.O. Box 20708, Houston, TX 77225, USA
RICKMAN, D.W.	Department of Anatomy and Cell Biology, University of Southern California, School of Medicine, Los Angeles, CA 90033, USA

ROBBINS, J.	Vision Research Unit, The Rayne Institute, St. Thomas's Hospital, London SE1 7EH, UK
SAVY, C.	Laboratoire de Neuro-Cytologie Oculaire, INSERM U-86, 15, rue de l'Ecole de Médecine, 75270 Paris 06, FRANCE
SEILER, M.	Eye Research Institute of Retina Foundation, 20 Staniford Str., Boston, MA 02114, U.S.A.
SHAH, M.A.	Department of Pharmacology, United Medical and Dental Schools of Guy's and St. Thomas's Hospitals (UMDS), St. Thomas's Hospital, London SE1 7EH, UK
SLAUGHTER, M.M.	Department of Biophysical Sciences, School of Medicine, State University of New York, Buffalo, NY 14214, USA
SPEKREIJSE, H.	The Netherlands Ophthalmic Research Institute, P.O. Box 12141, 1100 AC Amsterdam-Zuidoost, NL
STELL, W.K.	Department of Anatomy, The University of Calgary, Calgary, Alberta, T2N 4N1, CANADA
STERNINI, C.	Centre for Ulcer Research & Education. UCLA School of Medicine and VAMC-West, Los Angeles, CA 90073, USA
TACHIBANA, M.	National Institute for Physiological Sciences, Okazaki 444, JAPAN
TERANISHI, T.	Department of Neurophysiology, Neuroinformation Research Institute, University of Kanazawa, School of Medicine, 13-1 Takara-machi, Kanazawa 920, JAPAN
TURNER, J.E.	Bowman Gray School of Medicine, Wake Forest University, 300 S. Hawthorne Road, Winston Salem, NC 27103, USA
TUTTON, D.A.	Department of Anatomy, McMaster University, Hamilton, Ontario, L8N 3Z5, CANADA
VALLERGA, S.	Istituto di Cibernetica e Biofisica del Consiglio Nazionale delle Ricerche, 16146 Genova, ITALY
VANEY, D.I.	National Vision Research Institute of Australia, Carlton, Vic. 3053, and Vision Touch and Hearing Research Centre, University of Queensland, St. Lucia, Qld. 4067, AUSTRALIA
VOIGT, T.	Department of Neuroanatomy, Max-Planck-Institute for Brain Research, Deutschordenstr. 46, 6000 Frankfurt 71, FRG
WAGNER, H.J.	Institut für Anatomie und Zellbiologie, Philipps Universität Marburg, Robert Koch Str. 6, 3550 Marburg, FRG
WÄSSLE, H.	Department of Neuroanatomy, Max-Planck-Institute for Brain Research, Deutschordenstr. 46, 6000 Frankfurt 71, FRG
WATT, C.B.	Alice McPherson Laboratory of Retina Research, The Center for Biotechnology, Baylor College of Medicine, 4000 Research Forest Drive, The Woodlands, TX 77381, USA

WEILER, R.	Department of Neurobiology, University of Oldenburg, P.O. Box 2503, 2900 Oldenburg, FRG
WERBLIN, F.S.	Neurobiology Group, University of California, Berkeley, CA 94720, USA
WOLBURG, H.	Institute of Pathology, University of Tübingen, Liebermeisterstr. 8, 7400 Tübingen, FRG
YAMADA, M.	University of London, Department of Pure and Applied Biology, Neurobiology Group, London SW7 2BB, UK
YAZULLA, S.	Department of Neurobiology & Behavior, State University of New York, Stony Brook, NY 11794, USA
YELNIK, J.	INSERM U-3, Hôpital de La Salpêtrière, 47 Bd. de l'Hôpital, 75645 Paris 13, FRANCE
YOUNG, H.M.	National Vision Research Institute of Australia, Vic. 3053, and Vision Touch and Hearing Research Centre, University of Queensland, St. Lucia, Qld. 4067, AUSTRALIA
ZRENNER, E.	Laboratories of the Max-Planck-Institute for Physiological and Clinical Research, Mathildenstr. 8, 8000 München 2, FRG

more densely labelled, centering around 14 (Fig. 9). The median (Table 2) is significantly different from that of the Müller cells ($p < 0.01$). The labelled amacrine cells generally have a relatively clear and voluminous cytoplasm, and the nucleus is indented. They emit both horizontal and vertical processes into the inner plexiform layer. Unlabelled amacrine cell bodies with the same characteristics are common, and there appears to be no anatomical criterion that could be useful in singling out the labelled ones. Similarly, many of even the largest amacrine cell process profiles in the inner plexiform layer had zero or only one grain. However, a grain density distribution plot (data not shown) showed that there is also a large population of labelled amacrine process profiles. This agrees well with the observation of labelled amacrine cell bodies.

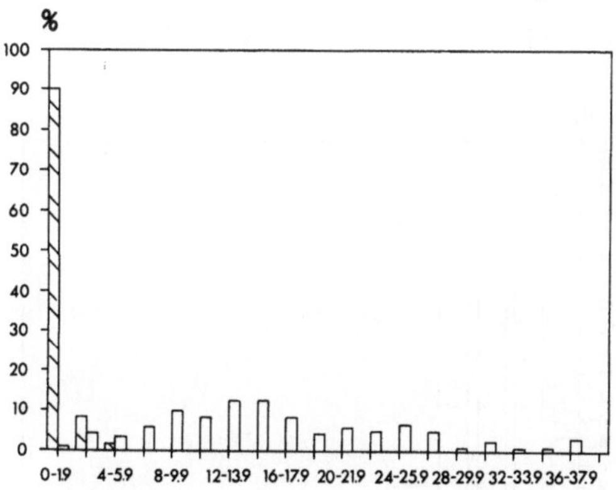

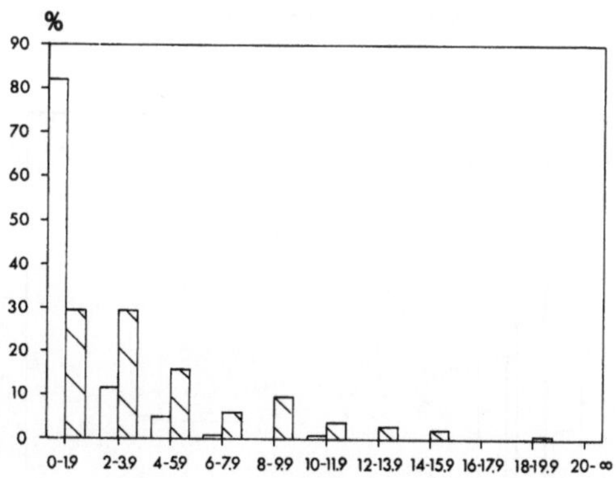

<u>Fig. 7 (left) and 8 (right).</u>
<u>Fig. 7:</u> Labelling densities in 121 bipolar cell processes in the inner plexiform layer (open columns) and 133 Müller cell profiles (cross hatched columns). There is some tendency for the bipolar cell columns to form two peaks, one around 14 and one around 25 grains per square micron.
X axis: grains per square micron. Y axis: percentage of the total count for each individual group.
<u>Fig. 8.</u> Comparison of the distribution of sizes of the processes in Fig. 7. Open columns: Müller cells; hatched columns: bipolar cell processes. The difference between the two populations is not significant.

Labelled amacrine profiles have been found to be presynaptic to labelled and unlabelled ganglion cell profiles, to be both pre- and postsynaptic to other labelled or unlabelled amacrine cell profiles and to be both pre- and postsynaptic to labelled bipolar cell profiles. Reciprocal synapses with bipolar cells have not been seen.

Most of the cell bodies in the ganglion cell layer show some moderate labelling, but many show only very little (Fig. 10). There is no morphological feature to distinguish the two types. Some of the labelled cells may be displaced amacrine cells, but there is also a significant number of moderately labelled nerve fibres (Table 2), which shows that some true ganglion cells are labelled.

Discussion

There is previous strong pharmacological and electrophysiological evidence that photoreceptors use an excitatory neurotransmitter, and glutamate has often been implied (see Ehinger and Dowling 1987). Correspondingly, glutamate immunoreactivity was seen in photoreceptors in this study. The labelling intensity (in turtle) was similar to that of other cells in the retina, with exception for the strongly labelled bipolar cells.

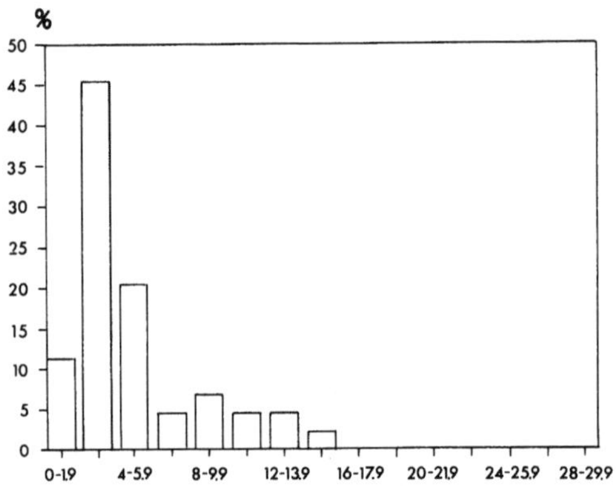

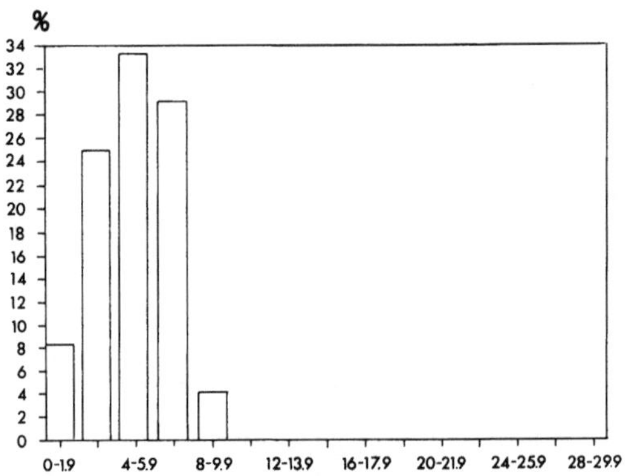

Figs. 9 (left) and 10 (right)
X axis: gold grains per square micron. Y axis: percentage of the total count.
Fig. 9. Labelling densities in 44 amacrine cell bodies. A significant proportion of the cell bodies had a grain count above 3.3.
Fig. 10. Labelling densities in 24 ganglion cell bodies. A significant proportion of the cell bodies had a grain count above 3.3.

Horizontal cells were for the most part unlabelled, but a moderate labelling was seen in a few. The turtle retina was analysed in the electron microscope, and whenever the type could be established, it was seen to be H1. There are no previous results that suggest glutamate as a neurotransmitter of horizontal cells. However, their neurotransmitter(s) is largely unknown. The current results suggest that glutamate should be investigated as a possible neurotransmitter in a minority of the horizontal cells.

The majority of the amacrine cells contain immunoreactive glycine or GABA (see Ehinger and Dowling 1987), and are usually regarded as inhibitory (Miller 1979). However, there is some electrophysiological evidence that amacrine cells may use an excitatory amino acid as their transmitter (Bloomfield and Dowling 1985). The present observations of high glutamate immunoreactivity in a small population and moderate immunoreactivity in a somewhat larger population suggest that some amacrine cells in the turtle retina may use glutamate as their neurotransmitter.

Bipolar cells receive input from amacrine cells in the inner plexiform layer. In many cases this input is at reciprocal synapses, and the transmitter of the amacrine process is then presumed to be inhibitory. Correspondingly, we have not been able to see any reciprocal synapses between bipolar cells and glutamate immunoreactive amacrine cell processes. However, the present observation of glutamate immunoreactive synapses onto bipolar cells suggests there may be excitatory influences by amacrine cells on bipolar cells, but not in reciprocal synapses.

The neurotransmitter of ganglion cells has not been established. However, glutamate is present in significant concentration in the ganglion cell layer of several species (Berger et al 1977, Voaden 1978, Voaden et al 1980). ^{3}H-D-aspartate is thought to label cells operating with glutamate or aspartate as their neurotransmitter, and will label cells in the ganglion cell layer (see Ehinger and Dowling 1987). More recently, Langdon and Freeman (1986) and Crunelli et al (1987) have found evidence that glutamate receptors are involved postsynaptically to the optic nerve. The previous results and the present observation of endogenous immunoreactive glutamate in many cells of the ganglion cell layer and in nerve fibres suggests that glutamate may be a neurotransmitter in many ganglion cells.

Strong glutamate immunoreactivity was found in all bipolar cells in all species investigated. All parts of the cells are about equally labelled. We have not been able to identify with certainty any unlabelled turtle bipolar cell in the electron microscope, despite efforts to do so. The result shows that there are significant concentrations of endogenous glutamate in bipolar cells. The observations therefore strongly support earlier suggestions, based on electrophysiological studies, that glutamate may be a neurotransmitter in bipolar cells (see, e.g., Slaughter and Miller 1983c, Kato, Teranishi and Negishi 1985, Bloomfield and Dowling 1985, Lukasiewics and McReynolds 1985). Further, our results suggest it is the main transmitter.

There are two main functional types of bipolar cells, ON cells and OFF cells, and the two can be distinguished pharmacologically (Slaughter and Miller 1983a-c). Nevertheless, both types must be glutamate immunoreactive because they appear to be at least roughly equal in numbers and there can at most be only a very small minority (if any at all) of bipolar cells which are not glutamate immunoreactive. The features determining whether a bipolar cell is of the ON or OFF type thus appear to reside in the receptors rather than in the type of neurotransmitter the bipolar cells use.

GABA, glycine and 5-hydroxytryptamine (5-HT) are now suspected to be neurotransmitters in bipolar cells (see Ehinger and Dowling 1987), and the number of, for instance, 5-HT bipolar cells in the turtle retina is fairly high (Weiler and Schütte 1985). Since we were unable to prove the existence of bipolar cells without glutamate immunoreactivity, we suspect it resides in bipolar cells which also contain GABA, glycine and 5-HT. The rationale for this coexistence is unclear. In the first case (GABA), glutamate might be present as a GABA precursor, but the explanation(s) must be more complex in the other two cases.

Acknowledgements

The antibody, 13 Glu, was kindly provided by Drs. OP Ottersen and J Storm-Mathisen, Anatomical Institute, University of Oslo. The study was supported by grants EY00811 to JE Dowling and by grants from the Swedish Medical Research Council (project no 0321), the Elsa och Torsten Segerfalks Stiftelse, the H och L Nilssons stiftelse and the Faculty of Medicine, University of Lund, to B Ehinger. Valuable statistical advice was given by dr. Armando Garsd.

References

Berger SJ, McDaniel ML, Carter JG, Lowry OH (1977) Distribution of four potential transmitter amino acids in monkey retina. J Neurochem 28:159-163

Bloomfield SA, Dowling JE (1985) Roles of aspartate and glutamate in synaptic transmission in rabbit retina. II. Inner plexiform layer. J Neurophysiol 53:714-725

Conover, WS, Practical Nonparametric Statistics (1980), second ed, J Wiley & Sons, New York-Chichester-Brisbane-Toronto-Singapore

Crunelli V, Kelly JS, Leresche N, Pirchio M (1987) On the excitatory post-synaptic potential evoked by stimulation of the optic tract in the rat lateral geniculate nucleus. J Physiol (London) 384:603-618.

Ehinger B, Dowling JE (1987) Retinal neurocircuitry and neurotransmission. In: Björklund A, Hökfelt T, Swanson LW (eds) Handbook of Chemical Neuroanatomy, vol. 5, Elsevier: Amsterdam pp 389-446

Ehinger B, Ottersen OP, Storm-Mathisen J, Dowling JE (1988) Bipolar cells in the turtle retina are immunoreactive to glutamate. Proc Natl Acad Sci USA, in press

Kato S, Teranishi T, Negishi K (1985) L-Glutamate depolarizes ON-OFF transient type of amacrine cells in the carp retina: an ionophoretic study. Brain Res 329:390-394

Kolb H, Jones J (1984) Synaptic organization of the outer plexiform layer of the turtle retina: an electron microscope study of serial sections. J Neurocytol 13:567-591

Langdon RB, Freeman JA (1986) Antagonists of glutaminergic neurotransmission block retinotectal transmission in goldfish. Brain Res 398:169-174

Lukasiewicz PD, McReynolds JS (1985) Synaptic transmission at N-methyl-D-aspartate receptors in the proximal retina of the mudpuppy. J Physiol (Lond) 367:99-115

Miller RF (1979) The Neuronal basis of ganglion-cell receptive-field organization of the physiology of amacrine cells. In: Schmitt FO, Worden FG (eds) The Neurosciences. Fourth Study Program, The MIT Press, Cambridge, Mass. and London, pp 227-246

Ottersen OP, Storm-Mathisen J, Madsen S, Skumlien S, Strømhaug J (1986) Evaluation of the immunocytochemical method for amino acids. Med Biol 64:147-158

Ottersen OP (1987) Postembedding light- and electron microscopic immunocytochemistry of amino acids: description of a new model system allowing identical conditions for specificity testing and tissue processing. Exp Brain Res 69:167-174

Slaughter MM, Miller RF (1983a) The role of excitatory amino acid transmitters in the mudpuppy retina: an analysis with kainic acid and N-methyl aspartate. J Neurosci 3:1701-1711

Slaughter MM, Miller RF (1983b) An excitatory amino acid antagonist blocks cone input to sign-conserving second-order retinal neurons. Science 219:1230-1232

Slaughter MM, Miller RF (1983c) Bipolar cells in the mudpuppy retina use an excitatory amino acid neurotransmitter. Nature 303:537-538

Somogyi P, Hodgson AJ (1985) Antiserum to gamma-aminobutyric acid. III. Demonstration of GABA in Golgi-impregnated neurons and in conventional electron microscopic sections of cat striate cortex. J Histochem Cytochem 33:249-257

Somogyi P, Hodgson AJ, Smith AD, Nunzi MG, Gorio A, Wu J-Y (1984) Different populations of GABAergic neurons in the visual cortex and hippocampus of cat contain somatostatin- or cholecystokinin-immunoreactive material. J Neurosci 4: 2590-2603

Sternberger, LA (1979) Immunocytochemistry, 2nd ed., J Wiley & Sons: Chichester, pp 1-354

Storm-Mathisen J, Leknes A K, Bore A T, Vaaland J L, Edminson P, Haug F M S, Ottersen O P (1983) First visualization of glutamate and GABA in neurones by immunocytochemitry. Nature 301:517-520

Storm-Mathisen J, Ottersen OP (1988) Localization of excitatory amino acid transmitters. In: Excitatory Amino Acids in Health and Disease (ed D Lodge), John Wiley & Sons Ltd pp 107-141

Voaden MJ (1978) The localization and metabolism of neuroactive amino acids in the retina. In: Fonnum F (ed) Amino acids as Chemical Transmitters NATO Advanced Study Institute Series, vol. 16, Plenum Press, New York, pp 257-274

Voaden MJ, Morjaria B, Oraedu ACI (1980) The localization and metabolism of glutamate, aspartate and GABA in the rat retina. Neurochem Int. 1:151-166

Weiler R, Schütte M (1985) Morphological and pharmacological analysis of putative serotonergic bipolar and amacrine cells in the retina of a turtle, Pseudemys scripta elegans. Cell Tissue Res 241:373-382

THE RELEASE OF ACETYLCHOLINE AND GABA BY NEURONS OF THE RABBIT RETINA

Richard H. Masland, Charles Cassidy and Donald M. O'Malley
Departments of Physiology and Neurosurgery
Harvard Medical School
Boston, MA 02115
U.S.A.

It has been known for a decade that the rabbit retina contains neurons that synthesize acetylcholine and neurons that synthesize GABA. Until recently, it was thought that they were separate groups of cells. However, several laboratories have now reported that the cholinergic neurons also contain GABA (Vaney & Young, 1988; Kosaka et al, 1988; Brecha et al, 1988; see also Agardh & Ehinger, 1983). If these cells release both an excitatory and an inhibitory neurotransmitter, sophisticated physiological mechanisms become possible. At the least, one can imagine excitation and inhibition of postsynaptic cells simultaneously resulting from their activation. Before proceeding far with such speculations, however, it seemed worthwhile to ask a simpler question: are the two neurotransmitters released together or separately? One can imagine a situation in which acetylcholine and GABA are released by a single cellular mechanism, in which case they would be secreted exactly in parallel. On the other hand, the two neurotransmitters could quite plausibly be released by different mechanisms, or from different sites within the the cell, so that they respond to different physiological inputs. The second possibility is strengthened by the demonstration of carrier-mediated release of GABA from horizontal cells; it provides an alternative to the vesicle-mediated mechanism by which acetylcholine is secreted (Schwartz, 1982, 1987; Yazulla, 1983).

The rabbit retina is a favorable one for studies of neurotransmitter release, because it is unusually thin. Except where the optic nerve fibers run together in the medullary rays, the retina is approximately 100 µm thick. This means that isolated rabbit retinas can be maintained *in vitro* at a physiological temperature, since the distance through which oxygen, substrates,

NATO ASI Series, Vol. H31
Neurobiology of the Inner Retina
Edited by R. Weiler and N.N. Osborne
© Springer-Verlag Berlin Heidelberg 1989

and metabolites must diffuse is relatively short. In electrophysiological experiments, vigorous ganglion cell responses to light have been recorded after more than two days incubation *in vitro* (Ames & Nesbett, 1981). The rabbit retina's thinness also means that compounds released within the inner plexiform layer reach the superfusing fluid quickly; this sharpens the time within which the effects of stimulation may be resolved.

Here we will survey some of our studies of the release of acetylcholine and GABA by isolated rabbit retinas. Our primary purpose is to make an overall comparison between the two neurotransmitters, and the experiments will not be described in detail. (For more detail see Masland et al. 1984a; Masland & Cassidy, 1987). We also will point out certain questions that remain unresolved.

SECRETION OF ACETYLCHOLINE BY THE CHOLINERGIC AMACRINE CELLS

The cholinergic neurons of the rabbit retina are amacrine cells, about half of which have cell bodies in the normal position of amacrine cells and half have cell bodies located in the ganglion cell layer. They form a single morphological class, with distinctive, radiating branches. The major dendritic arbors are confined to two precise strata within the inner plexiform layer, one for dendrites of the conventionally placed amacrine cells and one for the displaced cells (Masland & Mills, 1979; Tauchi & Masland, 1984; Masland et al. 1984b; Vaney, 1984).

If one incubates the retina in the presence of radioactive choline, the choline is transported into the cells on a hemicholinium-sensitive carrier and within the cell is incorporated into acetylcholine. Since choline labeled to a high specific activity is available, the radioactive acetylcholine subsequently released into a superfusing medium can be measured even at high flow rates (Masland et al. 1984a). We incubate the retina for periods ranging from 20 minutes to 2 hours in radioactive choline. The retina is then rinsed for a short period, to clear the labeling choline from the extracellular space and decrease the amount of free radioactive choline within the cells. It is mounted in a superfusion chamber. Samples of the superfusate are collected at timed intervals and their content of radiolabeled acetylcholine is determined by electrophoresis or liquid cation exchange.

Stimulation of the retina by flashing light causes an increased release of acetylcholine into the superfusing medium. The stimulation of release is extremely reproducible, both within a single preparation and from preparation to preparation (figure 1). This series of experiments now includes a total of 142 rabbit retinas; a light-evoked release of acetylcholine was observed in every preparation.

Within the limits of the temporal resolution of the superfusion system, the release of acetylcholine increases immediately after the beginning of the train of light flashes, and ceases immediately when they are turned off. The limits of resolution are about 5 seconds. This is slow compared to the operation of synapses, but is still fast compared to the actions of some peptides.

A train of flashes at 3 Hz is roughly optimal (Masland & Livingstone, 1976; Massey & Neal, 1979; Neal & Massey, 1980). Except under special conditions of superfusion, an effect of stimulation with steady light is barely detectable. The difference between steady light and flashing light suggests that it is the stimulus transients that trigger a release of acetylcholine by the cells. This was confirmed by fast-flow experiments. If the retina is incubated in a chamber that has a small internal volume, and the superfusing fluid is moved rapidly over the tissue, the release of acetylcholine at stimulus transients can be directly observed. A burst of acetylcholine is released at stimulus onset and a second burst at stimulus cessation (Masland et al. 1984a).

As one would expect from the functional subdivision of the inner plexiform layer, the separate on and off releases of acetylcholine come from the two mirror-symmetric cholinergic amacrine cells. The displaced cholinergic amacrine cells, which arborize in the part of the inner plexiform layer dominated by

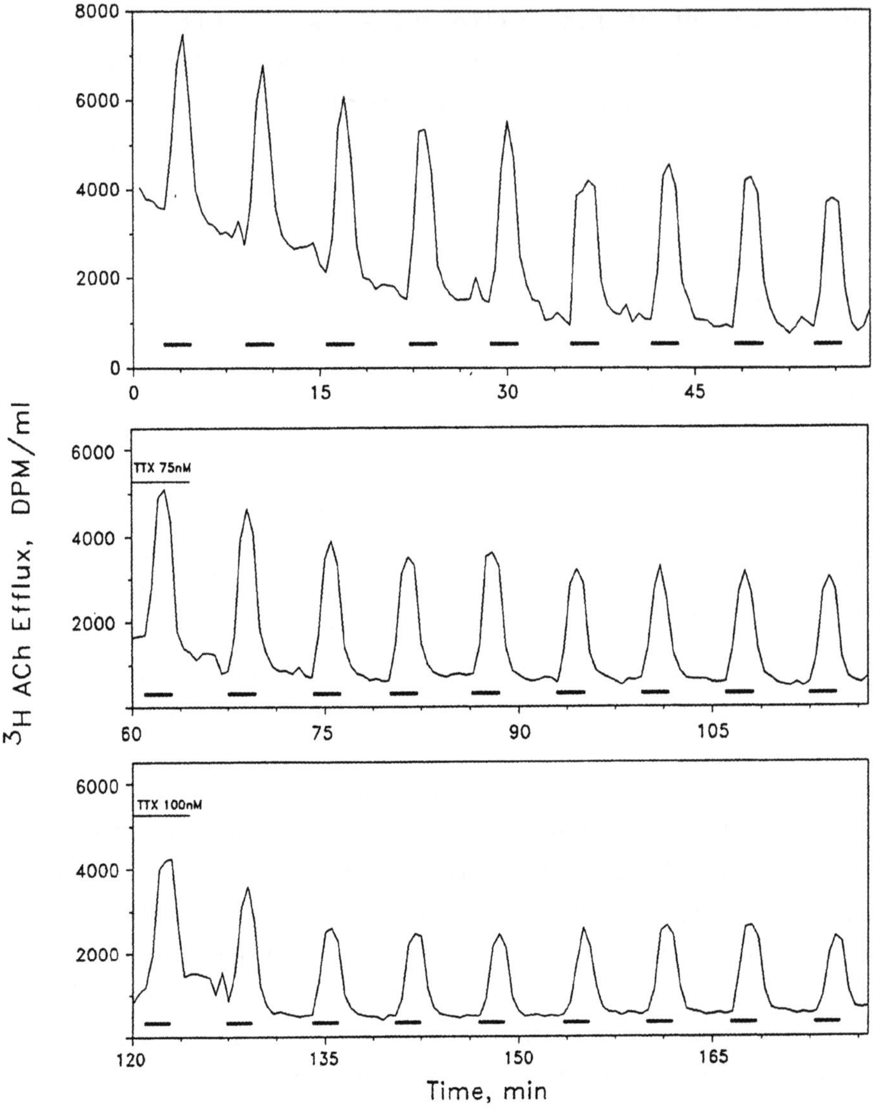

Figure 1. Release of acetylcholine by the rabbit retina. Acetylcholine was identified in the superfusate by liquid cation exchange. At the times indicated by bars the retina was stimulated by flashing light. An increased release of acetylcholine was reliably evoked. Replotted from Masland, Mills and Cassidy (1984a).

on activity, release acetylcholine at stimulus onset. The conventionally placed cholinergic amacrine cells, which arborize in the off-dominated part of the inner plexiform layer, release acetylcholine at stimulus cessation.

The light-stimulated release of acetylcholine depends on the presence of extracellular Ca^{++} ions. Lowering the concentration of Ca^{++} and raising the concentration of Mg^{++} entirely prevents the increased secretion of acetycholine that ordinarily results from stimulation with light. The release of acetylcholine in response to direct depolarization of the cells by K^{+} ions is also prevented.

In summary, the secretion of acetylcholine by the cholinergic amacrine cells appears entirely conventional: the cells respond to activation of the retina by secreting acetylcholine; secretion happens on the time scale usually associated with the retina's physiological activity; the release is phasic, as is much of the inner retina's activity; and the release depends on Ca^{++} ions, as one expects for the Ca^{++}-triggered exocytosis of synaptic vesicles (figure 2).

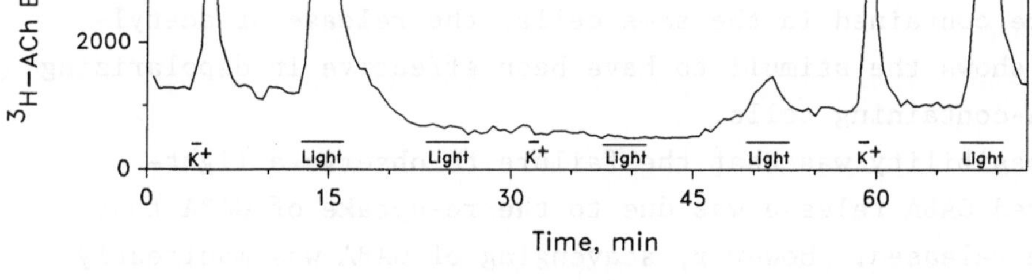

Figure 2. The release of acetylcholine in response to stimulation of the retina by flashing light or depolarization of the cells by 15 mM K+ ions. Where indicated, the incubation medium contained 0.2 mM Ca++ and 20 mM Mg++. Replotted from Masland and Cassidy (1987).

THE CO-RELEASE OF ACETYLCHOLINE AND GABA BY THE CHOLINERGIC AMACRINE CELLS.

To study the co-release of acetylcholine and GABA, we used a different labeling strategy: retinas were incubated in radioactive choline, as before, but were incubated at the same time in radioactive GABA. One compound was labeled with tritium and the other with ^{14}C (different combinations of labels were used for different purposes). The compounds released into the superfusing medium were separately identified by differential scintillation counting. In that way the release of acetylcholine and GABA could be studied simultaneously in a single preparation and directly compared.

The experiment shown in figure 3 summarizes our central findings. When the retina was stimulated by flashing light or by moving bars, the tissue's release of ^{3}H-acetylcholine was vigorously enhanced--just as in the earlier experiments. Stimulation with light never increased the rate of release of ^{14}C-GABA.

Stimuli tested included flashing lights at a wide series of rates and intensities; light and dark bars moving across the retina; and steady light at intensities ranging from scotopic to mid-photopic. The range and variety of the light stimuli tested would by itself make it seem unlikely that we missed the conditions that would have stimulated GABA release. More important, though, is the fact that the same preparations did release acetylcholine in response to light. Since ^{3}H-acetylcholine and ^{14}C-GABA were contained in the same cells, the release of acetylcholine shows the stimuli to have been effective in depolarizing the GABA-containing cells.

A possibility was that the failure to observe a light-stimulated GABA release was due to the re-uptake of GABA that had been released. However, scavenging of GABA was manifestly incomplete since GABA was recovered in the superfusate. As an additional control, further experiments were carried out in the presence of agents known to compete for the GABA carrier: nipecotic acid, SKF 89976-A, or an excess of unlabeled GABA.

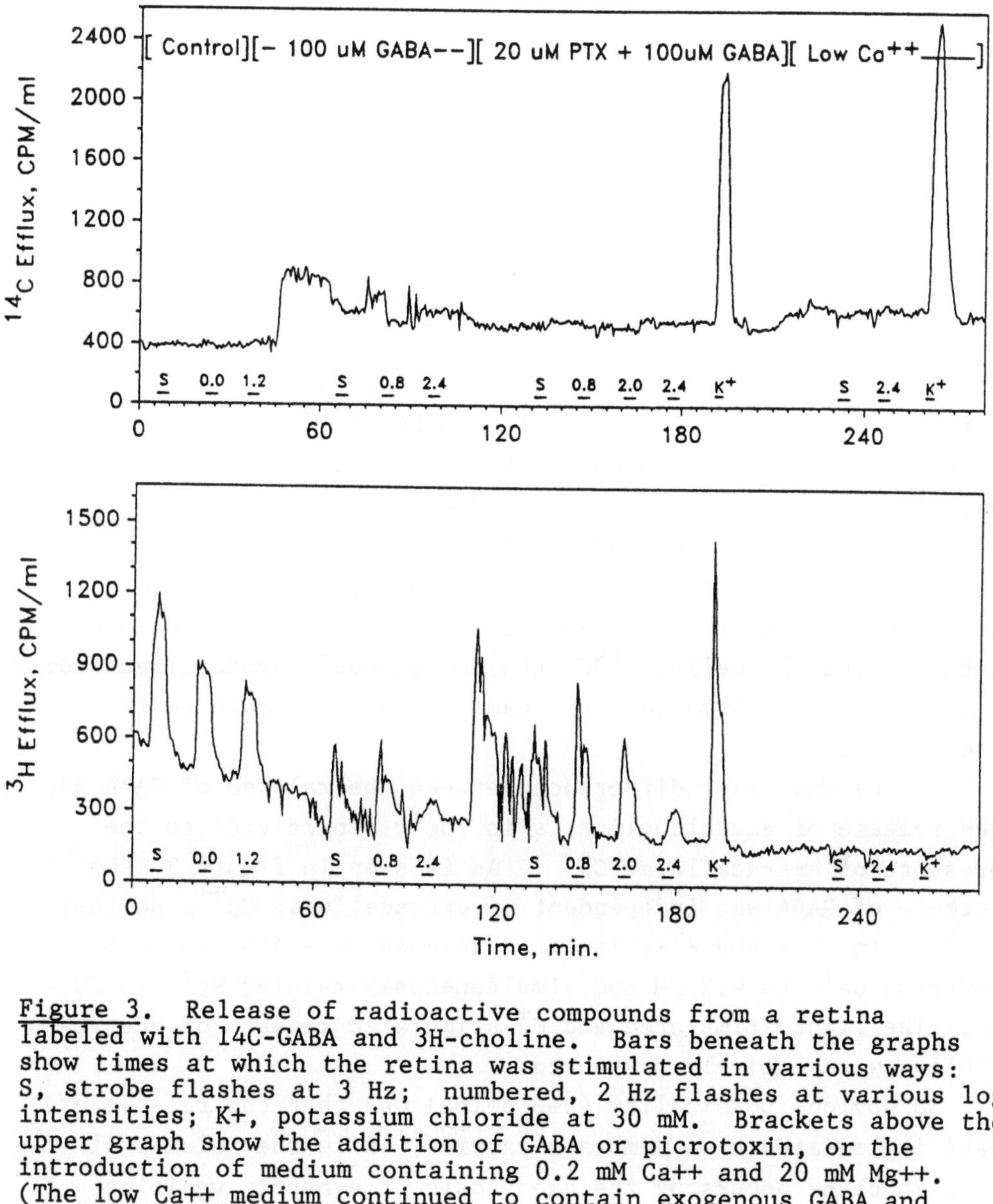

Figure 3. Release of radioactive compounds from a retina labeled with 14C-GABA and 3H-choline. Bars beneath the graphs show times at which the retina was stimulated in various ways: S, strobe flashes at 3 Hz; numbered, 2 Hz flashes at various log intensities; K+, potassium chloride at 30 mM. Brackets above the upper graph show the addition of GABA or picrotoxin, or the introduction of medium containing 0.2 mM Ca++ and 20 mM Mg++. (The low Ca++ medium continued to contain exogenous GABA and picrotoxin.)

Note that picrotoxin caused erratic and unstable activity in the cholinergic neurons, so that acetylcholine was released in spasmodic bursts. Such behavior has only been seen in response to picrotoxin; it shows the importance of GABA in maintaining the physiological stability of the inner retina.

All of them increased the basal release of GABA in the dark, indicating that they do interfere with GABA re-entry. None of them unmasked a light-evoked release of GABA.

The GABA contained in the cells was not sequestered in some sort of unresponsive pool, since the rate of GABA release did rise sharply when the cells were depolarized by K^+ ions. The release caused by depolarization could be demonstrated by an autoradiographic experiment to have come come from the same cells that contain acetylcholine. Labeled retinas were incubated in control medium or in medium that contained 20 mM K^+ (to depolarize the cells) and 1 mM unlabeled GABA (to prevent re-entry of GABA that had been released.) At the end of 1 hour the retinas were fixed and autoradiographed. The cholinergic cells of retinas incubated under GABA-releasing conditions were substantially depleted of radioactivity. Finally, the failure of light to stimulate GABA release was not due to the use of exogenous GABA to load the cells. ^{14}C-GABA endogenously synthesized from ^{14}C-glutamate behaved just the same as GABA accumulated from the medium.

The second major difference between the release of GABA and the release of acetylcholine is in their sensitivity to the presence of extracellular Ca^{++}. As is seen in figure 3, the release of GABA was independent of extracellular Ca^{++}; neither the resting nor the K^+-stimulated release were diminished by lowering Ca^{++} to 0.2 mM and simultaneously raising Mg^{++} to 20 mM. The light-stimulated and K^+-stimulated release of acetylcholine were entirely prevented.

An obvious possibility (see below) was that the release of GABA is mediated by a membrane carrier, while the release of acetylcholine occurs by the exocytosis of synaptic vesicles. If that is the case, the retina's acetylcholine would be stored substantially in vesicles, while the tissue's content of GABA would not. We purified synaptic vesicles from double-labeled retinas, by attaching them to beads coated with antibodies against a synaptic vesicle protein, one alternatively named P38 or synaptophysin (Buckley & Kelly, 1985; Lowe et al. 1988; De Camilli et al. 1988). We found that a substantial fraction

(15 to 33%, depending on the conditions) of the tissue's acetylcholine was pelleted with the synaptic vesicles, while no specific pelleting of radioactive GABA could be detected.

DISCUSSION

Some of our findings allow fairly certain conclusions, but for others the interpretation is less sure. The identity of the cholinergic neurons and the existence of a light-stimulated release of acetylcholine seem beyond much doubt. It also seems clear that those same cells contain GABA and release it into the medium. The two findings most open to question, if only because they clash with the general expectation, are the non-vesicular localization of GABA and the failure of stimulation by light to release GABA from the cells.

How certain can we be that the cholinergic amacrine cells do not have any GABA-containing synaptic vesicles? A possibility that is permitted by our experiments on immunopurified vesicles is that GABA-containing vesicles exist but are more permeable to GABA than the cholinergic vesicles are to acetylcholine. During purification their GABA would leak out. That explanation, however, does not account for the results of the physiological experiments: if GABA were released via vesicles of any kind there should be a Ca^{++}-dependent release of GABA from the living cells, which was not observed. There has in fact been some difficulty in isolating GABA-containing synaptic vesicles from any source (Rassin, 1972; Kontro et al. 1980). When they do appear to have been isolated (from brain), they are not particularly leaky (Fykse & Fonnum, 1988).

A second possibility is that retinal cells (amacrine cells and Muller cells, which both accumulate GABA) have a large pool of cytoplasmic GABA and a very few GABA-containing synaptic vesicles. The cytoplasmic GABA is in equilibrium across the cell on a carrier (see below) and the fluxes are large. They swamp out the much smaller vesicular release of GABA. This is a possibility that we cannot eliminate. If there is such a vesicular pool, however, it must be small indeed, because of the failure to detect GABA in the isolated vesicles and the absence of Ca^{++}-mediated GABA release. It is of course impossible to prove the non-existence of anything. The burden of proof can shift, however, and our inclination is if there are GABA-containing vesicles they mediate only a minor part of the GABAergic neurotransmission in the retina.

The second issue is the lack of effect of stimulation with light upon the rate of release of GABA. Here the amacrine cells contrast with the horizontal cells, where a light-induced inhibition of GABA release has been demonstrated (Yazulla, 1985). Since there is clear electrophysiological and biochemical evidence for the inhibition of mammalian inner retinal neurons by GABA, one must assume that it is a retinal neurotransmitter (Ames & Pollen, 1969; Ariel & Daw, 1982; Massey & Redburn, 1982; and Cunningham & Neal, 1983). A possibility is that its release is controlled by events quite different from those that control the release of acetylcholine; if the mechanisms of release of the two neurotransmitters are different they can be controlled independently, and the control of GABA release need not have anything to do with light. GABA might, for example, have a homeostatic role in controlling the overall excitability of the inner retina.

Alternatively, the different mechanisms of GABA and acetylcholine release could have differing "gains", so that a given amount of depolarization releases more acetylcholine than GABA. If the release of GABA is mediated by a carrier, GABA should be considered as being in equilibrium across the cell membrane, with a maintained concentration of GABA intracellularly and a maintained concentration extracellularly (Iversen, 1971; Wheeler, 1984; Kanner & Schuldiner, 1987). Since Na^{+} is electrogenically co-transported with GABA, the point of equilibrium depends directly on the membrane potential. The exocytosis of synaptic vesicles, however, results from a Ca^{++} influx caused by depolarization. At that stage amplification is possible, and a large effect on vesicle fusion can result. This is perhaps the simplest explanation of our results, since it requires postulation of no new mechanisms. The question that it leaves is why the amount of GABA that is released should be so much less--possibly several orders of magnitude less--than the amount of acetylcholine.

REFERENCES

Agardh E, Ehinger B (1983) Retinal GABA neuron labelling with [3H]isoguvacine in different species. Exp Eye Res 36: 215--229
Ames A, Nesbett FB (1981) In vitro retina as an experimental model of the central nervous system. J Neurochem 37:867-877.
Ames A, Pollen DA (1969) Neurotransmission in central nervous tissue: A study of isolated rabbit retina. J Neurophysiol 32:424-442.
Ariel M, Daw NW (1982) Pharmacological analysis of directionally sensitive rabbit retinal ganglion cells. J Physiol (London) 324:161-185.
Brecha N, Johnson D, Peichl L, Wassle H (1988) Cholinergic amacrine cells of the rabbit retina express GABA and GAD immunoreactivity. Proc Natl Acad Sci (in press).
Buckley K, Kelly RB (1985) Identification of a transmembrane glycoprotein specific for secretory vesicles of neural and endocrine cells. J Cell Biol 100:1284-1294.
Cunningham JR, Neal MJ (1983) Effects of GABA agonists, glycine, taurine and neuropeptides on acetylcholine release from the rabbit retina. J Physiol (London) 336:563-577.
De Camilli P, Vitadello M, Canevini MP, Zanoni R, Jahn R, Gorio A (1988) The synaptic vesicle proteins synapsin I and synaptophysin (protein P38) are concentrated both in efferent and afferent nerve endings of the skeletal muscle. J Neurosci 8:1625-1631.
Fykse EM, Fonnum F (1988) Uptake of GABA by a synaptic vesicle fraction isolated from rat brain. J Neurochem 50:1237-1242.
Iversen LL (1971) Role of transmitter uptake mechanisms in synaptic neurotransmission. Br J Pharmac 41:571-591.
Kanner BI, Schuldiner S (1987) Mechanism of transport and storage of neurotransmitters. CRC Crit Rev Biochem 22:1-38.
Kontro P, Marnela KM, Oja SS (1980) Free amino acids in the synaptosomes and synaptic vesicle fractions of different bovine brain areas. Brain Res 184:129-141.
Kosaka T, Tauchi M, Dahl JL (1988) Cholinergic neurons containing GABA-like and/or glutamic acid decarboxylase-like immunoreactivities in various brain regions of the rat. Exp Brain Res 70:605-617.
Lowe AW, Madeddu L, Kelly RB (1988) Endocrine secretory granules and neuronal synaptic vesicles have three integral membrane proteins in common. J Cell Biol 106:51-59.
Masland RH, Cassidy C (1987) The resting release of acetylcholine by a retinal neuron. Proc R Soc Lond B 232:227-238.

Masland RH, Livingstone CJ (1976) Effect of stimulation with light on synthesis and release of acetylcholine by an isolated mammalian retina. J Neurophys 39:1210-1219.
Masland RH, Mills JW (1979) Autoradiographic localization of acetylcholine in the rabbit retina. J Cell Biol 83:159-178.
Masland RH, Mills JW, Cassidy C (1984a) The functions of acetylcholine in the rabbit retina. Proc R Soc Lond B 223:121-139.
Masland RH, Mills JW, Hayden SA (1984b) Acetylcholine synthesizing amacrine cells: identification and selective staining by using radioautography and fluorescent markers. Proc R Soc Lond B 223:79-100.
Massey SC, Neal MJ (1979) The light evoked release of acetylcholine from the rabbit retina in vivo and its inhibition by GABA. J Neurochem 32:1327-1329
Massey SC, Redburn DA (1982) A tonic GABA-mediated inhibition of cholinergic amacrine cells in rabbit retina. J. Neurosci 11:1633-1643.
Neal MJ, Massey SC (1980) The release of acetylcholine and amino acids from the rabbit retina in vivo. Neurochem 1:191-208.
Rassin DK (1972) Amino acids as putative transmitters: failure to bind synaptic vesicles of guinea-pig cerebral cortex. J Neurochem 19:139-148.
Schwartz EA (1982) Calcium-independent release of GABA from isolated horizontal cells of the toad retina. J Physiol (London) 323: 211-227.
Schwartz EA (1987) Depolarization without calcium can release GABA from a retinal neuron. Science 238:350-355.
Tauchi M, Masland RH (1984) The shape and arrangement of the cholinergic neurons in the rabbit retina. Proc R Soc Lond B 223:101-119.
Vaney DI (1984) 'Coronate' amacrine cells in rabbit retina have the 'starburst' dendritic morphology. Proc R Soc Lond B 220:501-508.
Vaney DI, Young HM (1988) GABA-like immunoreactivity in cholinergic amacrine cells of the rabbit retina. Brain Res 438:369-373.
Wheeler DD (1984) The effect of membrane potential on initial velocity of GABA uptake and steady state distribution ratio in rat cortical synaptosomes. Neurochem Res 9:649-660.
Yazulla S (1983) Stimulation of GABA release from retinal horizontal cells by potassium and acidic amino acid agonists. Brain Res 275: 61-74.
Yazulla S (1985) Evoked efflux of [^{3}H]GABA from goldfish retina in the dark. Brain Res 325:171-180.

STUDIES ON THE LOCALIZATION OF SEROTONERGIC NEURONES AND THE TYPES OF SEROTONIN RECEPTORS IN THE MAMMALIAN RETINA

N.N. Osborne, N.L. Barnett, H. Ghazi, A. Calas* and M. Maitre**
Laboratory of Ophthalmology
Oxford University
Walton Str., Oxford
OX2 6AW England

Introduction

During the past decade there has been considerable debate about whether serotonin (5-hydroxytryptamine) plays a transmitter role in the vertebrate retina. There is now strong support for the idea that serotonin is a transmitter substance in non-mammalian species, i.e. that serotonin is present in significant amounts in these retinas and can be localized in specific populations of amacrine cells by the Falck and Hillarp method or with immunohistochemical techniques (Ehinger 1982, Osborne et al. 1982, Osborne 1984). However, it still remains to be shown that the amine has a neurotransmitter role in mammalian retina.

The main argument against the idea that serotonin exists in mammalian retina is that all attempts to localize the amine have failed (see Ehinger & Florén 1980, Florén & Hansson 1980). Osborne et al. (1982) have argued that this may be due simply to the low amounts of serotonin in neurones and that more sensitive procedures would be required to localize this amine. The amount of serotonin is indeed low in mammalian retina in comparison with non-mammalian retina (i.e. 9-18 ng/g fresh weight versus 100-500 ng/g).

Specific populations of amacrine cells in most mammalian retinas (an exception is the rat) take up exogenous serotonin. These "indoleamine-accumulating cells" as they have been called (Ehinger & Florén 1978) have been extensively studied in terms of amine transport (Osborne & Patel 1984), morphology (Sandell & Masland 1986, Vaney 1986) and synaptic connections (Brunken & Daw 1987). There still remains the challenge, however,

*Université de Bordeaux I, Bordeaux, France, **Centre de Neurochimie du CNRS and INSERM, Strasbourg, France.

NATO ASI Series, Vol. H31
Neurobiology of the Inner Retina
Edited by R. Weiler and N.N. Osborne
© Springer-Verlag Berlin Heidelberg 1989

of proving that these cells contain serotonin and that the amine is used as a neurotransmitter/neuromodulator. Evidence that serotonin has such a function in the retina is quite good. The amine can be synthesized from tryptophan in the retina (Osborne et al. 1984) and can also be released by potassium depolarization in a calcium-dependent manner. Binding sites for serotonin have been localized in crude membrane fractions of the retina (Mitchell & Redburn 1985, Osborne 1981). Furthermore, 5-HT_2 antagonists reduce the on-response of ON-centre cells as well as the surrounds of OFF-centre cells and enhance the centre off-responses of the latter cells in the rabbit retina (Brunken & Daw 1986, 1987).

Recent studies from our laboratory have been concerned with the identification of the serotonin-receptor types in the mammalian retina (Cutcliffe & Osborne 1987, Ghazi & Osborne 1988a). Classification of serotonin receptor subtypes is based on functional criteria, binding criteria or actions on 2nd messengers (see Osborne & Hamon 1988, Bradley et al. 1986, Göthert & Schlicker 1987) where it is generally accepted that three major subtypes exist in nervous tissues. The 5-HT_1 subtype is linked to adenylate cyclase/c-AMP cascade and has an affinity to the agonist 8-OH-DPAT. The 5-HT_2 subtypes are linked to the phospholipase C/phosphoinositide hydrolysis pathway and are specifically sensitive to the antagonist ketanserin. The 5-HT_3 receptor subtypes are generally located peripherally and are not linked to a second messenger system. One purpose of this article is to describe our recent studies concerned with characterizing the subtypes of serotonin receptors in the rabbit retina.

Previous studies by ourselves have included attempts to elevate cellular serotonin levels of the retina by injecting various substances (e.g. monoamine oxidase inhibitors, pteridine co-factors to activate the enzyme tryptophan-hydroxylase or tryptophan, the precursor for serotonin synthesis) intraocularly. These studies proved unsuccessful in that the amine level was nerver sufficiently elevated for serotonin to be visually localized by immunohistochemistry. However, recent studies with forskolin and dibutyryl c-AMP have proved of interest in this respect. These data, as well as some preliminary results on the localization of tryptophan-hydroxylase immunoreactivity will also be described, as they argue in favour of the existence of serotonin neurones in the mammalian retina.

Evidence for the occurrence of $5\text{-}HT_{1A}$ receptors

Florén and Hansson (1980) reported that serotonin had no effect on c-AMP levels in rabbit and chicken retinas. These results agreed with the data reported for calf and carp retinas (Makman et al. 1975, Dowling & Watling 1981). More recent studies by Blazynski et al. (1985) show clearly, however, that serotonin and other indoleamines significantly elevate intracellular levels of c-AMP in rabbit retina, suggesting the presence of a $5\text{-}HT_1$-like receptor. We have both confirmed and extended this observation.

As seen in fig. 1a, serotonin and the $5\text{-}HT_{1A}$ agonist 8-OH-DPAT stimulate c-AMP production in similar ways: both substances have comparable EC_{50} values. Furthermore, the shape of the curves is similar, suggesting that the two agonists mediate their effects through the same receptor sites. This is corroborated by the pharmacological characteristics of the two agonists in stimulating c-AMP production (Table 1). They are both inhibited by spiroxatrine, a known $5\text{-}HT_{1A}$ antagonist (Göthert & Schicker 1987). In contrast, neither $5\text{-}HT_2$ antagonists ketanserin, methysergide and miaserin, not the $5\text{-}HT_3$ antagonist MDL-72222 had an effect. Propranolol, a beta antagonist, but known to influence 5-HT receptors in the brain, was without effect. The dopaminergic antagonists spiroperidol and haloperidol were also without effect. Additional proof of the occurrence of $5\text{-}HT_{1A}$ receptors is shown by the effects of other known $5\text{-}HT_{1A}$ agonists. The true $5\text{-}HT_{1A}$ agonist, buspirone and ipsapirone and the mixed $5\text{-}HT_{1A/B}$ agonist, RU 24696 behave pharmacologically in a similar way to serotonin (Table 1).

Evidence for the occurrence of $5\text{-}HT_2$ receptors

Within the last few years an entire host of neurotransmitters (acetylcholine, noradrenaline, substance P, histamine etc.) have been shown to elicit effects through an inositol lipid receptor mechanism (Berridge 1987, Osborne et al. 1988). Activation of these receptors leads to a breakdown of phosphatidylinositol 4,5-bisphosphate (PIP_2) to form two second messengers, namely inositol 1,4,5-triphosphate (IP_3) and sn-1,2 diacylglycerol (DG). IP_3 releases calcium from a non-mitochondrial store, while one of the functions of DG is to activate the enzyme protein kinase C (PKC). In addition to mediating various intracellular biochemical effects, PKC has been shown in some studies to regulate the agonist-induced phospholipase C-linked phosphoinositide pathway (Gonzales et al. 1987). Our recent results show that in the rabbit retina serotonin stimulates IP_3 formation (Cutcliffe & Osborne 1987) and that this process is accompanied by a mobilization of intracellular calcium stores (Ghazi & Osborne 1988b), but we could not demonstrate that PKC activation influences these processes.

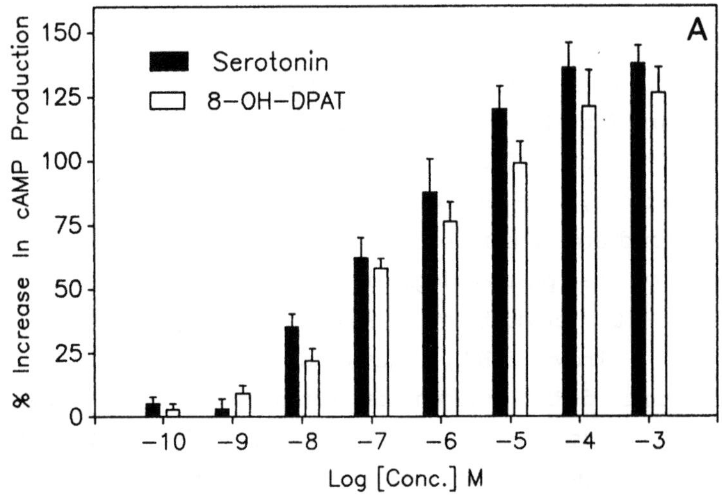

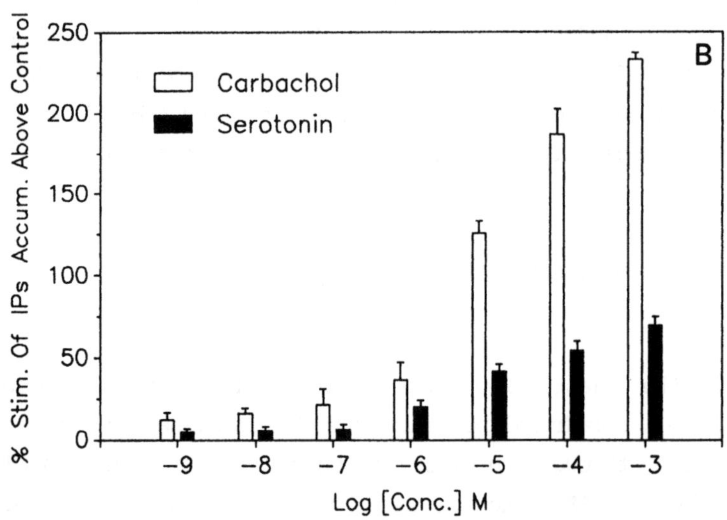

Fig. 1a show the dose-response relationship of serotonin and 8-OH-DPAT effects on c-AMP production in homogenates of rabbit retinas. Fig. 1b reveals the dose-response relationship of serotonin- and carbachol-induced stimulation of inositol phosphates in intact rabbit retinal pieces. Each point represent a mean ± SEM for 4 separate experiments.

TABLE 1

Agent	Antagonist Added	% Stimulation of cAMP Production Above Basal
Serotonin(500µM)		139.0±7.4
Serotonin	Spiroxatrine(1µM)	* 25.4±9.6
Serotonin	Ketanserin(10µM)	**130.0±9.1
Serotonin	Mianserin(10µM)	**136.9±10.1
Serotonin	Methysergide(10µM)	**132.0±13.6
Serotonin	Spiroperidol(10µM)	**135.4±12.0
Serotonin	MDL-72222(10µM)	**141.4±12.9
Serotonin	Propranolol(10µM)	**132.6±3.4
Serotonin	Haloperidol(10µM)	**135.0±5.8
8-OH-DPAT(500µM)		128.9±10.1
8-OH-DPAT	Spiroxatrine(1µM)	* 16.9±6.2
8-OH-DPAT	Methysergide(10µM)	**115.0±7.3
8-OH-DPAT	Ketanserin(10µM)	**125.6±12.4
8-OH-DPAT	MDL-72222(10µM)	**130.3±12.8
8-OH-DPAT	Propranolol(10µM)	**124.9±11.7
8-OH-DPAT	Haloperidol(10µM)	**126.3±8.2
Buspirone(500µM)		135.8±5.9
Buspirone	Spiroxatrine(1µM)	* 12.9±3.3
Buspirone	Ketanserin(10µM)	**129.4±8.0
Buspirone	Propranolol(10µM)	**124.2±13.9
Buspirone	Fluphenazine(10µM)	**133.7±4.0
Ipsapirone(500µM)		118.2±15.9
Ipsapirone	Spiroxatrine(1µM)	* 12.1±5.9
Ipsapirone	Ketanserin(10µM)	**108.7±14.6
Ipsapirone	Propranolol(10µM)	**112.1±12.3
Ipsapirone	Fluphenazine(10µM)	**117.0±16.4
RU24696(500µM)		104.9±8.0
RU24696	Spiroxatrine(1µM)	* 10.8±1.0
RU24696	Spiroxatrine(10µM)	* 4.6±0.5
RU24696	Methysergide(10µM)	**107.3±8.5
RU24696	MDL-72222(10µM)	**106.8±6.9
RU24696	Propranolol(10µM)	** 98.1±5.7
RU24696	Haloperidol(10µM)	**102.0±2.2
Spiroxatrine(1µM)		-1.2±3.7
Ketanserin(10µM)		1.4±0.8
Mianserin(10µM)		4.2±2.1
Methysergide(10µM)		0.7±1.7
MDL-72222(10µM)		2.6±1.1

Effect of various antagonists on stimulation of c-AMP production in rabbit retinal homogenates induced by serotonergic agonists. Results are expressed in terms of % stimulation of cAMP production above basal levels, and are given as means ± SEM where n is from 10-14 separate experiments, except for buspirone, ipsapirone and RU24696 where n=3. * and ** denote significantly and insignificantly different from agonist-induced stimulation, respectively.

Some of these results will now be summarised. Typical concentration response curves for serotonin and, as a comparison, for carbachol for the stimulation of ^{3}H-inositol phosphates (IPs) accumulation in rabbit retinal slices are shown in fig. 1b. Both substances elicit their effects maximally at 1mM but it is clear that carbachol is the more efficacious. Similarly, carbachol is more efficacious in mobilizing cytosolic calcium stores in primary rabbit retinal cultures (Table 2). It was necessary to use cultures in these experiments to load the cells with the fluorescent calcium indicator Quin 2. Substances which stimulate IPs accumulation in the retina were similarly effective in the cultures (Ghazi & Osborne 1988a). Only substances which stimulated IPs production increased intracellular levels of calcium. Furthermore, the 5-HT_2 antagonists like ketanserin, retanserin, miaserin and cyproheptadine were correspondingly effective in inhibiting the 5-HT-induced stimulation of inositol phosphates or the mobilization of intracellular calcium stores (Table 2). A number of other antagonists, in particular atropine and prazosin, were without effect. Atropine and prazosin block the carbachol and noradrenaline induced stimulation of IPs in rabbit retina, respectively (Ghazi & Osborne 1988a). From these results it is clear that the 5-HT_2 receptors in the retina are not unlike those found in other parts of the central nervous system (CNS) (Conn & Sanders-Bush 1986).

Table 2

Agent(s)	% Stim. of IPs accum. Above Control	Increase in $[Ca^{2+}]_i$ Above Basal Level
Carbachol (100µM)	186.8±15.9	311±26 nM
Serotonin (500µM)	71.3±3.2	142±25 nM
+ Ketanserin (1µM)	* 18.0±1.9	* 25±4 nM
+ Mianserin (1µM)	* 36.2±4.7	* 36±6 nM
+ Ritanserin (1µ)	* 41.6±3.6	
+ Methysergide (10µM)	** 68.7±11.2	
+ Cyproheptadine (10µM)	* 35.9±6.0	
+ Spiroxatrine (10µM)	** 70.1±1.8	** 148±12 nM
+ MDL72222 (10µM)	** 74.9±1.2	** 140±16 nM
+ Atropine (10µM)	** 67.8±5.1	** 146±19 nM
+ Prazosin (10µM)	** 70.5±2.9	** 136±14 nM
+ Hexamethonium (10µM)	** 73.1±7.2	
+ Propranolol (10µM)	** 71.9±6.9	
+ Haloperidol (10µM)	** 70.9±4.6	

Agonist-induced stimulation of IPs accumulation in rabbit retinal slices, and agonist-evoked rise in free intracellular calcium ($[Ca^{2+}]_i$) in primary rabbit retinal cultures containing both neurones and Müller cells. Results are mean values ± SEM where n=4-12. * and ** denote significantly and insignificantly different from serotonin-induced stimulation.

The effect of 4-β-phorbol 12-myristate 13-acetate (PMA), a phorbol ester which stimulates PKC, on receptor-mediated stimulation on IPs accumulation is shown in fig. 2. It can be seen that PMA on its own has no effect on the stimulation of IPs but reduced the efficacy of carbachol and noradrenaline in stimulating IPs production. However, in the case of serotonin, PMA has no effect in this respect. As shown in fig. 2, staurosporine, a PKC inhibitor (Watson et al. 1988) did not affect basal levels of IPs; however, it significantly reversed the inhibitory effects of PMA on the carbachol and noradrenaline-induced formation of IPs. These experiments suggest that a modulatory feedback "down regulation" role for PKC does not function at 5-HT_2 receptors, although it is effective at muscarinic and α_1-adrenergic receptors.

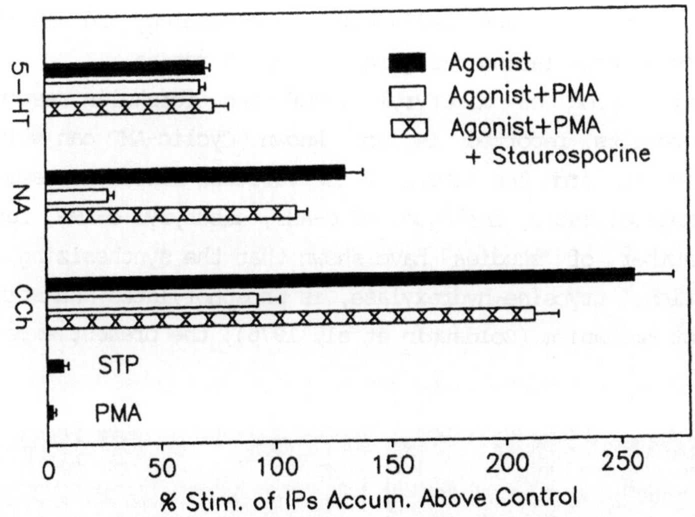

Fig. 2. Effect of 15 min preincubation with PMA (0.1µM) on the CCh(0.5mM)-, NA(0.1mM) and 5-HT(0.5mM)-mediated stimulation of IPs in rabbit retinal slices. Addition of 1µM staurosporine (STP) 5 min prior to PMA, completely blocked the inhibitory effects of the phorbol ester. PMA inhibited the CCh- and NA- but not the 5-HT-mediated response. Values are means of percentage stimulation above basal ± SEM from 3 separate experiments

Effect of intraocular injection of forskolin and dibutyryl c-AMP on serotonin-immunoreactivity

When paraformaldehyde-fixed rabbit retinal tissue is subjected to an analysis for the localization of serotonin immunoreactivity, negative results are obtained. However, retinas from animals which were either exposed in vitro to dibutyryl c-AMP (1µM for 30 min) or injected intraocularly with 50µl forskolin (100µM) for 6-8 hours and then analysed, a positive localization of serotonin immunoreactivity was revealed. Fig. 3a is a retinal section from an animal which had been treated with dibutyryl c-AMP where it can be seen that varicose-like serotonin-immunoreactive fibres bordering the inner plexiform and ganglion cell layers are apparent. In contrast, forskolin-treated tissues showed weak but clear serotonin-immunoreactive amacrine cells (Fig. 3b). These cells have the morphology and distribution of the known serotonin-accumulating cells in the rabbit retina.

The mode of action of dibutyryl c-AMP and forskolin responsible for providing the results reported is not known. Cyclic-AMP can stimulate the growth of cells and, from the nature of the varicose fibres it seems possible that the non-metabolizable analogue of c-AMP, dibutyryl c-AMP, functions in this way. A number of studies have shown that the synthesizing enzyme for dopamine formation, tryosine-hydroxylase, is phosphorylated and stimulated by a c-AMP dependent mechanism (Goldstein et al. 1976); the present experiments

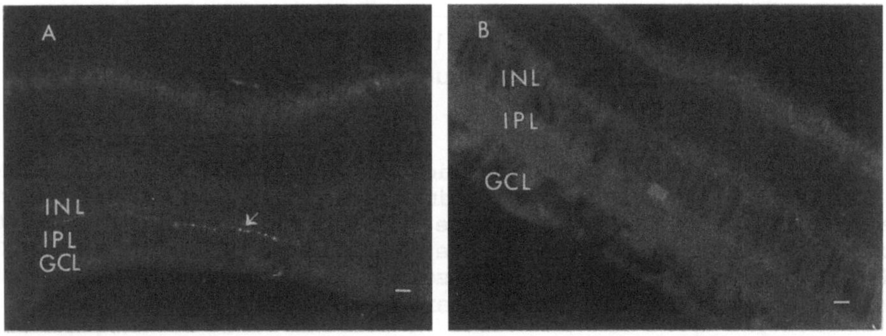

Fig. 3. Low power photomicrographs of rabbit retinal sections processed for the localisation of serotonin-immunoreactivity. Fig. 3a shows a section from a retina exposed in vitro to dibutyryl c-AMP where a varicose-like serotonin-immunoreactive fibre (arrow) can be seen. Fig. 3b is a section of a retina from an animal treated intraocularly with forskolin where weak serotonin immunoreactive amacrine cells are apparent. GCL= ganglion cell layer, IPL= inner plexiform layer, INL= inner nuclear layer. Scale bar= 15µm.

were therefore carried out to see whether the cyclic nucleotide can possibly stimulate the formation of retinal serotonin to a level sufficient for immunohistochemical localization. The dibutyryl c-AMP experiments support this insofar as serotonin-immunoreactivity can be localized. However, the distribution of the serotonin-immunoreactivity does not correspond with the distribution of the serotonin-accumulating cells. Forskolin can increase intracellular c-AMP levels via its known action on adenylate cyclase (Seamon & Daly 1986). It was therefore of significance to find that the serotonin-containing neurones can be localized to contain endogenous serotonin following forskolin treatment. Before we carried out the studies which show that c-AMP acts on tryptophan hydroxylase we were not aware of comparable results. A recent paper, however, by Garber & Makman (1987) suggests that tryptophan hydroxylase activity in the rat striatum is regulated by a c-AMP dependent process. This provides more substance to our present view therefore that serotonin accumulating cells in the retina are "serotonergic" in character.

Tryptophan-hydroxylase immunoreactivity in guinea pig retina

Tryptophan-hydroxylase is a cytosolic enzyme located in serotonin-producing cells and is the rate-limiting enzyme in the biosynthesis of the amine (Boadle-Biber 1982). We have recently used a polyclonal antiserum to tryptophan hydroxylase (Weissmann et al. 1985) for studies on the guinea pig retina. Details of the antiserum's specificity and production are described elsewhere (Weissmann et al. 1987, Cash et al. 1985). Tryptophan-hydroxylase immunoreactivity was observed in somata of amacrine cells located at the border of the inner nuclear and plexiform layers (Fig. 4a). No immunoreactive somata were observed in the outer nuclear and distal nuclear layer or the ganglion cell layer. Müller cells were not immunoreactive. Immunoreactive cell processes were observed only in the inner plexiform layer ramifying as a diffuse narrow band corresponding to lamina 1 of Cajal (Cajal 1893). While the number of tryptophan-hydroxylase immunoreactive amacrine perikarya in the guinea pig retina is small, they correspond both in number and morphology to the serotonin-accumulating cells in the guinea pig retina (Fig. 4b).

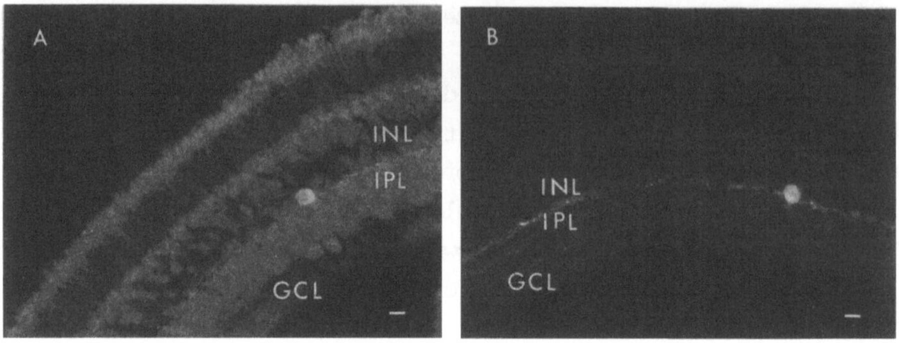

Fig. 4. Low power photomicrographs of a guinea pig retinal sections processed for the localisation of tryptophan-hydroxylase immunoreactivity (fig. 4a) and for the presence of serotonin-accumulating cells (fig. 4b). Serotonin-accumulating cells were observed by first incubating freshly dissected retinas in physiological solution containing serotonin and then processing for serotonin immunoreactivity as described elsewhere (Osborne & Patel 1984). Other details are as for fig. 3.

The serotonin-accumulating cells show two defined layers of processes in the inner plexiform layer as opposed to a single diffuse layer of tryptophan-hydroxylase immunoreactivity. It is possible that two layers of tryptophan-hydroxylase immunoreactivity actually exist, but because of low amounts of enzyme, one of the layers is not detectable.

Conclusions and Summary

The described recent, and to some extent, preliminary studies support the idea that serotonin actually exists in mammalian retinal neurones and that the amine therefore has a functional role in the retina. It is clearly demonstrated that serotonergic receptors exist which are linked to the second messengers c-AMP and inositol phosphates. The pharmacological characterization of these receptors shows that they are of the 5-HT_{1A} and 5-HT_2 subtypes as described for other parts of the mammalian CNS. Tryptophan-hydroxylase immunoreactive neurones occur quite clearly in the guinea pig retina and these neurones have a similar morphology to perikarya which take up exogenous serotonin. Finally, while immunohistochemistry cannot normally demonstrate the presence of serotonin neurones, such cells can be visualized by a prior intraocular injection of forskolin. The mechanism of action is not known whereby forskolin stimulates serotonin levels.

Acknowledgements

We are grateful to the Wellcome Trust for financial support.

References

Berridge MJ (1987) Inositol trisphosphate and diacylglycerol: Two interacting second messengers. Annual Review of Biochemistry 56: 159-194

Blazynski C, Ferrendelli JA, Cohen AI (1985) Indoleamine-sensitive adenylate cyclase in rabbit retina: Characterization and distribution. J Neurochem 45: 440-447

Boadel-Bider MC (1982) Biosynthesis of serotonin. In: Osborne NN (ed) Biology of serotonergic transmission John Wiley & Sons pp 63-94.

Bradley PB, Engel G, Feniuk W (1986) Proposal for the classification and nomenclature of functional receptors for 5-hydroxytryptamine. Neuropharmacology 5: 563-576

Brunken WJ, Daw NW (1986) 5-HT_2 antagonists reduce ON responses in the rabbit retina. Brain Res 384: 161-165

Brunken WJ, Daw NW (1987) The actions of serotonergic agonists and antagonists on the activity of brisk ganglion cells in the rabbit retina. J Neurosci 7: 4054-4065

Cajal SR (1893) La retine des vertebres. La cellule 9: 17-257

Cash CD, Vayer P, Mandel P, Maitre P (1985) Tryptophan-5-hydroxylase. Rapid purification from whole rat brain and production of specific antiserum. Eur J Biochem 149: 239-245

Conn PJ, Sanders-Bush E (1986) Regulation of serotonin-stimulated phosphoinositide hydrolysis: Relation to the serotonin 5-HT-2 binding site. Neurosci 6: 3669-3675

Cutcliffe N, Osborne NN (1987) Serotonergic and cholinergic stimulation of inositol phosphate formation in the rabbit retina. Evidence for the presence of 5-HT_2 and muscarinic receptors. Brain Res 421: 95-104

Dowling JE, Watling KJ (1981) Dopaminergic mechanisms in the teleost retina. II. Factors affecting the accumulating of cyclic AMP in pieces of intact carp retina. J Neurochem 36: 569-579

Ehinger B (1982) Neurotransmitter systems in the retina. Retina 2: 305-321

Ehinger B, Florén I (1978) Quantitation of the uptake of indoleamines and dopamine in the rabbit retina. Exp Eye Res 26: 1-11

Florén I, Hansson Ch (1980) Investigations into whether 5-hydroxytryptamine is a neurotransmitter in the retina of rabbit and chicken. Ophthalmol Vis Sci 19: 117-125

Garber SL, Makman MH (1987) Regulation of tryptophan hydroxylase activity by cyclic AMP dependent mechanism in rat striatum. Molec Brain Res 3: 1-10

Ghazi H, Osborne NN (1988a) Agonist-induced stimulation of inositol phosphates in primary rabbit retinal cultures. J Neurochem 50: 1851-1858

Ghazi H, Osborne NN (1988b) Receptors linked to phsophoinositide-specific PLC increase free intracellular calcium in rabbit retinal cultures. Neurochem Int (in press)

Goldstein M, Bronaugh RL, Ebstein B, Roberge C (1976) Stimulation of tyrosine hydroxylase activity by cyclic AMP in synaptosomes and in soluble striatal enzyme preparations. Brain Res 109: 563-574

Gonzales RA, Greger PHJr, Baker SP, Ganz NI, Bolden C, Raizada MK, Crews FT (1987) Phorbol esters inhibit agonist-stimulated phosphoinositide hydrolysis in neuronal primary cultures. Devel Brain Res 37: 59-66

Göthert M, Schlicker E (1987) Classification of serotonin receptors. J Cardiovascular Pharmacol 10: S3-S7

Makman MH, Mishira RK, Brown JH (1975) Drug interactions with dopamine-stimulated adenylate cyclase of caudate nucleus and retina: Direct agonist effect of a metabolites of ET 495. Adv Neurol 9: 213-222

Mitchell CK, Redburn DA (1985) Analysis of pre- and postsynaptic factors of the serotonin system in rabbit retina. J Cell Biol 100: 64-73

Osborne NN (1981) Binding of [^{3}H]serotonin to membranes of the bovine retina. Exp Eye Res 33: 371-380

Osborne NN (1984) Indoleamines in the eye with special reference to the serotonergic neurones of the retina. Progress in Retinal Research Vol 3 (eds NN Osborne, J Chader) Pergamon Press, Oxford pp 61-103

Osborne NN, Nesselhut T, Nicholas DA, Cuello AC (1982) Serotonin: Serotonin-containing neurones in vertebrate retinas. J Neurochem 39: 1519-1528

Osborne NN, Patel S (1984) Postnatal development of serotonin-accumulating neurones in the rabbit retina and an immunohistochemical analysis of the uptake and release of serotonin. Exp Eye Res 38: 611-620

Osborne NN, Patel S, Beaton DW (1984) Serotonin, a major metabolite of serotonin in the rabbit retina. In: Progress in tryptophan and serotonin research (eds HG Scholssberger, W Kochen, B Linzer, H Steinhart) de Gruyter, Berlin pp 275-284

Osborne NN, Hamon M eds (1988) Neuronal serotonin. John Wiley & Sons, Chicester

Osborne NN, Tobin AB, Ghazi H (1988) Role of inositol trisphosphate as a second messenger in signal transduction processes: An essay. Neurochem Res 13: 177-191

Sandell JH, Masland RH (1986) A system of indoleamine-accumulating neurones in the rabbit retina. J Neurosci 6: 3331-3347

Seamon KB, Daly JW (1986) Forskolin: Its biological and chemical properties. Adv Cyclic Nucleot Prot Phosphoryl Res 20: 1-150

Vaney DI (1986) Morphological identification of serotonin-accumulating neurons in the living retina. Science 233: 444-446

Watson S, McNally J, Shipton LJ, Godfrey PP (1988) The action of protein kinase C inhibitor, staurosporine, on human platelets. Biochem J 249: 345-350

Weissman D, Belin MF, Aguera M, Meunier C, Maitre M, Cash CD, Ehret M, Mandell P, Pujol JF (1987) Immunohistochemistry of tryptophan hydroxylase in the rat brain. Neurosci 23: 291-304

THE FUNCTION OF MULTIPLE SUBCLASSES OF GABA RECEPTORS IN RABBIT RETINA

Dianna A. Redburn, David L. Friedman and Stephen C. Massey[1]
Department of Neurobiology and Anatomy
University of Texas Medical School
P.O. Box 20708
Houston TX 77225 USA

INTRODUCTION

The presence of a functionally important GABAergic amacrine cell system in mammalian retina has been well-established (for review see Redburn and Massey, 1987). In most instances markers for GABAergic properties, such as high affinity uptake systems, endogenous stores of GABA and the presence of its synthetic enzyme, glutamic acid decarboxylase (GAD) are routinely observed within specific populations of amacrine cells. Labeling in the outer retina has been reported by some labs but not in others, especially in the case of immunocytochemical labeling of photoreceptor terminals (Nishimura et al., 1986) and horizontal cells in adult mammalian species (Brandon, 1987; Mosinger and Yazulla, 1987). It should also be noted that functional studies do not support a major role for GABAergic transmission in the outer retina of adult mammals. In contrast, electrophysiological experiments clearly support the autoradiographic and immunocytochemical localization of a functional GABAergic system within amacrine cells (Caldwell et al., 1978; Massey and Redburn, 1982).

A further understanding of the precise role(s) of GABA in visual information processing will require a more complete understanding of the functional subclasses of GABAergic cells and their unique contributions to ON and OFF circuits. Our own studies directed toward this goal have focused on the GABAergic circuit which regulates release from cholinergic amacrine cells. Our primary approach has been to monitor the release of specific neurotrans-

[1]Steve C. Massey
Sensory Sciences Center
University of Texas Graduate School of Biomedical Sciences
Houston, Texas 77030 USA

NATO ASI Series, Vol. H31
Neurobiology of the Inner Retina
Edited by R. Weiler and N.N. Osborne
© Springer-Verlag Berlin Heidelberg 1989

mitters as a measure of functional activity. In order to identify inputs to GABAergic neurons, retinas pre-loaded with 3H-GABA can be stimulated pharmacologically or physiologically (by exposure to light or darkness). The release of 3H-GABA can be monitored biochemically and the specific cellular source of the 3H-GABA released can be verified by autoradiography. A similar strategy is applied to ascertain outputs. Retinas can be preloaded with other neurotransmitter substances, in this case labeled choline, and the effect of exogenously applied GABAmimetics on release rates from cholinergic amacrines can be measured. These data represent an important supplement to other experimental approaches to circuit analysis of the inner retina, such as EM serial reconstruction of identified GABAergic synapses or intracellular recordings. With careful integration of all these approaches, a useful working model for the highly complex GABAergic amacrine cell system may emerge.

INPUT TO GABAERGIC NEURONS

Analysis of 3H-GABA release from mammalian retina is plagued by problems inherent in attempting to preload GABAergic neurons with labeled transmitter. Glial cells of mammalian retina possess an avid uptake system for GABA and thus represent formidable barriers to uptake by amacrine cells (Yazulla, 1986). By taking advantage of the higher metabolic rate of Muller cells as compared to GABAergic neurons, this problem can be partially overcome. The standard procedure utilized in our laboratory entails incubation of isolated retinas in vitro with 3H-GABA for 15 min, followed by a 30 min post-incubation in label-free medium, during which much of the 3H-GABA accumulated by Muller cells is metabolically cleared (Hampton and Redburn, 1983). Under these conditions, neuronal labeling predominates although some labeling of Muller cells is still observed.

Most investigators agree that release of 3H-GABA can be stimulated by depolarizing levels of potassium (Moran et al., 1986; O'Malley and Masland, 1988). In our studies with isolated rabbit retina which has been post-incubated as described above, the release rate is roughly doubled over basal rates in response to 56 mM potassium (Moran et al., in preparation). Under these conditions a significant proportion (approximately 70%) of the release can be inhibited by adding 10 mM magnesium to the medium to inhibit calcium-dependent release. However, Masland and co-workers (O'Malley and Masland, 1988) report that a much smaller component of 3H-GABA release from rabbit retina is calcium-dependent, and they suggest that release may occur through reverse transport rather than through calcium-dependent exocytosis.

It has not been possible for our lab or that of Masland (O'Malley and Masland, 1988) to demonstrate light or dark stimulated release of 3H-GABA. However, Bauer (1978) reported a very modest but significant increase of roughly 20% in 3H-GABA released from amacrine cells of the rabbit retina exposed to light flashes over dark release rates. The release was associated with labeled amacrines and was not observed when Muller cells were labeled. The physiological relevance of such a small increase in release rates is uncertain. In comparison 3H-ACh release from amacrines is strongly stimulated by flashing light. We routinely measure a 4-fold increase in 3H-ACh release (Massey and Redburn, 1985) under conditions similar to those used by Bauer (1978). With certain pharmacological manipulations which block the major inhibitory inputs to the cholinergic system, even the increase in response to a single flash of light can be measured (Massey and Redburn 1985).

3H-GABA release from isolated rabbit retina is also relatively insensitive to most retinal neurotransmitter agonists and antagonists. Even the most widespread receptor class in the retina, that of the kainic acid subclass of glutamate receptor, appears to have very modest effects on 3H-GABA release from amacrine cells. High concentrations (100μM) produce only a 25-30% increase in basal release rates (Moran et al., in preparation). In fact, autoradiographic analysis shows that amacrine cell bodies and distinct sublamina within the inner plexiform layer retain heavy labeling patterns in the presence of 100 mM kainic acid, in spite of the extensive excitotoxic damage incurred by many other types of amacrine cells including dopaminergic and cholinergic amacrines (Hampton and Redburn 1983). For example, kainic acid is particular potent in stimulating release of 3H-ACh from cholinergic amacrines in rabbit retina. The threshold dose of kainic acid was approximately 5 μM and a maximum effect was obtained between 20 and 50 μM. At these concentrations kainic acid caused a 30-fold increase in ACh efflux.

The relatively weak response of the GABA system to kainic acid seen in adult rabbit retina is in sharp contrast to the eight-fold increase in 3H-GABA release noted in neonatal tissues (Moran et al., in preparation). However the increase in release rates seen in neonates can be attributed to excitotoxic effects of kainic acid on horizontal cells which accumulate 3H-GABA at birth but lose this ability in the adult. Neonatal GABAergic amacrines are relatively insensitive to kainic acid as they are in the adult.

OUTPUT OF GABAERGIC NEURONS

The output from GABAergic amacrines to cholinergic amacrines has been intensively studied by monitoring ACh release directly (Masland and Livingstone, 1976; Neal and Massey, 1980). In our earlier work, we used the in situ rabbit eye cup preparation to examine the release of 3H-ACh in response to flashing light (Massey and Redburn, 1982; 1983; 1985; Massey et al., 1983). This has proven to be a highly successful method for monitoring the activity of the cholinergic amacrine cells and for identifying the synaptic mechanisms controlling them. The vitreal surface of the retina was perfused with 3H-choline, allowing cholinergic cells to accumulate the label and use it as a precursor to synthesize 3H-ACh. Light simulation at 3 Hz caused a prompt rise in the release of total radioactivity from the rabbit eyecup of about 4-5 fold over resting release. Virtually all of the light evoked release of radioactivity consisted of 3H-ACh rather than 3H-Ch. The light-stimulated release, the basal release and the release caused by GABA antagonists described below were all strongly calcium-dependent.

3H-ACH release was found to be highly sensitive to GABA agonists and antagonists (Massey and Redburn, 1982). GABA (1 mM) caused a marked depression in the light-evoked release of a ACh and drop in the resting release rate as well. The GABA agonist muscimol was extremely effective (approximately 2,000 times more potent than GABA) and caused a 90% inhibition of light-evoked ACh release at a concentration of 1 μM. The relatively high dose of GABA required to inhibit ACh release, when compared to the efficacy of muscimol, might be explained by the presence of neuronal and glial GABA uptake systems which would decrease the local concentration of GABA but only transport muscimol poorly. The GABA uptake system apparently regulates extracellular concentrations of endogenous GABA as well as exogenously applied GABA. Nipecotic acid (1 mM) , a substance known to block GABA transport without activating the GABA receptor, inhibited the light evoked release of ACh and decreased the basal efflux in a manner similar to the depression caused by GABA or muscimol.

In subsequent experiments we have determined that the ACh released in response to flashing light is a result of a transient response of the cholinergic neurons to both the ON and the OFF portion of the light stimulus (Massey and Redburn, 1985). Cholinergic neurons are divided into two functionally independent groups with conventionally placed cholinergic amacrine cells providing the OFF response and displaced cells providing the ON response (Masland et al., 1984). The ON and OFF response peaks are almost identical in size suggesting that both ON and OFF cells receive the same excitatory and inhibitor inputs in equal proportions. This also appears to be the case for GABA input to the system; in control medium a brief flash is not detectable but in the presence of picrotoxin both components are seen. In addition GABA antagonists potentiate the light evoked release of ACh in the presence or absence of AP4, a substance known to block ON pathways in the retina.

Recent experiments carried out with varying concentrations of picrotoxin in isolated rabbit retina, suggest an additional level of complexity within the GABA-ACh circuit (Friedman and Redburn, in preparation). As shown in Table I, the amount of 3H-ACh released by 20 μM picrotoxin applied during constant darkness is approximately equal to that seen in response to a one minute exposure to flashing light. However, the combination of picrotoxin and light results in a greater than additive release of 3H-ACh. A doubling of the concentration of picrotoxin causes an approximate doubling in the amount of 3H-ACh released during darkness. When 40 μM picrotoxin is added simultaneously with flashing light, the release of ACh is again greater than the sum of that release by either one alone. These results indicate that a tonic (light independent) inhibition of ACh release is mediated by GABA receptors. The release stimulated (disinhibited) by picrotoxin during flashing light is greater than either light or picrotoxin alone. Thus GABA receptors appear to be involved in both light independent (tonic) and light dependent control of ACh release. Furthermore these two GABA circuits appear to be functionally independent of each other.

Alternatively, the greater-than-additivity effect could well be the result of non-linear relationships between the GABA input and any one of the response mechanisms which are involved in determining the level of ACh release. The current data does not permit us to select between these two alternatives. However, the linear response of ACh release to varying durations of light flash exposure and to varying concentrations of several drugs, including picrotoxin may offer some argument against the possibility of non-linear elements in the response. In addition the small component of light evoked GABA release reported by Bauer (1978) would be consistent with the hypothesis of both a tonic and light/dark-stimulated GABAergic input to cholinergic amacrines.

Another interesting observation was that muscimol blocked the stimulation by bicuculline but not that caused by picrotoxin. These results are consistent with the difference in the mechanisms of action of these two antagonists. Bicuculline is known to be a competitive antagonist at the GABA recognition site on the receptor. Since muscimol competes for the same site, the antagonism between bicuculline and muscimol supports the notion that these two compounds are indeed acting on the GABA receptor to alter ACh release. On the other hand, picrotoxin blocks the chloride channel associated with the GABA receptor and therefore would remain effective even if the GABA recognition site were occupied with an agonist such as muscimol.

TABLE I

Effect of Flashing Light and GABA Analogues on ^{3}H-ACh Release from Isolated Rabbit Retina

STIMULUS	RELEASE	
	DPM Released Above Baseline	% of Total Radioactivity Released
Light (3Hz;60 sec)	8,235	0.331
20μM picrotoxin	9,550	0.377
20 μM picrotoxin plus light	24,900	0.984
40 μM picrotoxin	20,500	0.887
40 μM picrotoxin plus light	33,400	1.446
400 μM pentobarbital plus light	4,100	0.177
20 μM picrotoxin, 200 μM pentobarbital plus light	3,450	0.136
100 μM baclofen	2,470	0.087
200 μM baclofen	3,360	0.118
500 μM baclofen	5,105	0.179
1 mM baclofen	11,021	0.441
40 μM picrotoxin plus 1 mM baclofen	14,941	0.598

As an additional point, we can conclude from these experiments that the role of GABA is to provide inhibitory tone to a separate, excitatory input which results from light/dark exposure. It is clear that light/dark stimulation of ACh release is not driven solely by GABA disinhibition since the light/dark response was observed even when the GABA system was blocked by picrotoxin.

GABA RECEPTORS

The diversity and complexity of the GABA system in rabbit retina results not only from anatomically and functionally distinct classes of presynaptic GABAergic cells. In addition, post-synaptic elements show diversity as well. It appears that at least three subclasses of GABA receptors (A1, A2 and B) also contribute to different types of synaptic integration within the GABAergic system. Our strategy has been to examine the full complement of GABA receptor classes in rabbit retina using in vitro receptor binding studies. Functional studies were then used to establish which of the different receptor subclasses regulate release from cholinergic amacrine cells.

The GABA receptor consists of a membrane protein complex which may contain as many as three distinguishable components: 1) a recognition site for GABA, 2) an ionophore and 3) in some cases, peripheral regulatory sites which alter the properties of the recognition site (for review see Olsen and Venter 1986). Different subclasses of GABA receptors show variations in all three of these components. The receptor subclass which has been most intensively studied is the GABA-A site. It is characterized by the affinity of its recognition site for the antagonist bicuculline and the agonist muscimol. The GABA-A site contains a chloride ionophore which when activated leads to hyperpolarization in most post-synaptic cells. Two regulatory sites a benzodiazepine (BZD) site and a barbiturate site, are associated with one type of GABA receptors, designated as GABA-A1. Both regulatory sites cause an on overall increase in efficacy of GABAergic transmission. More recent evidence suggests that only one subcategory of GABA-A sites, designated as GABA-A1, are linked to regulatory sites. It is not clear whether or not all GABA-A1 sites are uniformly linked to both BZD and barbiturate sites. In addition there is reason to suspect that BZD and barbiturate sites may be linked to receptors for other neurotransmitters in addition to GABA-A1 sites. GABA-A sites not linked to regulatory sites are designated as GABA-A2.

Within the past few years an additional class of GABA receptors, classified as GABA-B has been recognized. They differ from GABA-A sites in the characteristics of all three receptor components. The GABA-B site has a low affinity for bicuculline and a high affinity for the agonist baclofen (para-chloro-phenyl-GABA) and the antagonist DAVA (delta-aminovaleric acid), (Bowery and Brown, 1974; Muhyddin et al., 1982; Watling and Bristow, 1986). Our binding studies in rabbit retinal membranes demonstrate the presence of all three subclasses of GABA receptors (Friedman and Redburn, in preparation). Approximately three fourths of the 3H-GABA binding sites were displaced by the GABA-A antagonist bicuculline and thus are GABA-A sites. In comparison one-fourth of the binding was displaced by the GABA-B agonist, baclofen and thus were GABA-B sites.

The presence of GABA-A1 or coupled sites was demonstrated by analyzing 3H-flunitrazepam binding in rabbit retinal membranes, in the presence and absence of GABA agonists and antagonists. The binding of 3H-flunitrazepam was more than doubled by the addition of 100μM GABA. Binding was also enhanced approximately 50% by 3 mM pentobarbital. Furthermore picrotoxin blocked this effect. These data suggest that a substantial portion of the total number of GABA receptors in rabbit retina are of the GABA-A1 type with recognition sites for GABA, BZD and barbiturates which are all physically and functionally linked.

In order to characterize the subtype of GABA-A receptor which mediates the inhibition of ACh release, we tested the effects of GABA-A1 receptor modulators on both basal and light/dark evoked 3H-ACh release. Since the response to endogenously released GABA is submaximal (Massey and Redburn, 1982; Cunningham and Neal, 1983), we reasoned that modulators should enhance the GABA mediated inhibition of light evoked release, if a coupled GABA-A receptor is present in this input pathway. Furthermore, if the response to the modulators is directly coupled to a GABA receptor, this effect should be blocked with picrotoxin. The benzodiazepine, flurazepam, produced no change in the basal release rate or release seen during flashing light. In contrast, pentobarbital in the range of 0.1 to 3 mM, caused an dose-dependent decrease in light/dark evoked release. However, when we attempted to block the inhibitory response with picrotoxin, we were unable even at high concentrations to reverse this inhibition. We conclude that the pentobarbital inhibition is independent of a GABA receptor mechanism. Studies in other parts of brain suggest that in addition to their effect on GABA receptors barbiturates may have a direct effect on release of several neurotransmitters including ACh (Haycock et al., 1977). Taken together, these results suggest that the GABA-A receptors which inhibit cholinergic amacrines are not coupled to BZD or barbiturate regulatory sites and thus are of the GABA-A2 subclass.

SUMMARY

We have proposed a working model for the the GABAergic circuit which regulates ACh release in rabbit retina. Our experiments suggest a relatively simple pathway with GABAergic amacrines interacting either directly with cholinergic amacrines or indirectly through bipolar cell inputs. It seems less likely that other amacrine cells are involved in the circuit since pharmacological agents which interact with other transmitter systems, such as glycine, serotonin and dopamine, have little effect on the GABA inhibition of ACh release. Likewise, it is possible to rule out two other pathways. Excitatory bipolar cell input to the cholinergic amacrines is not generated indirectly through disinhibition of GABAergic interneurons since light/dark evoked release is observed even in the presence of high concentrations of picrotoxin. A pathway involving cholinergic synapses feeding forward to GABAergic amacrine cells with a subsequent negative feedback may also be ruled out since cholinergic antagonists do not change the resting release of ACh, nor do they alter the effects of GABA antagonists.

The GABA receptors which regulate ACh release probably represent only a small portion of the total GABA receptor population. They do not appear to be linked to BZD or barbiturate regulatory sites, thus the inhibitory effect of GABA is the result of activation of GABA-A2 receptors. A small stimulatory effect of GABA via GABA-B receptors is also suggested from our experiments. The lack of additivity in the responses to a GABA-A2 antagonist and a GABA-B agonist suggests that there is an interaction between these to inputs to the cholinergic system. GABA-B sites could be located on pre-synaptic GABAergic terminals and act as they do in other systems, as a feedback mechanism to inhibit GABA release.

Inputs which regulate release from GABAergic amacrines are uncertain. A high level of tonic release activity is evident. However, we have little pharmacological evidence for input from any neurotransmitter systems known to reside in other types of amacrine cells. The degree of excitatory input from bipolar cells is also unclear. Glutamate is thought to be one of the major neurotransmitters used by bipolar cells yet glutamate receptor agonists are relatively weak stimulants of 3H-GABA release from amacrines. Three lines of evidence offer some counter argument. First, morphological studies suggest an extensive number of reciprocal synapses between GABAergic amacrines and rod and cone bipolars (Freed and Sterling, 1982; Pourcho and Goebel, 1983). Second, Bauer (1978) reported a modest increase in 3H-GABA release from amacrines in response to flashing light. Finally, our own data indirectly supports the possibility of a light/dark stimulated release of GABA which in turn inhibits light/dark stimulation of ACh.

The equipotent effects of GABA on the activity of ON and OFF cholinergic amacrine cells might suggest a symmetrial arrangement of GABAergic inputs to mimic the symmetrical arrangement of cholinergic amacrines. Given the prominent trilaminar banding pattern in the IPL of some (but not all) GABAergic markers, it seems reasonable to suggest the outermost band would influence OFF cholinergic amacrines and the innermost band would influence ON cholinergic amacrines. Masland et al. (1984) has shown that OFF and ON cholinergic processes are restricted to the inner and outer portions of the inner plexiform layer respectively and they do not cross the central region. A more complete reconciliation between our functional studies and current anatomical data is particularly difficult in light of the recent controversies over co-localization of various GABAergic markers. Perhaps the most remarkable co-localization studies published only this year (O'Malley and Masland, 1988; Vaney and Young, 1988), have shown that GABA immunoreactivity is co-localized within cholinergic markers in starburst amacrine cells. The possible functional consequences of this arrangement are not easily integrated into our current hypothesis about GABAergic or cholinergic function. These findings will no doubt demand serious consideration in future studies.

REFERENCES

Bauer B (1978) Photic release of radioactivity from rabbit retina preloaded with 3H-GABA. Acta Opthalmologica 56:270-283.

Bowery NG, Brown DA (1974) Depolarizing actions of -aminobutyric acid and related compounds on rat superior cervical ganglia in vitro. Brit. J. Pharmacol. 50:205-218.

Brandon C (1987) Cholinergic neurons in the rabbit retina: Immunocytochemical localization, and relationship to GABAergic and cholinesterase-containing neurons. Brain Res. 401:385-391.

Caldwell JH, Daw NW, Wyatt HJ (1978) Effects of picrotoxin and strychnine on rabbit retinal ganglion cells: Lateral interactions for cells with more complex receptive fields. J. Physiol., 276:277-298.

Cunningham, JR, Neal MJ (1983) Effect of γ-aminobutyric acid agonists, glycine, taurine, and neuropeptides on acetylcholine release from the rabbit retina. J. Physiol. 336:563-577.

Freed MA, Sterling P (1982) Amacrines reciprocal to the rod bipolar in cat retina are GABA-accumulating. Soc. Neurosci. Abstr 8:46.

Friedman DL, Redburn DA. Multiple GABA receptors in rabbit retina. Manuscript in preparation.

Hampton CK, Redburn DA (1983) Autoradiographic analysis of 3H glutamate. 3H dopamine and 3H-GABA accumulation in rabbit retina after kainic acid treatment. J. Neurosci. Res. 9:239-251

Haycock JW, Levy WB, Cotman CW (1977) Pentobarbital depression of stimulus - secretion coupling in brain -selective inhibition of depolarization - induced calcium dependent release. Biochem. Pharmacol. 26:159-161.

Masland RH, Livingstone CJ (1976) Effect of stimulation with light on the synthesis and release of acetylcholine by an isolated mammalian retina. J. Neurophysiol 39:1210-1219.

Masland RH, Mills JW, Hayden SA (1984) Acetylcholine-synthesizing amacrine cells: identification and selective staining by using radioautography and fluorescent markers. Proc. R. Soc. B.223:79-100.

Massey SC, Redburn DA (1982) A tonic γ-aminobutyric acid-mediated inhibition of cholinergic amacrine cells in rabbit retina. J. Neurosci. 2:1633-1643.

Massey SC, Redburn DA (1983) The cholinergic amacrine cells of rabbit retina receive ON and Off input: An analysis of [3H]-ACh release using 2-amino 4-phosphonobutyric acid (APB) and chloride free medium. Vis. Res. 23:1615-1620.

Massey SC, Redburn DA (1985) Light evoked release of acetylcholine in response to a single flash: Cholinergic amacrine cells receive ON and OFF input. Brain Res. 328:374-377.

Massey SC, Redburn DA, Crawford, MLJ (1983) The effects of 2-amino 4-phosphonobutyric acid (APB) on the ERG and ganglion cell discharge of rabbit retina. Vis. Res. 23:1607-1613.

Massey SC, Redburn DA (1987) Transmitter circuits in the vertebrate retina. Prog. in Neurobiol. 28:55-96.

Moran J, Pasantes-Morales H, Redburn DA (1986) Glutamate receptor agonists release 3H-GABA preferentially from horizontal cells. Brain Res. 398:276-287.

Moran J, Pasantes-Morales H, Redburn DA. Stimulatory effect of kainic acid and glutamic acid on 3H-GABA release from developing rabbit retina. Manuscript in preparation.

Mosinger J, Yazulla S (1987) Double-label analysis of GAD and GABA-like immunoreactivity in the rabbit retina. Vis. Res. 27:23-30.

Muhyddin M, Roberts PJ, Woodruff GN (1982) Presynaptic -aminobutyric acid receptors in rat anococcygeus muscle and their antagonism by 5-aminovaleric acid. Brit. J. Pharmacol. 77:163-168.

Neal MJ, Massey SC (1980) The release of acetylcholine and amino acids from the rabbit retina in vivo. Neurochem. 1:191-208.

Nishimura Y, Schwartz ML, Rakic P (1986) GABA and GAD immunoreactivity in photoreceptor terminals in primate retina. Nature 320: 753-756.

Olsen RW, Venter JC (eds) (1986) Benzodiazepine/GABA receptors and Chloride channels: Structural and functional properties, Alan R. Liss New York.

O'Malley DM, Masland RH (1988) Co-release of acetylcholine and GABA by a retinal neuron. Invest. Ophthal. and Vis. Sci. Suppl. 29:273.

Pourcho RG, Goebel DJ (1983) Neuronal subpopulations in cat retina which accumulate the GABA agonist 3H-muscimol: A combined Golgi and autoradiographic study. J. Comp. Neurol. 219:25-35.

Vaney DI, Young HM (1988) GABA-like immunoreactivity in cholinergic amacrine cells of the rabbit retina. Brain Res. 438:369-373.

Watling KJ, Bristow DR (1986) GABA-B receptor-mediated enhancement of vasoactive intestinal peptide-stimulated cyclic AMP production in slices of rat cerebral cortex. J. Neurochem. 46:1755-1762.

Yazulla S (1986) GABAergic mechanisms in the retina. Prog. Ret. Res. 5:1-52.

THE ANATOMY OF MULTIPLE GABAERGIC AND GLYCINERGIC PATHWAYS IN THE INNER PLEXIFORM LAYER OF THE GOLDFISH RETINA

Robert E. Marc
University of Texas HSC/Houston

INTRODUCTION

Glycine and γ-aminobutyric acid (GABA) are the dominant and perhaps the only **fast** inhibitory neurotransmitters for vertebrate interneurons. Why are even two such transmitters necessary? One compelling explanation is that a greater range of additivity of inhibitory currents is achieved by having two different pre-synaptic systems converge on the same kind of post-synaptic mechanism, as demonstrated by Gold and Martin (1984) for Muller cells in the lamprey medulla. Thus, in spite of tremendous molecular and biophysical similarity of the channels gated by glycine and GABA, their relative contributions to post-synaptic responses can be independent over a large range. In the vertebrate retina, and especially in the goldfish retina, there are multiple types of GABAergic and glycinergic amacrine cells (ACs) that constitute the fundamental assortment of inhibitory devices in the inner plexiform layer (IPL).

IDENTITIES OF GABAERGIC ACS

The first structural evidence for GABAergic ACs in the goldfish retina came from light microscope autoradiographs (LMARGs) of goldfish retinas incubated in [^{3}H] GABA (Lam and Steinman, 1971). Later, Stell, Bok, Lam and I were able to identify the dominant [^{3}H] GABA-accumulating AC as a variety of pyriform AC (Marc et al., 1978) which we named type Ab to designate its selective arborization in sublamina **b** of the IPL (Famiglietti et al., 1977). We also noted very weak labeling among many cells of the AC layer, but had no firm explanation for the phenomenon. The matter remained unexplored until Steve

NATO ASI Series, Vol. H31
Neurobiology of the Inner Retina
Edited by R. Weiler and N.N. Osborne
© Springer-Verlag Berlin Heidelberg 1989

Yazulla reported that the distribution of GABA receptors (Yazulla and Brecha, 1981) and the patterning of glutamic acid decarboxylase (GAD) immunoreactivity in the IPL (Zucker et al., 1984) did not match the unique labeling pattern of the Ab pyriform AC. We have found that all GAD immunoreactive pyriform ACs showed concordant [^{3}H] GABA labeling in LMARGs (Marc, Liu, Muller and Lam, in preparation). And though this result was consistent with the finding of others Bäll and Brandon (1986), Ball (1987) and Yazulla et al. (1987), these workers emphasized the existence of GABAergic ACs arborizing at all levels of the IPL, while in vitro autoradiography predominantly revealed the Ab AC. This difference between immunocytochemistry and in vitro autoradiography was due to **spatial buffering.** Powerful barriers to the rapid diffusion of GABA are demonstrable in retinas of several species (Marc, 1986) and in other areas of the central nervous system (Wood and Sidhu, 1986). When isolated goldfish retinas are incubated in micromolar levels of [^{3}H] GABA, the GABA encounters two barriers to penetration: (1) The nearly confluent surface of GABAergic H1 horizontal cells and (2) the dense mat of Ab AC processes in sublayer 5 of the IPL. Low concentrations of GABA have a low probability of making it past the gauntlet of transporters and little [^{3}H] GABA reaches sublayers 1-4 of the IPL during a brief in vitro experiment. One useful way to weaken this barrier effect (Marc, 1986) is partial competitive inhibition: Sufficient competitive inhibitor is added to the medium (e.g. unlabeled GABA or nipecotic acid) to shift the Km of the transport function to a higher concentration. This reflects an increase in the probability that a labeled molecule of GABA will get past the initial arrays of transporters. Competitive inhibition will diminish the overall rate of labeling, but after longer autoradiographic exposure, specific labeling will still be present in H1 horizontal cells and sublayer 5 of the IPL. However, a massive increase occurs in the amount of labeling in the distal IPL and the proportion of labeled cells in the AC layer increases from 2% to over 50%. This paradoxical increase in the number of labeled cells with increasing uptake inhibition is simply a consequence of overcoming spatial buffering of GABA uptake. Many aspects of previous instances of labeling discordance between [^{3}H] GABA uptake and GAD immunocytochemistry are attributable to spatial buffering of GABA.

How good is the match between [^{3}H] GABA uptake and other markers of GABAergic function? In goldfish retina, Ball and Brandon (1986) reported that ≈ 83% of GAD immunoreactive ACs displayed [^{3}H] GABA labeling. In a similar study, ≈ 75% of GABA immunoreactive ACs displayed uptake (Ball, 1987). It was not obvious why these markers failed to label identical sets. A 70%-80% percent concordance still leaves a high "false positive rate" for those interested in synaptology. By using partial competitive uptake inhibition autoradiography and silver intensified immunogold (SIG) post-embedding immunocytochemistry for GABA (Marc et al., 1988), we have shown that 96.6 % of [^{3}H] GABA labeled ACs were GABA immunoreactive and 94.8% of immunoreactive ACs were [^{3}H] GABA-labeled. The small mismatches arise because we are operating at the statistical limit of LMARG analysis. The relationship between [^{3}H] GABA uptake autoradiography and GABA immunocytochemistry in the goldfish retina is probably an identity. If there are GABAergic cells that lack GABA uptake, my experiments cannot provide any statistical evidence for their existence. Further, by combining GABA immunocytochemistry with [^{3}H] glycine autoradiography, we have observed that the populations labeled by these methods form two completely independent sets. No GABA immunoreactive cells ever showed any [^{3}H] glycine labeling.

Because GABAergic markers can be found at all levels of the IPL, many types of GABAergic ACs clearly exist. One can identify at least six forms of GABAergic ACs based on laminar or somatic features (Marc et al., 1988). From serial section SIG immunocytochemistry for GABA immunoreactivity, it has been possible to identify (in addition to type Ab) pyriform-like GABAergic ACs that arborize in sublayers 2 through 4, some of which are displaced ACs. A small fraction of the GABAergic AC population is composed of very large multipolar bistratified cells similar to dye-injected transient ACs described by Murakami and Shimoda (1975) and Teranishi et al. (1984). There are also some small multipolar cells that arborize in sublayer 1, although we do not know where else they may branch.

IDENTITIES OF GLYCINERGIC ACs

High affinity uptake of [^{3}H] glycine reveals a completely different set of ACs (Marc and Lam, 1981). As noted above, there is no overlap at all with the GABAergic population. This is convincingly reflected in the synaptologies of these cells, as we will see. Glycinergic ACs comprise over a third of all the somas in the AC layer and arborize broadly in the IPL with terminal density peaks in sublayers 2 and 4 (Marc, 1982, 1986). The glycinergic interplexiform cell constitutes a second variety of glycinergic interneuron (Marc and Lam, 1981; Marc and Liu, 1984). They are low frequency neurons ($\approx$ 50-100/mm^2) with somas situated in the inner nuclear layer and processes arborizing in both plexiform layers. The connections of glycinergic interplexiform cells in the IPL are still unknown.

I originally viewed the glycinergic AC population as homogeneous but that view is untenable. There are at least two autoradiographically identifiable soma types : (1) A small pyriform variety (Marc and Lam, 1981; Marc, 1982) and a multipolar type (Muller and Marc, 1985, 1988 ; Muller and Marc, in preparation). Studholme and Yazulla (1988) have identified at least four types of glycine immunoreactive ACs in the goldfish retina. Finally, most varieties of retinal neurons possess some local order to their distributions. The clumpy, apparently chaotic patterning of glycinergic and GABAergic AC populations is, no doubt, illusory. There are probably many superimposed orderly arrays of GABAergic and glycinergic ACs hidden in the mass labeling experiments.

SYNAPTIC CONNECTIONS OF INHIBITORY ACS

The first autoradiographic demonstration of neurochemically identified synapses in the retina were between GABAergic Ab ACs and Bb BCs (Marc et al., 1978). Although we inferred that Ab ACs made **reciprocal** synapses with type Bb BCs, we had only individual examples of ribbon inputs to Ab ACs and separate examples of conventional synapses onto BC terminals from Ab ACs. We now report that these synaptic relations are, in fact, largely reciprocal (Marc et al., 1988 and in preparation). We also find that the vast majority of all AC inputs to BCs at all levels of the IPL are GABAergic. Our combined samples for sublaminas a and b indicate that over 85% of all synapses and ≈ 90% of reciprocal synapses onto type Ba or Bb BC terminals are GABAergic (see Fig. 1, for example).

We have performed serial section reconstruction of portions of Bb BC terminals, which range from 4-8 μm in diameter. A typical Bb BC terminal contains at least 60 ribbon synapses and nearby reciprocal AC synapses are found at 2/3 of those contacts. By tracing the continuity of AC profiles, however, it is clear that many AC processes lacking reciprocal synapses near a ribbon will wrap around the BC terminal and make a synapse at another site. Thus each terminal receives at least as many feedback synapses as it has ribbons. But each Bb BC terminal also receives at ≈ 350 AC synapses and since we have not yet traced the paths of the fine inter-terminal AC dendrites around BC terminals, the true number of feedback synapses could be much higher than 60. In any case, we expect that each BC terminal, depending on size, will receive roughly fives times as much GABAergic innervation as there are ribbons in the terminal. This is a massive amount of inhibition and can certainly account for the large GABA-activated currents observed in the terminals of cyprinid BCs (Kondo and Toyoda, 1983; Tachibana and Kaneko, 1988). An additional observation is that feed-forward synapses are as frequent as feed-back synapses (Marc and Liu, in preparation), and that the targets of feed-forward synapses are rarely ACs, but rather small profiles similar to the processes observed in contact with ribbon synapses after HRP back-filling of GCs (Muller and Marc, 1985, 1988; Muller and Marc, in preparation). At least half of the feed-forward synapses can be shown to arise from AC terminals that also make feed-back synapses on the BCs. Thus one can view GABAergic signal flow as one which largely integrates inputs from BCs, providing negative feed-back for that particular stage of amplification, and negative feed-forward to the targets of those BCs, further reinforcing the strength of the surround signal. The roles of "lateral" inhibition may be several. In the case of the BC terminal, local feed-back may increase BC spatiotemporal bandpass and stability and decrease noise (Marmarelis and Marmarelis, 1978), while feed-forward to GCs may isolate strong spatial antagonism. The GABAergic domination of feed-back processes in the IPL seems to be shared with other vertebrate classes, including mammals (Pourcho and Goebel, 1983; Vaughn et al., 1981; Mariani and Caserta, 1986; Freed et al., 1987), so that this form of signal shaping may be a vertebrate generalization. This idea is further strengthened by the observation that GABAergic markers in all vertebrate retinas are highly stratified throughout the IPL (Marc, 1986), implying that some GABAergic systems are constrained to input-output operations within a stratum. In parallel with our observations on GABAergic inputs to BCs at all levels of the IPL, we have also found that GC dendrites have extensive GABAergic innervation at

---

Figure 1. An electron microscope autoradiograph of a Ba BC terminal in sublayer 1 of the IPL. The retina was incubated as described previously with [^{3}H] GABA and 1 mM nipecotic acid, yielding labeling throughout the IPL. With this technique, all BCs can be shown to (1) receive extensive GABAergic input from GABAergic ACs (arrowheads) and (2) engage in reciprocal synapses (R) with GABAergic ACs; BC ribbon synapse, (r). Type Ba and Bb BC terminals are virtually encased in a cage of GABAergic profiles (asterisks), reminiscent of the organization of mossy fiber terminals in cerebellum.

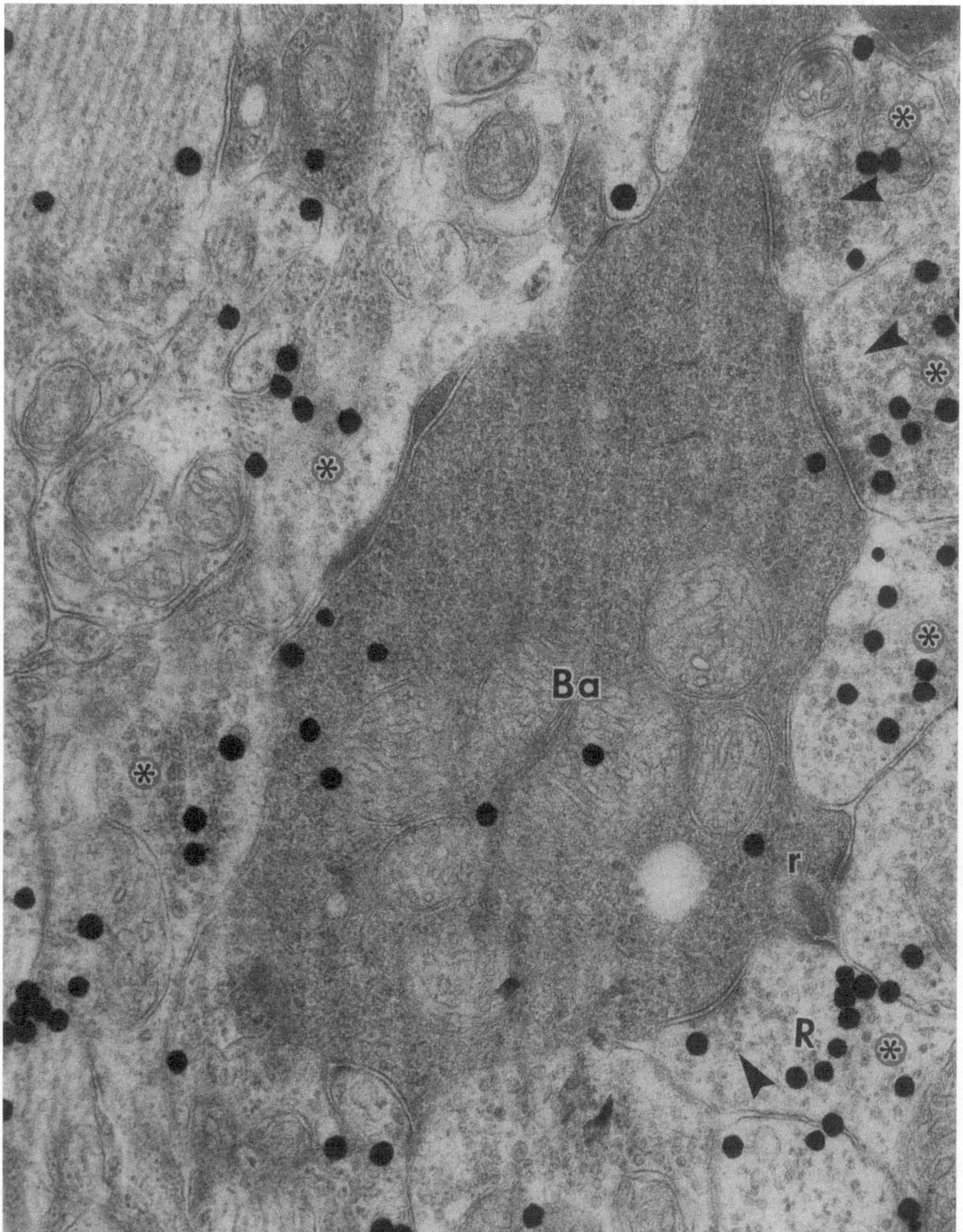
Ba
r
R

all levels of the IPL (Muller and Marc, 1985, 1988; Muller and Marc, in preparation). Thus, we can now conceive of a simple organizational rule that may account for the bulk of morphological diversity in GABAergic ACs: Each variety of BC, by virtue of its unique pattern of photoreceptor inputs, similarly requires a unique set of feed-back devices at its synaptic terminal for the generation of temporal, spatial and/or chromatic differentiations. This does not mean that different BCs cannot share some ACs, but rather that each kind of BC should have a unique pattern of inputs from such cells.

A very different picture emerges with [^{3}H] glycine autoradiography. As mentioned before, glycinergic ACs make no significant contribution to the control of BC functions in the goldfish retina, as far as we can discern (Marc and Lam, 1981; Marc, 1985). The input to glycinergic ACs comes primarily from other ACs. We do not know the neurochemical identities of those inputs yet, but it is very likely that they are partially GABAergic. And unlike GABAergic systems, we have not found any cases of glycine-glycine synapses. The major targets of glycinergic ACs are other ACs and GC dendrites. As with GABAergic systems, glycinergic synapses to GCs can be found at every level of the IPL. It is thus critical to define the varieties of glycinergic ACs, and this is in progress. Mammalian glycinergic ACs do interact with BCs (see review by Marc, 1989), but are not predominantly involved in reciprocal synapses. Glycinergic cells may be involved in disinhibitory processes (i.e. glycine-GABA interactions) where it is important to prevent occlusion. Furthermore, they may have a role in many retinas as a mechanism to control and perhaps dampen GC transients that occur at light onset or offset transitions (e.g. Belgum et al., 1985). This is very different from the presumed GABAergic transient AC whose role is probably to provide frequency-matched surround inhibition for transient GCs.

SUMMARY

In spite the the great morphological diversity of retinal ACs and the burgeoning list of neuroactive substances found in ACs, some fundamental, simplifying notions are beginning to emerge from our efforts in "structural neurochemistry." The immediate surround organization of BC to GC pathways in the goldfish is dominated by GABAergic mechanisms. Sign-inverting feed-back to BCs and sign-inverting feed-forward to GCs and other BCs is basically achieved by GABAergic neurons. The partitioning of GABAergic ACs into morphological subtypes simply reflects the signal specificities of the BC terminals they serve. Glycinergic cells probably participate in a variety of operations along with GABAergic cells, but behave as synaptically independent entities, even in the intimacy of the IPL, where synapses carrying differently "labeled" signals are often situated but a tenth of a micron apart on the same target process. Current physiological and pharmacological data regarding the partitioning of inhibitory visual processes in goldfish retina are sparse, but the anatomical findings are so far in accord with physiological results: That cyprinid BCs are very GABA sensitive and glycine insensitive (Tachibana and Kaneko, 1987; Kondo and Toyoda, 1983); that cyprinid GCs of most types typically show extensive sensitivity to both GABA and glycine (Negishi et al., 1978; Ishida and Cohen, 1988; Cohen and Fain, 1988). More attention to the numbers and kinds of ACs associated with each BC, the synaptic arrangements of GABAergic profiles around BC terminals and a better definition of glycinergic connectivity is necessary. It is satisfying to see some "order" begin to emerge from a decade of pathway tracing with neurotransmitter markers.

REFERENCES

Ball A (1987) Immunocytochemical and autoradiographic localization of GABAergic neurons in the goldfish retina. J Comp Neurol 255: 317-325

Ball A, Brandon C (1986) Localization of [^{3}H]-GABA, -muscimol, and -glycine in goldfish retinas stained for glutamate decarboxylase. J Neurosci 6: 1621-1627

Belgum JH, Dvorak DR, McReynolds JS (1984) Strychnine blocks transient but not sustained inhibition in mudpuppy retinal GCs. J Physiol (Lond) 354: 273-286

Cohen B, Fain GL (1988) GABA and glycine channels in isolated GCs from the goldfish retina. Invest Ophthal Vis Sci 29: 104

Famiglietti EV Jr, Tachibana M, Kaneko A (1977) Neuronal architecture of ON and OFF pathways to GCs in the carp retina. Science 198: 1267-1268

Freed MA, Smith R, Sterling P (1987) Rod bipolar array in the cat retina: Pattern of input from rods and GABA-accumulating ACs. J Comp Neurol 266: 445-455

Gold MR, Martin AR (1984) γ-Aminobutyric acid and glycine activate Cl^- channels having different characteristics in CNS neurones. Nature 308: 639-641

Ishida AT, Cohen B (1988) GABA-activated whole cell currents in isolated retinal GCs. J Neurophysiol: in press

Kondo H, Toyoda J-I (1983) GABA and glycine effects on the BCs of the carp retina. Vision Res 23: 1259-1264

Lam DMK, Steinman L (1971) The uptake of [^{3}H]-γ-aminobutyric acid in the goldfish retina. Proc natl Acad Sci USA 68: 2777-2781

Marc RE (1982) The spatial organization of neurochemically classified interneurons in the goldfish retina. I. Local patterns. Vision Res 22: 589-608

Marc RE (1985) The role of glycine in retinal circuitry. In: Morgan W (ed), Retinal Transmitters and Modulators: Models for the Brain, Vol 1, CRC Press, Boca Raton, Fla, pp 119-158

Marc RE (1986) Neurochemical stratification in the IPL of the vertebrate retina. Vision Res 26: 223-238

Marc RE (1989) The role of glycine in the mammalian retina. Prog Ret Res 8: in press

Marc RE, Lam DMK (1981) Glycinergic pathways in the goldfish retina. J Neuroscience 1: 152-165

Marc RE, Liu W-L S (1984) Horizontal cell synapses onto glycinergic interplexiform cells. Nature: 266-269

Marc RE, Liu W-LS (1985) (3H) Glycine-accumulating neurons of the human retina. J Comp Neurol 232: 241-260

Marc RE, Liu W-LS, Muller JF (1988) Multiple GABA-mediated surround channels in goldfish retina. Invest Ophthal Vis Sci 29: 272

Marc RE, Stell WK, Bok D, Lam DMK (1978) GABA-ergic pathways in the goldfish retina. J Comp Neurol 182: 221-246

Mariani AP, Caserta MT (1986) Electron microscopy of glutamate-decarboxylase immunoreactivity in the IPL of the rhesus monkey retina. J Neurocytol 15: 645-655

Marmarelis PZ, Marmarelis VZ (1978) Analysis of Physiological Systems: The white noise approach. Plenum Press, New York

Muller JF, Marc RE (1985) GABA-ergic and glycinergic synapses onto goldfish retinal GCs. Invest Ophthal and Visual Sci Suppl 26: 95

Muller JF, Marc RE (1988) GABAergic and glycinergic pathways in the IPL of the goldfish retina. Invest Ophthal and Visual Sci Suppl 29: 197

Murakami M, Shimoda Y (1975) Identification of amacrine and GCs in the carp retina. J Physiol (Lond) 264: 801-818

Negishi K, Kato S, Teranishi T, Laufer M (1978) Dual actions of some amino acids on spike discharges in the carp retina. Brain Res 148: 67-84

Pourcho RG, Goebel DJ (1983) Neuronal subpopulations in cat retina which accumulate the GABA agonist (3H) muscimol: A combined Golgi and autoradiographic study. J Comp Neurol 219: 25-35

Studholme KM, Yazulla S (1988) Localization of glycine immunoreactivity in the goldfish retina. Invest Ophthal Vis Sci 29: 273

Tachibana M, Kaneko A (1987) γ-Aminobutyric acid exerts a local inhibitory action on the axon terminal of BCs: Evidence for negative feedback from ACs. Proc natl Acad Sci USA 84: 3501-3505

Teranishi R, Negishi K, Kato S (1984) Dye coupling between ACs in carp retina. Neurosci Lett 51: 73-78

Vaughn JE, Famiglietti EV Jr, Barber RP, Saito K, Roberts E, Ribak C (1981) GABAergic ACs in rat retina: Immunocytochemical identification and synaptic connectivity. J Comp Neurol 197: 113-127

Wood JD, Sidhu HS (1986) Uptake of γ-aminobutyric acid by brain tissue preparations: a re-evaluation. J Neurochem 46: 739-744

Yazulla S, Brecha N (1981) GABAergic synapses in the goldfish retina: An autoradiographic study of [^{3}H]-muscimol and [^{3}H]-GABA binding. J Comp Neurol 200: 83-93

Yazulla S, Studholme KM, Wu J-Y (1987) GABAergic input to the synaptic terminals of mb_1 BCs in the goldfish retina. Brain Res 411: 400-405

Zucker C, Yazulla S, Wu J-Y (1984) Non-correspondence of [^{3}H]-GABA uptake and GAD localization in goldfish ACs. Brain Res 298: 154-158.

ACKNOWLEDGEMENTS

I wish to thank Ms. Sunny Liu and Dr. J.F. Muller for excellent technical assistance and long-standing collaborations. The research reported here was supported by PHS grant EY-02576 and Texas Higher Education Coordinating Board grant 1953.

DISTRIBUTION AND SPATIAL ORGANIZATION OF DOPAMINERGIC INTERPLEXIFORM CELLS IN THE RAT RETINA

J. Nguyen-Legros, C. Savy, E. Martin-Martinelli and J. Yelnik*
Laboratoire de Neuro-Cytologie Oculaire
INSERM U-86
15, rue de l'Ecole de Médecine
75270 Paris 06
France

Since the discovery of intrinsic retinal dopamine (DA)-neurons (Malmfors, 1963), a number of physiological roles in retinal neurotransmission have been established for DA (see Ehinger, 1983, for review). The best documented at the level of cellular and molecular biology is unquestionably the uncoupling action of DA on horizontal cell gap junctions via a D_1 receptor mechanism. The DA released by light stimulation (flashes or light adaptation from drakness) contributes to the regulation of horizontal cell receptive field size which is a mechanism coding for contrast sensitivity. The narrowing of receptive fields due to the diminished ion permeability of gap junctions results in an enhanced perception of contrast which is likely to improve visual acuity. The anatomical support for such an action of DA at the level of the outer plexiform layer would be the so-called interplexiform cells (Gallego, 1971), which contain DA in fish (Ehinger et al., 1969), because these cells, located in the amacrine cell layer together with DA-amacrines, send processes to both the inner (IPL) and outer (OPL) plexiform layers. Although the uncoupling action of DA on horizontal cells has been demonstrated in the retina of cold-blooded animals (Piccolino et al., 1984; Teranishi et al., 1983), there are supporting evidences for the occurence of such a mechanism in mammals and man whose retina also contains DA-interplexiform cells (DA-IPcs) (Hokoc and Mariani, 1987; Negishi et al., 1985; Nguyen-Legros et al., 1981, 1984; Oyster, 1985; Ryan and Hendrickson, 1987; Versaux-Botteri et al., 1984). Indeed, a decreased contrast sensitivity is observed in patients with diminished retinal DA content such as Parkinson's disease (Hamois et al., 1988; Skrandies and Gottlob, 1986), this dysfunction being reversed by treatment with L-DOPA. Moreover, dopaminergic drugs improve visual contrast sensitivity in healthy volunteers (Domenici et al., 1985). The present knowledge about the morphology and connectivity of mammalian

*INSERM U-3, Hôpital de La Salpêtrière, 47 Bd de l'Hôpital, 75645 Paris 13, France.

NATO ASI Series, Vol. H31
Neurobiology of the Inner Retina
Edited by R. Weiler and N. N. Osborne
© Springer-Verlag Berlin Heidelberg 1989

DA-IPcs is still bare. It has been supposed from Polyak descriptions (1941), that IPcs drive a recurrent input from the IPL to the OPL, and it was confirmed that a large number of DA-profiles are post-synaptic in the IPL while they are all pre-synaptic in the OPL in the monkey Cebus capucinus whose retina contains many DA-IPcs (Dowling et al., 1980).
However, since the internal processes of DA-IPcs are intermingled with those of DA-amacrines in the IPL, it is impossible to distinguish between the synapses of the two sorts of cells. The only analysis of single intracellularly injected cells could reveal their own synaptology. The intermingling of processes also renders difficult the evaluation of a percentage of DA-IPcs among the total DA-cell population, and their distribution.
The whole-mounted retina treated with anti-tyrosine hydroxylase antiserum for the demonstration of DA-neurons (Versaux-Botteri and Nguyen-Legros,1986) offers a unique opportunity for analysing single DA-cells. The immature retina represents an easier approach because the growing dendrites still bear growth cones which clearly indicate their terminals and because a lesser degree of overlapping in the IPL allows for a more accurate separation of single cells. Since this technique allows for a clear identification of DA-IPcs, their density and distribution can be observed. The arborization of individual cells can be reconstructed and analysed using a computer program developed by one of us (Yelnik et al., 1981).

Tridimensional analysis of single DA-IPcs

The present example concerns a DA-IPc chosen in the inferior temporal quadrant of a 9 day old rat retina. It is representative of the DA-IPc population of the immature retina. In a first step, the neuron is drawn using a camera lucida (Fig. 1A). Then, it is transformed into a graph by using a video camera attached to the microscope and which transmits an image onto a video monitor (Fig. 1B). Points (roots, nodes and tips) and segments (stems, internodes and terminal segments) are defined and the three coordinates are measured for each point. The X and Y coordinates are given by two orthogonal lines which may be positioned by the observator on the screen, the Z coordinate is measured by a circular potentiometer attached to the fine focus screw of the microscope. The neuron is digitalized by using the x100 oil objective. Rotations may be done around the initial X and Y axes (Fig. 1B). They correspond to cross sections of the retina for angles of 90°.

Fig. 1. A- Camera lucida drawing of one DA-IPc. (x100 oil objective). Some processes are thick, bear growth cones and branch at acute angles. Other are thin and branch at right angles.

B- Computer graph of the same IPc. The cell is represented in the plane of the retina (plan initial) and in two orthogonal planes (RX=90; RY=90) which correspond to the projection of the arborizations into the retinal layers.

A

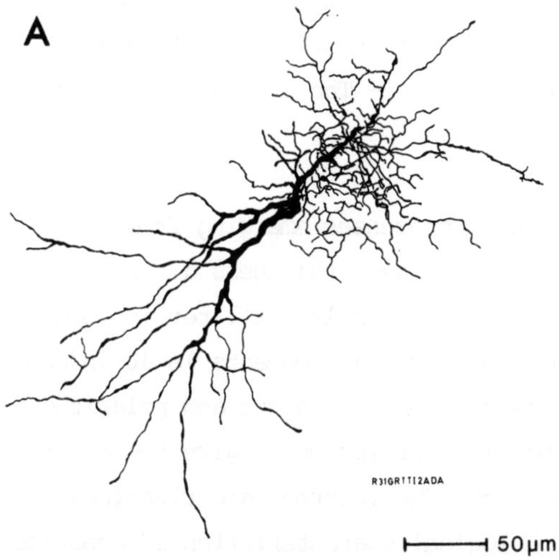

B

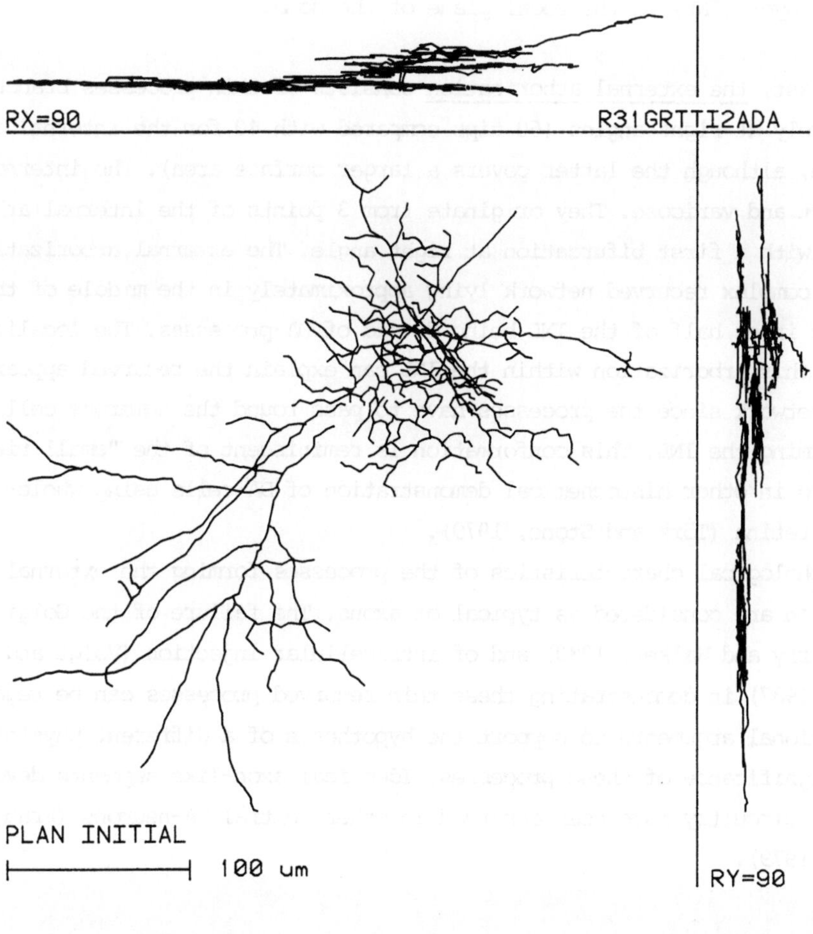

Two criteria which are the morphological appearance of the processes (thickness, branching angles, presence of varicosities) and their localization within the retinal layers allow to subdivide the arborization of DA-IPcs into three parts: 1- The internal arborization, 2- The external arborization, 3- The scleral process(es).

The internal arborization consists of thick processes emerging from the soma and branching at a short distance of it. Their thickness diminishes at each bifurcation which occurs at acute angle. The only small terminals of these processes are varicose. The number of terminal segments is 40 in comparison with 3 stems, but the internodes are long, straight and polarized in two major opposite directions so that it contributes to give the arborization a general poorly branched appearance. The internal arborization, which unquestionably exhibits the morphological characteristics of dendrites, is quite monostratified and lie at the margin between the IPL and the inner nuclear layer (INL) in the focal plane of the soma.

In contrast, the external arborization consists of thin processes branching extensively at right angles (60 tips compared with 40 for the internal arborization, although the latter covers a larger surface area). The internodes are short and varicose. They originate from 3 points of the internal arborization, with a first bifurcation at right angle. The external arborization forms a complex recurved network lying approximately in the middle of the INL, the inner half of the INL being devoid of DA-processes. The localization of this arborization within the INL can explain the recurved appearance of the network since the processes have to pass round the numerous cell bodies forming the INL. This conformation is reminiscent of the "small rings" described in other histochemical demonstration of DA-cells using whole-mounted retina (Törk and Stone, 1979).
The morphological characteristics of the processes forming the external arborization are considered as typical of axons. The failure of the Golgi method (Perry and Walker, 1980) and of intracellular injection (Voigt and Wässle, 1987) in demonstrating these thin recurved processes can be regarded as additional arguments to support the hypothesis of a different physiological significance of these processes. Identical axon-like segments devoted to local circuitry have been observed in other central DA-neurons (François et al., 1979).

Depending on the IPc, there is one or more scleral process(es). It also originates from the internal arborization, passes through the external network to reach the OPL level. Unfrequently, a scleral process can arise from the soma. The configuration of the present cell is the most frequent but the scleral process can sometimes be an internal dendrite itself which changes its direction to run sclerally. Consequently, the scleral processes exhibit rather morphological charateristics of dendrites, although some axon-like processes can emerge from their origin and contribute to the external arborization. The absence of terminal ramifications in the present example can be due to immaturity but, even in the adult, some scleral processes terminate abruptly with a bulky ending at the level of the OPL.

DA-IPc density and distribution

The precise counting of DA-IPcs is only possible in the immature retina, since the date of their first appearance (6th postnatal day). In the adult retina, the intermingling of DA processes at the level of the internal arborizations is so complex that it becomes impossible to trace the dendrites from which scleral processes arise to their parent cells. The only IPcs which can be clearly identified are those whose scleral process originates from the cell body. The density of scleral processes however, can be evaluated by focusing to the OPL level and mapping the processes either with a camera lucida or a X-Y-plotter.
The finest analysis has been made in a 9 day old rat retina in which the density and distribution of the total population of DA-cells is already close to those of adult. In this specimen, all the scleral processes have been mapped and counted. A total 73 DA-cells, randomly chosen on the retina (18±1 per quadrant) have been finely examined to established their relationship with scleral processes in order to determine if they were IPcs. Thirty three of these cells were identified as IPcs (43%). Their distribution shows a higher concentration in the temporal half of the retina (more than 50% of the DA-cells versus 30% in the nasal side). The percentage of DA-IPcs per quadrant can be compared with the number of scleral processes so that a theoretical mean number of scleral processes per IPc can be calculated (Table 1). This number is lower (2 processes/IPc) in the superior temporal quadrant than in the other parts of the retina (5-6 processes per IPc).

Finally, three important observations are derived from the parameters measured in this specimen retina:

1- The total number of IPcs per retina (43%) is slightly lower than half the total number of DA-cells. It corresponds to a mean dendity of 7-8 IPcs per square millimeter.

2- The distribution of IPcs shows a naso-temporal division of the retina, since the number of DA-IPcs is higher in the two temporal quadrants (> 50% of DA-cells) than in the two nasal quadrants (30% of DA-cells).

3- There is a regional specialization in the superior temporal quadrant, which can be related to the presence of an area centralis: the density of IPcs is the highest (15/mm^2) while they have the smallest number of scleral processes (2/IPc).

These results are different from those reported in cat retina (Oyster et al. 1985), the only mammalian species in which an evaluation of DA-IPcs is available for comparison.

	T	ST	IT	IN	SN
density of DA-cells :	18	30	16	7	18
percentage of DA-IPcs:	43%	50%	55%	30%	30%
density of DA-IPcs :	8	15	8	2	5
number of scleral processes :	825	202	272	111	240
processes/DA-IPc :		2	5	6	5

Table 1. Density and distribution of DA-IPcs compared with those of the total population of DA-cells in the 9 day old rat retina. T= total retina, ST= superior temporal, IT= inferior temporal, IN= inferior nasal, SN= superior nasal.

Developmental evolution of DA-IPcs

The first DA-IPcs can be identified by postnatal day 6 in the rat retina. They are very few at this date but are already more numerous in the temporal half retina. The figure 2B shows the scarcity of scleral processes in the mid temporal area of the 6 day old rat retina. The proportion of DA-IPcs/DA-cells is only 13% at this date. This proportion seems to regularly increase along the postnatal development. As for the total DA-cell population,

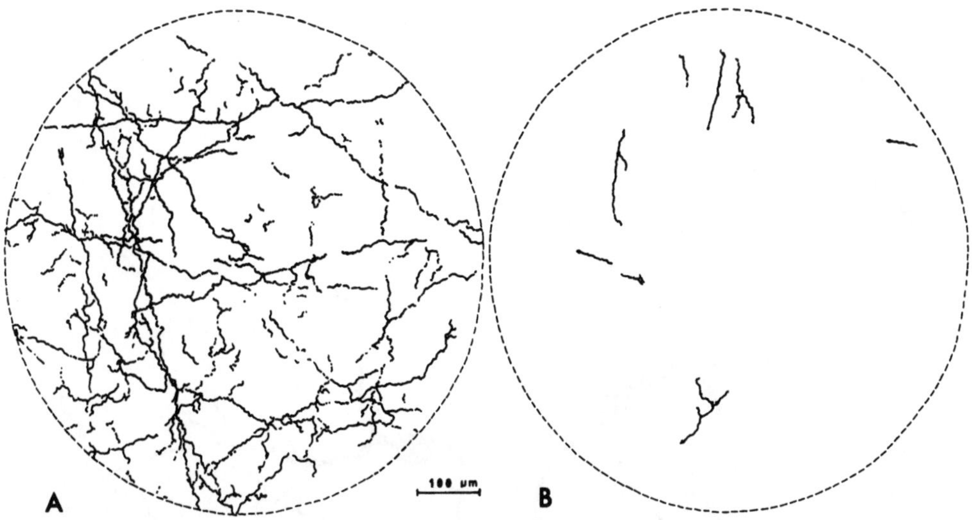

Fig. 2. Camera lucida drawing of the scleral processes in a surface area of retina corresponding to the field of the microscope (X25 objective) and chosen in the mid temporal part of the retina.
A- adult retina.
B- 6 day old retina (date of first appearance of DA-IPcs).

there is not a stage of proliferation followed by regulation but a constant growth of the number of cells and of the length and ramification of processes (Martin-Martinelli et al., in press). The figure 2A shows the density of scleral processes in an adult retina, for comparison.

A restriction must be made however concerning the external arborization described in the IPcs of the 9 day old retina. It is no more possible in the adult to clearly separate the dendritic internal arborization from the axon-like external one, because the empty space (corresponding to the inner part of the INL) observed between the two arborization planes progressively disappears, as a result of the thinning down of the INL. Consequently, the "small rings" formed by axon-like processes become in a focal plane very close to that of dendrites, rendering difficult the spatial analysis of arborizations with our computer program.

Computer analysis of DA-process network

A previous study using the nearest neighbour method in the adult rabbit retina (Brecha et al., 1984) established that the distribution of DA-cells is not completely aleatory. However in rat, it appears that some groups of

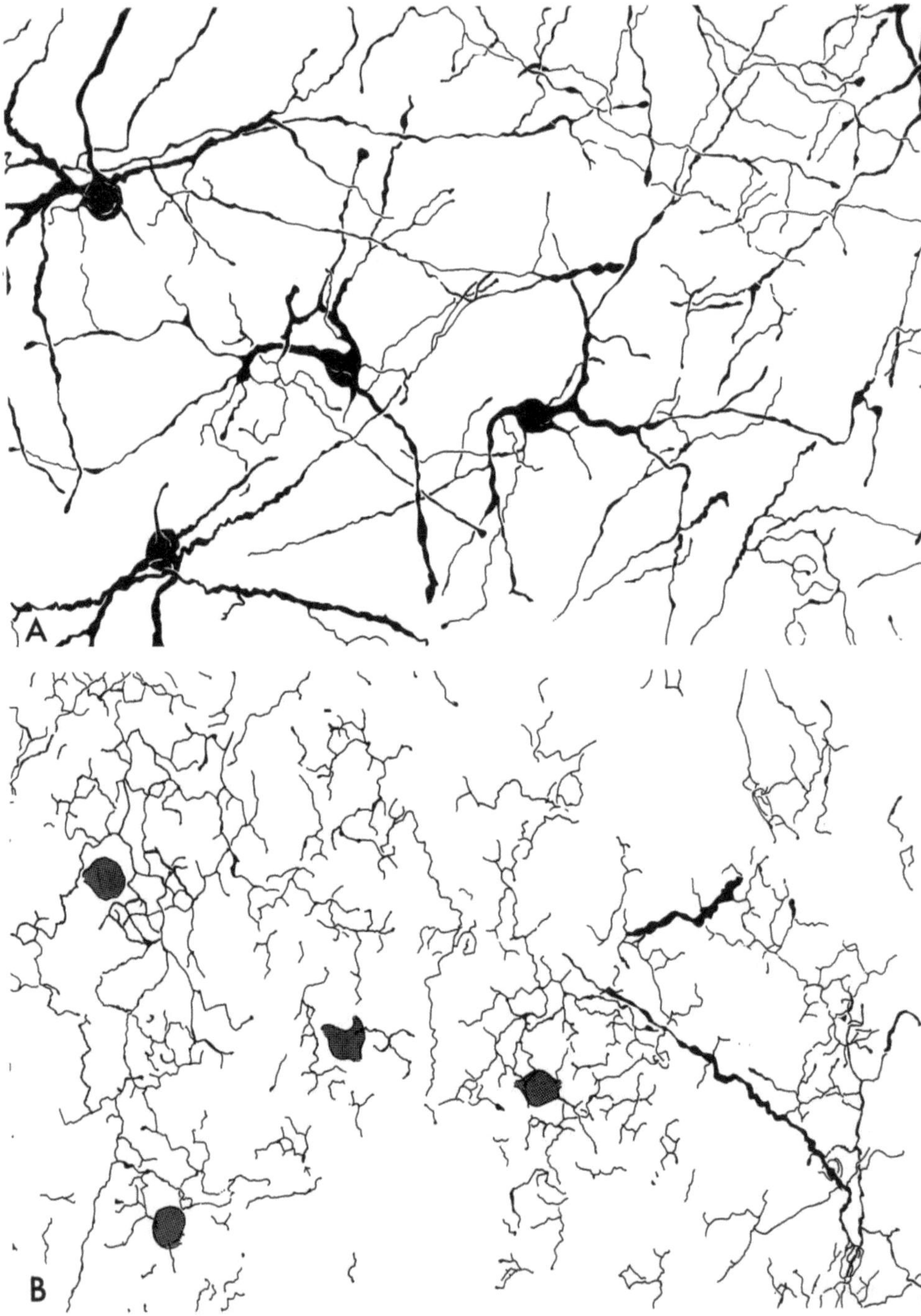

Fig. 3 . A- Camera lucida drawing (x100 oil objective) of DA internal arborizations in a field of 320 x 220 µm. Four cell bodies are represented among which one is of an IPc.

B- Drawing of the external arborizations together with scleral processes in the same field.

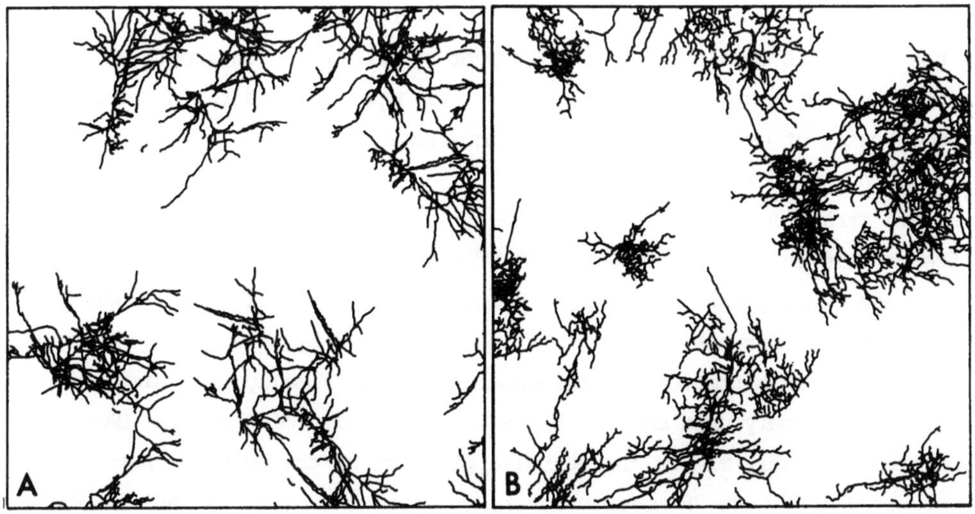

Fig. 4. Simulated network of internal arborizations (A) and external arborizations plus scleral processes (B), using 4 digitalized DA-IPcs randomly distributed on the basis of the total DA-cell density.

cells occur in contrast with large empty spaces. This appearance is most frequent in the inferior half of the retina where the cell density is the lowest and is still more evident in the immature retina. Figure 3 is a representation of one of these groups. Since it was suggested that the non random distribution of cell bodies in adult could be derived from a random one during development, we have used 4 digitalized cells, which are IPcs, to do a simulation of the network of internal and external arborizations, by using a computer program.

On the basis of the DA-cell density of the inferior temporal quadrant, where the 4 digitalized cells were chosen, and a random distribution of cell bodies, a simulation of the network of internal arborizations (Fig. 4A) and of external arborizations plus scleral processes (Fig. 4B) has been performed. With a random distribution of cells and a random orientation of arborizations, an extensive overlapping of processes of neighbouring cells ocuurs (Fig. 4) at the two levels of arborization. However, at the internal level, it is possible to minimize this overlapping by orienting manually the arborizations while keeping the same soma distribution. The network obtained by this artifice closely mimics the actual network observed in the 9 day old retina, i.e. there are groups of cells whose internal arborizations overlap (Fig. 3A) while empty spaces occur. It is thus likely that the cell bodies are actually randomly distributed during development and that the regular distribution of processes which results in a complete covering of the retina in the adult, is obtained by an adjustment of the processes growth and orientation. Another argument which supports this hypothesis is the observation of some pairs of cells whose arborizations have really complementary orientations in the adult.

The same stratagem can be used for the external arborizations. The only difference is that the actual network of external arborizations does not reveal any overlapping of processes (Fig. 3B). It is likely that, contrary to the internal network, the external one is formed by an adjusted juxtaposition of the arborizations of neighbouring cells. Finally, the simulation program, which only used DA-IPcs but with the density of the total DA-cell population, closely mimics the actual appearance of the external network as observed in figure 3B where only one cell is an IPc. It is thus likely that not only the IPcs have an external arborization in the INL but also the DA-amacrines in the 9 day old rat retina.

Brecha NC, Oyster CW, Takahashi S (1984) Identification and characterization of tyrosine hydroxylase immunoreactive amacrine cells. Invest Ophthalmol Vis Sci 25: 66-70

Domenici L, Trimarchi C, Piccolino M, Fiorentini A, Maffei L (1985) Dopaminergic drugs improve human visual contrast sensitivity. Human Neurobiol 4: 195-197

Dowling JE, Ehinger B, Floren I (1980) Fluorescence and electron microscopical observations on the amine accumulating neurons of the Cebus monkey retina. J Comp Neurol 192: 665-686

Ehinger B, Falck B, Laties AM (1969) Adrenergic neurons in teleost retina. Z Zellforsch mikr anat 97: 285-297

Ehinger B (1983) Functional role of dopamine in the retina. In: Osborne N, Chader G (eds) Progress in Retinal Research, vol 2. Pergamon Press, Oxford p 213

François C, Percheron G, Yelnik J, Heyner S (1979) Demonstration of the existence of small local circuit neurons in the Golgi stained primate substantia nigra. Brain Res 172: 160-164

Gallego A (1971) Celulas interplexiformes en la retina del gato. Arch Soc Esp Oftalm 31: 299-304

Hamois C, Di Paolo T, Marcotte G, Daigle M (1988) Retinal dopamine content in Parkinsonian patients. Ann meet Assoc Res Vis Ophthalm, Sarasota USA, p 200

Hokoc JN, Mariani AP (1987) Tyrosine hydroxylase immunoreactivity in the Rhesus monkey retina reveals synapses from bipolar cells to dopaminergic amacrine cells. J Neurosci 7: 2785-2793

Malmfors T (1963) Evidence of adrenergic neurons with synaptic terminals in the retina of rats demonstrated with fluorescence and electron microscopy. Acta Physiol scand 58: 99-100

Martin-Martinelli E, Simon A, Vigny A, Nguyen-Legros J (to be published) Postnatal development of tyrosine hydroxylase immunoreactive cells in the rat retina: Morphology and distribution. Dev Neurosci

Negishi K, Kato S, Teranishi H, Katayama Y, Tohyama M (1985) So-called interplexiform cells immunoreactive to tyrosine hydroxylase or somatostatin in rat retina. Brain Res 346: 136-140

Nguyen-Legros J, Berger B, Vigny A, Alvarez C (1981) TH-like immunoreactive interplexiform cells in the rat retina. Neurosci Lett 27: 255-259

Nguyen-Legros J, Botteri C, Le Hoang P, Vigny A, Gay M (1984) Morphology of primate's dopaminergic amacrine cells as revealed by TH-like immunoreactivity on retinal flat-mounts. Brain Res 295: 145-153

Oyster CW, Takahashi ES, Cilluffo M, Brecha NC (1985) Morphology and distribution of tyrosine hydroxylase-like immunoreactive neurons in the cat retina. Proc Nat Acad Sci USA 82: 6335-6340

Perry VH, Walker M (1980) Amacrine cells, displaced amacrine cells and interplexiform cells in the retina of the rat. Proc R Soc Lond B 208: 415-431

Piccolino M, Neyton J, Gerschenfeld H (1984) Decrease of gap junction permeability induced by dopamine and cyclic adenosine 3',5'-monophosphate in horizontal cells of turtle retina. J Neurosci 4: 2477-2488

Ryan MK, Hendrickson AE (1987) Interplexiform cells in macaque monkey retina. Exp Eye Res 45: 57-66

Skrandies W, Gottlob I (1986) Alteration of visual contrast sensitivity in Parkinson's disease. Human Neurobiol 5: 255-259

Teranishi T, Negishi K, Kato S (1983) Dopamine modulates S-potential amplitude and dye coupling between external horizontal cells in carp retina. Nature 301: 243-245

Törk I, Stone J (1979) Morphology of catecholamine-containing amacrine cells in the cat's retina as seen in retinal whole mounts. Brain Res 169: 261-273

Versaux-Botteri C, Nguyen-Legros J, Vigny A, Raoux N (1984) Morphology, density and distribution of tyrosine hydroxylase-like immunoreactive cells in the retina of mice. Brain Res 301: 192-197

Versaux-Botteri C, Nguyen-Legros J (1986) An improved method for tyrosine hydroxylase immunolabeling of catecholamine cells in whole-mounted rat retina. J Histochem Cytochem 34: 743-748

Voigt T, Wässle H (1987) Dopaminergic innervation of AII amacrine cells in mammalian retina. J Neurosci 7: 4115-4128

Yelnik J, Percheron G, Perbos J, François C (1981) A computer-aided method for the quantitative analysis of dendritic arborizations reconstructed from serial sections. J Neurosci Meth 4: 347-364

NEURONAL AND GLIAL RELEASE OF GABA FROM THE RAT RETINA

M J NEAL, J R CUNNINGHAM & M A SHAH

Department of Pharmacology, United Medical and Dental Schools of Guy's and St Thomas's Hospitals (UMDS), St Thomas's Hospital, London, SE1 7EH

There is strong evidence that in some retinas (eg fish, amphibia) GABA is a horizontal cell transmitter (for review see Yazulla, 1986) and that in most vertebrate retinas GABA is an inhibitory transmitter in about one third of amacrine cells. However, in contrast to horizontal cells, from which the release of GABA has been convincingly demonstrated (Schwartz, 1982; Yazulla & Kleinschmidt, 1983; Yazulla, 1983; Ayoub & Lam, 1985; Cunningham & Neal, 1985a; Schwartz, 1987), there have been few reports of GABA release from amacrine cells. Bauer (1978) demonstrated a Ca-dependent light-evoked release of radiolabelled GABA from the amacrine cells of the isolated superfused rabbit retina, but in similar experiments, O'Malley & Masland (1988) failed to show a light-evoked release of labelled GABA. Furthermore, O'Malley & Masland (1988) claim that the K-evoked release of labelled GABA from amacrine cells of the isolated rabbit retina is not Ca-dependent.

In the present study we have examined the release of endogenous GABA from amacrine cells of the rat retina. In this species, GABA is restricted mainly to amacrine cells, although a few interplexiform cells are also likely to be GABAergic (Brandon, 1985; Mosinger, Studholme & Yazulla, 1986).

For comparison, we have also examined the release of ^{3}H-GABA, which in the rat retina is accumulated mainly by the glial Muller fibres (Neal & Iversen, 1972). Thus, we have taken the release of endogenous GABA and ^{3}H-GABA to be an indication of neuronal and glial release respectively.

We have found that GABA release from amacrine cells can be stimulated by both potassium and glutamate in a Ca-dependent manner but glial ^{3}H-GABA release was stimulated only by glutamate, again by a Ca-dependent mechanism. No evidence was found for $GABA_B$ inhibitory autoreceptors on GABAergic amacrine cells. Some of the results of this study have been presented in a preliminary form (Neal, 1987).

METHODS

GABA Release Experiments

Male hooded Lister rats (weight 150-250g) were killed by cervical dislocation. The eyes

NATO ASI Series, Vol. H31
Neurobiology of the Inner Retina
Edited by R. Weiler and N. N. Osborne
© Springer-Verlag Berlin Heidelberg 1989

were rapidly removed and the retinas dissected into Krebs bicarbonate medium at room temperature. In most experiments the irreversible GABA-T inhibitor, γ-vinyl-GABA (GVG) (250mg/kg I.P.) was administered to the rats 18h before they were used in order to increase the storage levels of GABA.

Endogenous GABA Release

Two retinas (approximately 20mg tissue) were placed in a perspex chamber (0.5ml volume) containing Krebs bicarbonate medium at room temperature and bubbled with 5% CO_2 in oxygen. The tissue was washed with several changes of medium for 10 min and then the retinas were incubated in 250μl medium for 10 min 'collection' periods. Five such samples were collected and during the third period the tissue was exposed to medium containing a depolarizing agent. The amount of GABA released into each 250μl sample was measured by high performance liquid chromatography (HPLC). The samples were analysed on a Varian 5000 HPLC using pre-column derivatisation with o-phthalaldehyde (Turnell & Cooper, 1981).

^{3}H-GABA Release

In each experiment, four retinas were given a preliminary incubation at 25°C in 5ml Krebs solution containing aminooxyacetic acid (AOAA) (50 μM). ^{3}H-GABA was then added to give a final concentration of $2x10^{-8}$M and the incubation was continued with gentle shaking for 30 min. The retinas were removed, washed in Krebs medium (100ml) for 2 min and then each retina was placed between nylon grids in a separate superfusion chamber (volume 1ml). The retinas were superfused at 1ml min^{-1} with Krebs solution at 25°C and 2ml fractions were collected in counting vials using an automatic fraction collecter. Ecoscint (National Diagnostics) (10ml) was added to each vial and the radioactivity measured by liquid scintillation counting (Nuclear Chicago, Mk3). The counts were automatically corrected to d.p.m. using the external standard method.

Preliminary experiments in which samples of the superfusate and extracts of the tissue were subjected to chromatography (Goodchild & Neal, 1973) revealed that more than 95% of the radioactivity was present as authentic ^{3}H-GABA. Thus, the release of radioactivity was taken as a measure of ^{3}H-GABA release. The released ^{3}H-GABA was assumed to originate mainly from the Muller fibres because previous autoradiographical studies indicated that these cells accumulate most of the radioactivity taken up when rat retinas are incubated with ^{3}H-GABA in the light (Neal & Iversen, 1972).

Incubation and Superfusion Medium

Krebs Ringer bicarbonate of the following composition was used (mM): NaCl, 118; KCl, 4.8; $CaCl_2$, 2.4; $NaHCO_3$, 25; KH_2PO_4, 1.2; glucose, 9.5. The medium was continuously gassed with 5% CO_2 in O_2.

Materials

4-Amino-*n*-2,3-[^{3}H]butyric acid (specific activity 105 Ci mmol^{-1}) was obtained from Amersham International.
L-Aspartate, L-glutamate, kainate and quisqualate were obtained from Sigma Chemicals Co Ltd.
N-methyl-D-aspartate and (±)-*cis*-2,3-piperidine dicarboxylic acid were obtained from Cambridge Research Biochemicals Ltd.

RESULTS

Endogenous GABA Release

Effect of Glutamate and High-K on GABA Release

The spontaneous resting release of GABA was 11.2 ± 2.60 pmol/min/mg (mean ± SEM of 6 experiments) but exposure of the retinas to either KCl (50mM) or glutamate (10mM) did not produce any detectable increase in GABA release (Fig 1).

Effect of γ-vinyl-GABA (GVG)

In order to increase the storage levels of GABA, rats were injected with GVG (250mg/kg I.P.) 18h before death. Treatment with this irreversible GABA-T inhibitor increased retinal GABA levels 5-fold (saline injected controls: 2.9 ± 0.31, GVG injected: 14.4 ± 1.06μmol/g wet wt, means ± SEMs of 6 experiments).

In retinas isolated from GVG treated rats, the spontaneous resting release of GABA was doubled, but more importantly, glutamate (10mM) and high-K (50mM) now evoked a striking increase in GABA release (Fig 1). Thus, in subsequent experiments, retinas from GVG treated rats were used.

Calcium Dependence of Glutamate and K-evoked GABA Release

When the retinas from GVG treated rats were exposed to high-Mg/low Ca medium the spontaneous resting release of GABA was reduced by approximately 75%. In addition, the glutamate-evoked and K-evoked release of GABA was greatly reduced or abolished respectively (Fig 2). Thus, the glutamate and K-evoked release of GABA from rat amacrine cells was clearly Ca-dependent.

Effect of Excitatory Amino Acid Analogues on GABA Release

The release of GABA from the retina was increased by kainate but only at very high (toxic) concentrations (Fig 3). Thus, the maximum evoked release occurred at a concentration of 250μM. In contrast 10μM kainate produces a large increase in ^{3}H-GABA release from frog horizontal cells (Cunningham & Neal, 1985a) and ^{3}H-ACh release from rabbit amacrine cells (Cunningham & Neal, 1985b).

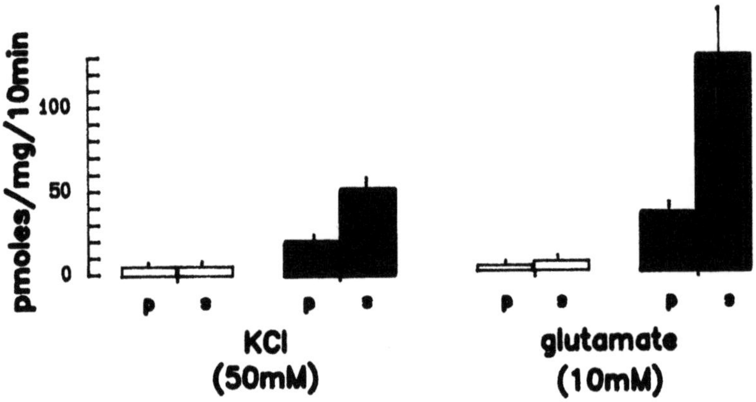

Fig. 1 Effect of glutamate (10mM) and KCl (50mM) on GABA release from retinas isolated from rats injected 18h previously with either saline (open histobars) or GVG (250mg/kg I.P.) (solid histobars). In this, and in subsequent figures, p indicates the GABA released into the sample immediately preceding exposure of the retina to drug and s indicates GABA release during exposure to drug. Each experiment is the mean ± SEM of 4-8 experiments.

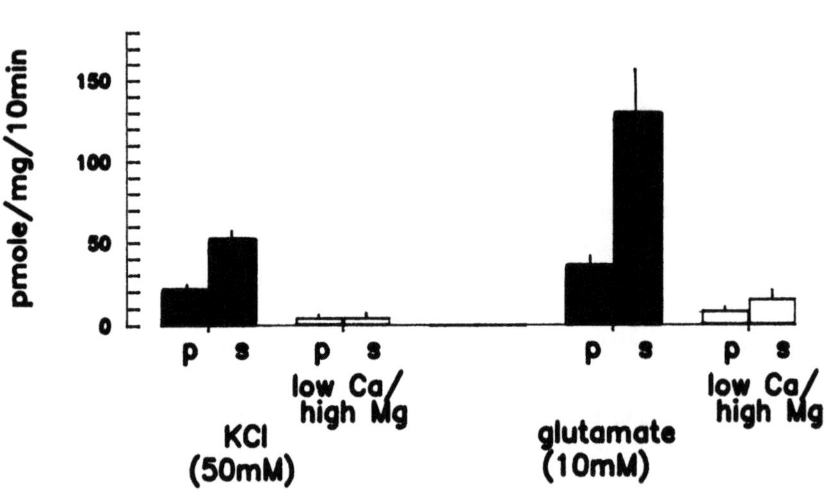

Fig. 2 Effect of high-Mg, low-Ca medium on the evoked release of GABA from the rat retina. Retinas from GVG treated rats were exposed to either KCl (50mM) or glutamate (10mM) in normal medium (solid histobars) or in medium containing $MgCl_2$:20mM, $CaCl_2$ 0.2mM (open histobars). Each result is the mean ± SEM of 4-8 experiments. The high-Mg, low-Ca medium significantly reduced the spontaneous resting release of GABA (P< 0.001) and almost abolished the K and glutamate evoked release.

The effects of quisqualate and NMDA are illustrated in Fig 4. These analogues had little effect on GABA release, even at high concentrations. Thus, only glutamate was able to evoke GABA release from rat amacrine cells at concentrations similar to those required to evoke horizontal cell GABA release and amacrine cell ACh release.

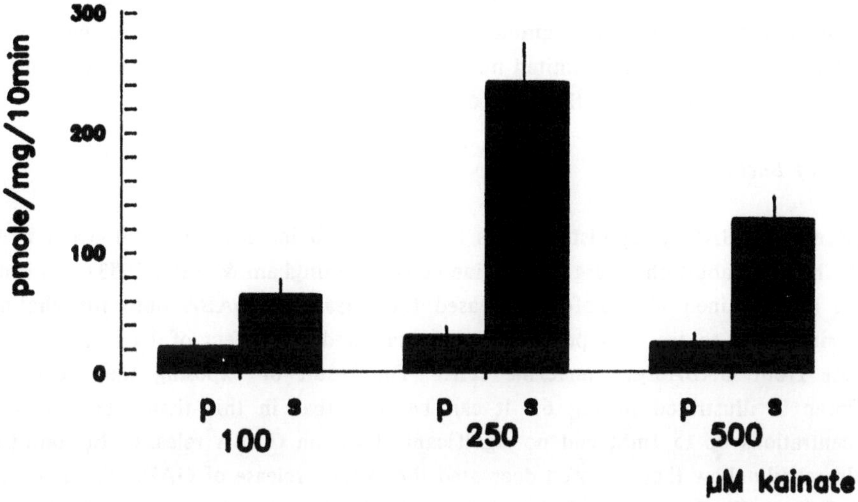

Fig. 3 Effect of kainate on GABA release from the retinas of GVG treated rats. Kainate increased GABA release but only at high concentrations (250 and 500 μM, P<0.01). Each result is the mean ± SEM of 4-6 experiments.

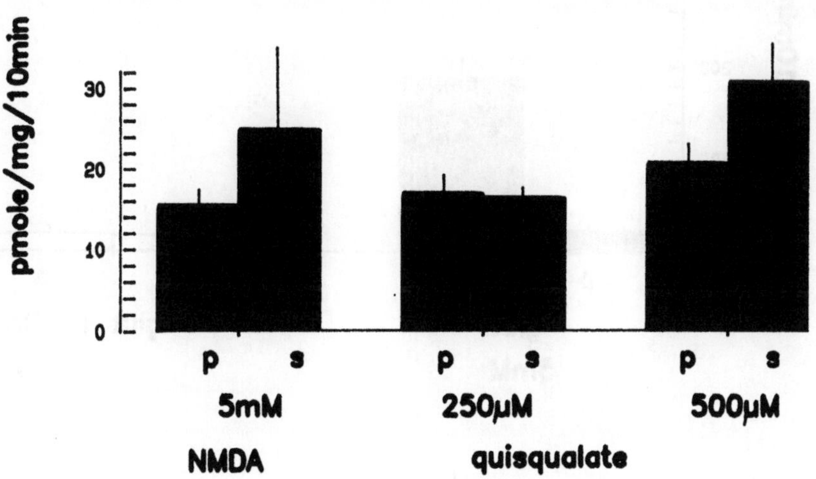

Fig. 4 Effect of N-methyl-D-aspartate (NMDA) and quisqualate on GABA release from the retinas of GVG treated rats. Each result is the mean ± SEM of 4 experiments. Neither drug produced a significant increase in GABA efflux.

Effect of PDA on Glutamate Evoked GABA Release

In order to test whether the glutamate-evoked release of GABA was mediated by some form of coupled transport system or by activation of excitatory amino acid receptors, the effect of cis-2,3-piperidine dicarboxylic acid (PDA) on the glutamate-evoked release of GABA was studied. As illustrated in Fig 5, this general excitatory amino acid antagonist abolished the effect of glutamate on retinal GABA release indicating the involvement of an excitatory amino acid receptor. This receptor has not yet been characterized, but from our limited number of experiments with agonists, it seems more likely to be a kainate type rather than a NMDA or quisqualate type.

Effect of Baclofen

Baclofen is a $GABA_B$ agonist that has been shown to increase the light-evoked release of ACh from rabbit cholinergic amacrine cells (Cunningham & Neal, 1983). This effect could be explained if baclofen decreased the release of GABA onto the cholinergic amacrines, and to test this possibility, we examined the effect of baclofen on GABA release from GABAergic amacrine cells. The result of exposing the rat retina to baclofen is illustrated in Fig 6. It can be seen that in this tissue, (±) baclofen at concentrations up to 1mM had no significant effect on GABA release. In contrast, we could readily show that baclofen decreased the evoked release of GABA from cortical or spinal slices (Fig 6) presumably by an action on inhibitory autoreceptors. In the case of cortical slices, similar results were obtained in both GVG treated and untreated animals (not shown).

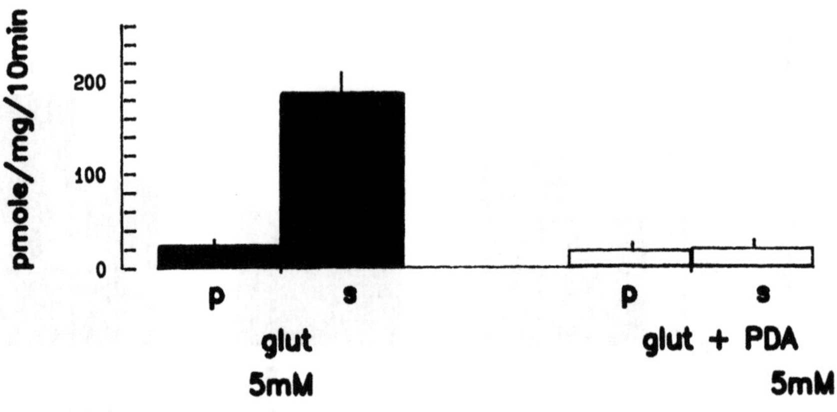

Fig. 5 Effect of PDA on the glutamate-evoked release of GABA from the retinas of GVG treated rats. In control experiments, glutamate (5mM) evoked a large release of GABA (closed histobars) but in the presence of PDA (5mM) the effect of glutamate was abolished (open histobars). Each result is the mean ± SEM of 4 experiments.

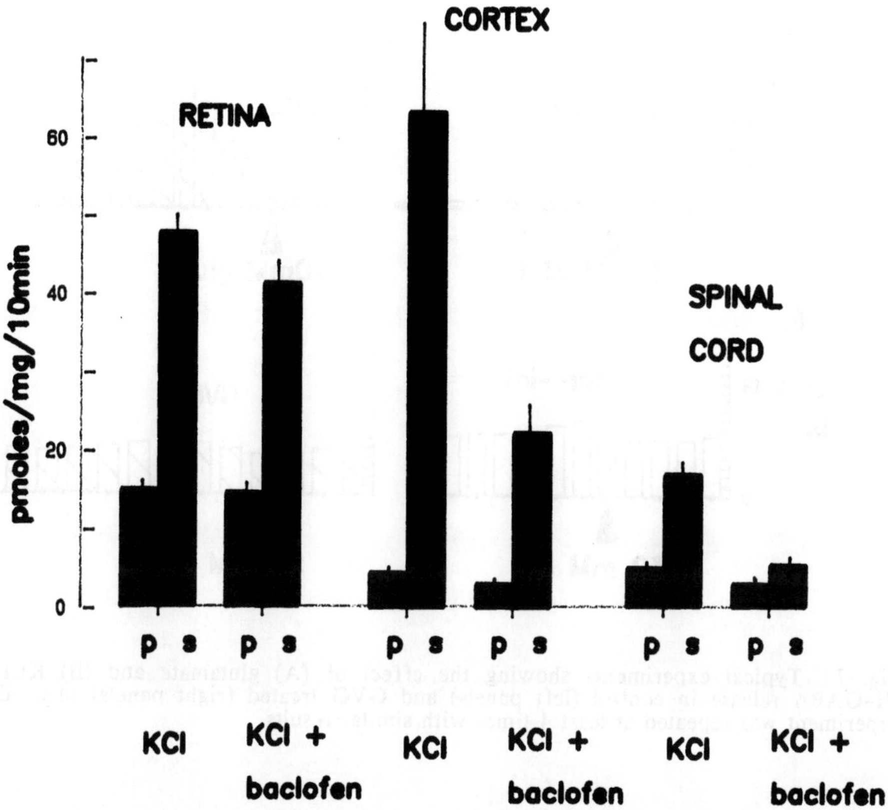

Fig. 6 Effect of baclofen on the release of GABA from GVG treated rats. Each result is the mean ± SEM of 4-8 experiments. 1mM (±) baclofen had no effect on KCl evoked release from the retina, but 10 μM (-) baclofen significantly reduced the evoked release from cortex ($P < 0.01$) and spinal cord ($P < 0.05$).

^{3}H-GABA Release

Effect of Glutamate and High-K on ^{3}H-GABA Release

When the retina was exposed to KCl (50mM) or glutamate (10mM) the evoked release of ^{3}H-GABA was very small. However, when retinas from animals treated with GVG were used, a large glutamate evoked release of ^{3}H-GABA was observed (Fig 7a). In contrast to the release of endogenous GABA, GVG did not affect the K-evoked release of ^{3}H-GABA (Fig 7b).

Calcium Dependence of Glutamate Evoked ^{3}H-GABA Release

The glutamate evoked release of ^{3}H-GABA from Muller fibres was greatly reduced in the presence of low-Ca/high-Mg medium (Fig 8).

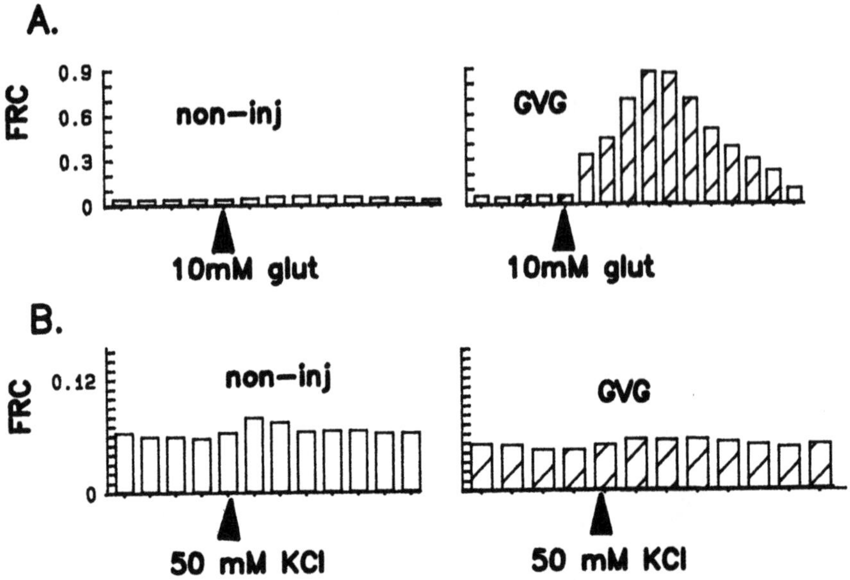

Fig. 7 Typical experiments showing the effect of (A) glutamate and (B) KCl on ^{3}H-GABA release in control (left panels) and GVG treated (right panels) rats. Each experiment was repeated at least 4 times with similar results.

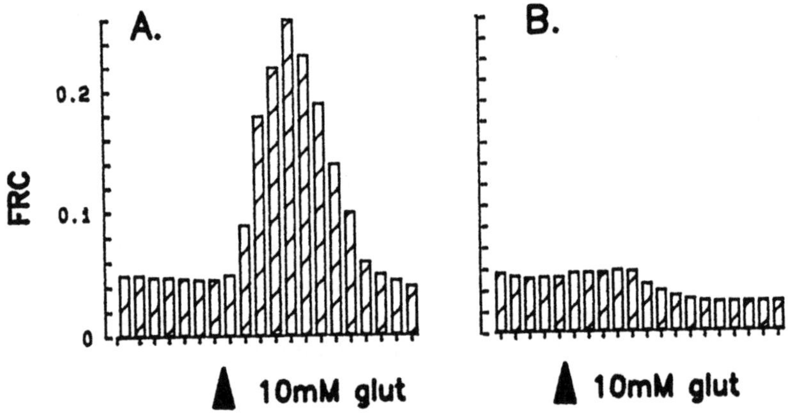

Fig. 8 Typical experiments showing glutamate evoked release of ^{3}H-GABA in GVG treated rats in (A) normal medium and (B) low Ca/high Mg medium.

Effect of Kainate on ^{3}H-GABA Release

Kainate (50-500μM) did not significantly affect the release of ^{3}H-GABA from retinas taken from rats untreated with GVG. However, when retinas from animals treated with GVG were exposed to kainate (500μM) there was a large increase in ^{3}H-GABA release. A typical experiment illustrating this effect is shown in Fig 9.

Effect of PDA on Glutamate and Kainate Evoked ^{3}H-GABA Release

The glutamate and kainate evoked release of ^{3}H-GABA from the retinas of GVG treated rats was almost abolished by PDA (5mM). This is illustrated for typical experiments in Fig 9.

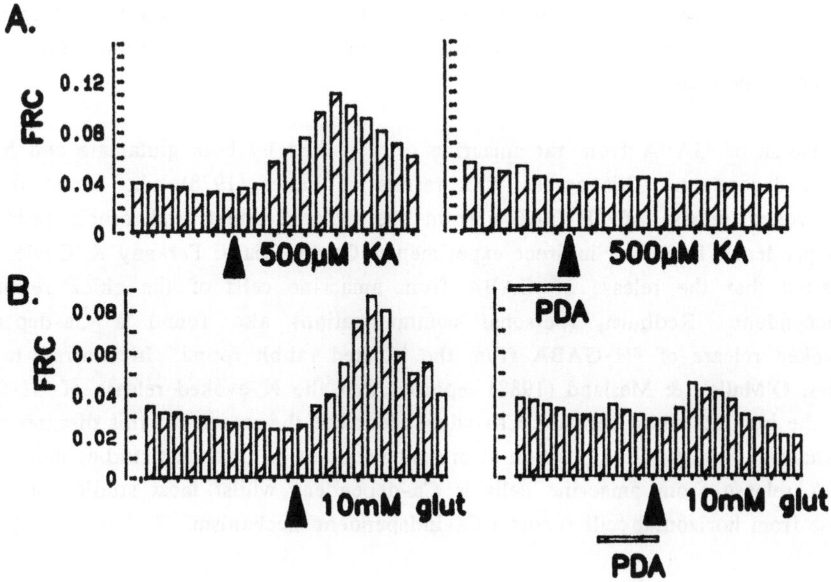

Fig. 9 Typical experiments showing that PDA (5mM) blocks both (A) kainate and (B) glutamate evoked ^{3}H-GABA release. Left panels show control responses. Right panels show responses in presence of PDA. Each experiment was repeated four times with similar results.

DISCUSSION

In the present study we have shown that endogenous GABA release from the rat retina can be stimulated by glutamate and high-K, but only if the storage levels of GABA are increased by pretreatment of the animal with the GABA-T inhibitor, γ-vinyl-GABA (GVG). Since GABA-T occurs in Muller fibres as well as in GABAergic neurones we considered it important to see if inhibition of the enzyme with GVG altered the distribution of GABA within the retina. Preliminary immunocytochemical studies using a GABA antiserum indicated that retinas from rats treated with GVG, but not controls, exhibited GABA-IR in the Muller fibres (Cunningham, Neal & Yazulla, unpublished results).

This raises the possibilty that the evoked endogenous GABA release measured in this study was due to GABA originating in the glial Muller fibres rather than the amacrine cells. Whilst we cannot totally exclude this possibility it seems unlikely for two reasons (a) the GABA-IR in the Muller fibres was very slight compared with the heavy staining in the amacrine cells and innerplexiform layer and (b) when the Muller fibres were loaded with radiolabelled GABA little GABA release occurred with K-depolarization even from retinas of rats treated with GVG. Since endogenous and radiolabelled GABA are taken up into glial cells and presumably stored by the same mechanisms, it is difficult to believe that K-depolarization could selectively cause an increase in only endogenous GABA release from these cells and this strongly suggests that in retinas from GVG treated rats the endogenous GABA released into the medium must originate from amacrine cells.

The release of GABA from rat amacrine cells evoked by both glutamate and high-K was Ca-dependent. This agrees with results of Bauer (1978) who found that the light-evoked release of ^{3}H-GABA from the isolated superfused rabbit retina was Ca-dependent. In rather indirect experiments, Campochiaro, Ferkany & Coyle (1985) suggested that the release of GABA from amacrine cells of the chick retina was Ca-dependent. Redburn, (personal communication) also found a Ca-dependent, K-evoked release of ^{3}H-GABA from the isolated rabbit retina. In contrast to these studies, O'Malley & Masland (1988) reported that the K-evoked release of ^{3}H-GABA from the isolated rabbit retina is Ca-independent but the reason for this disagreement is not known. However, the position at present seems to be that most studies indicate that GABA release from amacrine cells is Ca-dependent whilst most studies on GABA release from horizontal cells report a Ca-independent mechanism.

The release of GABA from rat amacrine cells was very insensitive to the excitatory amino acid analogues, kainate, quisqualate and NMDA. This raised the possibility that glutamate might be acting by reversing the GABA uptake mechanism as suggested for horizontal cells (Schwartz, 1982; Yazulla & Kleinschmidt, 1983; Schwartz, 1987) or perhaps by exchanging for GABA (Nasimento & de Mello, 1985) However, the clear Ca-dependency of the glutamate evoked GABA release and the fact that the action of glutamate was abolished by the antagonist PDA argues against such a mechanism in rat amacrine cells.

The characteristics of ^{3}H-GABA release from the rat retina were very similar to those of endogenous GABA. Thus, the glutamate-evoked release of ^{3}H-GABA was observed only in retinas from rats treated with GVG: It was Ca-dependent and was blocked by PDA, indicating that activation of excitatory amino acid receptors was necessary to evoke the release of ^{3}H-GABA. However, in spite of these similarities it seems clear that endogenous GABA and ^{3}H-GABA originate from different pools because high-K stimulated the release of only endogenous GABA and did not significantly increase the release of ^{3}H-GABA.

In a previous study, we found that high-K produced a small Ca-independent release of ^{3}H-GABA from the rat retina (Neal & Bowery, 1979) but a more detailed study by Sarthy (1983) showed that the K-evoked release of ^{3}H-GABA from the rat retina was Ca-dependent. This Ca-dependency raises the possibility that the ^{3}H-GABA released by glutamate may originate from the relatively tiny pool of labelled GABA in the amacrine cells.

It has been shown that an initial localisation of GABA in the Muller cells can change with time to be mainly in the amacrine cells (Ehinger, 1977). However, this 'redistribution' depends on the rapid metabolism of GABA in Muller cells by GABA-T and the subsequent loss of labelled metabolites (Ehinger, 1977). In the present experiments, AOAA was present to prevent the metabolism of ^{3}H-GABA and although we have not performed autoradiography on the tissue at the end of a superfusion experiment, it is unlikely that any redistribution of GABA occurred during the experiment. Furthermore, in studies of ^{3}H-GABA from the rat retina, Sarthy (1983) showed that the presynaptic agent, α-latrotoxin which depletes neuronal stores of transmitter by increasing the membrane permeability to Na and Ca, did not affect ^{3}H-GABA release although it abolished the (neuronal) release of ^{14}C-glycine. These experiments strongly suggest that ^{3}H-GABA released from the rat retina does indeed originate mainly from Muller cells rather than amacrine cells.

The insensitivity of both endogenous and ^{3}H-GABA release to excitatory amino acid analogues is surprising in view of the ability of glutamate to evoke large releases in retinas from GVG treated rats. A similar resistance to kainate in the rat retina was reported by Moran, Pasantes-Morales and Redburn (1986) who studied ^{3}H-GABA release from a number of species. It is possible that analogues such as kainate cause swelling of the retina, preventing access of the drug to the inner retina. However, this does not seem very likely, because in the present experiments, glutamate itself was able to evoke large releases of both GABA and ^{3}H-GABA. At present we have no information on the type of excitatory amino acid receptor involved in the glutamate-evoked release of GABA or ^{3}H-GABA although kainate was the most potent analogue to evoke release, especially of ^{3}H-GABA.

The role of $GABA_B$ receptors in the retina is also unknown. Baclofen increases the light-evoked release of ^{3}H-ACh from the rabbit retina (Cunningham & Neal, 1983; Friedman & Redburn, 1987) and since there is a tonic GABA input onto the cholinergic amacrine cells (Cunningham & Neal, 1983), it seemed possible that baclofen might modulate ACh release indirectly by decreasing the release of GABA. Inhibitory $GABA_B$ autoreceptors have been postulated in other areas of the central nervous system (Waldmeier et al, 1988) and in the present experiments we found that low concentrations of baclofen greatly reduced the K-evoked release of GABA from rat cortical and spinal slices, an effect that was blocked by the $GABA_B$ receptor antagonist phaclofen (Shah & Neal, 1988). Futhermore, phaclofen alone increased the resting release of GABA from cortical slices suggesting that GABA was inhibiting its own release. However, baclofen did not reduce the K-evoked release of GABA from the retinas of GVG treated rats. It might be argued that the release of GABA from tissues of GVG treated animals was abnormal. However, this does not seem likely since baclofen reduced the K-evoked release of GABA from cortical slices taken from rats which had and hadn't been treated with GVG.

In preliminary experiments we have found that phaclofen also has no effect on the release of GABA from the rat retina suggesting that the failure of baclofen to reduce the release of GABA from GABAergic amacrines was not due to maximal stimulation of the $GABA_B$ sites by endogenous GABA.

In summary, our results with baclofen and phaclofen provide no evidence for the presence of inhibitory $GABA_B$ autoreceptors on GABAergic amacrine cells in the rat retina, although evidence for such receptors in cortical slices was readily obtained.

ACKNOWLEDGEMENTS

M A Shah is supported by the Government of Pakistan. We are grateful to Merrell Dow Pharmaceuticals Ltd for a gift of γ-vinyl-GABA (Vigabatrin).

REFERENCES

Ayoub GS & Lam DM (1985) The content and release of endogenous GABA in isolated horizontal cells of the goldfish retina. *Vision Res.* **25**, 1187-1193

Bauer B (1978) Photic release of radioactivity from rabbit retina preloaded with ^{3}H-GABA. *Acta Ophthalmol.* **56**, 270-282

Brandon C (1985) Retinal GABA neurones: localization in vertebrate species using an antiserum to rabbit brain glutamate decarboxylase. *Brain Res.* **344**, 286-295

Campochiaro P, Ferkany JW & Coyle JT (1985) Excitatory amino acid analogs evoke release of endogenous amino acids and acetylcholine from chick retina *in vitro*. *Vision Res.* **25**, 1375-1386

Cunningham JR & Neal MJ (1983) Effect of γ-aminobutyric acid agonists, glycine, taurine and neuropeptides on acetylcholine release from the rabbit retina. *J. Physiol.* **336**, 563-577

Cunningham JR & Neal MJ (1985a) Effect of excitatory amino acids and analogues on ^{3}H-GABA release from frog horizontal cells. *J. Physiol.* **362**, 51-67

Cunningham JR & Neal MJ (1985b) Effect of excitatory amino acids and analogues on ^{3}H-acetylcholine release from amacrine cells of the rabbit retina. *J. Physiol.* **366**, 47-62

Ehinger B (1977) Glial and neuronal uptake of GABA, glutamic acid, glutamine and glutathione in the rabbit retina. *Exp. Eye. Res.* **25**, 221-234

Friedman DL & Redburn D (1987) Regulation of ACh release from rabbit retina by multiple GABA receptors. *Invest Ophthal.* **28**, suppl. p279

Goodchild M & Neal MJ (1973) The uptake of ^{3}H-GABA by the retina. *Brit. J. Pharmac.* **47**, 529-542

Moran J, Pasantes-Morales H & Redburn DA (1986) Glutamate receptor agonists release ^{3}H-GABA preferentially from horizontal cells. *Brain Res.* **398**, 276-287

Mosinger JL, Studholme K & Yazulla S (1986) Immunocytochemical localization of GABA in the retina: a species comparison. *Exp. Eye Res.* **42**, 631-644

Nascimento JLM & de Mello FG (1985) Induced release of γ-aminobutyric acid by carrier-mediated, high-affinity uptake of L-glutamate in cultured chick retina cells. *J. Neurochem.* **45**, 1820-1827

Neal MJ & Iversen LL (1972) Autoradiographical localization of ^{3}H-GABA in rat retina. *Nature New Biology* **235**, 217-218

Neal MJ & Bowery NG (1979) Differential effects of veratridine and potassium depolarization on neuronal and glial GABA release. *Brain Res.* **167**, 337-343

Neal MJ (1987) Effects of gamma-vinyl-GABA on the GABAergic system in the rat retina. *Invest. Ophthal.* **28**, Suppl. p352

O'Malley DM & Masland RH (1988) Co-release of acetylcholine and GABA by a retinal neuron. *Invest. Ophthal.* **29**, Suppl. p273

Sarthy PV (1983) Release of ^{3}H-γ-aminobutyric acid from glial (Muller) cells of the rat retina: efects of K^{+}, veratridine, and ethylenediamine. *J. Neurosci.* **3**, 2494-2503

Schwartz EA (1982) Calcium-independent release of GABA from isolated horizontal cells of the toad retina. *J. Physiol.* **323**, 211-227

Schwartz EA (1987) Depolarization without calcium can release γ-aminobutyric acid from a retinal neuron. *Science* **238**, 350-355

Shah MA & Neal MJ (1988) Effect of baclofen on the release of GABA and other amino acids from rat cortex, spinal cord and retina. *Brit. J. Pharmac.* (In press)

Turnell DC & Cooper JDH (1981) Rapid assay for amino acids in serum or urine by precolumn derivatization and reversed phase liquid chromatography. *Clin. Chem.* **28**, 101-118

Waldmeier PC, Wicki P, Feldtrauer JJ & Baumann PA (1988) Potential involvement of a baclofen-sensitive autoreceptor in the modulation of the release of endogenous GABA from rat brain slices *in vitro.* Naunyn-Schmiedebergs Arch. *Pharmacol.* **337**, 289-295

Yazulla S & Kleinschmidt J (1983) Carrier-mediated release of GABA from retinal horizontal cells. *Brain Res.* **263**, 63-75

Yazulla S (1983) Stimulation of GABA release from retinal horizontal cells by potassium and acidic amino acid agonists. *Brain Res.* **275**, 61-74

Yazulla S (1986) GABAergic mechanisms in the retina. *Prog. Retinal Res.* **5**, 1-52

TRANSMITTER-SPECIFIC SYNAPTIC CONTACTS INVOLVING MIXED ROD-CONE BIPOLAR CELL TERMINALS IN GOLDFISH RETINA.

Stephen Yazulla
Department of Neurobiology & Behavior
State University of New York
Stony Brook, NY 11794 USA

Retinal bipolar cells are the neurons largely responsible for conveying information from photoreceptors to ganglion cells. They are also subject to lateral inputs from horizontal cells in the outer plexiform layer (OPL), amacrine cells in the inner plexiform layer (IPL), and centrifugal input from interplexiform cells (IPC). Bipolar cells in teleost retinas receive either pure cone input or a mixture and rod and cone input. The morphology of mixed rod-cone bipolar cells have been correlated with their response polarity to illumination over the receptive field center. There are two major types: type Mb are depolarizing (ON-center), with synaptic terminals in the proximal 40% of the IPL (sublamina b), and type Ma are hyperpolarizing (OFF-center) with synaptic terminals in the distal 60% of the IPL (sublamina a) (Fig. 1, Stell, 1967; Famiglietti et al 1977; Stell et al 1977; Iśhida et al 1980; Saito et al 1983). The synaptic terminals of Mb and Ma cells are relatively large, ranging from 4-10 um in diameter, and are surrounded by amacrine cell synapses. Despite the elegant work describing the photoreceptor input to bipolar cell dendrites (Stell and Kock, 1984, for review), there is less information on the transmitter-specific inputs to their synaptic terminals.

Most data indicate massive GABAergic input to Mb terminals (Marc et al, 1978; Ishida and Kaneko, 1987; Yazulla et al, 1987). However, there is only scattered information regarding transmitter-specific input to Ma terminals and non-GABAergic input to Mb terminals (Marc and Lam, 1981; Yazulla, 1981; Zucker and Yazulla, 1982; Marshak et al, 1984; Yazulla et al, 1985; Yazulla and Zucker, 1988). In this paper, data on GABA, glycine, acetylcholine (ACh), dopamine, somatostatin (SRIF) and substance P from previous publications will be incorporated

NATO ASI Series, Vol. H31
Neurobiology of the Inner Retina
Edited by R. Weiler and N.N. Osborne
© Springer-Verlag Berlin Heidelberg 1989

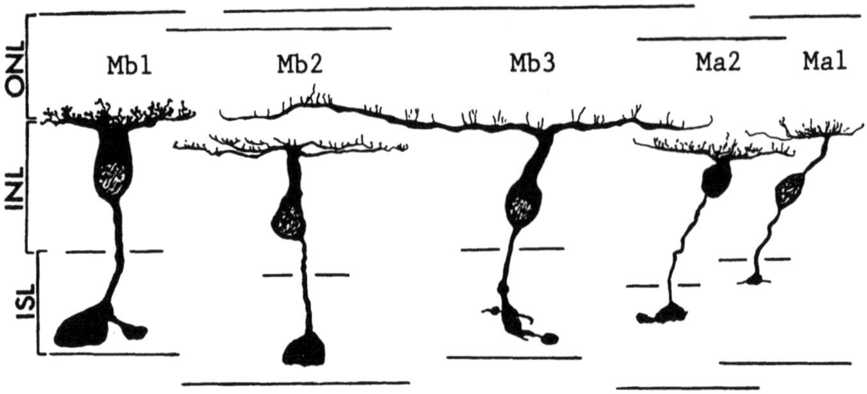

Fig. 1. Camera lucida drawing of mixed rod-cone bipolar cells of goldfish retina. Three types of ON-center cells (Mb1, Mb2, Mb3) ramify in the proximal IPL. Three types of OFF-center cells (Ma1, Ma2, Ma3, not illustrated) ramify in the distal IPL. One subtype each (Ma1, Mb1) contacts rods and red cones, whereas two subtypes each (Ma2, Ma3, Mb2, Mb3) contact rods and red and green cones (Stell and Kock, 1984). Used with permission (Fig. 9 from Stell, 1979). ONL, outer nuclear layer; INL, inner nuclear layer; ISL, inner synaptic layer.

with new data on GABA, glycine and somatostatin to present a more complete picture of transmitter-specific inputs to the synaptic terminals of mixed rod-cone bipolar cells. The smaller pure cone bipolar cells will not be dealt with in this paper, although it is possible that some of the contacts to be described involved them. Also, we could not differentiate between the subtypes of Mb and Ma terminals without serial reconstruction. Techniques included single and double-label electron microscopic autoradiography, pre- and post embedding immunocytochemistry, and ligand binding localization. Technical details are to be found in the cited publications.

GABA

Earlier studies suggested GABAergic amacrine cell input to Mb terminals by localizing high affinity ^{3}H-GABA uptake (Marc et al 1978) and to Ma terminals by localizing ^{3}H-muscimol binding (Yazulla, 1981). These suggestions have been corroborated in subsequent studies using a combination of high affinity ^{3}H-GABA uptake, GAD-immunoreactivity (IR) and GABA-IR (Yazulla et al, 1986, 1987; Studholme et al, 1988). It is clear now that putative GABAergic amacrine cells comprise several morphological types whose combined processes span the entire depth of the IPL (Ball and Brandon, 1986; Yazulla et al 1986). Except for the easily identified pyriform Ab amacrine cells (Marc et al 1978), the morphology and hence the contacts of the other types of "GABA" amacrine cells have not been differentiated.

GAD-IR - Thirty-two Mb terminals and 34 Ma terminals were analyzed for the percentage of GAD-IR amacrine cell contacts. Thirty-one of the 34 Ma terminals were in the distal 25% of the IPL. Both Mb and Ma terminals receive extensive contact from GAD-IR processes and also were found to make ribbon synaptic contacts onto GAD-IR processes (Fig. 2). There was a higher percentage of GAD-IR contacts onto Mb terminals than onto Ma terminals (77% vs 61%). If Mb terminals are divided into proximal and distal halves, a higher percentage of GAD-IR processes was around the proximal half than the distal half (82% vs 68%). Thus, the distal portion of Mb terminals and Ma terminals are similar in the proportion of GAD-IR amacrine cell synapses, which is considerably less than that onto the proximal portion of Mb terminals. A likely explanation for this difference is that the proximal portion of Mb terminals receives almost exclusive contact from a class of pyriform Ab amacrine cell that is unistratified in the proximal 25% of IPL and whose processes show a very high degree of ^{3}H-GABA uptake, GAD-IR and GABA-IR (Marc et al 1978, Yazulla et al 1987; Studholme and Yazulla, 1988). The other "GABAergic" amacrine cells may have their dendrites restricted to the distal 75% of the IPL and a less encompassing projection to Mb and Ma terminals compared with Ab amacrine cells.

GABA-IR - Analysis of 10 Mb terminals and 24 Ma terminals by postembedding immunocytochemistry of GABA-IR showed extensive labeling of amacrine cell processes onto bipolar cell terminals (Fig. 3). There was a higher percentage of GABA-IR processes around Mb terminals than Ma terminals (84% vs 75%). Also, there was a higher percentage of GABA-IR processes than GAD-IR processes around bipolar cell terminals: 75% vs 61% for

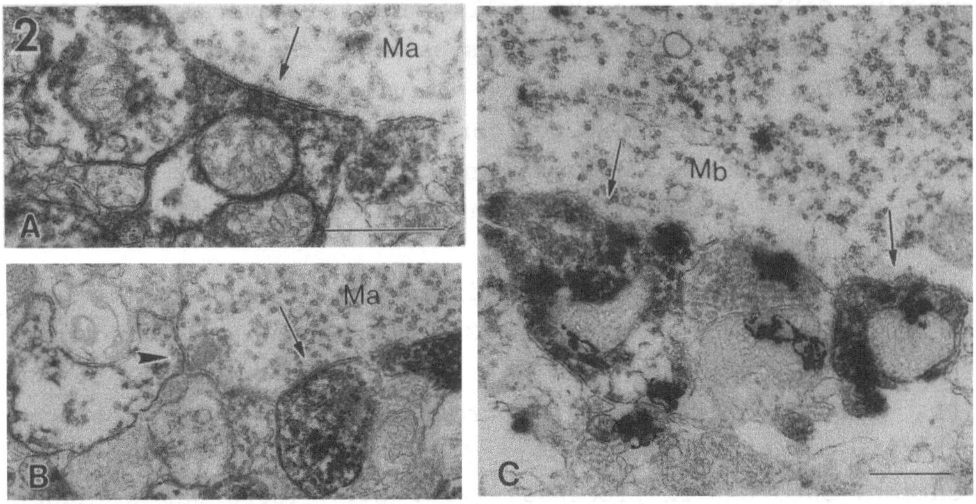

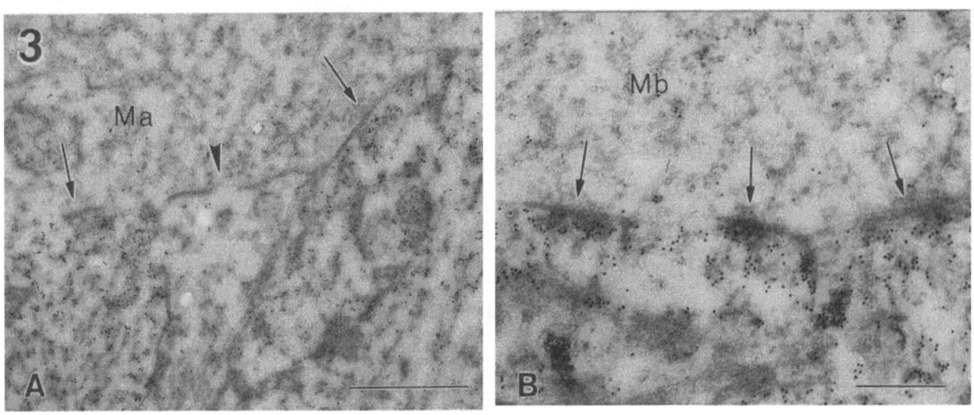

Fig. 2. GAD-IR amacrine cell processes in contact with bipolar cell terminals in goldfish retina. A and B) GAD-IR processes are presynaptic to Ma terminals (arrows). In B, a GAD-IR process is postsynaptic to a ribbon synapse (arrowhead). C) GAD-IR double-labeled with ^{3}H-GABA uptake; amacrine cell contacts onto an Mb terminal, all 3 of which took up ^{3}H-GABA but only 2 were GAD-IR (arrows). Scale bars for Figs. 2 & 3 = 0.5 um. Methods from Yazulla et al (1987).

Fig. 3. GABA-IR amacrine cell processes contacting Ma (A) and Mb terminals (B) in goldfish (arrows); LR White resin, postembedding with 10 nm colloidal gold. Note that the center process in A is not GABA-IR (arrowhead). Method from Studholme and Yazulla (1988).

Ma terminals, and 84% vs 77% for Mb terminals. The higher proportion of GABA-IR could be due, in part, to differences in the criteria used to determine what constitutes "labeling" for the pre-embedding (GAD-IR) and post embedding (GABA-IR) procedures. However, these data are consistent with earlier double-label studies of ^{3}H-GABA uptake and GAD-IR (Yazulla et al, 1986, 1987) that showed amacrine cell bodies and processes that were labeled only by ^{3}H-GABA uptake (Fig. 2C). There are reports in different species of a higher proportion of labeling with either GABA-IR or ^{3}H-GABA uptake compared with GAD-IR (Hendrickson et al 1985; Ball and Brandon, 1986; Mosinger and Yazulla, 1985, 1987). This raises the thorny question as to whether GAD-IR results in an underestimate or do ^{3}H-GABA uptake and GABA-IR result in a overestimate of GABAergic neurons. This dilemma will be difficult to resolve until the subtypes of "GABA"-positive amacrine cells are identified both morphologically and electrophysiologically.

GLYCINE

^{3}H-glycine is taken up by about 36% of the amacrine cells and a class of IPC in goldfish retina (Marc and Lam, 1981). Glycine-accumulating cells make extensive synaptic contact with amacrine cells and occasional contact with cone bipolar cell terminals (Marc and Lam, 1981). In our preliminary studies, utilizing a pre-embedding immunocytochemical procedure for glycine-IR, we confirmed the findings of Marc and Lam (1981). In addition, examples were found in which glycine-IR processes were pre- and post synaptic to presumed Ma terminals (Fig. 4A) accounting for perhaps 10% of their synaptic input. In regard to Mb terminals, small glycine-IR processes occasionally were postsynaptic at ribbon synapses (Fig. 4B) but, in agreement with Marc and Lam (1981) glycine-IR processes were not observed as a presynaptic element to Mb terminals.

ACETYLCHOLINE

Tumosa and Stell (1986) used choline acetyltransferase-IR to identify two types of putative cholinergic neurons in goldfish retina: an amacrine cell and displaced amacrine cell with

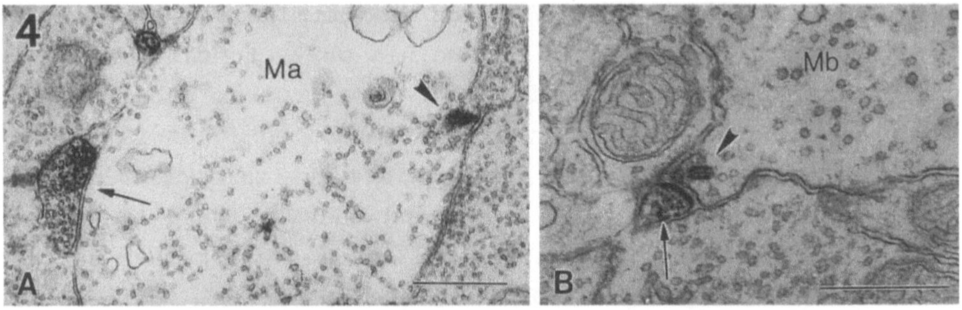

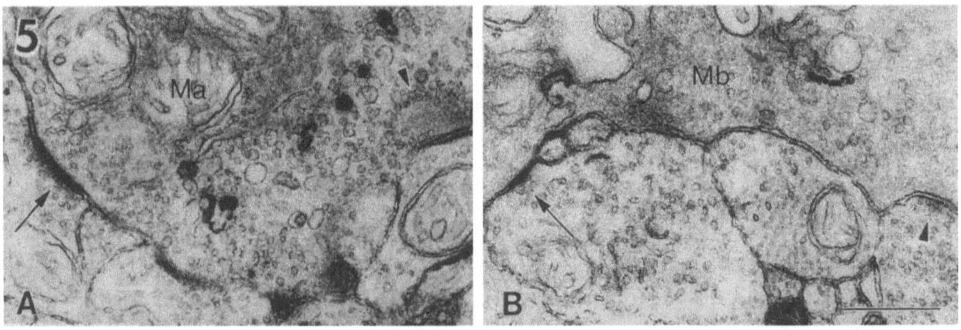

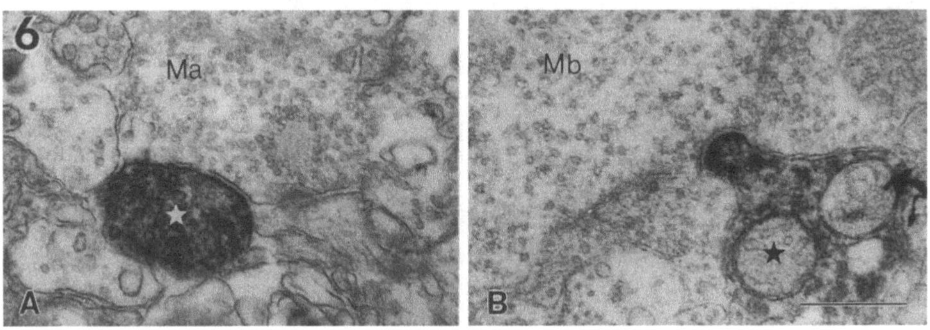

Fig. 4. Glycine-IR amacrine cell process that is presynaptic to an Ma terminal (A, arrow) and postsynaptic at a ribbon synapse of an Mb terminal (B, arrow). Note the synaptic ribbons in both bipolar cell terminals (arrowheads). Scale bars for Figs. 4,5,6 = 0.5 um. Method from Yazulla et al 1987.

Fig. 5. Putative nicotinic cholinergic synapses, indicated by binding of an HRP- -bungarotoxin conjugate, onto Ma (A) and Mb terminals (B) (arrows); a synaptic ribbon in the Ma terminal (arrowhead in A) and an unlabeled amacrine cell synapse onto the Mb terminal (arrowhead in B) are indicated. From Zucker and Yazulla (1982, figs. 9 and 10).

Fig. 6. DA-IPC processes, indicated by TOH-IR (stars), in contact with Ma (A) and Mb (B) terminals. The polarity of these contacts is uncertain. From Yazulla and Zucker (1988, figs. 10 and 11).

processes ramifying in sublamina 2 and 4. There is evidence for nicotinic cholinergic input from amacrine cells onto Ma and Mb synaptic terminals in the above mentioned sublamina (Fig. 5, Zucker and Yazulla, 1982). These contacts were labeled with an HRP- bungarotoxin conjugate and represented a very small percentage of the amacrine cell contacts onto the bipolar cell terminals. In view of the questions about the specificity of -bungarotoxin for nicotinic synapses in the central nervous system, it remains to be determined by electrophysiological analysis whether bipolar cells receive cholinergic input.

DOPAMINE

Dopaminergic interplexiform cells (DA-IPC) make extensive synaptic contact with other amacrine cells in the IPL (Dowling and Ehinger, 1978; Yazulla and Zucker, 1988). Analysis of tyrosine hydroxylase-IR revealed a small percentage of ribbon synapses from Ma and Mb terminals onto DA-IPC boutons. In addition, numerous DA-IPC boutons were observed in the IPL, ganglion cell and optic fiber layers that were filled with synaptic vesicles but lacked the membrane specializations characteristic of conventional synapses. It was suggested that these boutons were sites of transmitter release (Yazulla and Zucker, 1988). On occasion such DA-IPC boutons were adjacent to Ma and Mb terminals (Fig. 6); however, the direction of transmission at these sites is uncertain.

SUBSTANCE P

There is a type of substance P-IR amacrine cell (SP-1, Li et al 1986) that has a unistratified projection to layer 3 of the IPL. We found that about 50% of the synaptic input to SP-1 cells was derived from bipolar cells (Fig. 7). Similar to our analysis of DA-IPCs, SP-1-IR processes often were seen adjacent to bipolar cell terminals, but seemed to lack synaptic specializations. Although it was not suggested at the time, in context of our studies of DA-IPCs, it is possible that bipolar cells receive substance P input at these non-conventional sites of contact. The identity of the bipolar cells in contact with SP-1 cells is uncertain. The most likely possibility is the distal swelling of Mb3 terminals which are in layer 3 of the IPL, although proximally located Ma terminals cannot be excluded (Fig. 1). Li et al (1986) showed that substance P and neurotensin colocalized in SP-1 cells. Thus, synaptic interactions of Mb3 bipolar cells in sublamina 3 probably involve both substance P and neurotensin.

SOMATOSTATIN (SRIF)

There are two types of SRIF-IR amacrine cells in goldfish retina, one unistratified in sublamina 1 and the other in sublamina 5 (Yamada et al 1980). Analysis of the synaptic contacts of SRIF-IR amacrine cells showed that about 5-10% of their synaptic input and output involved bipolar cell terminals (Marshak et al 1984) in both sublaminae. Although cone bipolar cell terminals were not differentiated from Ma terminals, it is possible that Ma terminals were involved with SRIF-IR contacts.

MODULATION OF GABA INPUT AT Mb TERMINALS?

The vast majority of the synaptic input to mixed bipolar cells in the IPL comes from GABAergic amacrine cells. Yet there is pharmacological evidence that GABAergic transmission in the IPL is modulated by dopamine and enkephalin (O' Brien and Dowling, 1985; Su et al 1986). We reported morphological evidence in support of dopaminergic modulation of GABA inhibition of Mb terminals (Yazulla and Zucker, 1988). Numerous examples of a serial arrangement of a DA-IPC process, ^{3}H-GABA labeled amacrine cell process and an Mb terminal (Fig. 8A,B) were observed. Enkephalin-IR and ^{3}H-GABA uptake colocalize in an unidentified subtype of amacrine cell in goldfish retina (Su et al 1986). It is not known whether this cell type contacts bipolar cell terminals.

^{3}H-GABA Ab amacrine cells appear similar to Ab amacrine cells that are SRIF-IR (Yamada et al 1980), but they are separate types (Yazulla et al 1984). SRIF-IR Ab amacrine cells make a small percentage of their contacts onto Mb terminals (Marshak et al 1984). Upon re-examination of our micrographs, we found examples of a serial arrangement between SRIF-IR synaptic processes, amacrine cell processes and Mb terminals (Fig. 8C),

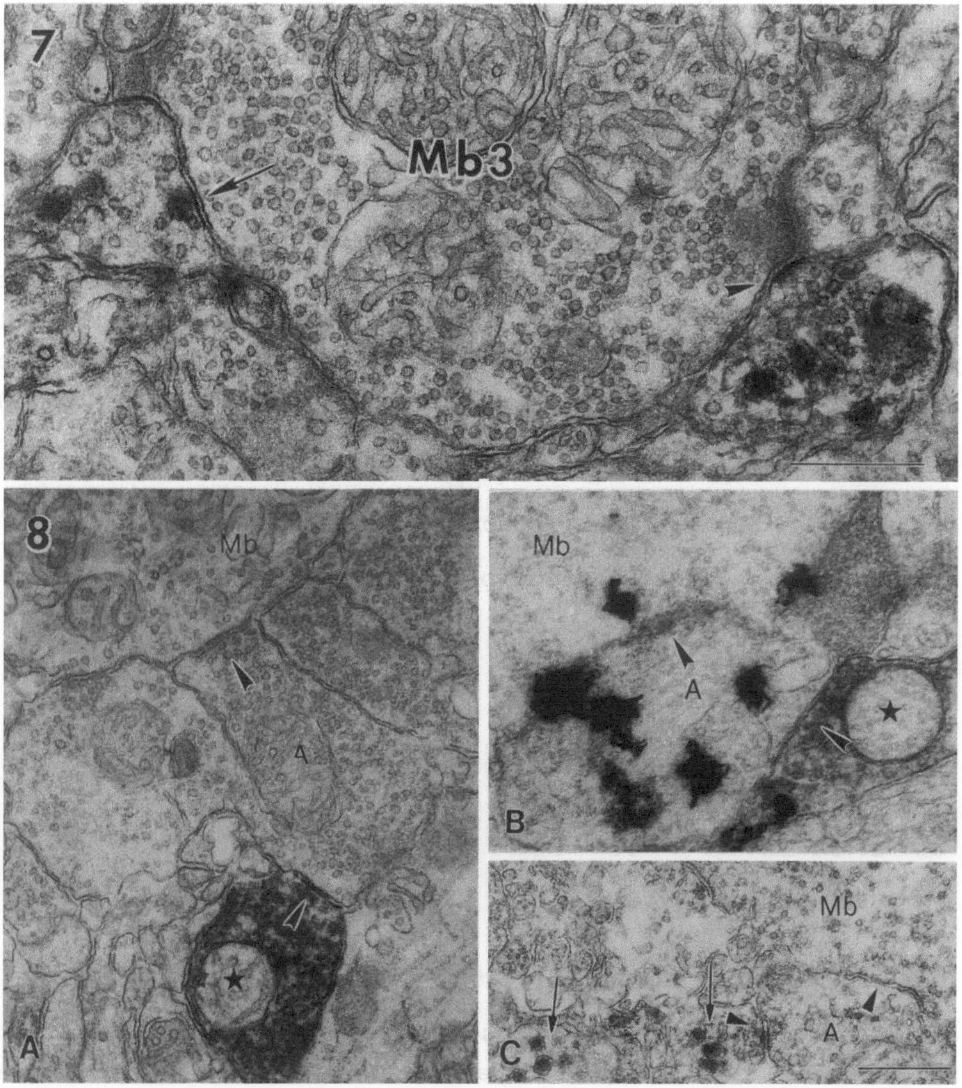

Fig. 7. Substance P-IR amacrine cell processes that are postsynaptic to a ribbon (arrowhead) and in apparent presynaptic contact (arrow) with an probable Mb3 terminal in layer 3 of the IPL. Scale bar = 0.5 um. (Yazulla et al, 1985; fig. 6).

Fig. 8. A,B) Serial arrangement in sublamina b involving a DA-IPC process (TOH-IR, star), amacrine cell process [A] (labeled by ^{3}H-GABA uptake in B) and Mb terminal. C) Two SRIF-IR processes (arrows); a serial arrangement between an SRIF-IR synaptic process, amacrine cell process [A] and Mb terminal. Arrowheads indicate the direction of presumed transmission. scale bar = 0.5 um. A,B) from Yazulla & Zucker (1988, fig. 7), C) from Yazulla et al (1984).

similar to that described for DA-IPCs. One of the two SRIF-IR processes is in synaptic contact with an amacrine cell process, which does not show a synaptic specialization. However, we have found by serial sectioning that all such amacrine cell processes eventually make synaptic contact with Mb terminals. Note that the SRIF-IR dense-cored vesicles are not adjacent to the presynaptic membrane; small agranular vesicles are. Perhaps SRIF colocalizes with a conventional transmitter and these substances are coreleased. Since 84% of the amacrine cell processes in this region stain for GABAergic markers and are likely to be GABAergic, these SRIF-IR terminals may be involved in the modulation of GABAergic inhibition of Mb bipolar cells.

Table 1. Summary and approximate percentage of various transmitter-specific inputs to Ma and Mb bipolar cell terminals. Identification of presynaptic contacts of DA-IPC boutons is equivocal.

	Ma Terminals	Mb Terminals	(ref.)
GABA			
GABA-IR	75%	84%	(1)
GAD-IR	61%	77%	(1,4)
Glycine-IR	<10%	0%	(1)
ACh	<10%	< 5%	(6)
SP/NT-IR	0%	Mb3 - <10%	(3)
Dopamine	< 5%?	?	(5)
SRIF-IR	< 5%	< 2%	(2)

1)-New Data; 2)-Marshak et al '84; 3,4)-Yazulla et al '85, '87; 5) Yazulla and Zucker '88; 6) Zucker and Yazulla '82.

Ma and Mb terminals synapse with dendritic processes that are stained for probes indicating GABA, glycine, substance P/neurotensin, dopamine and SRIF; contact with cholinergic processes has not been yet reported. Thus, bipolar cell output appears varied in regard to the transmitter types of amacrine cells and IPCs. Input to the synaptic terminals of Mb (ON) and Ma (OFF) mixed rod-cone bipolar cells (Table 1) seems to be overwhelmingly GABAergic, with a higher proportion onto Mb terminals than Ma terminals. Ma terminals also appear to receive a wider variety of transmitter-specific inputs than do Mb terminals, most notably glycine and perhaps dopamine. Despite the anatomical evidence for major GABAergic and minor glycinergic input to Ma terminals, Kondo and Toyoda (1983) reported that GABA and glycine had no effect on Ma bipolar cells when applied in the IPL. Ishida and Kaneko (1987), however, reported high sensitivity to GABA at the synaptic terminal of isolated mixed rod-cone bipolar cells but did not differentiate between Mb or Ma types. There is also the intriguing possibility that the distal swelling of Mb3 terminals is the sole bipolar cell target of SP-1 amacrine cells. Thus, Ma and Mb terminals appear to differ regarding the proportion and variety of potential synaptic inputs. We conclude that modulation of signal transmission from bipolar cells to ganglion cells via amacrine cells is different for the ON and OFF pathways, suggesting functional differences between the parallel pathways in visual processing.

I wish to acknowledge Drs. Judith Mosinger, Charles Zucker and Mr. Keith Studholme for their technical and intellecutal collaboration in the many studies described in this paper. Also, thanks to Drs. J.Y. Wu, R. Wenthold, T. Joh and I. Black for supplying us with antisera to GAD, GABA, glycine, tyrosine hydroxylase and substance P. Support by NIH Grant EY01682.

Ball AK and C Brandon (1986) Localization of ^{3}H-GABA, ^{3}H-muscimol, and ^{3}H-glycine uptake in goldfish retina stained for glutamate decarboxylase. J Neurosci 6:1221-1227.

Dowling JE and B Ehinger (1978) The interplexiform cell system I. Synapses of the dopaminergic neurons of the goldfish retina. Proc Roy Soc Lond B, 201:7-26.

Famiglietti EV Jr, A Kaneko and M Tachibana (1977) Neuronal architecture of on and off pathways to ganglion cells in carp retina. Science 198:1267-1269.

Hendrickson A, M Ryan, B Noble and JY Wu (1985) Colocalization of ^{3}H-muscimol and antisera to GABA and glutamic acid decarboxylase within the same neurons in monkey retina. Brain Res 348:391-396.

Ishida AT, WK Stell and DO Lightfoot (1980) Rod and cone inputs to bipolar cells in goldfish retina. J Comp Neurol 191:315-335.

Kondo H and J-I Toyoda (1983) GABA and glycine effects on bipolar cells of carp retina. Vision Res 23:1259-1264.

Li HB, DW Marshak, JE Dowling and DMK Lam (1986) Colocalization of immunoreactive substance P and neurotensin in amacrine cells of the goldfish retina. Brain Res 366:307-313.

Marc RE and DMK Lam (1981) Glycinergic pathways in the goldfish retina. J Neuroscience 1:152-165.

Marc RE, WK Stell, D Bok and DMK Lam (1978) GABAergic pathways in the goldfish retina, J Comp Neurol 182: 221-246.

Marshak DW, T Yamada and WK Stell (1984) Synaptic contacts of somatostatin-immunoreactive amacrine cells in goldfish retina, J Comp Neurol 225:44-52.

Mosinger JL and S Yazulla (1985) Co-localization of GAD immunoreactivity and ^{3}H-GABA uptake in amacrine cells of rabbit retina. J Comp Neurol 240:396-406.

Mosinger JL and S Yazulla (1987) Double-label analysis of GAD- and GABA-like immunoreactivity in the rabbit retina Vision Res 27:23-30.

O'Brien DR and JE Dowling (1985) Dopaminergic regulation of GABA release from the intact goldfish retina. Brain Res 360:41-50.

Saito T, T Kujiraoka and T Yonaha (1983) Connections between photoreceptors and horseradish peroxidase-injected bipolar cells in the carp retina. Vision Res 23:353-362.

Stell WK (1967) The structure and relationships of horizontal cells and photoreceptor-bipolar synaptic complexes in goldfish retina. Am J Anat 120:401-424.

Stell WK, AT Ishida and DO Lightfoot (1977) Structural basis for On- and Off-center responses in retinal bipolar cells. Science 198:1269-1271.

Studholme KM and S Yazulla (1988) Localization of GABA and glycine in goldfish retina by postembed immunocytochemistry: Improved visualization of synaptic structures with LR White resin. J. Neurocytol. (in press).

Su YYT, KR Fry, DMK Lam and CB Watt (1986) Enkephalin in the goldfish retina. Cell & Mol Neurobiol 6:331-347.

Tachibana M and A Kaneko (1987) -aminobutyric acid exerts a local inhibitory action on the axon terminal of bipolar cells: evidence for negative feedback from amacrine cells. Proc Nat Acad Sci USA 84:3501-3505.

Tumosa N and WK Stell (1986) Choline acetyltransferase immunoreactivity suggests that ganglion cells in the goldfish retina are not cholinergic. J Comp Neurol 244:267-275.

Yamada T, D Marshak, S Basinger, J Walsh, J Morley and WK Stell (1980) Somatostatin-like immunoreactivity in the retina. Proc Nat Acad Sci, USA 77:1691-1695.

Yazulla S (1981) GABAergic synapses in the goldfish retina: an autoradiographic study of ^{3}H-muscimol and ^{3}H-GABA binding. J Comp Neurol 200:83-93.

Yazulla S and CL Zucker (1988) Synaptic organization of dopaminergic interplexiform cells in the goldfish retina. Visual Neurosci 1:13-29.

Yazulla S, J Mosinger and C Zucker (1984) Two types of pyriform Ab amacrine cells in the goldfish retina: and EM analysis of [^{3}H]GABA uptake and somatostatin-like immunoreactivity. Brain Res 321:352-356.

Yazulla S, K Studholme and J-Y Wu (1986) Comparative distribution of ^{3}H-GABA uptake and GAD immunoreactivity in goldfish retinal amacrine cells: A double-label analysis. J Comp Neurol 244:149-162.

Yazulla S, KM Studholme and J-Y Wu (1987) GABAergic input to mb_1 bipolar cells in the goldfish retina. Brain Res 411:400-405.

Yazulla S, K Studholme and C Zucker (1985) Synaptic organization of substance-P like immunoreactive amacrine cells in the goldfish retina. J Comp Neurol 231:232-238.

Zucker C, and S Yazulla (1982) Localization of synaptic and nonsynaptic nicotinic-cholinergic receptors in the goldfish retina. J Comp Neurol 204:188-195.

EFFERENT PROJECTIONS TO THE GOLDFISH RETINA

Alexander K. Ball, William K. Stell* and **Diane A. Tutton**

Department of Anatomy,
McMaster University,
Hamilton, Ontario, L8N 3Z5

*Department of Anatomy,
The University of Calgary
Calgary, Alberta, T2N 4N1, Canada.

INTRODUCTION

The centrifugal innervation of the retina has been extensively described in birds (Cowan, 1970), but only recently has the existence of efferents been demonstrated in a variety of other species, including several species of fish (Witkovsky, 1971; Ebbesson and Meyer, 1981; Munz and Claas, 1981; Munz et al, 1982; Gerwerzhagen et al, 1982; Crapon de Caprona and Fritzsch, 1983; Meyer et al, 1983; Springer, 1983). Retrograde tracing methods have shown that there may be as many as five central sources of fish retinal efferents (Ebbesson and Meyer, 1981). Reports on the number and location of these efferent sources are inconsistent, but there is agreement that one source of retinal efferents is the terminal nerve (TN) (Springer, 1983).

The vertebrate TN was initially described by Fritsch in 1878 and designated a supernumary cranial nerve. The fish TN was later characterized by Sheldon (1909) and Brookover (1910). Somata of the TN are scattered along its course on the ventromedial surface of the olfactory bulb. Processes from these neurons extend rostrally into the olfactory epithelium, and caudally to the medial and lateral septal nuclei, olfactory tubercle, and preoptic areas (Demski and Schwanzel-Fukuda, 1987).

IMMUNOREACTIVE STAINING AND MORPHOLOGY OF TN EFFERENTS

A common feature among vertebrates is the staining of a proportion of TN cells with antisera directed against GnRH-like peptides (Demski, 1987) which can be blocked by appropriate pre-absorption controls (Stell et al, 1984). Antisera directed against luteinizing hormone releasing hormone (LHRH), molluscan cardioexcitatory peptide (FMRFamide) and one antisera against Substance P (SP) stained no more than 25-30% of the 60-65 ganglion cells in the goldfish TN (Stell et al, 1987). Labelling of the TN by retrograde

NATO ASI Series, Vol. H31
Neurobiology of the Inner Retina
Edited by R. Weiler and N.N. Osborne
© Springer-Verlag Berlin Heidelberg 1989

transport of True Blue injected into one eye and Lucifer Yellow into the other demonstrated that 75% of TN cells project to the contralateral retina, while the remaining project ipsilaterally or bilaterally (Stell et al, 1984, 1987). These retrogradely labelled cells were also shown to be immunoreactive for GnRH-like peptides and immunoreactive fibres from the TN cells could be traced to the optic nerve in serial frozen sections (Stell et al, 1984). A schematic drawing of the TN projection to the goldfish retina is shown in Fig. 1.

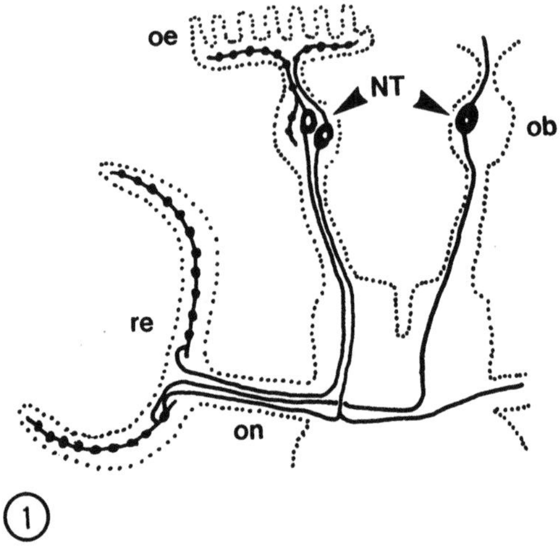

FIGURE 1 Schematic drawing of a dorsal view of the goldfish brain showing the retinopetal LHRH-IR *nervus terminalis* (NT) cells in the olfactory bulb (ob) which sends processes rostrally to the olfactory epithelium (oe) and caudally to the optic nerve (on).

Two antisera were used in studies of the efferent terminations in the retina: one, directed against mammalian LHRH, has been shown to conformationally specific (Ellinwood et al, 1985), while the other, directed against GnRH recognizes internal aa sequences (Fig. 2). In frozen-section and flatmount preparations, retinas stained with either antisera contained immunoreactive fibres in the optic nerve. These fibres could be followed to the mid-periphery in the optic nerve layer. They then crossed the inner plexiform layer (IPL) to ramify as a dense beaded plexus at its distal border with the inner nuclear layer (INL)(Fig. 3a). The beaded plexus extended over the entire retina in IPL substrata 1 (S1). The thin fibres crossed over one another frequently and varicosities were located at 8 um intervals (Fig. 3b).

The intensity of LHRH immunoreactive staining depended on the light condition in which fish were kept prior to fixation and subsequent processing for pre-embedding immunocytochemistry (Eldred et al, 1983). Staining was weakest in light-adapted retinas and densest in dark-adapted retinas (Fig.4).

mLHRH

pGlu-His-Trp-Ser-Tyr-Gly-Leu-Arg-Pro-Gly-NH_2

Salmon I GnRH

pGlu-His-Trp-Ser-Tyr-Gly-Trp-Leu-Pro-Gly-NH_2

(2)

FIGURE 2 Amino acid (aa) sequences of mammalian LHRH and Salmon I GnRH decapeptides. One antisera used in this study is specific for aa at both ends of mLHRH (line), suggesting that this antisera requires conformational specificity for recognition. Another antisera produced against GnRH precursor is specific for the internal to N terminal aa sequences 6 - 10 (line).

RETINAL CELLS CONTACTED BY EFFERENTS

Because of its location in IPL S1, the efferent fibre plexus could potentially contact bipolar, amacrine, and interplexiform cells as their neurites extend into the IPL. To determine if a pharmacologically identified cell type was specifically contacted by TN efferents, dopaminergic I1 interplexiform cells (IPCs) were labelled by uptake of [^{3}H]-dopamine, glycinergic amacrine cells (ACs) and I2 IPCs were labelled by uptake of [^{3}H]-glycine, or GABAergic amacrine cells were labelled by uptake of [^{3}H]-GABA or [^{3}H]-muscimol (Marc, 1986) prior to pre-embedding LHRH immunocytochemistry (Eldred et al, 1983) and light or electron microscope autoradiography (Fig. 5a, 6a). To increase the number of LHRH-IR fibres encountered in each section, serial horizontal sections were cut in the plane of IPL S1 (Fig. 5b, 6b).

Light microscopic observations demonstrated likely contacts between the efferent fibres and ACs identified using these methods, as well as other cell types (Fig. 7 - 12). Several beaded portions of efferent fibres were seen adjacent to GABAergic cells labelled by [^{3}H]-GABA uptake, [^{3}H]-muscimol uptake, or double-label immunocytochemistry (Fig. 7). In addition, efferent fibres were seen in the proximal IPL (Substrata 5) and varicosities were located next to displaced GABAergic amacrine cells (Ball and St. Denis, 1986)(Fig. 8). Close contact between efferent fibres in IPL S4/5 and large lightly-stained structures in the proximal IPL were also observed (Fig. 9a). These structures have been shown to correspond to the terminals of MBb1 bipolar cells (Fig. 9b). Occasionally, efferent fibres were observed leaving

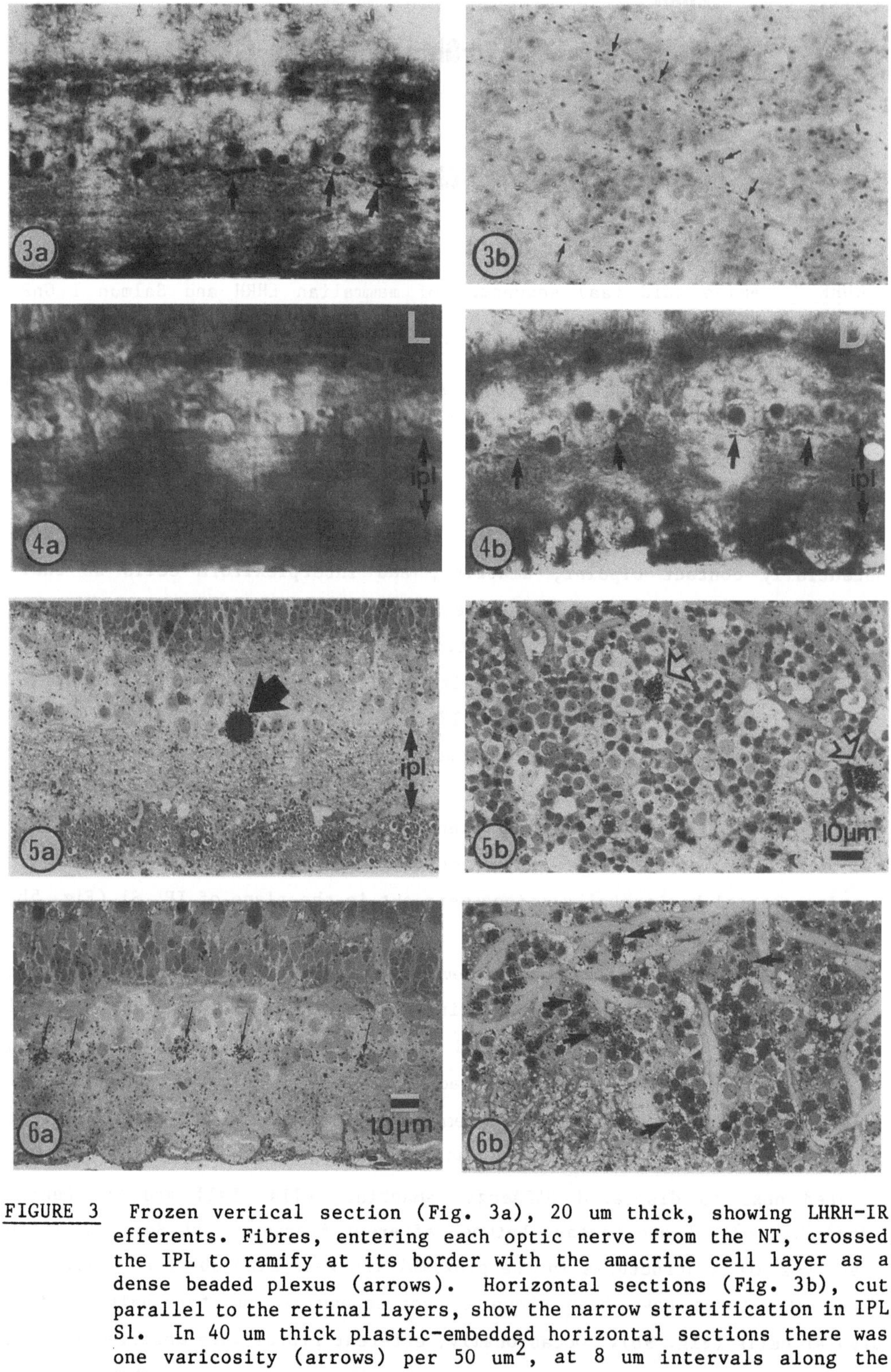

FIGURE 3 Frozen vertical section (Fig. 3a), 20 um thick, showing LHRH-IR efferents. Fibres, entering each optic nerve from the NT, crossed the IPL to ramify at its border with the amacrine cell layer as a dense beaded plexus (arrows). Horizontal sections (Fig. 3b), cut parallel to the retinal layers, show the narrow stratification in IPL S1. In 40 um thick plastic-embedded horizontal sections there was one varicosity (arrows) per 50 um^2, at 8 um intervals along the fibres.

FIGURE 4 Vertical plastic sections (Figs. 4a & 4b), 40 um thick, of retinal slices processed simultaneously using pre-embedding immunocytochemistry with mLHRH antisera. Fig. 4a shows immunoreactive staining in a 2hr light adapted goldfish retina and Fig. 4b shows staining in a 2hr dark adapted retina. No staining of light adapted retinas was observed, but LHRH-IR in dark adapted retinas was intense. Such staining was evident in fish during late summer and fall.

FIGURE 5 Light microscope autoradiographs showing the uptake of [^{3}H]-dopamine (DA) by goldfish I1 interplexiform cells (IPCs, arrows) in 0.5 um vertical (Fig. 5a) and horizontal (Fig. 5b) sections.

FIGURE 6 Light microscope autoradiographs showing the uptake of [^{3}H]-glycine by goldfish Aa amacrine cells (ACs, arrows) and I2 interplexiform cells in 0.5 um vertical (Fig. 6a) and horizontal (Fig. 6b) sections.

the main plexus in IPL S1 and ascending to the distal retina. Fibres have been observed making baskets around small somas in the INL (Fig. 10a) which may correspond to small bipolar cell somas (Fig. 10b). Beaded portions of the fibres travelled horizontally in the outer plexiform layer (OPL), where they could potentially contact bipolar cell dendrites, horizontal cells, or photoreceptor synaptic terminals (Fig. 11a). Rarely, fibres were seen making baskets around photoreceptor nuclei (Fig. 11b).

Electron microscopic observations showed evidence of synaptic contacts between efferent fibre varicosities and the cells observed by light microscopy. Although it has been shown that bipolar cells receive a considerable input from GABAergic amacrine cells (Marc, 1978; Yazulla, this volume), LHRH-IR efferents contact bipolar cell synaptic terminals as well (Fig. 13). Amacrine cell somata were usually contacted by the efferents as the cell narrowed at the INL-IPL border, before giving off primary processes (Fig. 13). Multiple contacts between efferents and amacrine cells were frequently observed. Efferent contact with retinal cells was most frequently made as the efferents ran along with their neurites in IPL S1. Close contacts between efferent fibre varicosities (probably synaptic) and retinal cell neurites in the IPL were identified in retinas double-labelled with [^{3}H]-glycine, [^{3}H]-dopamine, or [^{3}H]-GABA, and LHRH pre-embedding immunocytochemistry. These studies showed that each of these cell types received input from TN efferents (Fig. 14). Contacts between a similar efferent system and dopaminergic IPCs has previously been identified in the white perch retina (Zucker and Dowling, 1987). However, it is evident that TN efferents also contact many other cell types in the retina and the system must therefore have widespread influence on the retina.

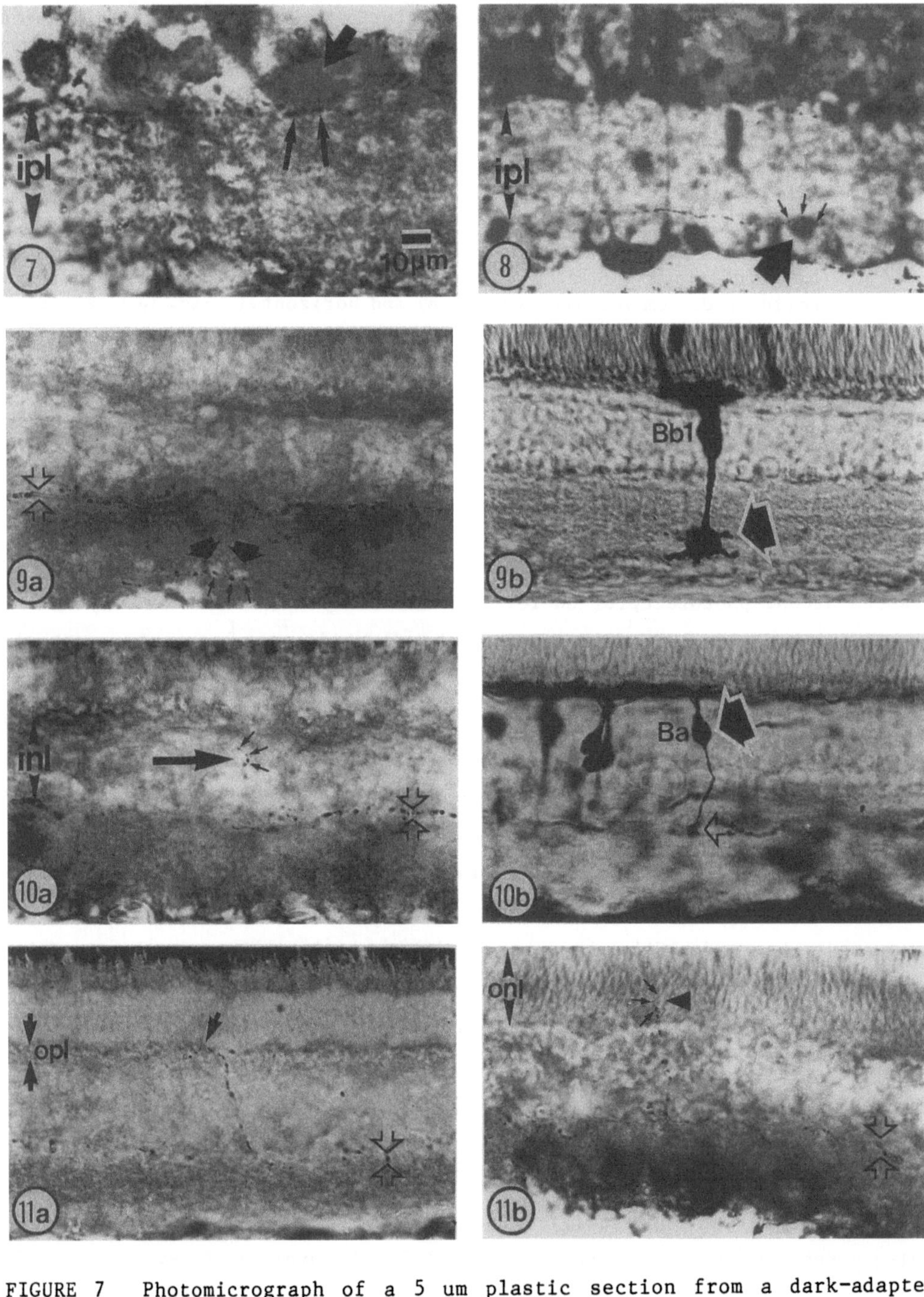

FIGURE 7 Photomicrograph of a 5 um plastic section from a dark-adapted goldfish retina which was double labelled for LHRH immunoreactivity (using DAB as the chromogen- brown diffuse) and subsequently GABA immunoreactivity (using BDHC as the chromogen- black needles). Beaded

portions of LHRH-IR processes in IPL S1 (arrows) make close contact with a GABA-IR amacrine cell (large arrow).

FIGURE 8 Autoradiograph of a dark-adapted goldfish retina which was exposed to [^{3}H]-muscimol and then processed for LHRH-immunoreactivity. Focus is at the level of LHRH-IR and silver grains over GABAergic ACs are out of focus. Short lengths of LHRH-IR fibres were sometimes located in IPL sublamina b. Beaded portions of these fibres made close contact with [^{3}H]-muscimol-labelled displaced GABAergic amacrine cell (large arrow).

FIGURE 9 Occasional LHRH-IR fibres left the main LHRH plexus in IPL S1 (open arrows) and travelled for short distances in IPL sublamina b. In addition to possible contacts with displaced amacrine cells or ganglion cells in this layer, beaded portions of these fibres came close to the large bulbous axon terminals of b1 bipolar cells (Fig. 9b, Golgi impregnation, arrow).

FIGURE 10 LHRH-IR processes were infrequently observed leaving the main plexus (open arrows) and ascending through the INL to reach the distal retina. Beaded portions of the fibres (arrows) were observed making baskets around possible bipolar cell somas (large arrow)(see Fig. 10b, Golgi impregnation).

FIGURE 11 Photomicrographs showing distal terminations of LHRH-IR ascending processes which left the primary plexus in IPL S1 (open arrows). Beaded processes (Fig. 11a, arrow) were seen travelling laterally in the outer plexiform layer (opl) and also ascending to the ONL (Fig. 11b), to make beaded baskets around photoreceptors (arrowhead).

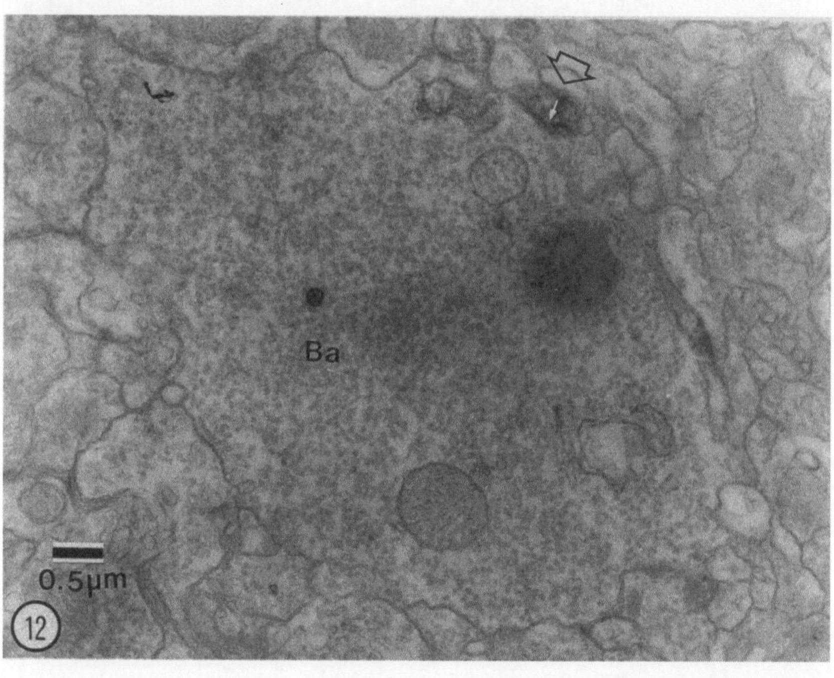

FIGURE 12 Electron micrograph of a horizontal section through IPL sublamina a. LHRH-IR efferent fibres (open arrow) made synaptic contact (white arrow) with a bipolar cell axon terminal (see Fig. 10b, open arrow).

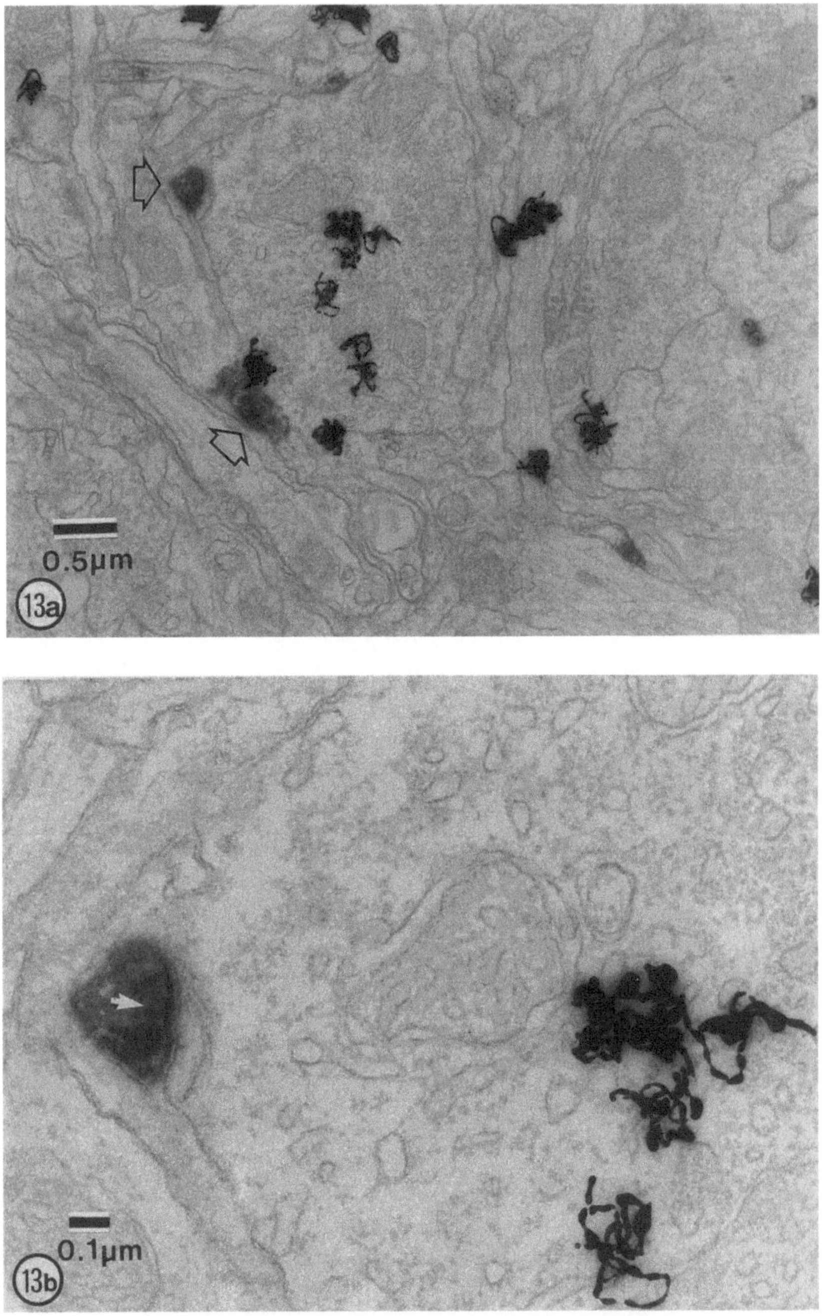

FIGURE 13 Combined electron microscope immunocytochemistry/ autoradiography showing an invaginating synapse of an efferent fibre onto a [^{3}H]-glycine accumulating amacrine cell soma (arrows). Horizontal section at the border of the IPL and IPL. At high magnification (Fig. 13b) small vesicles, a dense-cored vesicle and presynaptic density were evident (arrow).

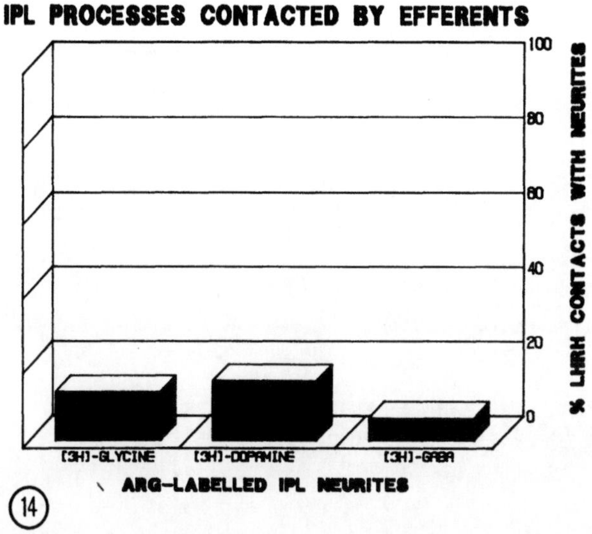

FIGURE 14 Graph summarizing frequency of LHRH contacts with [^{3}H]-glycine-[^{3}H]-dopamine-, and [^{3}H]-GABA-labelled IPL processes in combined EM immunocytochemistry/ autoradiographs. These results suggest that glycinergic amacrine cells or I2 interplexiform cells, dopaminergic I1 interplexiform cells and GABAergic amacrine cells are all contacted by the efferent system.

DIURNAL AND SEASONAL CHANGES

Although dense immunoreactive staining was present in dark, but not light-adapted retinas, staining of efferents was difficult to obtain in the spring with various pre-embedding immunocytochemical methods designed to preserve antigenicity and enhance antibody penetration (Eldred et al, 1983). However, in formalin-fixed frozen sections, light/dark differences in staining could be observed at this time of year as well (Fig. 15). As in the late summer, fall, and winter (Fig. 4), staining of efferents was dense in dark-adapted retinas (Fig. 15a, b) and absent in light-adapted retinas (Fig. 15c, d) using antisera against either GnRH or LHRH (Fig. 2). Immunocytochemistry is not a quantitative technique, but marked changes in staining suggest that there are higher concentrations of GnRH-like peptide present in the dark than in the light. Processing for pre-embedding immunocytochemistry causes a greater loss of antigenicity than conventional processing (4% formalin fixation, frozen sections). Staining was rarely evident in light-adapted retinas processed using this method (Fig. 4a), but staining was usually present in conventionally processed light-adapted retinas (Fig. 15c, d). Regardless of the immunocytochemical method used, staining was always more intense in dark-adapted retinas.

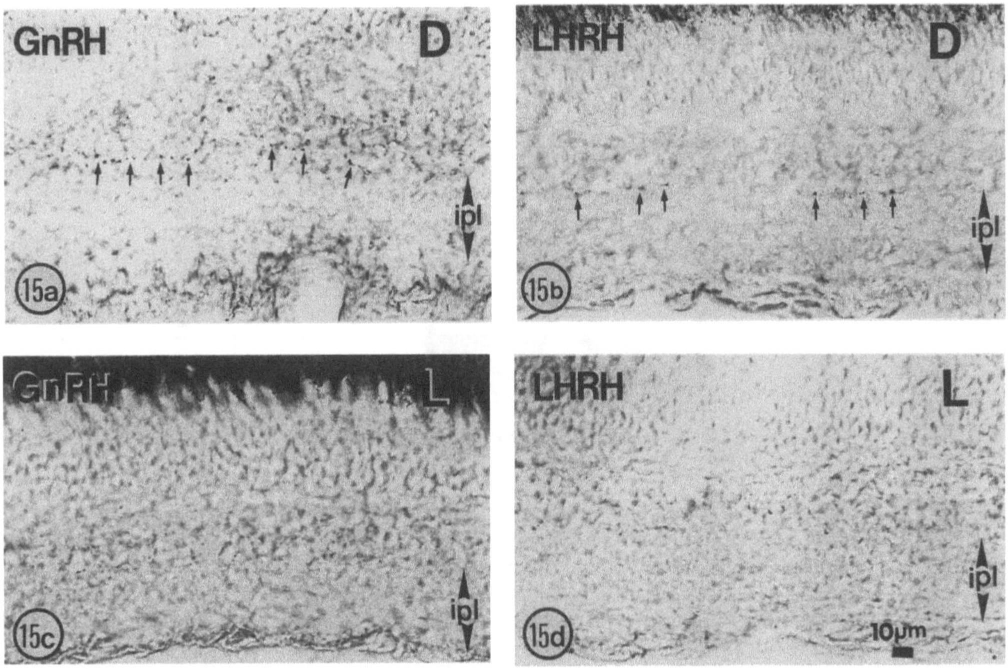

FIGURE 15 Light micrographs of 20 um formalin-fixed frozen sections from dark (D)(Figs. 15a & 15b) or light (L)(Figs. 15c & 15d) adapted retinas and incubated in antisera directed against internal sequence of Salmon I GnRH (Figs. 15a & 15c) or conformationally specific mLHRH (Figs. 15b & 15d). Retinas were from animals fixed during the reproductive season. With both antisera, less immunoreactive staining was found in light-adapted retinas.

Such light/dark differences could not be detected in the spring using pre-embedding immunocytochemistry, but were still evident in conventionally processed tissue. This finding suggests that less GnRH-like peptide is present in the retinal efferents in the spring and that a seasonal variation is superimposed on the light/dark variation in staining. Radioimmunoassay has recently shown that significant decreases in GnRH concentrations in certain brain regions, including olfactory bulb, occur in fish undergoing spontaneous ovulation compared to non-ovulating fish (Yu et al, 1987). It was found that changes in GnRH concentrations in different brain regions occur concomittantly with spontaneous ovulation and that elements of the GnRH neuronal system function as an integrated unit. The decreased staining of GnRH-IR retinal efferent fibres in the spring suggest that they are a part of this GnRH unit.

To study the time-course of the appearance of LHRH-immunoreactive staining in the dark and disappearance in the light, dark-adapted fish were killed 5, 30, 60, or 120 min after lights OFF in the middle of the day and light-adapted fish were killed 5, 30, 60 or 120 min after lights ON at the beginning of the light cycle. Immunoreactive staining in the dark-adapted fish was diminished 30 min after lights ON and decreased thereafter. Immunoreactive staining in the light-adapted fish began to appear 30 min after lights OFF and increased in intensity until 120 min after lights OFF. These findings suggest that the release of GnRH-like peptide may occur at lights ON and may continue to be released at a greater rate than can be replenished by transport from the TN. After lights OFF, release is discontinued and GnRH may reappear in the fibres. The continual release of GnRH in the light suggests a modulatory role for this peptide in the retina.

EFFECT OF LIGHT OR DARK ON LHRH-IR

TIME IN LIGHT		TIME IN DARK	
5 min	++	5 min	–
30 min	+	30 min	+
60 min	+–	60 min	++
120 min	–	120 min	+++

(16)

FIGURE 16 Table showing the effect of time in the light or dark on immunoreactive staining with antisera directed against mLHRH from animals fixed during the winter season (see Fig. 4). Immunoreactive staining was markedly diminished 30 min after lights ON. Staining intensity increased during dark adaptation, reaching a maximum at 120 min after lights OFF.

CONCLUSION

The function of the GnRH system remains obscure, but it has been suggested that the TN component may be involved in triggering sexual responses to pheromones (Demski and Northcutt, 1983). This is consistent with the observation of seasonal variations in the GnRH system. The retinal terminations of the TN may be associated with the visual part of the regulation and coordination of reproductive behaviours (Stell et al, 1987). Sandeman and Rosenthal (1974) have demonstrated that a centrifugal fibre system in fish may be facilitatory to ganglion cell responses. Although the effect of exogenous GnRH on ganglion cell activity has been shown to be rapidly desensitizing (Walker and Stell, 1986), it is not clear if the observed effects could be attributed to the system described by Sandeman and Rosenthal (1974). The consequences of such seasonal effects on visual processing remain speculative.

The light/dark variations in GnRH-immunoreactive staining suggests that aspects of adaptation may also be under efferent control. The fish dopaminergic interplexiform cell (DA IPC) system (Dowling, 1986) has been shown to have multiple modulatory roles in the retina related to light adaptation. Dopamine released in the light has been shown to regulate photoreceptor retinomotor movements (Dearry and Burnside, 1986), horizontal cell dendritic spinule formation (Weiler et al, 1988), and horizontal cell coupling through gap junctions (Negishi and Drujan, 1978). The demonstration that the efferent system contacts dopaminergic cell neurites (Fig. 14) and somata (Zucker and Dowling, 1987) suggests that this DA IPC system may be under efferent control. Further evidence supporting this suggestion comes from the observation of Arey (1916) that retinomotor responses depend on the integrity of the optic nerve, and that sectioning the optic nerve eliminates light/dark differences in horizontal cell gap junction particle density (Wolburg and Kurtz-Isler, 1985). Release of GnRH in the light could regulate the release of dopamine, which results in the cellular changes associated with light adaptation. Such indirect effects may also explain the observed desensitizing effect of GnRH on ganglion cell activity.

The significance of the efferent contacts on other retinal cells remain unknown, but may be related to other adaptational mechanisms. Activation of GABA B receptors have been shown to change the receptive field properties of ganglion cells, giving rise to a retinal model of selective attention (Slaughter, this volume). The demonstration of a large efferent input to GABAergic amacrine cells, including displaced GABAergic amacrines, could provide a basis for central control in such a model. The widespread terminations on bipolar cells, different types of amacrine cells, and photoreceptors, suggest diffuse and nonspecific actions of the system on visual processing, however, the significance of all these retinal contacts remains to be explained.

REFERENCES

Arey LB (1916) The function of the afferent fibres of the optic nerve of fishes. J Comp Neurol 26:213-245

Ball AK, Stell WK (1983) Structure of LHRH- and FMRFamide-Like immunoreactive fibres in the goldfish retina. Supp Invest Ophthal Vis Sci 24:66

Ball AK, St.Denis J (1986) Displaced GABAergic amacrine cells in the ganglion cell layer of the goldfish retina. Supp Invest Ophthal Vis Sci 27:332

Ball AK, St.Denis J (1987) LHRH-immunoreactive efferent fibres contact glycinergic and dopaminergic cells in the goldfish retina. Supp Invest Ophthal Vis Sci 28:277

Brookover C (1910) The olfactory nerve, the nervus terminalis and the pre-optic sympathetic system in amia calva L.. J Comp Neurol 20:49-119

Cowan MW (1970) Centrifugal fibres to the avian retina. Br. Med. Bull. 26:112-118

Crapon de Crapona M-D, Fritzsch B (1983) The development of the retinopetal nucleus olfacto-retinalis of two cichlid fish as revealed by horseradish peroxidase. Dev Brain Res 11:281-301

Dearry A, Burnside B (1986) Dopaminergic regulation of cone retinomotor movement in isolated retinas: I. Induction of cone contraction is mediated by D2 receptors. J Neurochem 46:1006-1021

Demski L, Northcutt RG (1983) The terminal nerve: A new chemosensory system in vertebrates? Science 220:435-437

Demski L (1987) Phylogeny of luteinizing hormone-releasing hormone systems in protochordates and vertebrates. Ann NY Acad Sci 519:1-11

Demski L, Schwanzel-Fukuda M (1987) Introduction: The Terminal Nerve; structure, function and evolution. Ann NY Acad Sci 519:ix

Dowling JE (1986) A dopaminergic neuromodulatory system. TINS 9:236-240

Ebbesson SOE, Meyer DL (1981) Efferents to the retina have multiple sources in teleost fish. Science 214:924-926

Eldred WD, Zucker C, Karten HJ, Yazulla S (1983) Comparison of fixation and penetration enhancement techniques for use in ultrastructural immunocytochemistry. J Histochem Cytochem 31:285-292.

Ellinwood WE, Ronnekleiv OK, Kelly MJ, Resko JA (1985) A new antiserum with conformational specificity for LHRH: Usefulness for radioimmunoassay and immunocytochemistry. Peptides 6:45-52.

Fritsch G (1878) Untersuchungen uber den feineren Bau des Fischgehirns mit besonderer Berucksichtigung der Homologien bei anderen Wirbelthierklassen. Verlag Gutmann'sche Buchhand. Berlin.

Gerwerzhagen K, Rickmann MJ, Meyer DL, Ebbesson SOE (1982) Optic tract cells projecting to the retina in the teleost, Pantodon bucholzi. Cell Tiss Res 225:23-28

Marc RE, Stell WK, Bok D, Lam DMK (1978) GABA-ergic Pathways in the Goldfish Retina. J Comp Neurol 182:221-245.

Marc RE (1986) Neurochemical stratification in the inner plexiform layer of the vertebrate retina. Vision Res. 26:223-238.

Meyer DL, Gerwerzhagen K, Fiebig E, Ahlswede F, Ebbesson SOE (1983) An isthmo-optic system in a bony fish. Cell Tiss Res 231:129-133

Munz H, Claas, B (1981) Centrifugal innervation of the retina in cichlid and poecilid fishes. A horseradish peroxidase study. Neurosci Lett 22:223-226

Munz H, Claas B, Stumpf WE, Jennes L (1982) Centrifugal innervation of the retina by luteinizing hormone releasing hormone (LHRH)-immunoreactive telencephalic neurons in teleostean fishes. Cell Tiss Res 222:313-321

Negishi K, Drujan B (1978) Effects of catecholamines on the horizontal cell membrane potential in the fish retina. Sens Proc 2:388-395

Sandeman DC, Rosenthal NP (1974) Efferent axons in the fish optic nerve and their effect on the retinal ganglion cell. Brain Res 68:41-54

Sheldon RE (1909) The nervus terminalis in the carp. J Comp Neurol 19:192-201

Springer AD (1983) Centrifugal innervation of goldfish retina from ganglion cells of the Nervus Terminalis. J Comp Neurol 214:404-416

Stell WK, Walker SE, Chohan KS, Ball AK (1984) The goldfish nervus terminalis: A luteinizing hormone-releasing hormone and molluscan carioexcitatory peptide immunoreactive olfactoretinal pathway. Proc Nat'l Acad Sci USA 81:940-944

Stell WK, Walker SE, Ball AK (1987) Functional-anatomical studies on the terminal nerve projection to the retina of bony fishes. Ann NY Acad Sci 519:80-96

Weiler R, Kohler K, Kirsch M, Wagner H-J (1988) Glutamate and dopamine induce synaptic plasiticity in horizontal cell dendrites of fish retina. Neurosci Lett, in press

Walker SE, Stell WK (1986) Gonadotropin-releasing hormone (GnRF), Molluscan Cardioexcitatory peptide (FMRFamide), enkephalin and related neuropeptides affect goldfish retinal ganglion cell activity. Brain Res 384:262-273

Wolburg H, Kurtz-Isler G (1985) Dynamics of gap junctions between horizontal cells in the goldfish retina. Exp Brain Res 60:397-401

Witkovsky P (1971) Synapses made by myelinated fibres running to the teleost and elasmobranch retinas. J Comp Neurol 142:205-222

Zucker C, Dowling J (1987) Centrifugal fibres synapse on dopaminergic interplexiform cells in the teleost retina. Nature 330:166-168

Yu KL, Nahorniak CS, Peter RE, Corrigan A, Rivier JE, Vale WE (1987) Brain distribution of radioimmunoassayable gonadotropin-releasing hormone in female goldfish: Seasonal variation and periovulatory changes. Gen. Comp. Endocrin. 67:234-24

CORRELATION BETWEEN ELECTROPHYSIOLOGICAL RESPONSES AND MORPHOLOGICAL CLASSES OF TURTLE RETINAL AMACRINE CELLS.

J. Ammermüller[1] and R. Weiler[2].

[1]Zoolog. Institute, University of Munich, Luisenstr. 14, 8 München 2, FRG;

[2]Dept. of Neurobiology, University of Oldenburg, 29 Oldenburg, FRG.

INTRODUCTION

Turtle retinal amacrine cells have been classified into twenty-nine morphological types (Kolb 1982; Kolb et al. 1988). Like in other species, four basic physiological types of amacrine cell responses (ON, OFF, slow ON-OFF and fast ON-OFF; e.g. Teranishi et al. 1987) have been described in turtle retina which can be correlated to some of the morphological types (Marchiafava and Weiler 1982; Weiler and Marchiafava 1981; Jensen and DeVoe 1982; Ammermüller and Weiler 1988). However, there still exists a large discrepancy between the number of physiologically and the number of morphologically classified amacrine cell types.

In the outer plexiform layer it is now well accepted, that the different morphological types of photoreceptors, horiziontal cells and probably also bipolar cells reflect distinct physiological properties (Miller et al. 1973; Normann and Kolb 1981; Weiler 1981; Ohtsuka 1985). In this study we try to answer the question, whether this is also true for the great morphological variety of amacrine cells in the inner plexiform layer (IPL). Therefore, we recorded and stained approximately 300 amacrine cells in the retina of *Pseudemys scripta elegans* during the last years and stained most of them with Lucifer yellow or horseradish peroxidase, using conventional recording and staining techniques. Onehundret and five of these amacrine cells were stained unequivocally and could therefore be classified morphologically, using the criteria and nomenclature introduced by Kolb (1982). The number of identified amacrine cells seems now high enough to present a first, preliminary 'catalogue' correlating most of the morphological amacrine cell types in turtle retina with electrophysiological responses to various light stimuli.

In this paper we will concentrate on amacrine cell types which so far have not been identified physiologically and which were recorded and stained at least three times during the course of this study. In a summarizing figure we will present all of the classified amacrine cell types schematically.

NATO ASI Series, Vol. H31
Neurobiology of the Inner Retina
Edited by R. Weiler and N.N. Osborne
© Springer-Verlag Berlin Heidelberg 1989

SUSTAINED, HYPERPOLARIZING AMACRINE CELLS (OFF-centre).

Figure 1 is a camera lucida drawing of two wide field, monostratified amacrine cells ramifying in sublayer S1. The cells resemble very much A 17, described by Kolb et al. (1988). The responses to various light stimuli are shown in Fig. 2.

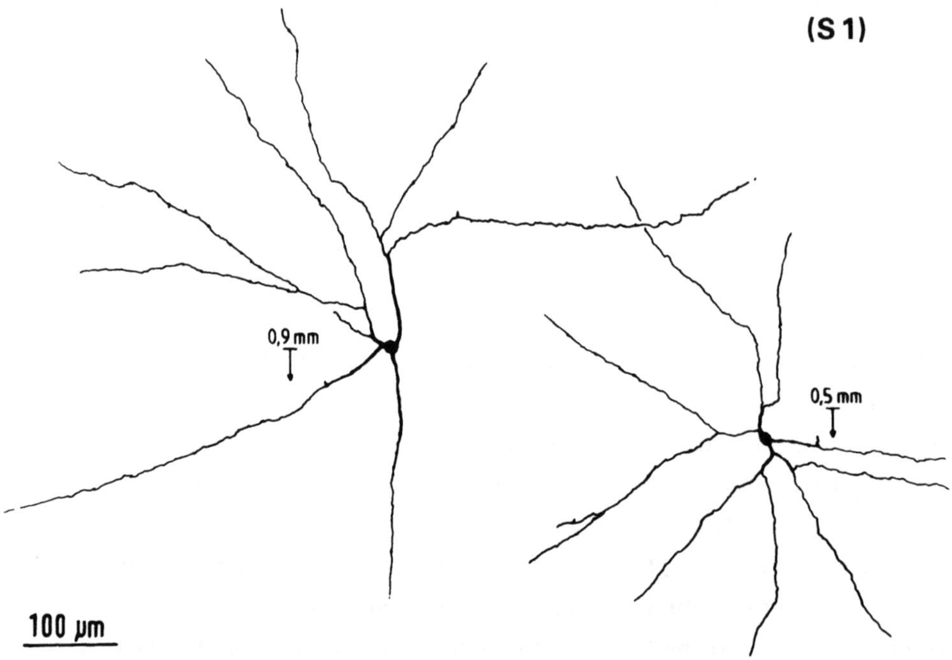

Fig. 1: Camera lucida drawing of two A 17 type amacrine cells. In this and all following drawings the scale bar represents orientation of the longitudinal visual streak. Distance of the soma from the visual streak is indicated by the numbers above the arrows. Ramification level(s) in IPL is shown in brackets.

During light-ON the response was a sustained hyperpolarization, which became larger for increasing spot sizes. There the hyperpolarization was also followed by a small depolarizing component after light OFF (Fig. 2A). No response was detectable during stimulation with an annulus. With the intensity used, all stained A 17 were less sensitive to blue than to red or green spot stimulation (Fig.2B). The response amplitudes increased with increasing intensities (Fig.2C). 'Noisy' voltage fluctuations occured throughout the whole recording of this amacrine cell.

A camera lucida drawing of two examples of another type of sustained hyperpolarizing amacrine cell is shown in Fig. 3. The elongated, asymmetric areas covered by their processes measure about 500 x 300 µm in diameter. Most branches ramify in sublayers S1 and S3. However, parts of the branches extend into

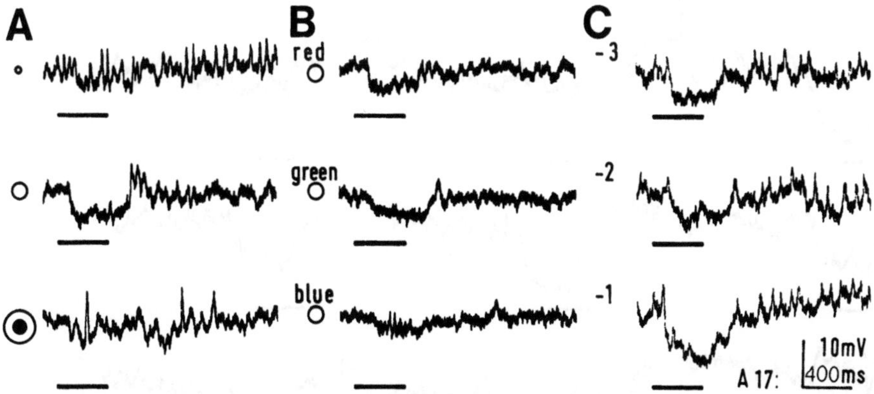

Fig. 2: Photoresponses of a A 17 type amacrine cell. In this and the following figures showing photoresponses, spot stimulation is symbolized by circles, annulus stimulation by annuli. The spot sizes were always 200 µm, 800 µm and 3000 µm, respectively, unless otherwise indicated. Annulus dimension was always 3000/1000 µm outer/inner diameter. The wavelengths used for colour stimulation were: 625 nm (red), 520 nm (green) and 450 nm (blue). Relative light intensity was always 10^{-3} for colour stimulation. Bars indicate duration of light stimuli. A: Photoresponses to spots and an annulus. Relative intensity 10^{-2}. Spot sizes; 200 µm (top), 800 µm (middle). B: Responses to colours. Spot size; 800 µm. C: Responses to increasing light intensities, indicated by the numbers near the recordings (white light, spot size 400 µm).

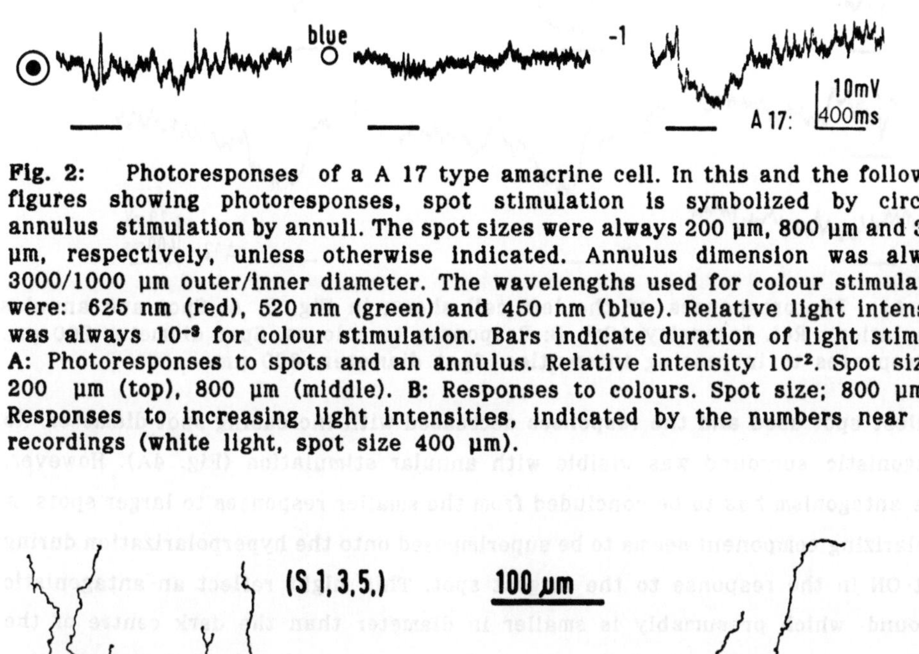

Fig. 3: Two tristratified amacrine cells, presumably A 12.

sublayer S5. The soma is decentered with respect to the processes. The ramification levels are equal to those of Kolb's (1982) small field A 12 amacrine cell, but the field sizes seem to be larger. Therefore, we are not completely sure whether these cells correspond to A 12. The responses of one of the stained cells are shown in Fig. 4. It responded with the largest hyperpolarization to the

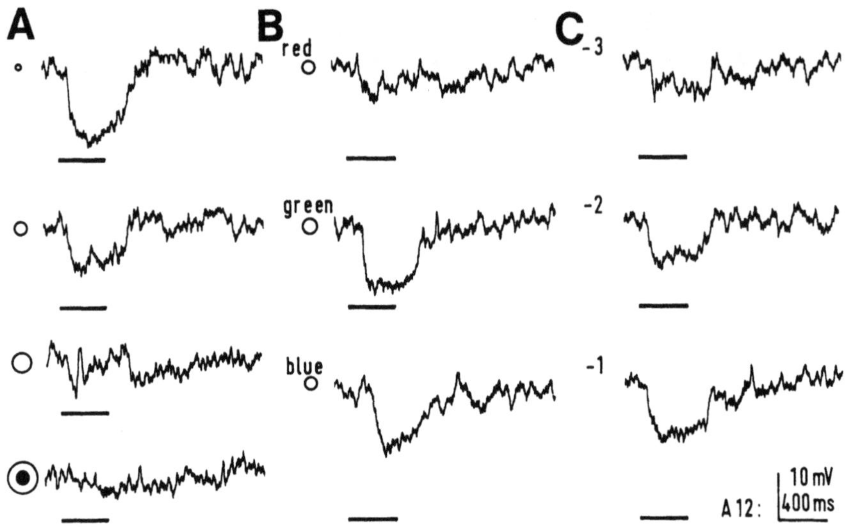

Fig. 4: Photoresponses of the left cell shown in Fig. 3. A: Spot and annulus stimulation. Rel. intensity 10^{-2}. B: Responses to colours. Spot diameter 200 µm. C: Responses to increasing intensities. Spot diameter; 800 µm.

smallest spot used and the responses decreased with increasing spot diameter. No antagonistic surround was visible with annular stimulation (Fig. 4A). However, some antagonism has to be concluded from the smaller responses to larger spots. A depolarizing component seems to be superimposed onto the hyperpolarization during light ON in the response to the largest spot. This might reflect an antagonistic surround which presumably is smaller in diameter than the dark centre of the annulus used and therefore, the surround was not stimulated adequately by this annulus. This amacrine cell type was always less sensitive to red light than to green or blue light at the intensity tested (Fig. 4B). Increasing intensities elicited larger hyperpolarizations (Fig. 4C).

TRANSIENT, HYPERPOLARIZING AMACRINE CELLS.

A micrograph and a camera lucida drawing of another tristratified, medium field sized amacrine cell type is shown in Fig. 5. These amacrine cells ramify in sublayers S1, S2 and at the S4/5 border. The main branching field is approximately circular, however, always one or two long processes are running radially. The ramification levels and the central branching field resemble the putative dopaminergic A 28 (Kolb et al. 1987, 1988). However, the long processes have not been described in these studies. It might be possible that these

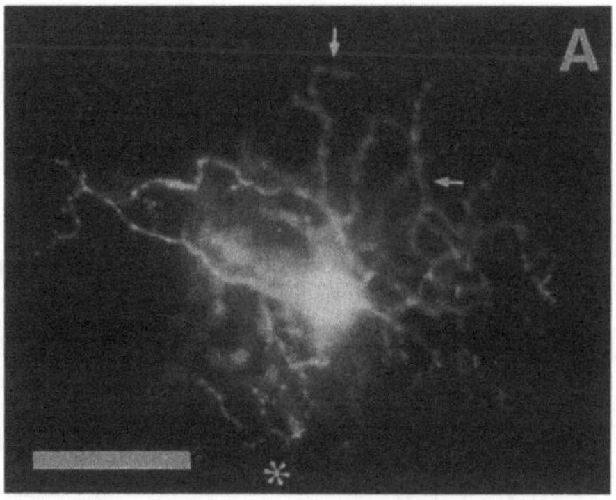

Fig. 5: A: Micrograph of presumptive A 28. Level of focus was on S2. Branches in S1 are only slightly out of focus (arrows). Branches in S4/5 border are largely defocused. One long process is running in S2 (asterisk), the other in S4/5 (not visible). Scale 100 µm.
B: Camera lucida drawing of two presumptive A 28. The cell on the right side is shown in the micrograph of A.

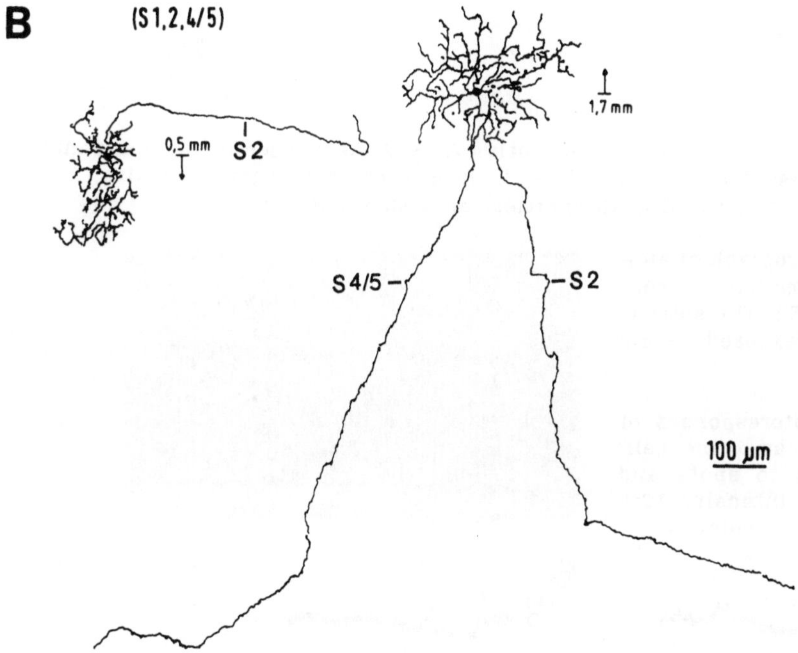

processes are too difficult to see in immunocytochemically or Golgi stained material, leaving the possibility open that this amacrine cell type really is A 28. The responses of this amacrine cell type were always transient hyperpolarizations at light ON and after light OFF, especially for medium sized spots. With increasing spot size the responses became more sustained and for annulus stimulation, the responses were pure sustained hyperpolarizations (Fig. 6A). The amplitude of the hyperpolarizations at light ON as well as at light OFF were smallest for blue light stimuli (Fig. 6B). At high intensities the responses were sustained hyperpolarizations similar to annulus stimulation, which is most probably due to strong straylight, stimulating a larger area (Fig. 6C).

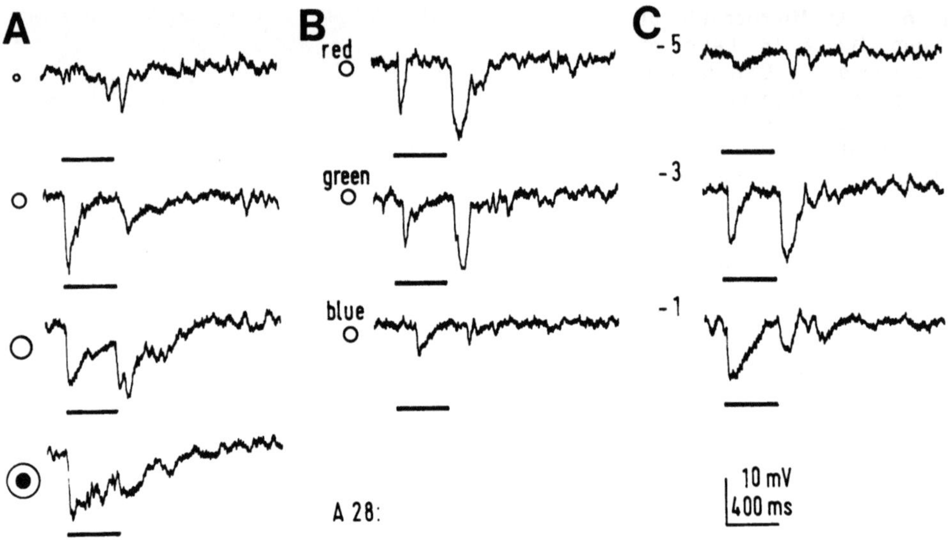

Fig. 6: Photoresponses of presumptive A 28. A: Responses to spots and annulus. Spot diameters; Rel. intensity 10^{-2}. B: Responses to colours. Spot size; 800 µm. C: Responses to increasing intensities. Spot size; 800 µm.

Fig. 7: Micrograph of an A 2 type amacrine cell, ramifying in S2. The soma is heavily overexposed. Scale 100 µm.

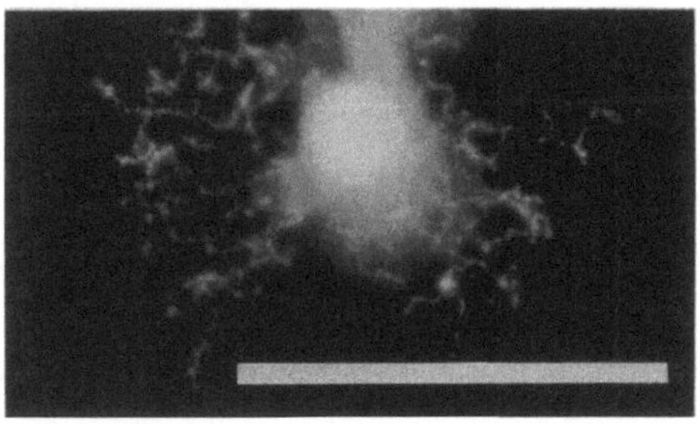

Fig. 8: Photoresponses of an A 2 type amacrine cell. A: Responses to spots and annulus. Rel. intensity 10^{-2} B: Responses to colours.

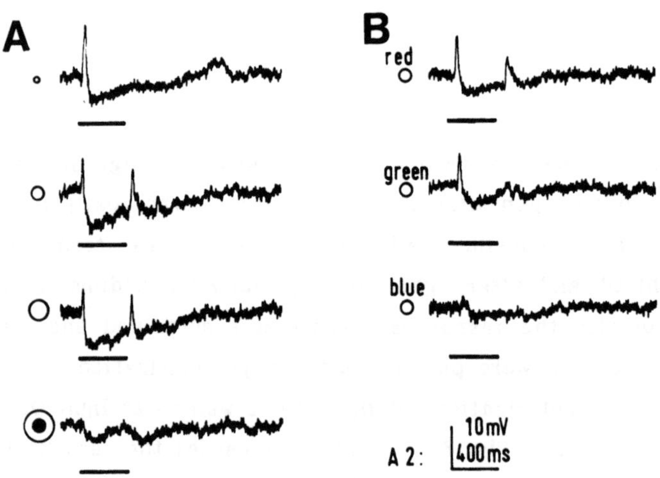

TRANSIENT, DEPOLARIZING AMACRINE CELLS (ON-OFF).

Figure 7 shows a narrow field, monostratified amacrine cell ramifying with very fine, varicose processes in sublayer S2 and resembling A 2 of Kolb's (1982) classification. The responses to small spots were fast, transient depolarizations, followed by a more sustained and long lasting hyperpolarizing component. Stimulation with larger spots introduced a second, fast depolarization after light OFF superimposed onto the decay of the hyperpolarizing component (Fig. 8A). No antagonistic surround was detectable. The sensitivity to colours was the typical one for amacrine cells, namely a smaller response to blue than to green than to red (Fig. 8B).

ON-OFF depolarizations in response to spots were also obtained from an amacrine cell type resembling A 24 of Kolb's (1982) classification (Fig. 9). This wide field amacrine cell type has numerous long, and some shorter, radiating processes in sublayer S3. All branching points are near the soma and in the flat mount view this amacrine cell type belongs to the 'stellate' or 'radiate' type, seen in many vertebrate retinas (Cajal 1892). The responses to spots were fast, transient

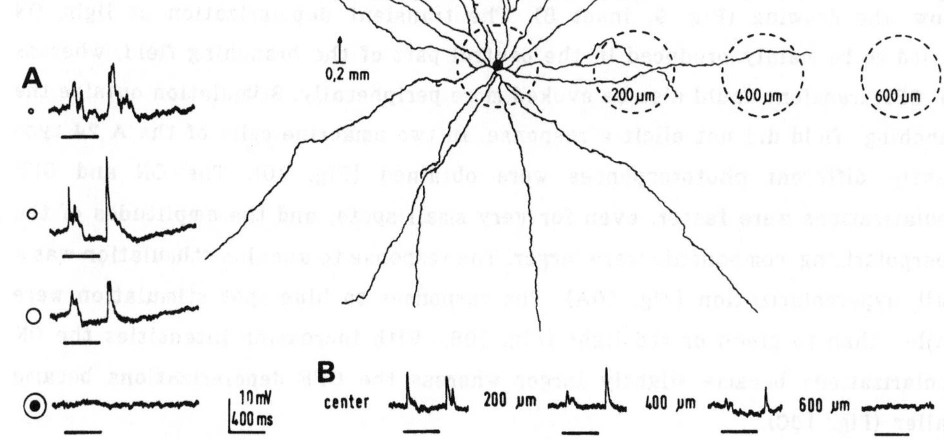

Fig. 9: An A 24 type amacrine cell. Inset A: Responses to centered spots and annulus. Rel. intensity 10^{-2}. Inset B: Responses to a spot of 100 µm diameter while stimulating the retina at the areas indicated in the camera lucida drawing. The distances to the soma (=center) are indicated beneath the recordings. Rel. intensity 10^{-4}.

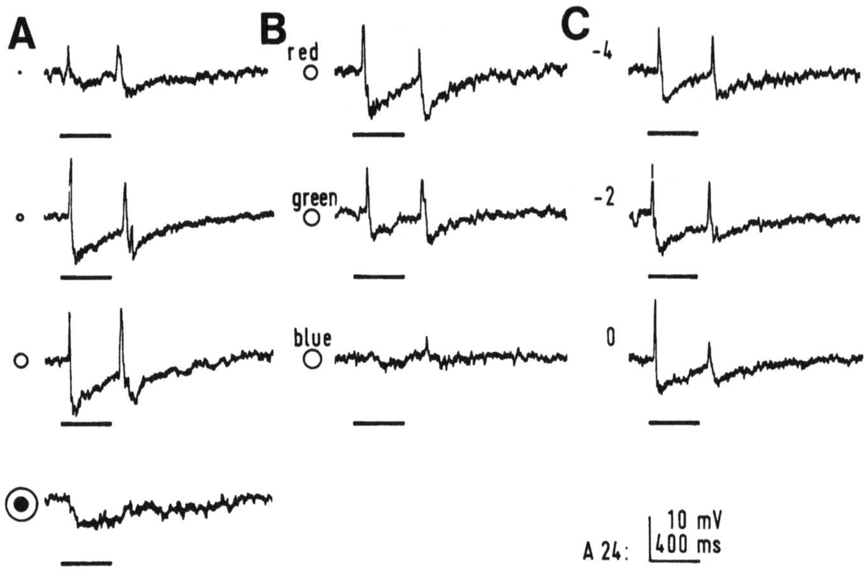

Fig. 10: Slightly differing A 24 photoresponses. A: Responses to spots and annulus. Spot sizes; 100 µm, 200 µm and 800 µm. Rel. intensity 10^{-2}. B: Responses to colours. Spot size; 3000 µm. C: Responses to increasing intensities. Spot size; 200 µm.

depolarizations at light ON and OFF, accompanied by small hyperpolarizing components. No significant response to annulus stimulation was detectable (Fig. 9, inset A). With the cell shown in Fig. 9, a rough test of the receptive field size was possible. Stimulation with small-diameter spots at the retinal areas indicated in the camera lucida drawing of Fig. 9B resulted in the responses shown below the drawing (Fig. 9, inset B). The transient depolarization at light ON seemed to be mainly produced in the central part of the branching field, whereas the OFF transient could also be evoked more peripherally. Stimulation outside the branching field did not elicit a response. In two amacrine cells of the A 24 type slightly different photoresponses were obtained (Fig. 10). The ON and OFF depolarizations were faster, even for very small spots, and the amplitudes of the hyperpolarizing components were larger. The response to annulus stimulation was a small hyperpolarization (Fig. 10A). The responses to blue spot stimulation were smaller than to green or red light (Fig. 10B). With increasing intensities the ON depolarizations became slightly larger whereas the OFF depolarizations became smaller (Fig. 10C).

A very characteristic turtle amacrine cell type is A 16, which cannot be confused with any other type. The main process bifurcates in sublayer S5 and the two only branches run into the opposite directions, crossing nearly the entire retina.

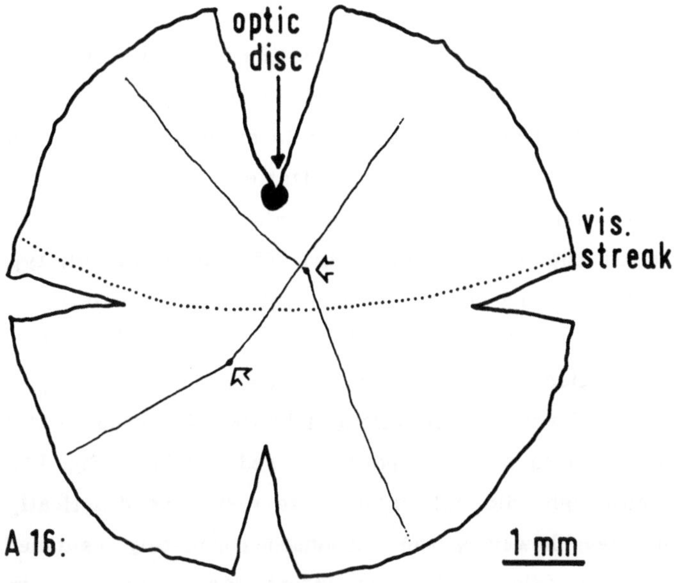

Fig. 11: Schematic drawing of a retina with two A 16 type amacrines. Somas are indicated by arrows.

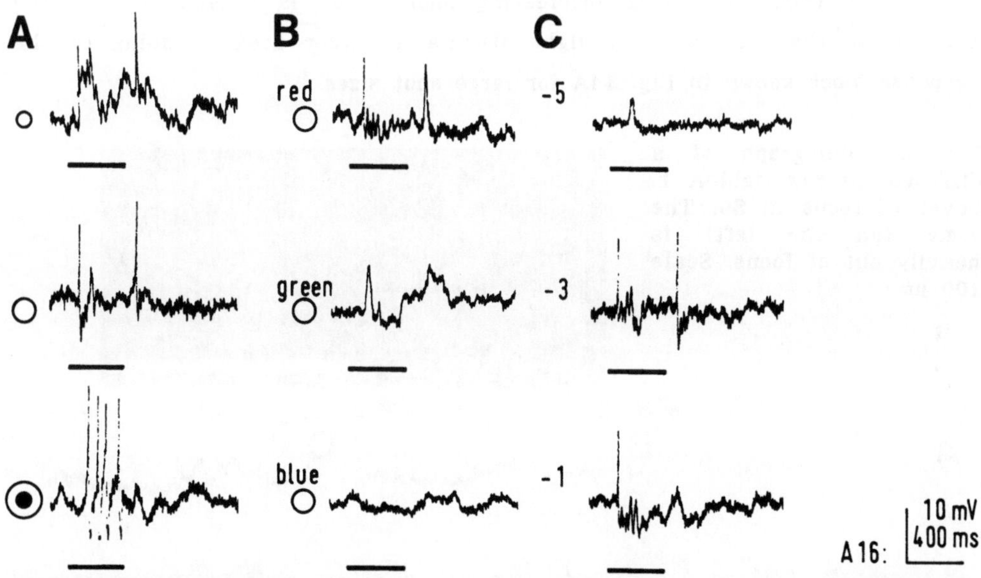

Fig. 12: Photoresponse of an A 16 type amacrine cell. A: Responses to spots and annulus. Spot sizes; 400 µm and 800 µm. Rel. intensity 10^{-2}. B: Responses to colours. Spot size 800 µm. C: Responses to increasing intensity. Spot size 800 µm

Figure 11 shows a semi-schematic drawing of a retina with two stained and recorded A 16, indicating the large extension of these strange cells. The responses of A 16 were, especially for larger spots, fast spikelike depolarizations at light ON and OFF. The response to an annulus was a 'burst' of 'spikes' during light ON (Fig. 12A). The amplitude to blue light stimulation was very small with the intensity used (Fig. 12B). With high intensity stimulation the fast OFF transient vanished (Fig. 12C).

Another example of a transient, depolarizing amacrine cell type is shown in the micrograph of Fig. 13. The processes of this narrow field amacrine cell branch diffusely throughout the whole IPL - identifying it as an A 1 type amacrine cell (Kolb 1982). The light responses to stimulation with a medium sized spot were basically slow transient depolarizations at light ON and OFF. However, small hyperpolarizing components were also visible. Stimulation with a large spot inhibited the response (Fig. 14A). This amacrine cell type is the first example for colour coding in turtle amacrine cells. Stimulation with red light elicited mainly a small hyperpolarization during light ON, followed by a small depolarization after light OFF. Stimulation with blue light evoked mainly a more sustained depolarization during light ON. The response to green light seems to be a superposition of the responses to red and blue (Fig. 14B). Increasing intensity of white light did not change the responses drastically for low and medium intensities, however, the responses completely vanished with high intensity stimulation (Fig. 14C). This interesting phenomenon is consistent with the assumption that strong stray light stimulates larger areas, leading to the response block shown in Fig. 14A for large spot sizes.

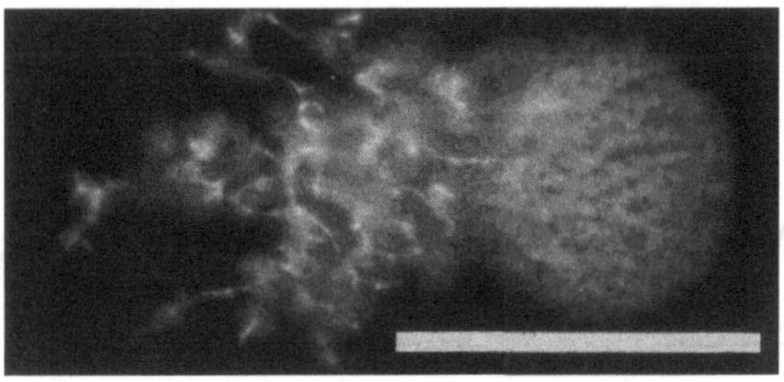

Fig. 13: Micrograph of a diffused, narrow field A 1. Level of focus in S5. The soma (on the left) is heavily out of focus. Scale 100 μm.

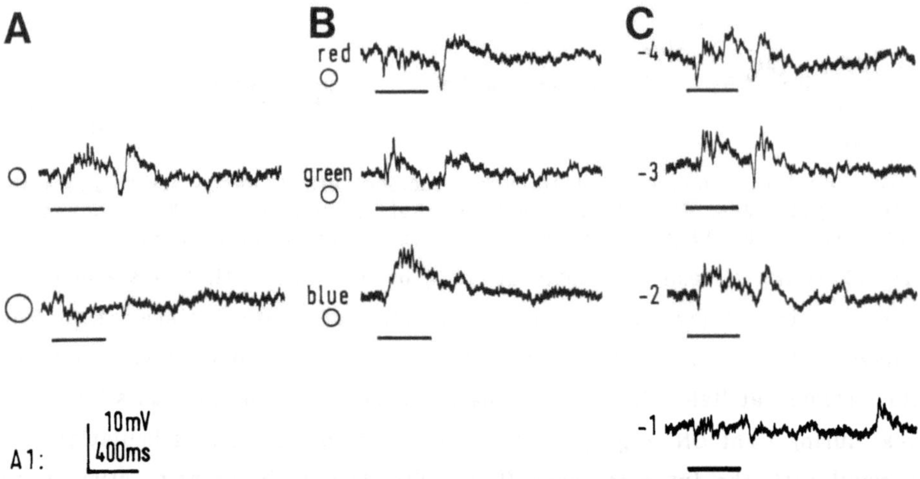

Fig. 14: Photoresponses of A 1. A: Responses to spots (800 μm and 3000 μm). No response was obtained with an annulus. Rel. intensity 10^{-2}. B: Responses to colours. C: Responses to increasing intensities. Spot size; 800 μm for B and C

SUSTAINED, DEPOLARIZING AMACRINE CELLS (ON-centre).

The small to medium field amacrine cell shown in the micrograph of Fig. 15A resembles A 10 (Kolb et al. 1988). These cells ramify through sublayers S3 to S5 with numerous curved, varicose processes. The responses to spots and annuli were sustained depolarizations during light ON (Fig. 15B). The amplitude to blue light

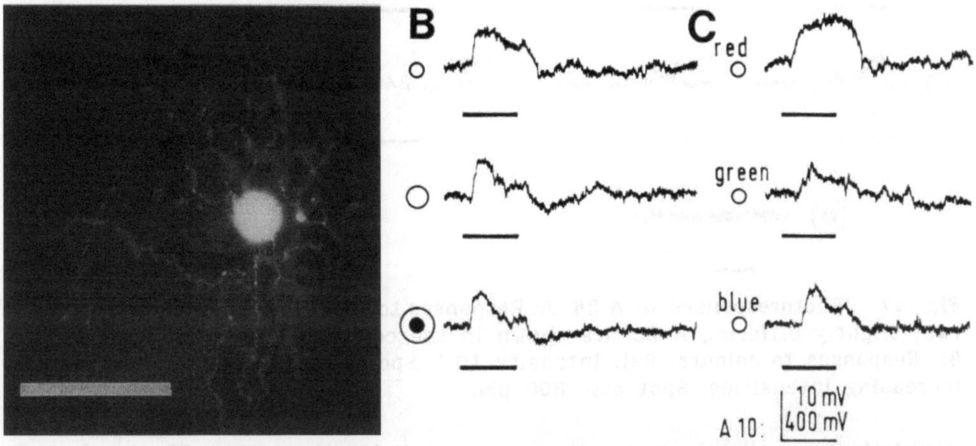

Fig. 15: A: Micrograph of an A 10, branching S3 to S5. Level of focus is on S3. Scale 100 µm. B: Photoresponses to spots and annulus. Spot sizes; 800 µm and 3000 µm. Rel. intensity 10^{-4}. C: Responses to colours. Spot size; 800 µm.

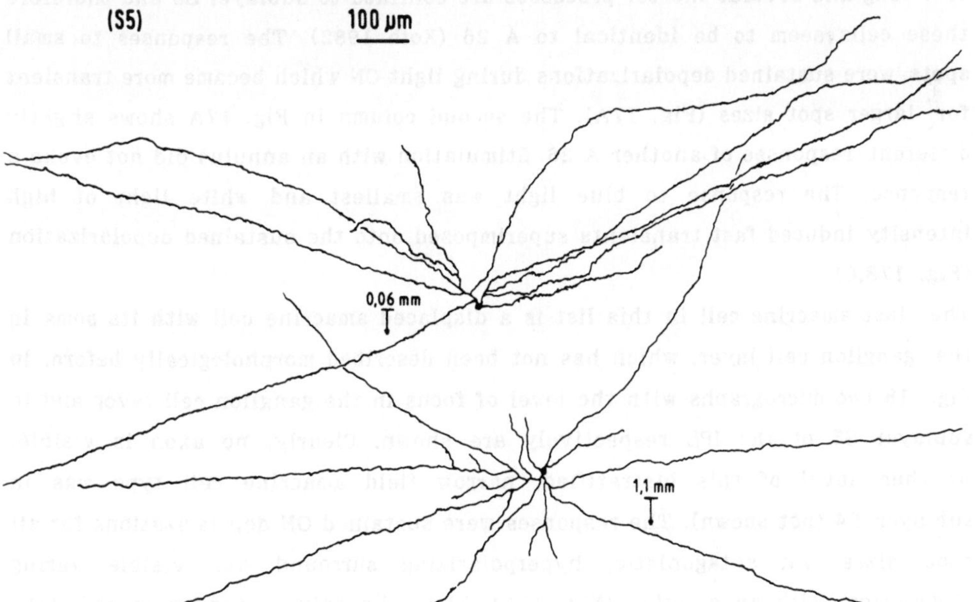

Fig. 16: Camera lucida drawings of two A 26.

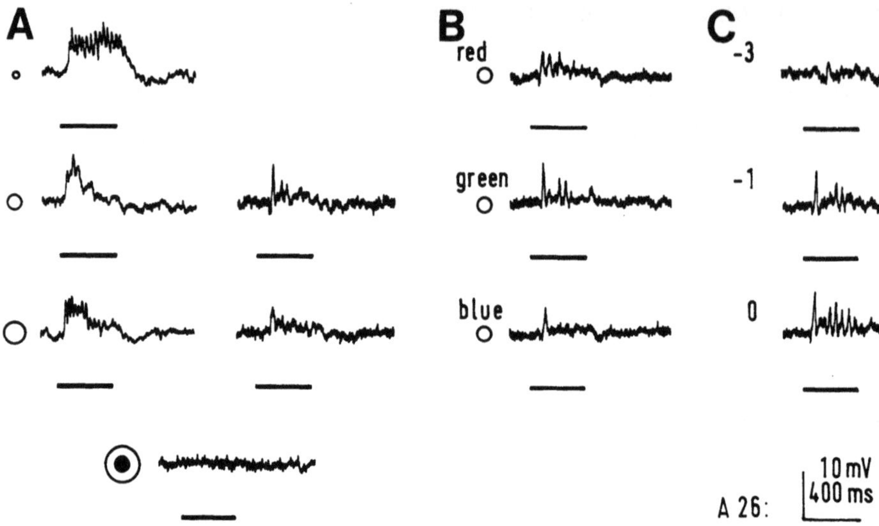

Fig. 17: Photoresponses of A 26. A: Responses to spots and annulus. Responses of two, slightly differing A 26 are shown in the columns. Rel. intensity 10^{-2}. B: Responses to colours. Rel. intensity 10^{-1}. Spot size; 800 µm. C: Responses to increasing intensities. Spot size 800 µm.

stimulation was slightly larger than to green light stimulation. The response to red light was largest (Fig. 15C).

Two examples of amacrine cells with a very wide field are shown in Fig. 16. The 5 to 7 long and several shorter processes are confined to sublayer S5 and therefore these cells seem to be identical to A 26 (Kolb 1982). The responses to small spots were sustained depolarizations during light ON which became more transient for larger spot sizes (Fig. 17A). The second column in Fig. 17A shows slightly different responses of another A 26. Stimulation with an annulus did not evoke a response. The response to blue light was smallest and white light of high intensity induced fast transients superimposed onto the sustained depolarization (Fig. 17B,C).

The last amacrine cell in this list is a displaced amacrine cell with its soma in the ganglion cell layer, which has not been described morphologically before. In Fig. 18 two micrographs with the level of focus in the ganglion cell layer and in sublayer S5 of the IPL respectively are shown. Clearly, no axon is visible. Another level of this bistratified, narrow field amacrine cell type was in sublayer S4 (not shown). The responses were sustained ON depolarizations for all spot sizes. An antagonistic, hyperpolarizing surround was visible during stimulation with an annulus (Fig. 19A). The cells only responded to red light with the intensity tested (Fig. 19B).

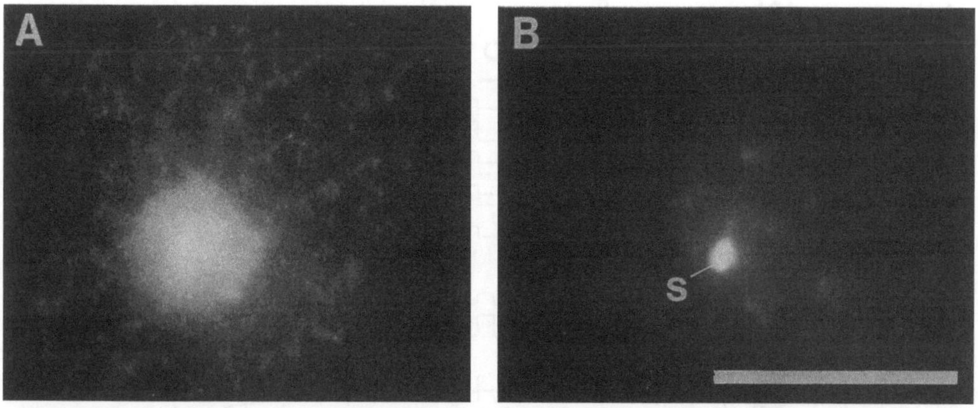

Fig. 18: Micrographs of displaced amacrine cell with the levels of focus on S5 (A) and ganglion cell layer (B). No axon is visible. S = soma. Scale 100 µm.

Fig. 19: Photoresponses of a displaced amacrine cell. A: Responses to spots and annulus. Rel. intensity 10^{-2} B: Responses to colours. Spot size; 200 µm.

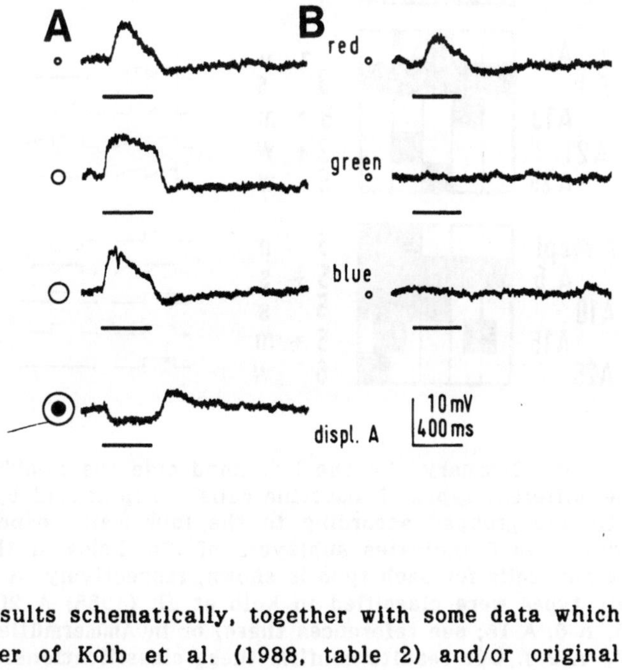

SUMMARY AND CONCLUSIONS.

Figure 20 summarizes the results schematically, together with some data which were adopted from the paper of Kolb et al. (1988, table 2) and/or original publications. The ramification levels of the different amacrine cell types in the IPL are indicated by the black areas on the left hand side. Schematic representations of the photoresponses elicited by spot and annular stimulation are shown below the spot and annulus symbols. Dotted lines in these schematic responses indicate variations in the responses seen sometimes in amacrine cells of a specified morphological type. The relative responses to red, green and blue light stimulation are indicated on the right hand side.

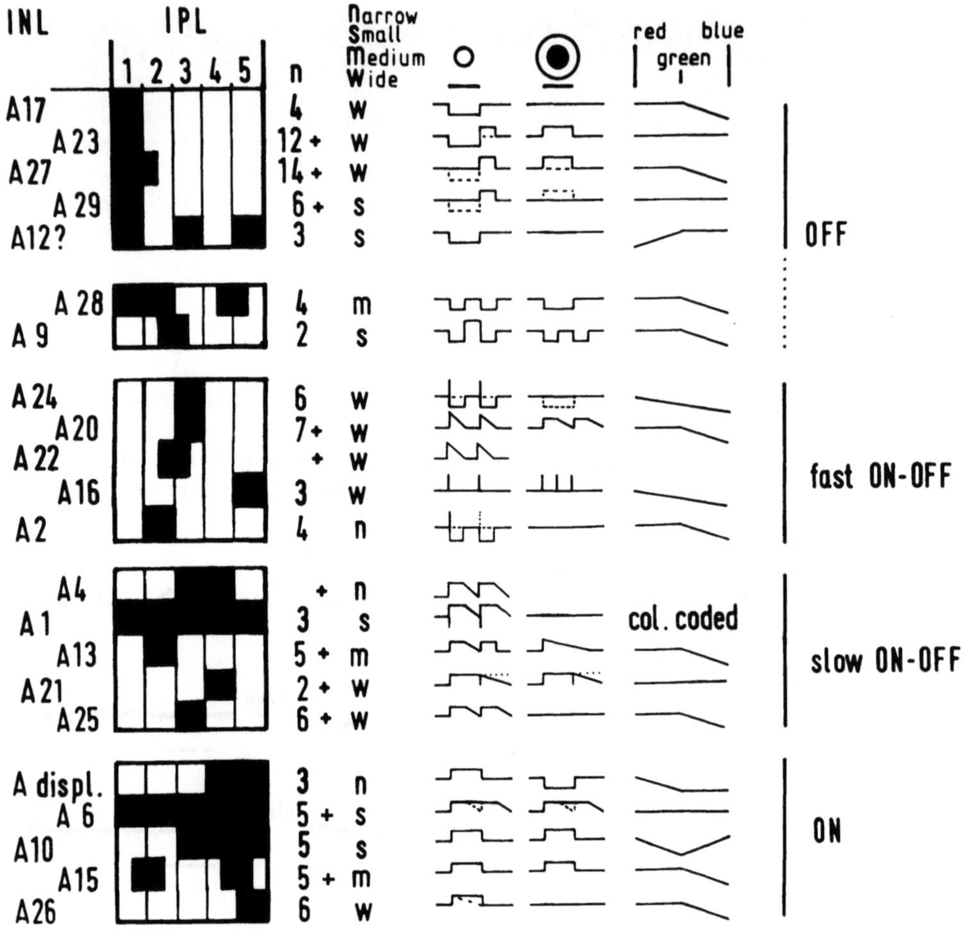

Fig. 20: Summary. On the left hand side the ramification levels within IPL of the different types of amacrine cells are indicated by black areas. The amacrine cells are grouped according to the four main response types (see right hand side). 1 to 5 indicates sublayers of IPL. Below n the number of recorded and stained cells for each type is shown, respectively. A + sign indicates that these cell types were classified in Kolb et al. (1988; A 20, A 22, A4, A 13, A 21, A 25, A 6, A 15; see references there) or in Ammermüller and Weiler (1988; A 23, A 27, A 29). Our results confirm these classifications. Below the spot symbol and the annulus symbol schematic representations of responses are drawn for every morphological type (see the original recordings above for comparison). For A 22 and A 4, which were not recorded in this study, data were taken from Jensen and DeVoe (1982; A 22) and Marchiafava and Weiler (1982; A 22, A 4). The relative responses to red, green and blue stimuli are indicated on the right. It has to be noted, however, that these responses were only tested with one intensity.

Several conclusion can be drawn from the results above.

- A first, clear evidence for colour coding in amacrine cells of the turtle retina was obtained.
- Every turtle amacrine cell ramifying exclusively in sublayer S1 of the IPL is an OFF-centre amacrine cell. However, not every hyperpolarizing amacrine cell type is ramifying exclusively in S1.
- On-centre amacrine cells always have processes in S4 and/or S5. However, not every amacrine cell with processes in S4 and/or S5 is an ON-centre amacrine cell.
- With the exception of A 16, ON-OFF amacrine cells are restricted to the middle (sublayers S2 to S4) or are diffused throughout IPL. 'Fast' ON-OFF amacrine cells seem to be always monostratified in sublayers S2 or S3.
- Every morphological amacrine cell type is also a distinct physiological type. Combining the responses to spots, annuli and different colours, is becomes evident, that cells showing the same responses to spots may differ in their receptive field properties and/or in their responses to colours. Additional stimulus parameters (like e.g. oriented slits, moving bars or gratings) may lead to a further physiological subdivision (which might be reflected by the variability of some response types, seen in many examples of Fig.20).

Comparing OPL and IPL, it is evident, that the morphological and electrophysiological complexity is greatly enhanced in the IPL. This is accompanied by increased neurochemical complexity (Massey and Redburn 1987). However, not all of these amacrine cell types may be really important for a general understanding of visual processing. Many of these types are shurely species specific adaptations for turtle vision and some may have accessory functions. Therefore, criteria reducing this complexity have to be developed (Masland 1988).

Achnowledgements:

This work was supported by a grant from the Deutsche Forschungsgemeinschaft to RW

REFERENCES:

Ammermüller J, Weiler R (1988) Physiological and morphological characterization of OFF-center amacrine cells in the turtle retina. J Comp Neurol 273:137-148

Cajal SR y (1892) La rétine des vertébrés. La Cellule 9

Jensen RJ, DeVoe RD (1982) Ganglion cells and (dye-coupled) amacrine cells in the turtle retina that have possible synaptic connection. Brain Res 240:146-150

Kolb H (1982) The morphology of the bipolar cells, amacrine cells and ganglion cells in the retina of the turtle *Pseudemys scripta elegans.* Phil Trans Roy Soc Lond B 298:355-393

Kolb H, Cline C, Wang HH, Brecha N (1987) Distribution and morphology of dopaminergic amacrine cells in the retina of the turtle *(Pseudemys scripta elegans).* J Neurocytol 16:577-588

Kolb H, Perlman I, Normann RA (1988) Neural organization of the retina of the turtle *Mauremys caspica:* a light microscope and Golgi study. Vis Neurosci 1:47-72

Marchiafava PL, Weiler R (1982) The photoresponses of structurally identified amacrine cells in the turtle retina. Proc R Soc Lond B 214:403-415

Masland RH (1988) Amacrine cells. TINS 11:405-410

Massey SC, Redburn DA (1987) Transmitter circuits in the vertebrate retina. Progress Neurobiol 28:55-96

Miller WH, Hashimoto Y, Saito T, Tomita T (1973) Physiological and morphological identification of L- and C-type S-potentials in the turtle retina. Vision Res 13:443-447

Normann RA, Kolb H (1981) Anatomy and physiology of the horizontal cells of the visual streak region of the turtle retina. Vision Res 21:1585-1588

Ohtsuka T (1985) Spectral sensitivities of seven morphological types of photoreceptors in the retina of the turtle *(Geoclemys reevesii).* J Comp Neurol 237:145-154

Teranishi T, Negishi K, Kato S (1987) Functional and morphological correlates of amacrine cells in carp retina. Neurosci 20:935-950

Weiler R (1981) The distribution of center-depolarizing and center-hyperpolarizing bipolar cell ramifications within the inner plexiform layer of turtle retina. J Comp Physiol 144:459-464

Weiler R, Marchiafava PL (1981) Physiological and morphological study of the inner plexiform layer in the turtle retina. Vision Res 21:1635-1638

DENDRITIC MORPHOLOGY OF A CLASS OF INTERSTITIAL AMACRINE CELLS IN CARP RETINA

K. Negishi and T. Teranishi
Department of Neurophysiology
Neuroinformation Research Institute
University of Kanazawa School of Medicine
13-1 Takara-machi, Kanazawa 920
Japan

Introduction

A fluorescent dye, 4,6-diamino-2-phenylindole (DAPI) has been known to be taken up preferentially by cholinergic amacrine cells in the rabbit retina (Vaney, 1984; Tauchi and Masland, 1985) and by AII amacrine cells in the cat retina (Vaney, 1985; Voigt and Wässle, 1987). However, since DAPI appears to be accumulated non-specifically by many classes of cells in the fish retina, it has been hard to discriminate a particular cell class among DAPI-labeled amacrine and ganglion cells. Beyond this difficulty, we have found it possible to select one special class of cells among DAPI-labeled cells for intracellular injection with Lucifer yellow CH (LY). That class is the interstitial amacrine cell. Since the somata of these cells are located in the inner plexiform layer (IPL), their DAPI-labeled nuclei can be isolated in visual field of the wholemounted retina by adjusting the focal level of the fluorescence microscope. The present paper reports a unique dendritic morphology of the interstitial amacrine cells, which have been thus isolated and injected with LY in the carp retina. A similar intracellular LY-injection has been applied to reveal the dendritic morphology of dopamine and serotonin cells in the carp retina (Teranishi et al, 1986, 1987, 1988).

NATO ASI Series, Vol. H31
Neurobiology of the Inner Retina
Edited by R. Weiler and N.N. Osborne
© Springer-Verlag Berlin Heidelberg 1989

Materials and Methods

Adult carp (Cyprinus carpio) were used; their body length and eyeball diameter were about 33 cm and 13 mm, respectively. The eyes were removed 24 h after intravitreal injection of 10 µl saline containing 0.05-0.10 µg DAPI. For intravitreal injection and eye removal, the fish were anesthetized with MS222 (Sankyo, Tokyo). The retinas were isolated and the pigment epithelium was detached with small pieces of filter paper. The isolated retina was wholemounted on a slide glass, either living or after a brief period (15-20 min) of fixation in a mixture of 4% paraformaldehyde and 0.1% glutaraldehyde in 0.1 M phosphate buffer (PB) at pH 7.4. Prior to wholemouting, the isolated retina was cut into 4-6 triangle pieces so that each piece contained an area from the margin to the central region. The vitreous body was carefully removed, and the retinal pieces were kept in 0.1 M PB and then placed one by one, vitreous side up, in a tissue chamber on the fixed-stage of a fluorescence microscope (Nikon UM-2) fitted with a long focal distance objective (Nikon M-plan, x40). With the UV filter set, the nuclei of DAPI-labeled cell were observed.

Glass micropipettes were filled with 4-5% LY. The resistance of the LY-filled pipettes was 100-200 MΩ. The tip of the micropipette was directed toward a blue fluorescent nucleus in retinal flatmounts under microscopic control; the iontophoretic injection technique was the same as in our earlier studies (Teranishi et al, 1986, 1987). For intracellular injection of LY, only one class of cells (interstitial type), located in an intermediate area between the optic disc and the retinal margin, was selected among numerous DAPI-labeled cells. This class was the only type of cell which could be discriminated, because of their soma position in the IPL. After injection, 10-20 photomicrographs of single LY-injected cells were taken through the V filter with different magnifications.

On the photomicrographs, the dendritic tree of each cell was traced, and the dendritic field size and the distance

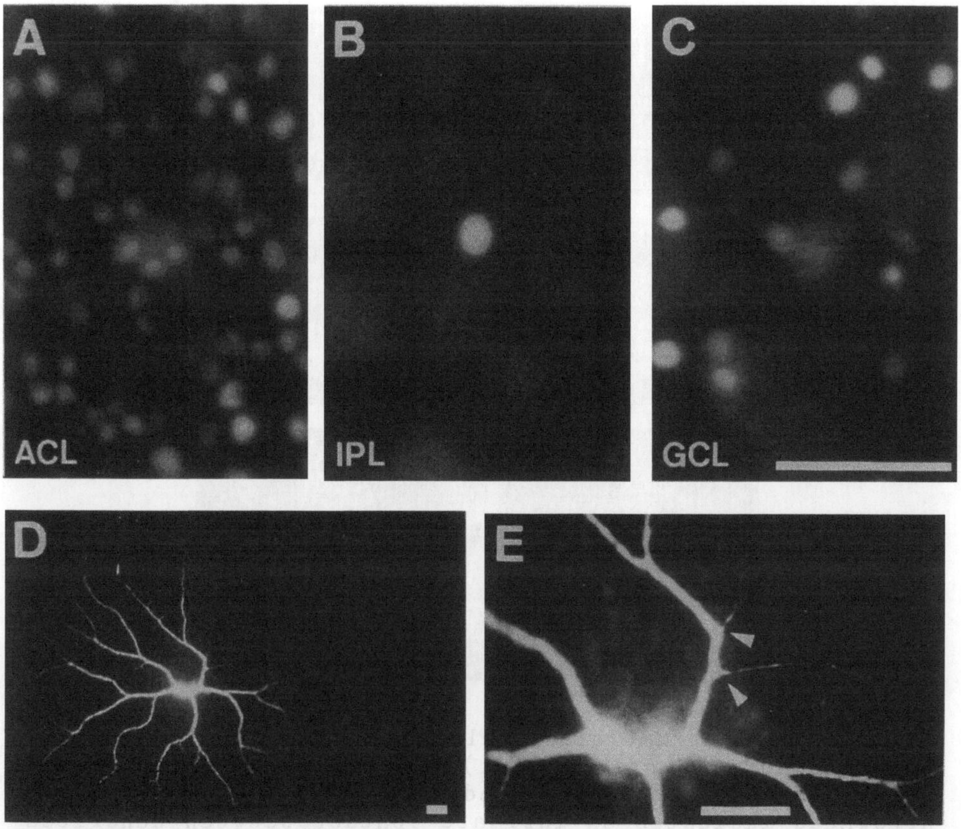

Fig. 1. DAPI-accumulated cells at different foacl levels and a LY-injected interstitial amacrine (ISA) cell in wet wholemounts of carp retina. A, DAPI-accumulated cells at the amacrine cell layer (ACL); B, one cell in the inner plexiform layer (IPL); C, cells at the ganglion cell layer (GCL); D, one LY-injected ISA cell; E, enlarged part of the same cell. Pictures A-C were taken through the UV filter system, while D and E were taken through the V system. In E, arrowheads indicate the branch points of long axon-like processes arising from one of the primary dendrites. Scale bars, 50 µm.

between the neighboring cells were measured. The LY-injected cells exhibited a relatively large-sized fusiform soma. After the intracellular injection experiment, some retinal pieces were further fixed in a mixture of 4% paraformaldehyde and 30% sucrose in 0.1 M PB (pH 7.4) overnight, and processed for frozen section (20 µm thickness) to examine the LY-injected cells in radial view.

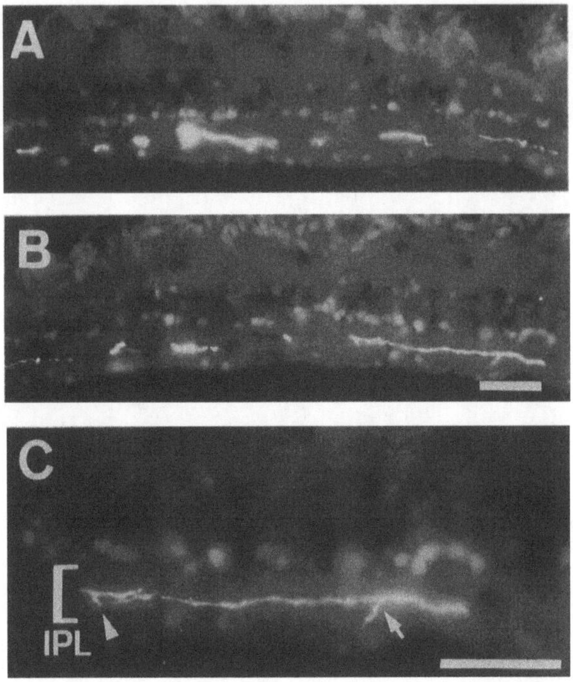

Fig. 2. DAPI-accumulated cells and one LY-injected ISA cell in radial sections (wet). Pictures A and B are sequential, and part of the dendrites in B is enlarged in C; arrowhead indicates an intimate contact between dendrites possibly arising from different ISA cells, and arrow shows a telodendrite directing proximally. Vertical bar approximately corresponds to the IPL. Scale bars, 50 µm.

Results

Figure 1 shows how we could differentiate a cell located in the IPL by adjusting the focal levels at the amacrine cell layer (ACL; A), the IPL (B) and at the ganglion cell layer (GCL; C) in the same area of a wholemounted retina. At both the ACL and GCL, numerous nuclei with various sizes and fluorescent intensities are seen, suggesting that some cells have uptaken DAPI more than other cells. However, it would be hard to select one type of cells among these DAPI-labeled nuclei for intracellular injection with LY. In the IPL, on the other hand, the nucleus of an interstitial amacrine (ISA) cell is isolated, and the tip of the LY-filled glass-micropipette can

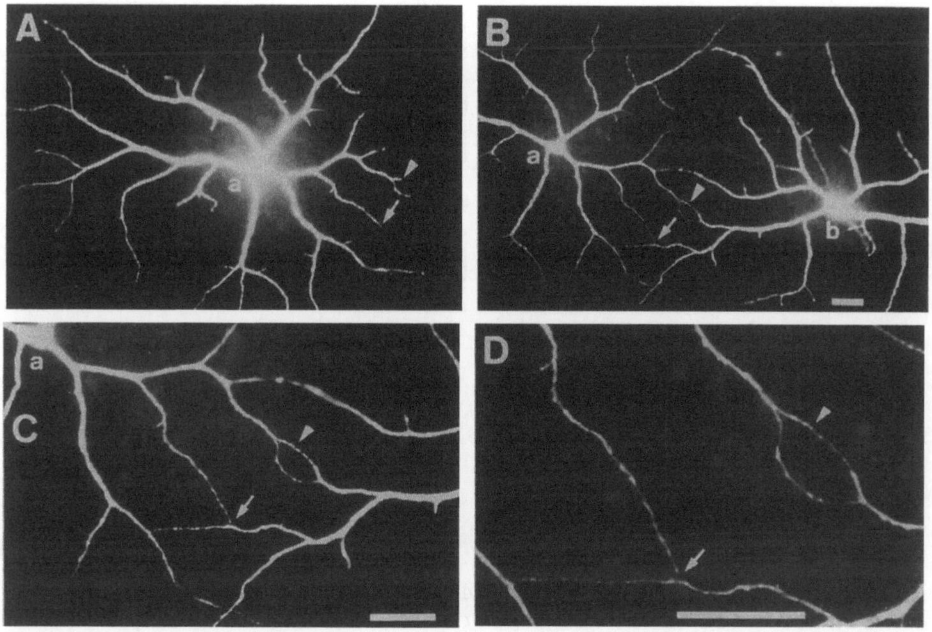

Fig. 3. Two neighboring ISA cells injected with LY. Cell 'a' (A) was first and cell 'b' (B) was then injected. Arrows and arowheads indicate corresponding points in pictures A-D, respectively. Arrows point one of tip-tip contacts between dendrites arising from cells 'a' and 'b', while arrowheads point a dendritic tip (from cell 'a') on the side of dendrite (from cell 'b'). Scale bars, 50 µm.

be directed to this cell. In raidal section, it would be further ascertained that the cell body was located in the IPL (Fig. 2).

Representative photomicrographs of LY-injected ISA cells are illustrated in Figs. 1D and 3A (single) and Fig. 3B (2 neighboring cells). The dendritic morphology of these cells was unique and homogenous; the soma shape was fusiform, 3-4 primary stout dendrites branched on the same plane (non-overlapping), and 2-5 distal fine axon-like processes emerged from the primary dendrites (arrowheads indicate their branch points in Fig. 1E). In radial sections, these dendrites were found to be monostratified at the middle level of the IPL (Fig. 2), probably sending their dendritic ends into the proximal layer of the IPL (sublamina _b_) (arrowhead in Fig. 3C).

The most attractive finding was cellular contacts between

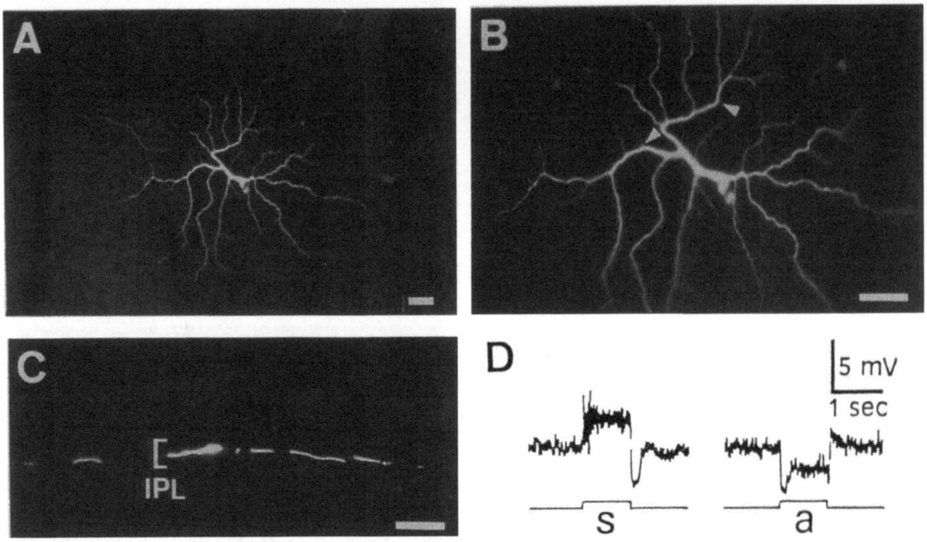

Fig. 4. One ISA cell injected with LY after light-induced on-center response was recorded. A, the cell in whole-mount (dry); B, the same cell was enlarged; C, the radial view of the same cell in a frozen section (dry); D, on-center response (depolarizing to a centered spot 's' and hyperpolarizing to an annular light 'a'. In B, arrowheads indicate the branch points of long axon-like processes. In C, vertical line approximately indicates the IPL. Scale bars, 50 μm.

their dendrites. After some neighboring cells were injected with LY, we could not judge which dendrites come from which cell bodies, because the dendrites were intimately apposed. Therefore, we had to take photomicrographs of each cell one by one when we planned to successively inject with LY into some neighboring ISA cells. In such a manner of successive injections, dendritic contacts could be clarified. In Fig. 3, one cell 'a' was first injected with LY (A), then the neighboring cell 'b' was injected and photographed together with cell 'a' (B). Part of this picture was enlarged photographically (Fig. 3C and D). Marks of arrow and arrowhead in these pictures indicate two corresponding spots, respectively. The arrow points one of "tip-tip" contacts between dendrites from cells 'a' and 'b', while the arrowhead point one of "tip-side" contact between dendrites from the cells. The axon-like processes also made similar but less frequent contacts between them or ended on dendrites (not shown).

The spatial parameters measured on ISA cells are following on the average: density, 8 cells/mm^2; intercellular distance, 400 µm (n=561); dendritic field size, 0.27 mm^2 (n=508); estimated dendritc coverage, 2.2. We could not trace the total course of long axon-like processes of these cells.

Discussion

In our earlier experiments (Teranishi et al, 1985, 1987), functional and morphological correlates of amacrine cells in the carp retina were systematically investigated. From the collection of many LY-injected amacrine cells, we could find a few cells whose dendritic morphology (Fig. 4) is exactly the same as that described in the present paper. Detailed re-examination of the earlier photomicrographs in radial view revealed that those cells are ISA cells, their somata being in the IPL (Fig. 4C). The light-evoked response of the cells was an on-center type (cell type Fnb), as can be expected from their monostratified dendrites in sublamina b of the IPL (Fig. 4D). Although their dendrites were intimately apposed to each other (Fig. 3), dye diffusion was not observed in living retinas in our present and earlier experiments (Teranishi et al, 1987), suggesting that there are no gap junctions between these apposed dendrites. An ISA cell marked with Procion yellow has been illustrated in an earlier paper, and an on-center type response was said to be recorded from it (Murakami and Shimoda, 1977). The frequent tip-tip and tip-side intimate contacts between dendrites or with axon-like processes of ISA cells are assumed to form a kind of reverberating circuits within the IPL.

It is well known that in the rabbit retina cholinergic starburst amacrine cells are classified into two subgroups; the somata of one group are located at the ACL and those of the other group are at the GCL (displaced amacrine cell type)

(Masland and Mills, 1979; Famiglietti, 1983; Masland et al, 1984; Vaney, 1984; Tauchi and Masland, 1985). The dendrites of the starburst cells located at the ACL monostratify in sublamina a and those of the displaced starburst cells monostratify in sublamina b. Such dual layers of soma localization and dendritic stratification of cholinergic amacrine cells have been found commonly in the retina of various vertebrates (Millar et al, 1983; Voigt, 1986), including the fish retina (Tumosa et al, 1984; Tumosa and Stell, 1986; Morgan et al, 1986). The ISA cells revealed with intracellular LY injection in the present study appeared to extend their dendrites into sublamina b. It is reasonably assumed that there is another subtype of cells belonging to a class similar to ISA cells, whose dendrites may monostratify in sublamina a, however. Indeed, we have recorded an off-center type response from a class of amacrine cells (cell type Fna) located at the amacrine cell layer (Teranishi et al, 1987; see Fig. 8). In the present study, we paid special attention to detecting a similar shape of nuclei in the ACL to that of the ISA cells, whose nuclei were relatively oval and large in shape. The intracellular LY injection into several such cells revealed that they were the same Fna type amacrine cell previously described (Teranishi et al, 1987). Since no displaced cells, whose dendritic morphology is similar to that of the ISA cells, were found at the GCL (although not yet systematically explored), the cell class in the present study appear to be located both at the ACL and in the middle of the IPL. Dendrites of the former ramify in sublamina a, while dendrites of the latter monostratify in sublamina b.

Dopaminergic cells have been classified into three categories depending upon their soma locations; junctional cells located at the ACL, eremite (interstitial) cells placed within the IPL as in the case of cells explored in the present work, and alloganglion (displaced) cells located at the GCL (Ehinger, 1976). A series of immunohistochemical experiments were carried out with no success to identify what a kind of neuroactive substance is contained in LY-injected ISA cells. We did not show any immunoreactivity to antibodies against, for example, tyrosine hydroxylase, glucagon, neuropeptide Y and somatostatin. Some classes of ISA cells may exist; for instance, it would be cholinergic (Tumosa et al, 1984; Tumosa and Stell, 1986) or GABAergic (Ball and Brandon, 1986; Ball, 1987). Indeed, we seleted for intracellular LY-injection one class of ISA cells, whose nuclei were largest among cells located in the IPL. It is certain that the LY-injected ISA cells are the Fnb type amacrine cells generating an on-center type response. These cells appear not to be misplaced but genetically designed to be positioned in the middle of the IPL, and may represent a specific morphofunctional system in the fish retina.

Summary

The dendritic morphology of a class of interstitial amacrine (ISA) cells in the carp retina was investigated by identifying their fluorescent nuclei preloaded with 4,6-diamino-2-phenylindole (DAPI) and followed by iontophoretic injection of Lucifer yellow (LY) in isolated, living or formaldehyde-fixed wholemounts under microscopic control. The ISA cell nuclei were discriminated only by microscopic focal levels, because DAPI appeared to be accumulated non-specifically by amacrine and ganglion cells. The LY-injected fusiform cell bodies were located at the middle of the inner plexiform layer (IPL) and their dendrites monostratified in sublamina b of the IPL. The cells extended 3-4 primary stout dendrites which were decorated with fine telodendrites and some long-extending axon-like processes. The spatial properties of these cells were on the average: density, 8 cells/mm^2; intercellular distance, 400 μm; dendritic field size, 0.27 mm^2; and dendritic coverage, 2.2. The dendritic morphology was unique, in particular their dendrites and axon-like processes made intimate appositions between dendrites, between axon-like processes or between both ("tip-tip" or "tip-side" contacts). These cells generate an on-center type response, but the neuroactive substance contained in the LY-injected ISA cells is unknown at present. The ISA cells appear to represent a certain morphofunctional system.

Acknowledgments

The authors thank Dr. William K. Stell for his suggestion and correction of the text English at the early stage of manuscript and Mrs. Tami Urano for her secretarial assistance. Supported in part by the Ministry of Education of Japan (63570054, 63304031, 62221014), Japan Society for Promotion of Science, and from the Johnan Hospital in Takaoka. Part of traveling expenses was given to K.N. by the Yamada Science Foundation to attend the NATO workshop at Oldenburg.

References

Ball AK (1987) Immunocytochemical and autoradiographic localization of GABAergic neurons in the goldfish retina. J Comp Neurol 255:317-325

Ball AK, Brandon C (1986) Localization of ^{3}H-GABA, -muscimol, and -glycine in goldfish retinas stained for glutamate decarboxylase. J Neurosci 6:1621-1627

Ehinger B (1987) Biogenic monoamines as transmitters in the retina. In: Bonting SL (ed) Transmitters in the visual process. Pergmon, Oxford, pp.145-152

Famiglietti EV (1983) On and off pathways through amacrine cells in mammalian retina: the synaptic connections of "starburst" amacrine cells. Vision Res 23:1265-1279

Masland RH, Mills JW (1979) Autoradiographic identification of acetylcholine in the rabbit retina. J Cell Biol 83:159-178

Masland RH, Mills JW, Hayden SA (1984) Acetylcholine-synthesizing amacrine cells: identification and selective staining using radioautography and fluorescent markers. Proc R Soc Lond B223:79-100

Millar TJ, Ishimoto I, Boelen M, Epstein ML, Johnston CD, Morgan IG (1987) The toxic effects of ethylcholine mustad aziridinium ion on cholinergic cells in the chicken retina. J Neurosci 7:343-356.

Morgan IG, Ishimoto I, Negishi K, Millar T (1986) Cholinergic-choinergic interactions may be a phylogenetically-conserved feature of the vertebrate retina (abstract), Colloque INSERM on structure et fonction dans le systeme visual

(Lyon):20
Murakami M, Shimoda Y (1977) Identification of amacrine and ganglion cells in the carp retina. J Physiol (Lond) 246: 801-818
Tauchi M, Masland RH (1984) The shape and arrangement of the cholinergic neurons in the rabbit retina. Proc R Soc Lond B223:101-119
Teranishi T, Negishi K (1986) Dendritic morphology of dopaminergic cells revealed by intracellular injection of Lucifer yellow in fixed carp retina. Brain Res 170:196-199
Teranishi T, Negishi K (1988) Regional difference in the dendritic morphology of dopamine cells in carp retina. Dev Brain Res 39:9-17
Teranishi T, Negishi K, Hidaka S, Naka K-I (1987) Dendritic morphology of indoleamine cells revealed by intracellular injection of Lucifer yellow in fixed carp retina. Neuroscience 22:323-329
Teranishi T, Negishi K, Kato S (1985) Correlations between photoresponse and morphology of amacrine cells in the carp retina. Neurosci Res Suppl 2:S211-S226
Teranishi T, Negishi K, Kato S (1987) Functional and morphological correlates of amacrine cells in carp retina. Neuroscience 20:935-950
Tumosa N, Eckenstein F, Stell WK (1984) Immunocytochemical localization of putative cholinergic neurons in the goldfish retina. Neurosci Lett 48:255-259
Tumosa N, Stell WK (1986) Choline acetyltransferase immunoreactivity suggests that ganglion cells in the goldfish retina are not cholinertgic. J Comp Neurol 244:267-275
Vaney DI (1984) 'Coronate' amacrine cells in the rabbit retina have a 'starburst' dendritic morphology. Proc R Soc Lond B220:501-508
Vaney DI (1985) The morphology and topographic distribution of AII amacrine cells in the cat retina. Proc R Soc Lond B244: 475-488
Voigt T (1986) Cholinergic amacrine cells in the rat retina. J Comp Neurol 248:19-35
Voigt T, Wässle H (1987) Dopaminergic innervation of AII amacrine cells in mammalian retina. J Neurosci 7:4115-4128

HOW MANY AMACRINE CELLS DOES A RETINA NEED ?

A COMPARATIVE ANALYSIS OF AMACRINE CELL MORPHOLOGY

H.-J. Wagner
Institut für Anatomie und Zellbiologie
Philipps Universität Marburg
Robert Koch Str. 6
3550 MARBURG F.R.G.

Among all retinal neurones, amacrine cells (ACs) have always held the greatest thrill for students of morphology and, more recently cytochemistry. This is due to their name-giving chacracteristic, the apparent lack of an axon (Cajal, 1893), as well as to their astounding structural diversity (Cajal,1893; Kolb, 1982; Kolb et al., 1981; Linberg et al., 1986; Wagner and Wagner, 1988), and to the wealth of neurotransmitters and neuromodulators they contain (Brecha et al, 1984; Stell, 1985; Marc, 1986; Massey and Redburn, 1987), which, in many cases, colocalize (Weiler and Ball, 1984; Lam et al, 1986). Electrophysiologists, on the other hand, have succeeded in recording only two basic types of light-evoked responses from ACs: transient and sustained; the latter ones could be depolarizing or hyperpolarizing (for review see Djamgoz and Wagner, 1987)

In spite of the great amount of data on ACs, understanding of AC function within the retinal network is far less advanced than in other retinal cell types. Correlation of the various facets and determination of the precise function of individual cell types would seem to be the principal approach towards the elucidation of their role. While in lower vertebrates, especially fish, several correlative attempts using microinjection techniques have not yielded clear concepts about the function of single cell types (Teranishi et al, 1987; Djamgoz et al, 1989), in mammals such a characterization has been successful for individual AC types (cat: AII, A4, A8, A17, Kolb and Nelson, 1984; starburst AC, Masland and Tauchi, 1986). However, it is still difficult to derive a general concept of AC function.

Remembering the traditional morphological dogma, that form determines function (or vice versa), one may ask whether it is reasonable to expect ACs to work according to a unifying concept in view of their enormous structural and neurochemical diversity. If differentiation of the various cell types may be regarded as the result of adaptation to functional requirements it may be useful to distinguish special, species-specific adaptations on the one hand from other, more basic forms of differentiation reflecting more general demands of visual function, on the other. Consequently, a comparative, interspecific analysis of ACs in several classes of vertebrates may help to identify cell types with the same morphology in many species and other more species-specific cell types.

Teleosts are most renowned for the adaptive radiation of their eyes and their retinas, in particular; the extent of photorecpetor differentiation including the visual pigments is best documented and best reflected in the concept of visual ecology (Lythgoe, 1979). Therefore, ACs will first be compared in a

NATO ASI Series, Vol. H31
Neurobiology of the Inner Retina
Edited by R. Weiler and N. N. Osborne
© Springer-Verlag Berlin Heidelberg 1989

number of teleost species to identify the types they have in common; subsequently, AC of other vertebrates will also be considered.

The unequivocal definition of a cell "type" is critical to such a comparative approach; this, in turn, relies on a common set of criteria used for the identification of cell types. While most authors agree that the morphological parameters used should be of functional significance and should also be quantifiable, the relative importance of the various features is often treated differently. For the roach, a teleost, we developed a system of AC type definition (Wagner and Wagner, 1988) which is also used here to analyse ACs in several other teleosts not described previously in this respect, and also to reclassify cells recorded by other workers in order to make them compatible for comparison.

In this sense, the question of how many cell types a retina might need - however provocative it is formulated - may lead to the formulation of a group of AC types common to most vertebrates and thought to subserve basic visual functions; consequently, species-specific adaptations may then be responsible for the major part of AC variability.

Material and Methods

Isolated retinas (dark adapted) were obtained from five species of teleosts and prepared for rapid Golgi impregnation (Stell, 1975). Details about sectioning, evaluation and classification procedures are described by Wagner and Wagner (1988). The following figures indicate the numnber of retinae successfully stained, the total number of ACs analysed and the number of cell types identified for the respective species: roach, Rutilus rutilus, Cyprinidae: 35, 250, 43; Nannacara anomala, Cichlidae: 46, 180, 32; Aequidens pulcher, Cichlidae: 24, 128, 28; goldeye, Hiodon alosoides, Notopteridae: 24, 98, 23; catfish, Kryptopterus bicirrhis, Siluridae: 38, 73, 21. None of these species shows cell density gradients exceeding a factor of three; therefore, the excentricity of the respective cells was ignored.

Briefly, the classification was mainly based on the wholemount aspect of the dendritic field (size, symmetry and orientation; branching pattern, number of ramification points; process diameter, ocurrence and location of spines and varicosiites); cells with identical features in this respect were regarded as types. Furthermore, the radial aspect was analysed; although, in the roach, seven sublamiae can be distinguished, for the present comparative aspect, only a stratification in two sublaminae (a and b) is used here. Monostatified cell types occurring at more than a single level of the ipl are referred to as subtypes. Among the diffuse cell types, narrow-diffuse (20% total width), broad-diffuse (50%) and all-diffuse (50%) are distinguished.

Observations

For the present comparative purposes the ACs of the European roach are used as a reference. According to the above criteria 43 cell types including 70 subtypes have been identified (Wagner and Wagner, 1988) in this species. Fig. 1 shows an overview of roach ACs visualized after Golgi impregnation or

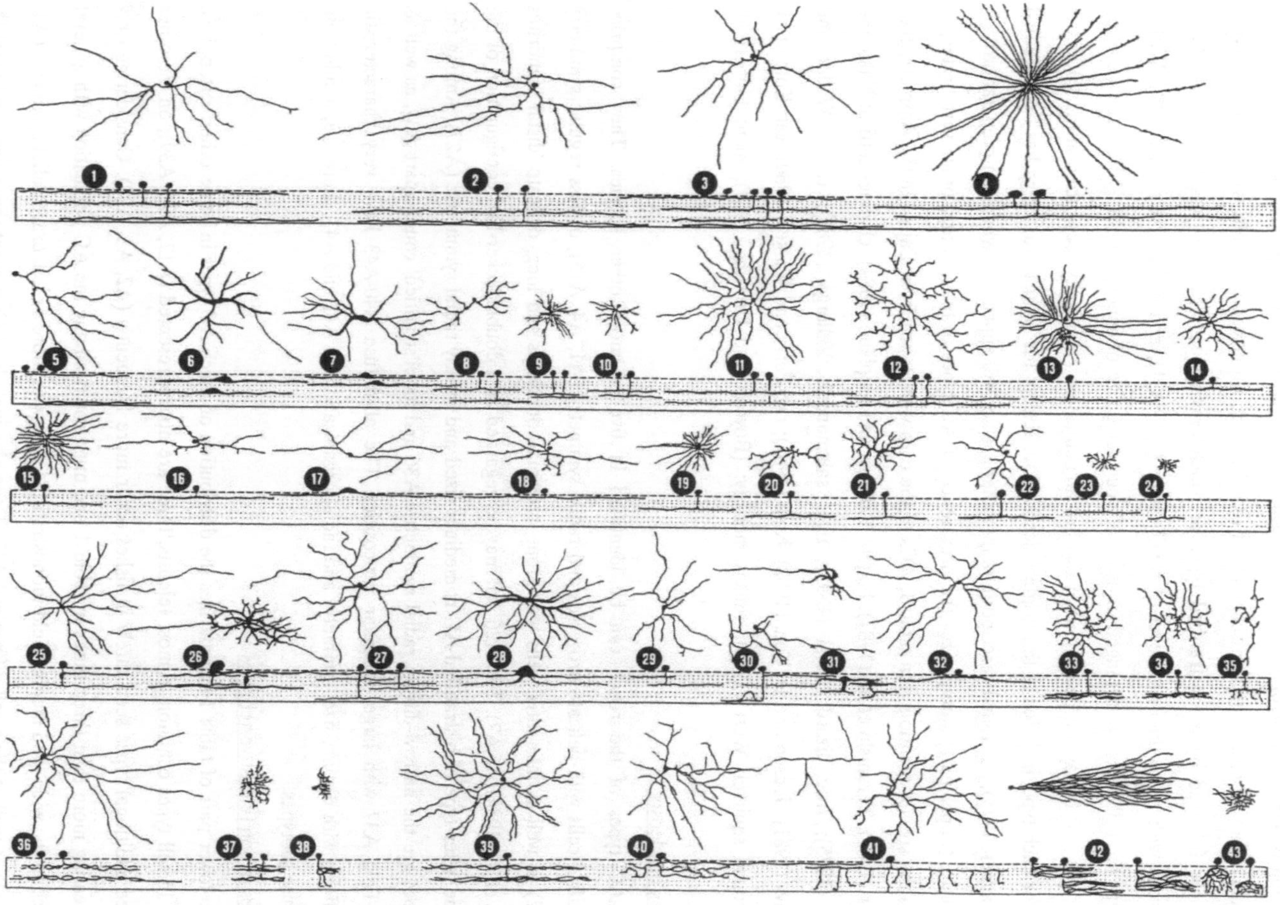

Fig. 1: Summary diagram of AC types and subtypes in the roach, slightly modified from Wagner and Wagner (1988) including microinjected cells from Djamgoz et al.(1989); scale bar: 100 µm; reproduced with permission of the Royal Society

microinjection of HRP (Djamgoz et al., 1989); each cell type is shown in tangential view according to a camera lucida drawing and in its semischematical radial aspect. Monostratified cells with several subtypes (A1-A12) are followed by monostratified types in which only a single subtype was found (A13-A24), stratified cells with more than one level of ramification (A25-A31), and diffuse ACs (A32-A43). Within these groups the cell types are arranged in order of decreasing size of dendritic field. Differences between some cell types which may not be obvious at this low magnification are readily recognized in the ID charts prepared for each type, included in the original description. Except for Nannacara, which has been studied before (Wagner, 1973), a detailed description of the other species will appear elsewhere (in preparation).

Table 1 (upper part) summarizes the occurrence of AC types in teleosts and other vertebrates; in addition to the species mentioned in table 1, ACs of the carp were taken from Cajal (1893), Ammermüller and Weiler (1981), Teranishi et al. (1987); bogue: Vallerga and Deplano (1984); channel catfish: Naka and Carraway (1975); frog, lizard: Cajal (1893); tiger salamander: Vallerga (1981); turtle: Weiler and Marchiafava (1981); Jensen and DeVoe, (1982); Kolb (1982); rat: Perry and Walker (1980); cat: Kolb et al. (1981); ground squirrel: West (1976); rhesus monkey: (Boycott and Dowling, 1969); man: Linberg et al.(1986).

1. AC types in teleosts

16 AC types of the roach can be identified in five or more other species. They comprise monostratified cells with delicate processes and radial symmetry (A1, A4, A15), or less regular symmetry (A12, A22) as well as interstitial cells with stout 'proximal' dendrites and long, delicate 'distal' dendrites (Djamgoz et al., 1984)(A6,A7). The A26 cell may be regarded as a doubledecker-like configuration of the A6 cell; the other type of bistratified AC is medium sized and mostly radial symmetric (A27). Among the diffuse types are the narrow-diffuse radial symmetric A36 and its bi'stratified' counterpart A39, as well as the small-field A37 with large, irregular varicosities. The all-diffuse A40-A43 have very characteristic dendritic fields with distinct size, symmetry and radial organization; in A42 and A43, there are, in addition, broad diffuse subtypes.

2. AC types in non teleost vertebrates

The lower part of table 1 summarizes the distribution of roach AC types in some other vertebrates. From the 16 cell types common to most teleosts, three are not represented (A12, A26, A39); on the other hand, three additional types seen rarely in fishes occur more frequently (A2, A3, A38). Considering only those appearing more than three times in table 1 one must count among the AC population with general visual function the wide- and mediumfield monostratified cells with more or less exact radial symmetry (A1, A2, A4, A15) as well as the small diffuse cell A43; if one also includes ACs present three times in the table, the stout A6, and the widefield A40 and A41 also belong to this group.

Looking more closely at A4 cells in lower vertebrates, on the one hand, and in mammal 'starburst'cells, on the other, it becomes apparent that judged strictly by our criteria they may not form an identical type; differences concern mainly the location of branching points, size constancy and the fact that

roach	Golgi	A1	A2	A3	A4	A5	A6	A7	A8	A9	A10	A11	A12	A13	A14	A15	A16	A17	A18	A19	A20	A21	A22	A23	A24	A25	A26	A27	A28	A29	A30	A31	A32	A33	A34	A35	A36	A37	A38	A39	A40	A41	A42	A43
roach	micro-inj.	+		+	+		+	+					+													+	+		+								+			+	+		+	
carp	Golgi	+	+	+	+		+	+		+				+		+				+							+		+			+		+							+	+	+	+
carp	micro-inj.	+	+		+	+	+	+							+								+				+							+			+			+		+		
Nannacara	Golgi	+		+	+			+				+	+			+				+	+		+			+	+	+			+						+	+		+		+	+	+
Aequidens	Golgi	+			+		+	+				+	+						+				+				+	+			+						+	+			+	+	+	+
goldeye	Golgi	+			+		+			+			+		+	+									+		+	+			+						+			+	+	+	+	
bogue	Golgi	+			+		+		+							+						+			+							+					+	+					+	+
catfish	Golgi	+	+	+	+		+	+	+				+				+				+		+				+	+	+					+			+	+		+	+	+	+	+
catfish	micro-inj.	+	+	+	+					+										+							+		+														+	+
frog	Golgi	+	+		+											+												+									+		?		+			+
tiger salaman.	micro-inj.	+																																							+			
turtle	Golgi	+	+	+	+		+																+																			+		+
lizard	Golgi	+	+		+		+									?																					+		?		+	+		+
rat	Golgi				+											+																												+
cat	Golgi	+	+		?											+																						+				+		+
ground squirrel	Golgi				+			+																																				+
rhesus monkey	Golgi	+	+				+									+																												+
man	Golgi	+																																										+

Table 1: Vertebrate AC types visualized by Golgi and microinjection techniques. Roach ACs were used as reference (Wagner and Wagner, 1988); additional cell types in other species not shown for greater clarity

instead of the mirror symmetric organization of paramorphic pairs (Famiglietti, 1983a, b; Kolb and Nelson, 1981) in teleosts, A4 cells occur as two subtypes with perikarya not clearly separated. However, considering the average covering factor of mammal 'starburst' cells and of A4 cells (Wagner and Wagner, 1988) as well as the localization of the transmitter content (ACh Tumosa et al., 1984; Marc, 1986; Masland and Tauchi, 1986) it can be assumed that they subserve equivalent functions in spite of the morphological differences.

Discussion

1. Methodical problems of staining techniques

The attempt to identify a population of AC types of possible general importance for visual function is fraught with methodical problems when based on Golgi impregnation or microinjection techniques. The major drawback of both these methods for the present purpose lies in the statistical nature with which cells are stained. Clearly, therefore, the above list is not likely to represent the complete inventory of AC types. Consequently, the population of cell types shared by all species may well be larger than established by the present comparison. Furthermore, these methods do not allow an estimate of the frequency and hence, maybe, of the relative of the various AC types within a given retina.

It is of course essential for the evaluation of these findings to know whether identical morphological types have identical functions in different species. Therefore, detailed physiological data are urgently required to answer this question. In the absence of such data, some considerations about the functional implications of AC morphology, of the distribution of subtypes, and of the known transmitter content may help to shed some light on this problem.

2. General considerations on AC form and function

The structural differentiation of the various AC types is mainly determined by the organization of bipolar terminals, ganglion cell dendrites and other AC dendrites with which they synapse. The specificity of these interactions may be supposed to be the cause of the large number of dendritic field patterns and of surface specializations (spines and varicosities). At the same time, spatial restraints may also be responsible for the mode of ramification of a given dendrite: Considering, for example the almost straight course of teleost A4 cells, it is obvious that they can only occur in ipl sublayers where there are no other voluminous structures such as large bipolar terminals which they would have to circumvent. Conversely, the T-like branching pattern of A12 cells may be due to the coexistence of just two such elements within the confined space of a given sublayer.

Disregarding the fine structural details, a number of basic principles of organization of amacrine dendritic fields can be determined. Among the monostratified cells, the most common pattern of arborization consists of a stem process giving off radiating secondary dendrites (A1, A3, A4, A9-A15, A19, A22-A24). Depending on the number and location of branching points different degrees of coverage of the dendritic field are achieved. The mostly radial symmetry would suggest that the main function of these types is to provide a homogeneous density of synaptic interactions. and would make these radiate cells candidates for the processing of local contrast and chromatic information.

Cells with asymmetric or polar dendritic fields (A2, A5, A8, A16-A18, A20), on the other hand, may serve more specialized functions such as mediating motion and directional sensitivity (Naka, 1980).

A third group of stratified cells showing some common features are those AC types with stout proximal dendrites and long delicate d-type dendrites (A6, A7, A26, A28, A31). In view of their coarse structure and their intercellular coupling (Teranishi et al., 1987), it is tempting to speculate that their role might be to influence the resting potential of neighbouring cells over large distances and thus to control the threshold of their sublayer. As to the d-type dendrites, their function certainly is enigmatic, especially since no special terminal structure was ever detected. Interestingly, Mariani (1982) described an 'association' AC resembling the roach A31 cell in the pigeon, which shows a large field of terminal arborization at the end of a d-type dendrite-like process.

Some of these considerations may not only apply to monostratified, but also to multistratified and diffuse cells. In these latter cells, however, additional functional properties must be taken into account if they are activated as a whole. In these cases, they may fulfill either transfer roles between individual sublayers, or simultaneous activation of different combinations of sublayers. They may also collect information from e.g. sublamina a or b and relay them to a specific class of ganglion cell in only one sublamina (cat A8 and A17 cells: Kolb and Nelson, 1985). In this context, cell types which have different dendritic fine structures in different sublaminae (A38, A42) are especially interesting since they might have differential functions in individual sublaminae. In cat AII cells, electron microscopic analysis has suggested such a role (Kolb and Nelson, 1984).

3. The role of subtypes

In numerous types of monostratified, and several of the narrow- or broad-diffuse cells, subtypes are observed, i.e. cells with identical whole mount dendritic field morphology occur at different levels of the ipl. This is sometimes correlated with perikarya localized in the nuclear layer next to the stratification level (displaced, paramorphic cells: Vaney et al., 1981; Kolb and Nelson, 1981). This paramorphism is so widespread that, in the roach, only a single of the general teleost ACs type is present in only one sublamina. However, the frequency of subtypes observed varies greatly among different ACs and among different species (roach: A3: 2 subtypes in each sublamina; A42: 4 subtypes; A1, A7: 3 subtypes; Nannacara: A3: 4 subtypes). In catfish and goldeye, on the other hand, never more than two subtypes for a given type were found; structurally, this correlates with an ipl of reduced width (goldeye: 20-30 m; catfish: 10-15 m). Functionally, these are monochromatic species which may be considered essentially colour blind. By contrast, in species with well developed colour vision, like Nannacara (Ali et al., 1978) or the roach (Downing et al., 1986), the number of AC subtypes is markedly higher. It is therefore tempting to speculate that the greater number of AC subtypes may serve the parallel processing of chromatic information within the different bipolar colour channels (Scholes, 1975). This argument implies that the various AC subtypes serve analogous functions on the different sublayers of a given retina. Consequently, it may be assumed that identical AC types also serve analogous functions in different species.

4. Transmitter content and morphological cell types

Apart from the specific synaptology, as derived from the respective dendritic field morphology, the content of neurotransmitters or neuromodulators is an important key to the function of ACs (Stell, 1985). Therefore, if the same transmitter were found within a given cell type across species boundaries, this could be taken as another indication for the constancy of their function.

The cell type most extensively studied in this respect, is the starburst cell (Masland and Tauchi, 1986; Voigt, 1986; Pourcho and Osman, 1986) as an example of "the" retinal cholinergic neuron. As shown above, A4 cells in lower vertebrates and starburst cells in mammals are not identical, morphologically; yet, their Ach or ChaT content would make these cells equivalent. In cat, however, the cholinergic cells (A14 and dA14; Schmidt et al., 1985; Pourcho and Osman; 1986) only have a faint resemblance to starburst cells. Furthermore, in chicken, in addition to the starburst cells a third type of ACh AC with a different morphology has been described not seen outside non-avian retinas (Millar et al., 1987).

A second AC population that has been studied in this respect contains neurotensin-like immunoreactivity. In roach, Aeqidens and catfish, these correspond to A22 and stratify in the middle of the ipl (Wagner and Zeutzius, 1987). In turtle, by contrast, neurotensin containing cells come in two different types (Weiler and Ball, 1984), one of which is similar to the teleost A22; it is, however, localized in sublamina b (Eldred and Carraway 1987).- Finally, somatostatin-like immunoreactivity is present in A18-like ACs in cyprinids, cichlids in catfish ; furthermore, in goldfish and roach, at least two additional cell types were observed (Marshak et al., 1984; Wagner and Zeutzius, 1987). In other vertebrates, the variability of somatostatin containing AC types is even greater (Brecha et al., 1981; Spira et al., 1984; Sagar et al., 1985).

In conclusion, while it does not appear possible, on a general level, to correlate transmitter content and cell type morphology, in some specific cases, such a correlation can indeed be made.

5. Light-evoked responses and cell type morphology

In cyprinid teleosts, two recent studies have attempted to relate the various kinds of membrane potentials to their cellular origin using microinjection of HRP or Lucifer Yellow (Teranishi et al.,1987; Djamgoz et al., 1989). The results in both experiments agree in that morphologically different types give identical electrophysiological responses. The best correlation possible was between subtypes and "on" or "off" centre reactions; consequently, the radial position within the ipl appears to be more important for the respective type of response than the pattern of ramification. In accordance with the functional subdivision of the ipl (Famiglietti and Kolb, 1976; Famiglietti et al., 1977), sustained cells in sublamina a were hyperpolarizing, and those in sublamina b were depolarizing; bistratified or diffuse cells occupying both sublaminae, on the other hand, had transient responses. One other structural detail that was reflected in the membrane responses concerned the diameter of processes: AC types with stout proximal dendrites (A6, A7,

A26, A28) had fast responses, whereas cell types with thin or delicate processes had slow membrane responses (Teranishi et al., 1987).

Conclusion

ACs with identical wholemount morphology (called cell types) are observed (i) within the same retina at different ipl sublayers (as subtypes), (ii) among different species of teleosts and (iii) even in representatives of different vertebrate classes. Structural constraints leading to the differentiation of identical ramification patterns would suggest similar principles of functional organization at a general level for these cell types. In some cases, this is supported by the correlation of transmitter and morphological cell type; a similar correlation for electrophysiological responses and cell type is less obvious; however, the role of subtypes is well defined by this technique.

From the 43 types of ACs found in roach, 16 could be identified in five of the seven teleost species considered here. Furthermore, five of these types were also present in at least five non-teleost vertebrate species. Thus, there appears to be a population of cell types with identical form and, possibly, function throughout vertebrates. A majority of cell types, however, seems to have evolved in response to the species-specific requirements on the visual system.

Subtypes, on the other hand, may serve complementary roles in the two ipl sublaminae. An increased number of subtypes of a given cell type within the same retina is thought to indicate a high degree of parallel processing required in colour vision, for example.

Acknowledgements: The support of the Deutsche Forschungsgemeinschaft is gratefully acknowledged. Thanks are also due to E.Wagner for preparing most of the camera lucida drawings; to M.B.A.Djamgoz (London) for making available HRP-injected cells and the NATO, which supplied travel funds for this collaboration; and to R.H. Douglas (London) for discussion and revision of the manuscript. The goldeye material was obtained through the help of M.A.Ali (Montréal); it was sectioned by M.Schneider.

References

Ali MA, Harosi F and Wagner H-J (1978) Photoreceptors and visual pigments in a cichlid fish, Nannacara anomala.Sensory Processes 2: 130-145.

Ammermüller J and Weiler R (1981) The ramification pattern of amacrine cells within the inner plexiform layer of the carp retina. Cell Tissue Res. 220: 699-723.

Boycott BB and Dowling JE (1969) Organization of the primate retina: Light microscopy. Phil. Trans. R. Soc. Lond. B 255: 109-184.

Brecha N Retinal neurotransmitters: Histochemical and biochemical studies. In: (P.C. Emson, ed.) Chemical Neuroanatomy. pp 85-129 Raven Press, New York 1983.

Brecha NC, Sharma SC and Karten HJ (1981) Localization of Substance P-like immunoreactivity in the adult and developing goldfish retina. Neuroscience 6: 2737-2746.

Brecha NC, Eldred W , Kuljis RO and Karten HJ (1984) Identification and localization of biologically active peptides in the vertebrate retina. Progress in Retinal Research 3: 185-226.

Cajal Ramon y, S (1893) La retine des vertebres. La Cellule 9: 121-225.

Djamgoz MBA and Wagner H-J (1987) Intracellular staining of retinal neurones: Applications to studies of functional organization. Progress in Retinal Research 6: 85-150.
Djamgoz MBA, Downing JEG and Wagner H-J (1984) The dendritic fields of functionally identified amacrine cells in a cyprinid fish retina. J. Physiol. Lond. 349: 21P.
Djamgoz MBA, Downing JEG and Wagner H-J (1989) An intracellular horseradish peroxidase study of amacrine cells in the retina of a cyprinid fish. Cell Tissue Res. in press: .
Downing JEG, Djamgoz MBA and Bowmaker JK (1986) Photoreceptors of a cyprinid fish, the roach: morphological and spectral characteristics. J. comp. Physiol. A 159: 859-868.
Eldred, WD and Carraway RE (1987) Neurocircuitry ot two types of neurotensin - containing amacrine cells in the turtle retina. Neuroscience 21: 603-618.
Famiglietti E V Jr and Kolb H (1976) Structural basis for 'on' and 'off'-center responses in retinal ganglion cells. Science, Wash. 194: 193-195.
Famiglietti E V Jr, Kaneko A and Tachibana M (1977) Neuronal architecture of on and off pathways to ganglion cells in carp. Science, Wash. 198: 1267-1269.
Famiglietti E V Jr (1983a) 'Starburst' amacrine cells and cholinergic neurons: mirror-symmetric on and off amacrine cells of rabbit retina. Brain Res. 261: 138-144.
Famiglietti E V Jr (1983b) On and off pathways through amacrine cells in mammalian retina: The synaptic connections of 'starburst' amacrine cells. Vision Res. 23: 1265-1279.
Famiglietti E V Jr (1985) Starburst amacrine cells: morphological constancy and systematic variation in the anisotropic field of rabbit retinal neurons. J. Neurosci. 5: 562-577.
Jensen RJ and DeVoe RD (1982) Ganglion cells and (dye coupled) amacrine cells in the turtle retina that have possible synaptic connection. Brain Res. 240: 146-150.
Kolb H (1982) The morphology of bipolar cells, amacrine cells and ganglion cells in the retina of the turtle Pseudemys scripta elegans. Phil. Trans. R. Soc. Lond. B 298: 355-393.
Kolb H and Nelson R (1981) Amacrine cells of the cat retina. Vision Res. 21: 1625-1633.
Kolb H and Nelson R (1984) Neural architecture in the cat retina. Progress in Retinal Research 3: 21-60.
Kolb H and Nelson R Functional neurocircuitry of amacrine cells in the cat. In: (A. Gallego and P. Gouras, eds.) Neurocircuitry of the Retina. A Cajal Memorial. pp 215-232 Elsevier, New York 1985.
Kolb H, Nelson R and Mariani A (1981) Amacrine cells, bipolar cells and ganglion cells of the cat retina: A Golgi study. Vision Res. 21: 1081-1114.
Lam DM-K, Su Y-YT and Watt CB (1986) The self-regulating synapse: A functional role for the co-existence of neuroactive substances. Brain Res. Rev. 11: 249-257.
Linberg KA, Fisher SK and Kolb H (1986) A Golgi study of amacrine cells in the human retina. Invest. Ophthalmol. Vis. Sci. Suppl. 26: 203.
Lythgoe, JN The ecology of vision. Clarendon Press, Oxford 1979.
Mariani A P (1982) Association amacrine cells could mediate directional selectivity in pigeon retina. Nature, Lond. 298: 654-655.
Marc RE (1986) Neurochemical stratification in the inner plexiform layer of the vertebrate retina. Vision Res. 26: 223-238.
Marshak D W, Yamada T and Stell W K (1984) Synaptic contacts of somatostatin-immunreactive amacrine cells in goldfish retina. J. comp. Neurol. 225: 44-52.
Masland R H and Tauchi M (1986) The cholinergic amacrine cell. Trends in NeuroSciences 95: 218-223.
Massey SC and Redburn D A (1987) Transmitter circuits in the vertebrate retina. Progress in Neurobiology 28: 55-96.
Millar TJ, Ishimoto I, Chubb IW, EpsteinML Johnson CD and Morgan IG (1987) Cholinergic amacrine cells of the chicken retina: a light and electron microscope immunocytochemical study. Neuroscience 21: 725-743.
Naka K-I (1980) A class of catfish amacrine cells responds preferentially to objects which move vertically. Vision Res. 20: 961-965.
Naka K-I and Carraway NRG (1975) Morphological and functional identification of catfish retinal neurons I.Classical morphology J. Neurophysiol. 38: 53-71
Perry V H and Walker M (1980) Amacrine cells, displaced amacrine cells and interplexiform cells of the rat. Proc. Roy. Soc. Lond. B 208: 415-431.

Pourcho RG and Osman K (1986) Cytochemical identification of cholinergic amacrine cells in cat retina. J. comp. Neurol. 247: 497-504.

Sagar SM, Marshall PE and Landis DMD (1985) Immunoreactive somatostatin in the rat retina; light microscopic immunocytochemistry and chromatographic characterization. Brain Res. 336: 235-242.

Schmidt MH, Wässle H, and Humphrey M (1985) Number and distribution of putative cholinergic neurons in the cat retina. Neuroscience Letters 59: 235-240.

Scholes JH (1975) Colour receptors, and their synaptic connexions in the retina of a cyprinid fish. Phil. Trans. R. Soc. Lond. B 270: 61-118.

Spira AW, Shimizu Y and Rorstad OP (1985) Localization, chromatographic characterization, and development of somatostatin-like immunoreactivity in the guinea pig retina. J. Neurosci.: 3069-3079.

Stell WK (1975) Horizontal cell axons and axon terminals in goldfish retina. J. comp. Neurol. 159: 503-520.

Stell WK Putative peptide transmitters, amacrine cell diversity and function in the inner plexiform layer. In: (A.Gallego and P.Gouras, eds.) Neurocircuitry of the Retina. A Cajal Memorial. pp 171-181 Oxford, Elsevier 1985.

Tauchi M and Masland RH (1984) The shape and arrangement of the cholinergic neurons in the rabbit retina. Proc. R. Soc. Lond. B 223: 101-119.

Teranishi T, Negishi K and Kato S (1987) Functional and morphological correlates of amacrine cells in carp retina. Neuroscience 20: 935-950.

Tumosa N, Eckenstein F and Stell WK (1984) Immunocytochemical localization of putative cholinergic neurons in the goldfish retina. Neuroscience Letters 48: 255-259.

Vallerga, S (1981) Physiological and morphological identification of amacrine cells in the retina of the larval tiger salamander. Vision Res. 21: 1307-1317.

Vallerga S and Deplano S (1984) Differentiation, extent and layering of amacrine cell dendrites in the retina of a sparid fish. Proc. R. Soc. Lond. B 221: 465-477.

Vaney DI, Peichl L and Boycott BB (1981) Matching populations of amacrine cells in the inner nuclear and ganglion cell layers of the rabbit retina. J. comp. Neurol. 199: 373-391.

Voigt T (1986) Cholinergic amacrine cells in the rat retina. J. comp. Neurol. 248: 19-35.

Wagner H-J and Wagner E (1988) Amacrine cells in the retina of a teleost fish, the roach (Rutilus rutilus). Phil. Trans. R. Soc. (Lond.) in press: .

Wagner H-J and Zeutzius I (1987) Amacrine cells with neurotensin- and somatostatin-like immunoreactivity in three species of teleosts with different colour vision. Cell Tiss. Res. 248: 663-673.

Wagner H-J (1973) Die nervösen Netzhautelemente von Nannacara anomala (Cichlidae, Teleostei).I. Darstellung durch Silberimprägnation. Z. Zellforsch. mikrosk. Anat. 137: 63-84.

Weiler R and Ball AK (1984) Co-localization of neurotensin-like immunoreactivity and 3H-glycine uptake system in sustained amacrine cells of the turtle retina. Nature, Lond. 311: 759-761.

Weiler R and Marchiafava PL (1981) Physiological and morphological study of the inner plexiform layer in the turtle retina. Vision Res. 21: 1635-1638.

West RW (1976) Light and electron microscopy of the ground squirrel retina: Functional considerations. J. comp. Neurol. 168: 355-378.

Dendritic Relationships between Cholinergic Amacrine Cells and Direction-Selective Retinal Ganglion Cells

David I. Vaney*, Shaun P. Collin and Heather M. Young

National Vision Research Institute of Australia, Carlton, Vic. 3053, and
Vision Touch & Hearing Research Centre, University of Queensland, St Lucia, Qld. 4067

Although the apparent simplicity of direction selectivity has fascinated retinal physiologists since the process was first characterized by Barlow & Levick (1965), its synaptic mechanism still eludes us. There is substantial cholinergic input to direction-selective (DS) retinal ganglion cells (Masland & Ames 1976, Ariel & Daw 1982) but the functions of cholinergic amacrine cells in complex visual processing are poorly understood. Direction selectivity is believed to arise from non-linear interactions between cholinergic and GABAergic inputs, either in the ganglion cell itself or at a presynaptic level. The recent finding that the cholinergic amacrines also contain γ-aminobutyric acid (GABA; Vaney & Young 1988) and its synthetic enzyme (GAD; Kosaka et al 1988, Brecha et al 1988), is compatible with a single type of interneuron mediating both the excitation and inhibition to DS ganglion cells (Dowling 1970).

The input and output synapses of cholinergic amacrines are segregated centrifugally (Famiglietti 1983b) and we have proposed that this arrangement could provide the asymmetric inhibition necessary for direction selectivity (Vaney & Young 1988). Our hypothesis makes a testable prediction: ganglion cells that respond to different preferred directions should receive synaptic input from different segments of the cholinergic dendritic field. We are testing this hypothesis directly by injecting dye under visual control into neighbouring cholinergic amacrines and DS ganglion cells, and then correlating their dendritic branching patterns and synaptic connectivity. Preliminary findings are presented here.

Cholinergic amacrine cells: 'starburst' morphology and dendritic fasciculation

Retinal cholinergic neurons comprise two matching populations of mirror-symmetric amacrine cells (Masland & Mills 1979, Vaney et al 1981). The cell bodies of the type Ca amacrines are conventionally located in the inner nuclear layer and their dendrites are narrowly stratified in sublamina a of the inner plexiform layer; the type Cb amacrines are displaced to the ganglion cell layer and stratify in sublamina b. Rabbit cholinergic amacrines can be selectively and reliably labelled with the fluorescent nuclear stain, diamidino-phenylindole (DAPI; Masland et al 1984a), enabling the cells to be visualized microscopically in an isolated living preparation.

---

* NH&MRC Research Fellow

NATO ASI Series, Vol. H31
Neurobiology of the Inner Retina
Edited by R. Weiler and N.N. Osborne
© Springer-Verlag Berlin Heidelberg 1989

Injection of Lucifer yellow into these DAPI-stained cells revealed that they have a distinctive unistratified morphology (Vaney 1984, Tauchi & Masland 1984) that resembles a 'starburst' firework (Famiglietti 1983a). Amacrine cells with this morphology were first described from Golgi-stained wholemounts of the rat retina (Perry & Walker 1980).

A Lucifer-filled Cb amacrine from central rabbit retina is shown in Figure 1: each cell usually produces three to five primary dendrites that branch regularly and repeatedly, forming a monostratified arborization with as many as 80 terminal dendrites. The dendritic field diameter of Cb amacrines ranges from 180 μm on the visual streak (900 cells/mm^2) to 640 μm in the superior periphery (200 cells/mm^2; Vaney 1984). Cholinergic amacrines receive bipolar input over the whole dendritic field, but their output to retinal ganglion cells is confined to an annular zone that is marked by small varicosities on the terminal dendrites and which covers the distal third of the tree (Famiglietti 1983b, Brandon 1987b). In addition, it has been reported that cholinergic-cholinergic synapses are also common (Millar & Morgan 1987).

The relatively high density of cholinergic amacrines combined with their wide-field morphology results in substantial dendritic overlap, with each point on the rabbit retina covered by the dendritic fields of 25 to 70 Cb amacrines (Vaney 1984, Tauchi & Masland 1984). However the cholinergic dendrites do not blanket the retina as originally proposed, but show a fasciculated topology with lacunae in the cholinergic meshwork of 10 to 50 μm diameter (Tauchi & Masland 1985, Brandon 1987b). The cholinergic fasciculation is particularly pronounced in cat retina (Figure 2) where the dendrites are more tightly bundled and the lacunae larger, perhaps reflecting the sparser dendritic branching of cat cholinergic amacrines (Pourcho & Osman 1986, Schmidt et al 1987, Vaney unpublished).

Direction-selective ganglion cells: looping dendrites follow cholinergic fascicules

DS neurons respond optimally to movement in one 'preferred' direction, but show little or no response to movement in the reverse 'null' direction. In the rabbit retina, there are two distinct types of DS ganglion cells which respond to flashed illumination with an On-Off or an On-centre response respectively; the two types also differ in their specificity for stimulus size and speed (Barlow et al 1964, Oyster 1968, Wyatt & Daw 1975). Amthor et al (1984) have recorded intracellularly from On-Off DS ganglion cells in eyecup preparations and then injected the cells with horseradish peroxidase: they have a characteristic bistratified morphology with looping dendrites that cover the field uniformly. The dendrites of the bistratified ganglion cells branch in the same strata as the two populations of cholinergic amacrines (Famiglietti 1987), but their dendritic relationships have not been examined. Does the fasciculated meshwork of the cholinergic dendrites reflect the looping morphology of the DS ganglion cells (Tauchi & Masland 1985, Brandon 1987b, Vaney & Hughes 1988), or is it secondary to developmental

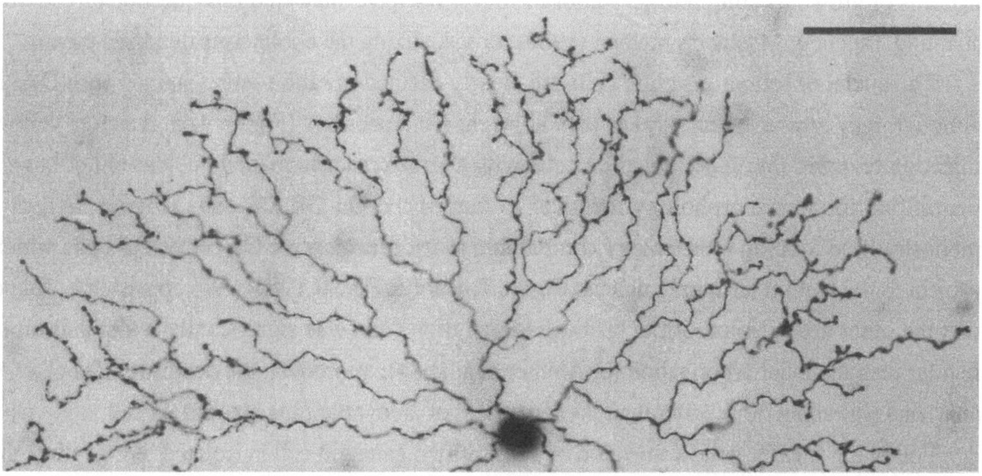

Figure 1. Type b cholinergic amacrine cell from the visual streak of rabbit retina. The neuron was injected with the fluorescent dye, Lucifer yellow, which was later converted to an opaque reaction product by photo-oxidation in the presence of DAB. Scale bar = 50 μm

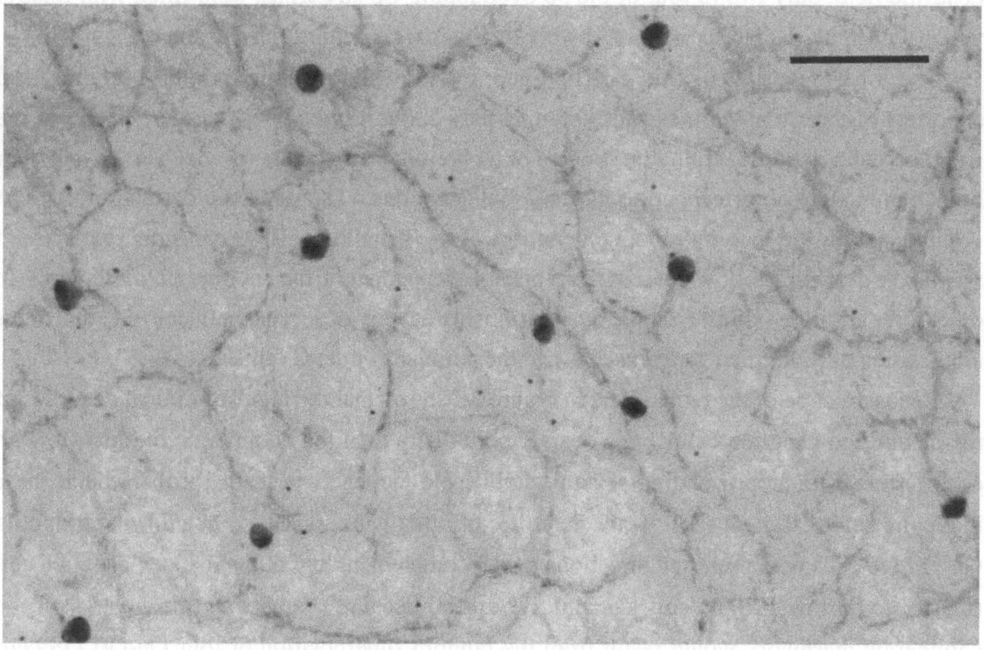

Figure 2. Cholinergic amacrine cells in a wholemount preparation of peripheral cat retina reacted with a monoclonal antibody against choline acetyltransferase and then processed by the streptavidin-biotin-peroxidase method. The irregular fasciculation of cholinergic dendrites in the inner plexiform layer is apparent. Scale bar = 50 μm

patterns of growth (Famiglietti & Tumosa 1987)? We have answered this question directly by adapting Tauchi & Masland's method (1985) for visualizing the cholinergic dendritic plexus.

The nuclei of retinal ganglion cells are clearly defined in rabbit retina stained with DAPI, although they appear much weaker than adjacent Cb amacrines (Figure 4A). Lucifer yellow injection revealed that those ganglion cells with a large crescent-shaped nucleus either have a bistratified looping morphology identical to that of the On-Off DS cells (Figure 3A), or a monostratified looping morphology comparable to the presumptive On-centre DS cells which project to the medial terminal nucleus (Figure 7; Buhl & Peichl 1986). We consistently found that the outer arborization of the bistratified ganglion cells was significantly wider and more regular than the inner arborization (cf Amthor et al 1984); moreover, the dendritic field of a Cb amacrine was about 50% wider than the inner field of an overlapping bistratified cell.

Following Lucifer yellow injection of a bistratified ganglion cell in isolated living retina, the incubation medium was drained from the tissue chamber, the retina briefly coverslipped and the cell's dendritic morphology recorded in a focal series of micrographs (Figure 3A). Up to ten Cb amacrine cells outside the dendritic field of the ganglion cell were then injected with Lucifer yellow (Figure 3B), the retina fixed and the dye in the filled processes converted to an opaque reaction product by photo-oxidation in the presence of DAB. At each point in the ganglion cell's dendritic field, only two to four of the 30 overlapping Cb amacrines were filled with Lucifer yellow; even so, the fasciculated meshwork of cholinergic dendrites is apparent (Figure 3C).

The inner dendrites of this ganglion cell are stratified in the same plane as the distal dendrites of the Cb amacrines and they may pass above or below each other. Many of the ganglion cell's dendrites follow the course of a cholinergic fascicule (or vice versa), and their branch points often correspond to fascicule intersections. The dendritic correlation is most pronounced for the looping terminal dendrites of the ganglion cell, whereas the primary and secondary dendrites follow courses that appear independent of the cholinergic meshwork. In other preparations, similar dendritic relationships were observed between the the outer arborization of bistratified ganglion cells and the terminals of the Ca amacrines.

Reconstruction of the retinal field in Figure 3C shows that the regular terminal dendrites of the bistratified ganglion cell have a coarser periodicity than the cholinergic meshwork, with each terminal loop surrounding several cholinergic lacunae. It seems probable that these interconnecting fascicules follow the dendrites of other ganglion cells, including overlapping On-Off DS cells with preferred directions which complement that of the mapped ganglion cell (Oyster & Barlow 1967). This has intuitive appeal because the apparent irregularity of the cholinergic meshwork would result from the random superposition of four On-Off DS cells with regular arborizations, but it does not explain how the pattern of cholinergic fascicules in sublamina a is largely duplicated in sublamina b (Brandon 1987b, Famiglietti & Tumosa 1987). Cholinergic dendrites skirt around vertical retinal elements such as Müller cell processes, but these local deviations account for only the small cholinergic lacunae.

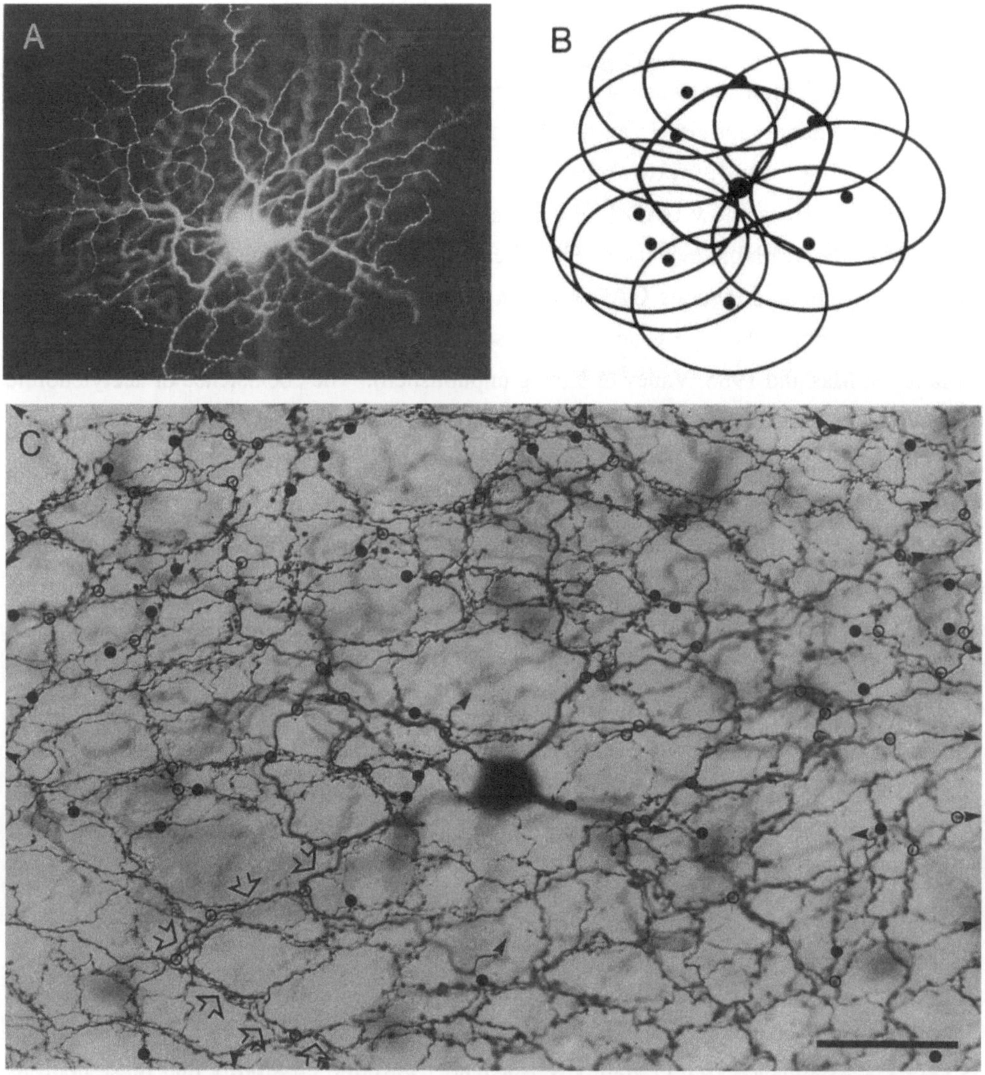

Figure 3A. A bistratified retinal ganglion cell from the visual streak of rabbit retina which was injected with Lucifer yellow: the focus is on the inner stratification in sublamina b of the inner plexiform layer. The cell has a looping morphology that corresponds to the dendritic form of On-Off direction-selective ganglion cells.

Figure 3B. Schematic diagram showing the cell positions and dendritic field overlap of the bistratified ganglion cell and of ten surrounding cholinergic amacrines which were subsequently injected with Lucifer yellow.

Figure 3C. Central portion of the bistratified ganglion cell and overlapping cholinergic amacrines as viewed with DIC optics. Circles mark the dendritic branch points of the ganglion cell, dots mark their dendritic terminations, and arrow-heads mark ganglion cell dendrites that extend beyond the micrograph or descend to sublamina a. In many places the ganglion cell dendrites follow cholinergic fascicules, as indicated by the open arrows. Scale bar = 50 μm

Localization of GABAergic markers in cholinergic amacrine cells

We recently demonstrated that the cholinergic amacrine cells in rabbit retina show strong GABA-like immunoreactivity (Vaney & Young 1988); independent confirmation of this surprising finding came from two other laboratories, both of which also reported that the cholinergic amacrines show GAD-like immunoreactivity (Kosaka et al 1988; Brecha et al 1988). Thus, the available evidence now indicates that the cholinergic amacrines contain an abundant store of endogenous GABA which is synthesized within the cells; like other GABAergic neurons, the type C amacrines also selectively accumulate extracellular GABA (O'Malley & Masland 1988, Vaney & Young unpublished). The coexistence of acetylcholine and GABA in these interneurons was foreshadowed by Agardh & Ehinger's observation (1983) that those neurons in the ganglion cell layer which accumulate isoguvacine, a selective GABA agonist, appear to overlap the population of choline-accumulating amacrines.

In a rabbit retinal wholemount, DAPI staining revealed the regular array of Cb cholinergic amacrine cells in the ganglion cell layer (Figure 4A); all of the Cb amacrines showed strong GABA-like immunoreactivity and they accounted for 65-75% of the GABA-positive neurons in this layer (Figure 4B). The remaining GABA-positive/DAPI-negative cells included some strongly labelled neurons in the displaced amacrine size range, and a heterogenous population of moderate and weakly labelled neurons, many in the ganglion cell size range (Vaney & Young 1988). In the cat retina, we have been able to show colocalization of cholinergic and GABAergic markers directly, by using a rat antibody against choline acetyltransferase and a rabbit antiserum against GABA (Figure 5; Vaney, Whitington & Young unpublished).

The synaptic mechansim of direction selectivity: cotransmission models

Although light-evoked cotransmission of acetylcholine and GABA from type C amacrines has yet to be established (see O'Malley & Masland 1988), the possibility that these neurons play an inhibitory role in direction-selective processing calls for reinterpretation of their synaptic organization. In the simplest case, GABAergic inhibition at cholinergic-cholinergic synapses or at autoreceptors may underlie the transient nature of acetylcholine release at light Onset and Offset (Masland et al 1984b), consistent with reports that the light-evoked release of acetylcholine is potentiated by GABA antagonists (Massey & Redburn 1982). Such isotropic inhibition could be accomodated by existing models of direction selectivity which regard the cholinergic amacrines as mediating only the excitatory input (Ariel & Daw 1982). The type C amacrines which contact a single On-Off DS cell cover a larger area than the ganglion cell's excitatory receptive-field (Tauchi & Masland unpublished), supporting the concept that the dendritic branches of a cholinergic amacrine are functionally separate (Miller & Bloomfield 1983, Masland et al 1984b). The distal location of the output synapses is thus rationalized in

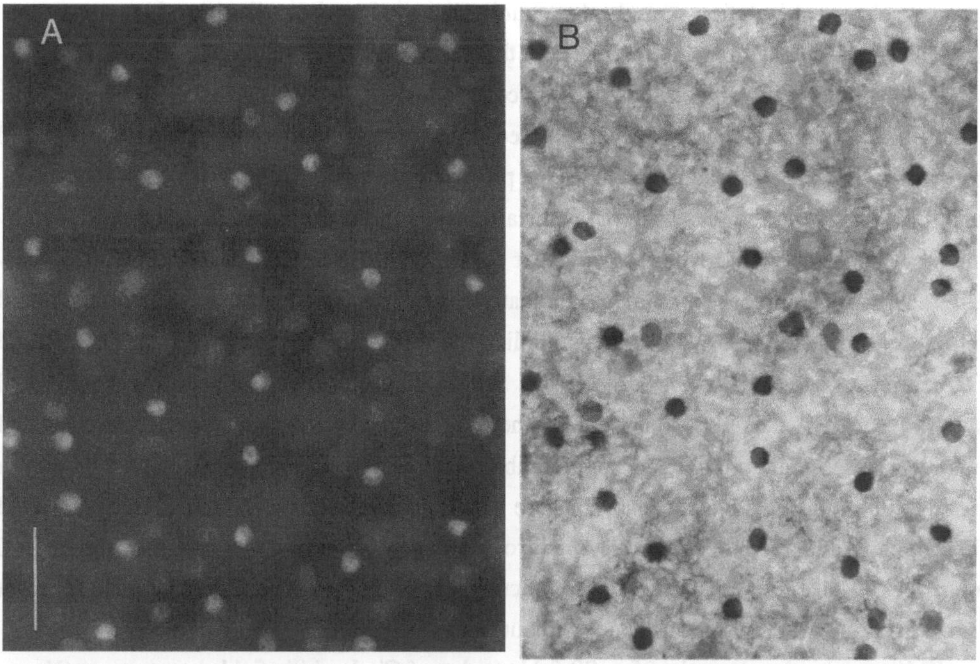

Figure 4A. The fluorescent dye DAPI selectively stains the array of displaced cholinergic amacrines in the ganglion cell layer of rabbit retina: the field contains 31 Cb amacrines at a density of 350 cells/mm^2. Scale bar = 50 μm

Figure 4B. Following immunolabelling, all of the Cb amacrine cells show strong GABA-like immunoreactivity. Differential shrinkage and cover-slip compression have distorted the field during processing. (From Vaney & Young 1988)

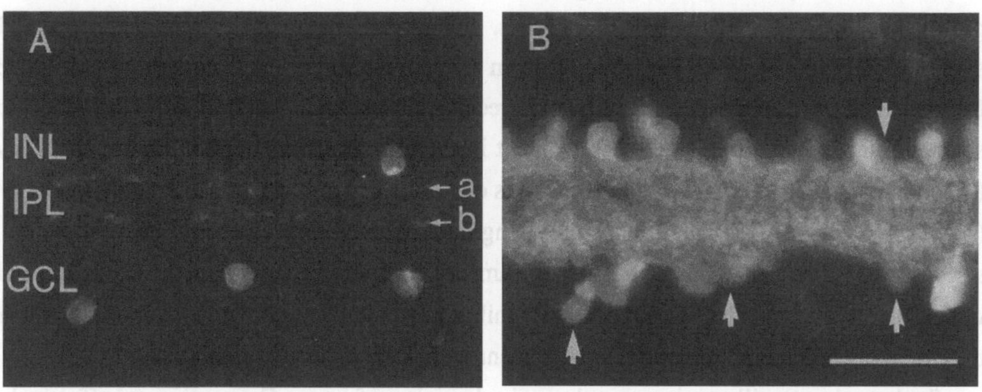

Figure 5. Frozen 10 μm transverse section of cat retina double-labelled with a rat monoclonal antibody against choline acetyltransferase (A) and a rabbit polyclonal antiserum against GABA (B). The Ca amacrine in the inner nuclear layer (INL) and three Cb amacrines in the ganglion cell layer (GCL) all show GABA-like immunoreactivity. The cholinergic strata in sublaminae a and b of the inner plexiform layer (IPL) are marked with small arrows. Scale bar = 50 μm

terms of electrical isolation of multiple subunits, but the broad distribution of input synapses seems incompatible with the requirements of these models.

Two-spot experiments on DS ganglion cells indicate that each part of the receptive-field receives an excitatory input from overlying receptors and an inhibitory input through a delayed pathway from a larger area that excludes the preferred direction; for null-direction movement, multiplicative interaction between the pathways results in the excitatory input being vetoed by the slow inhibitory input from the adjacent region (Barlow & Levick 1975, Wyatt & Daw 1975). Pharmacological experiments indicate that GABAergic neurons mediate this lateral inhibition, because GABA antagonists abolish the direction selectivity of both On-Off and On-centre DS cells (Caldwell et al 1978). The finding that type C amacrines contain and synthesize GABA, leads to the novel hypothesis that the centrifugal segregation of input and output synapses could provide the lateral inhibition necessary for direction selectivity (Vaney & Young 1988); the asymmetry arises because ganglion cells that respond to different preferred directions would receive synaptic input from different dendritic quadrants of the type C amacrines. The concept that pronounced functional asymmetries may be derived from the most symmetrical neuron in the retina is not without irony.

This model rationalizes the 25 to 70-fold overlap of Cb dendritic fields in two ways (Vaney 1985). First, only the outer varicose zone in one quadrant of each overlying amacrine would provide input to an On-Off DS cell: on the visual streak, fewer than four Cb amacrines would contact local regions of the dendritic field. Second, wide dendritic fields staggered regularly at short intervals would ensure positional sensitivity while preserving responsiveness to high velocities. A necessary corollary of this arguement is that the type C amacrines would not provide a suitable inhibitory substrate for the On-centre DS cells.

The inhibitory veto could take place at the level of the ganglion cell membrane or presynaptically in the afferent type C dendrites. In the latter case, a cholinergic dendrite would symmetrically receive GABAergic input from all neighbouring type C amacrines (Millar & Morgan 1987), but an On-Off DS cell would receive cholinergic input only from the dendritic segment that is first stimulated by movement in the null direction. The inhibitory pathway (bipolar → type CI → type CE → DS cell) has one more synapse than the excitatory pathway (bipolar → type CE → DS cell), thus providing the asymmetric delay necessary for direction selectivity. Although it is postulated that one amacrine system mediates both the excitation and inhibition, the spatial extent of GABAergic inhibition would be asymmetric to, and larger than the extent of cholinergic excitation, in agreement with Ariel & Daw's findings (1982).

Such a scheme still requires that the major branches of a type C amacrine are effectively isolated from each other. Otherwise, a wave of inhibition would hyperpolarize the whole amacrine, preventing appropriate release of acetylcholine to preferred-direction movement and of GABA to null-direction movement. In support of this, two-spot experiments indicate that the receptive-field diameter of On-Off DS cells is 2.5 to 3 times greater than the extent of the

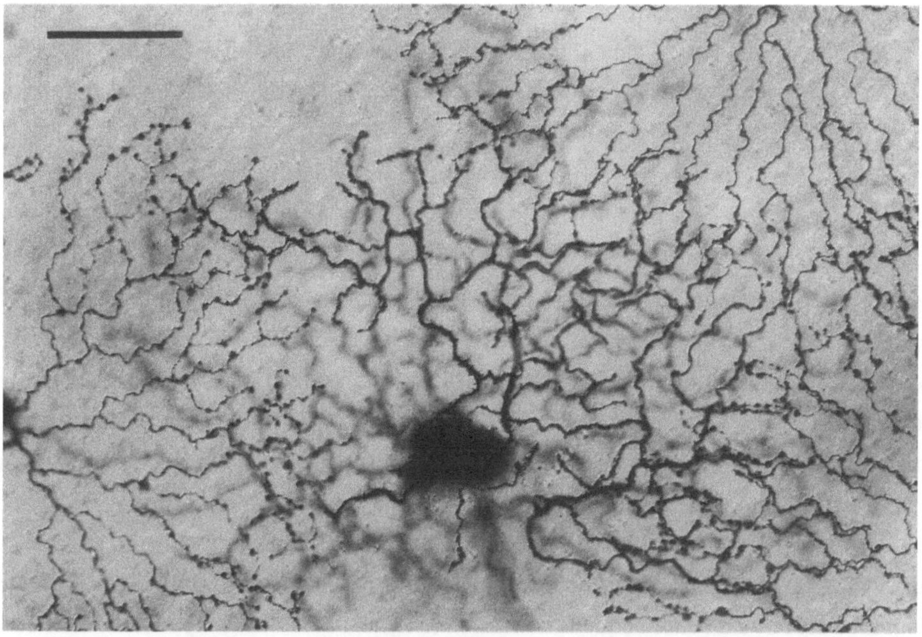

Figure 6. A bistratified ganglion cell and three displaced cholinergic amacrines as viewed with DIC optics: terminals from each of the cholinergic amacrines closely follow the looping dendrites of the ganglion cell. The out-of-focus arborization in sublamina a is more extensive than the sublamina b arborization. Scale bar = 50 μm

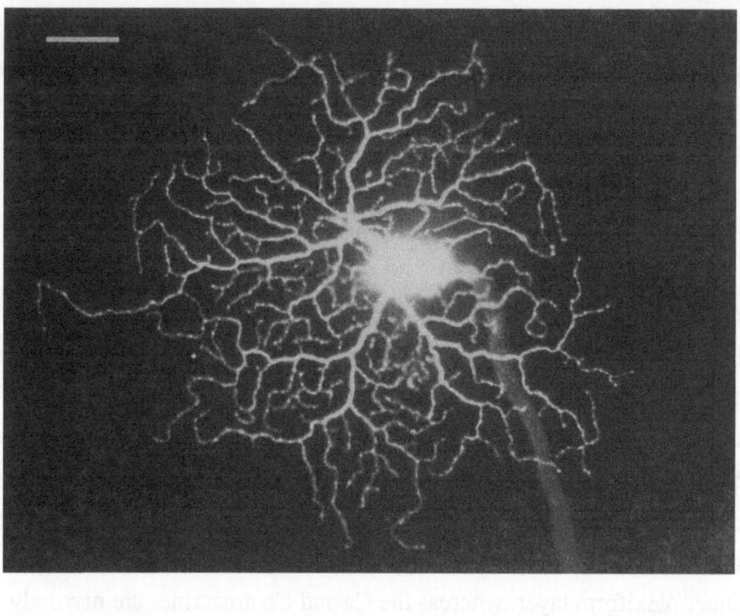

Figure 7. Presumptive On-centre direction-selective ganglion cell from the visual streak of rabbit retina. Scale bar = 50 μm

asymmetric inhibition (Barlow & Levick 1965, Wyatt & Daw 1975); given that the DS receptive-field is about 1.25 times larger than the dendritic field of a Cb amacrine at each eccentricity (Tauchi & Masland unpublished), we conclude that the inhibition extends for only half the width of a type C amacrine.

Different constraints are placed on cotransmission models of direction selectivity if the underlying non-linear interactions are mediated by shunting inhibition in the ganglion cell dendrites (Torre & Poggio 1978, Koch et al 1982). The DS cell would then be expected to receive type C-mediated GABAergic input only from the dendritic segment that is first stimulated by movement in the preferred direction, whereas the excitation could be mediated by cholinergic input from all type C dendrites. This would require that some synapses between type C dendrites and DS ganglion cells are GABAergic whereas others are not, which seems implausible; moreover, a mechanism other than synaptic delay would be necessary to account for the slower time course of the inhibitory input. If the excitation came directly from bipolar cells however (Famiglietti 1983b, Brandon 1987b) these factors would not be relevant: pharmacological experiments indicate that about half the light-evoked excitation of DS ganglion cells is mediated by non-cholinergic transmission (Masland & Ames 1976, Ariel & Daw 1982).

The common feature of these cotransmission models is the requirement that either the excitatory or inhibitory synapses between type C amacrines and a DS ganglion cell should arise from a restricted segment of each amacrine's dendritic field. We have examined whether such dendritic relationships are apparent under the light microscope by injecting Lucifer yellow into a bistratified ganglion cell and into adjacent Cb amacrines whose terminal dendrites partially overlap the ganglion cell's dendritic field (Figure 6). Following several experiments, we conclude that cholinergic dendrites on all sides closely follow the ganglion cell's arborization in sublamina b and, therefore, it will be necessary to examine this material under the electron microscope to test our hypothesis.

Presumptive On-centre direction-selective cells do not contact cholinergic amacrines

While searching for On-Off DS cells, we frequently injected a type of monostratified ganglion cell that had a dendritic morphology which resembled that of the bistratified ganglion cells, although their dendritic loops were smaller and more numerous (Figure 7). Their cell body size, dendritic morphology and stratification level matched that of the presumptive On-centre DS ganglion cells which project to the medial terminal nucleus (Buhl & Peichl 1986). Injection of overlapping Cb amacrines revealed that these monostratified ganglion cells cannot make contact with the terminal dendrites of type C amacrines. They stratify at about 60% of the depth of the inner plexiform layer, whereas the Ca and Cb amacrines are narrowly stratified at 23% and 70% respectively (Brandon 1987a, Famiglietti & Tumosa 1987): this corresponds to a separation of 1.5 to 2 μm.

When the acetylcholinesterase inhibitor, physostigmine, is infused into the retinal circulation, the potentiated cholinergic excitation overcomes the null-direction inhibition of both On-Off and On-centre DS cells, leading Ariel & Daw (1982) to conclude that similar excitatory and inhibitory pathways are involved for both types of ganglion cells. How can these results be reconciled with our finding that presumptive On-centre DS cells cannot receive a direct synaptic input from type C amacrines? Three possibilities come to mind. First, the ganglion cells in question may not be On-centre DS cells. Second, the light-evoked release of acetylcholine may have paracrine effects both within and beyond the cholinergic strata, as proposed for the light-independent resting release (Masland et al 1984b). Third, acetylcholine which would normally be hydrolyzed in the synaptic cleft by acetylcholinesterase may diffuse into surrounding neuropil with unpredictable results. In support of this hypothesis, On-centre DS cells respond well to Off stimuli in the presence of physostigmine (Ariel & Daw 1982), perhaps reflecting diffuse release of acetylcholine from the Ca amacrines.

Physostigmine symmetrically potentiates cholinergic excitation in both the null and preferred directions for On-Off DS cells, suggesting that the release of acetylcholine is not direction selective, as would be expected if the GABAergic inhibition occurred presynaptically (Ariel & Daw 1982). If the acetylcholine was not confined to the synaptic cleft, however, other ganglion cells within the fascicle may be excited by diffuse transmitter release, blurring the direction-selective relationship between a cholinergic dendrite and its target ganglion cell.

References

Agardh E, Ehinger B (1983) Retinal GABA neuron labelling with [^{3}H]isoguvacine in different species. Exp Eye Res 36:215-229

Amthor FR, Oyster CW, Takahashi ES (1984) Morphology of on-off direction-selective ganglion cells in the rabbit retina. Brain Res 298:187-190

Ariel M, Daw NW (1982) Pharmacological analysis of directionally sensitive rabbit retinal ganglion cells. J Physiol 324:161-185

Barlow HB, Levick WR (1965) The mechanism of directionally selective units in rabbit's retina. J Physiol 178:477-504

Barlow HB, Hill RM, Levick WR (1964) Retinal ganglion cells responding selectively to direction and speed of image motion in the rabbit. J Physiol 173:377-407

Brandon C (1987a) Cholinergic neurons in the rabbit retina: immunocytochemical localization, and relationship to GABAergic and cholinesterase-containing neurons. Brain Res 401:385-391

Brandon C (1987b) Cholinergic neurons in the rabbit retina: dendritic branching and ultrastructural connectivity. Brain Res 426:119-130

Brecha N, Johnson D, Peichl L, Wässle H (1988) Cholinergic amacrine cells of the rabbit retina contain glutamate decarboxylase and γ-aminobutyrate immunoreactivity. Proc Natl Acad Sci USA 85:6187-6191

Buhl EH, Peichl L (1986) Morphology of rabbit retinal ganglion cells projecting to the medial terminal nucleus of the accessory optic system. J Comp Neurol 253:163-174

Caldwell JH, Daw NW, Wyatt HJ (1978) Effects of picrotoxin and strychnine on rabbit retinal ganglion cells: lateral interactions for cells with more complex receptive fields. J Physiol 276:277-298

Dowling JE (1970) Organization of vertebrate retinas. Invest Ophthalmol 9:655-680
Famiglietti EV (1983a) 'Starburst' amacrine cells and cholinergic neurons: mirror-symmetric on and off amacrine cells of rabbit retina. Brain Res 261:138-144
Famiglietti EV (1983b) On and off pathways through amacrine cells in mammalian retina: the synaptic connections of "starburst" amacrine cells. Vision Res 23:1265-1279
Famiglietti EV (1987) Starburst amacrine cells in cat retina are associated with bistratified, presumably directionally selective, ganglion cells. Brain Res 413:404-408
Famiglietti EV, Tumosa N (1987) Immunocytochemical staining of cholinergic amacrine cells in rabbit retina. Brain Res 413:398-403
Koch C, Poggio T, Torre V (1982) Retinal ganglion cells: a functional interpretation of dendritic morphology. Phil Trans R Soc Lond B 298:227-264
Kosaka T, Tauchi M, Dahl JL (1988) Cholinergic neurons containing GABA-like and/or glutamic acid decarboxylase-like immunoreactivities in various brain regions of the rat. Exp Brain Res 70:605-617
Masland RH, Ames A (1976) Responses to acetylcholine of ganglion cells in an isolated mammalian retina. J Neurophysiol 39:1220-1235
Masland RH, Mills JW (1979) Autoradiographic identification of acetylcholine in the rabbit retina. J Cell Biol 83:159-178
Masland RH, Mills JW, Hayden SA (1984a) Acetylcholine-synthesizing amacrine cells: identification and selective staining using radioautography and fluorescent markers. Proc R Soc Lond B 223:79-100
Masland RH, Mills JW, Cassidy C (1984b) The functions of acetylcholine in the rabbit retina. Proc R Soc Lond B 223:121-139
Massey SC, Redburn DA (1982) A tonic γ-aminobutyric acid-mediated inhibition of cholinergic amacrine cells in rabbit retina. J Neurosci 2:1633-1643e
Millar TJ, Morgan IG (1987) Cholinergic amacrine cells in the rabbit retina synapse onto other cholinergic amacrine cells. Neurosci Lett 74:281-285
Miller RF, Bloomfield SA (1983) Electroanatomy of a unique amacrine cell in the rabbit retina. Proc Natl Acad Sci USA 80:3069-3073
O'Malley DM, Masland RH (1988) Co-release of acetylcholine and GABA by a retinal neuron. Invest Ophthamol Suppl 29:273
Oyster CW (1968) The analysis of image motion by the rabbit retina. J Physiol 199:613-635
Oyster CW, Barlow HB (1967) Direction-selective units in rabbit retina: distribution of preferred directions. Science 155:841-842
Perry VH, Walker M (1980) Amacrine cells, displaced amacrine cells and interplexiform cells in the retina of the rat. Proc R Soc Lond B 208:415-431
Pourcho RG, Osman K (1986) Cytochemical identification of cholinergic amacrine cells in cat retina. J Comp Neurol 247:497-504
Schmidt M, Humphrey MF, Wässle H (1987) Action and localization of acetylcholine in the cat retina. J Neurophysiol 58:997-1015
Tauchi M, Masland RH (1984) The shape and arrangement of the cholinergic neurons in the rabbit retina. Proc R Soc Lond B 223:101-119
Tauchi M, Masland RH (1985) Local order among the dendrites of an amacrine cell population. J Neurosci 5:2494-2501
Torre V, Poggio T (1978) A synaptic mechanism possibly underlying directional selectivity to motion. Proc R Soc Lond B 202:409-416
Vaney DI (1984) 'Coronate' amacrine cells in the rabbit retina have the 'starburst' dendritic morphology. Proc R Soc Lond B 220:501-508
Vaney DI (1985) News and views: fireworks in the retina. Nature 314:672-673.
Vaney DI, Hughes A (1988) Is there more than meets the eye? In: Blakemore C (ed) Vision: coding and efficiency. Cambridge University Press, Cambridge (in press)
Vaney DI, Young HM (1988) GABA-like immunoreactivity in cholinergic amacrine cells of the rabbit retina. Brain Res 438:369-373
Vaney DI, Peichl L, Boycott BB (1981) Matching populations of amacrine cells in the inner nuclear and ganglion cell layers of the rabbit retina. J Comp Neurol 199:373-391
Wyatt HJ, Daw MW (1975) Directionally sensitive ganglion cells in the rabbit retina: specificity for stimulus direction, size, and speed. J Neurophysiol 38:613-626

STRUCTURAL ORGANIZATION AND DEVELOPMENT OF DORSALLY-DIRECTED (VERTICAL) ASYMMETRICAL AMACRINE CELLS IN RABBIT RETINA

E. V. Famiglietti

Department of Anatomy
and Lions' Sight Centre
University of Calgary
Faculty of Medicine
Calgary Alberta Canada
T2N 4N1

INTRODUCTION

Amacrine cells are key elements in the inner retina, interposed between ganglion cells and neurons of simpler organization in the outer retina. Because ganglion cells exhibit a diversity of form and function (Cajal, '93; Cleland and Levick, '74; Famiglietti and Kolb, '76; Caldwell and Daw, '78; Famiglietti and Siegfried, '79; Famiglietti, '87), the roles of amacrine cells are necessarily diverse. This diversity is reflected in variations of morphology (Cajal, '93; Boycott and Dowling, '69; Famiglietti and Kolb, '75; Famiglietti and Siegfried, '80), neurotransmitter localization (Brecha, '83), synaptic connectivity (Famiglietti and Kolb, '75; Famiglietti, '83), and neurotransmitter function (Daw et al., '88,'89).

Some ganglion cells have "simple", centre-surround receptive field properties (Kuffler, '53). Amacrine cell input to these cells is essential in mammals (Famiglietti and Kolb, '75; Famiglietti, '83), and may affect the balance between centre and surround (Jensen and Daw, '84), but the effects seem to be radially symmetrical within the receptive field. Some types of ganglion cell with "complex" receptive field properties have radially asymmetrical receptive field organization, the most well-known of which are directionally selective (DS) and orientation selective (OS) ganglion cells (Barlow and Levick, '65; Levick, '67; Caldwell and Daw, '78). These receptive field properties are synthesized in the retina's inner plexiform layer (IPL), and it is of interest to

NATO ASI Series, Vol. H31
Neurobiology of the Inner Retina
Edited by R. Weiler and N.N. Osborne
© Springer-Verlag Berlin Heidelberg 1989

determine which amacrine cells are involved, and to ask how they create receptive field asymmetries in some ganglion cells.

It is natural to think that amacrine cells with specialized dendritic field organization are responsible for creation of the complex receptive fields of DS and OS ganglion cells. A search for amacrine cells with the appropriate features has revealed a number of amacrine cells in the rabbit retina with asymmetrical dendritic field organization (Famiglietti and Siegfried, '80). The most conspicuous of these is the morphologically heterogeneous set of vertically-oriented and dorsally-directed amacrine cells, which is the subject of this communication.

MATERIALS AND METHODS

Rabbit retinas were taken whole from pigmented or albino rabbit adults and from immature rabbits in the first 10 postnatal days, and mounted flat on a slide between pieces of filter paper for fixation in 3% glutaraldehyde. They were subjected to a Golgi impregnation regime of 3 days immersion in 2-5% glutaraldehyde and 4-6% potassium dichromate, followed by 1% silver nitrate for 2 days, with an intermediate wash in 0.1% silver nitrate. Retinas were brushed free of crystals, and some were postfixed in 1% osmium, dehydrated and mounted in thin sheets of epoxy, while others were dehydrated. cleared in xylene, and mounted in DPX. Cells were studied in a dozen adult retinas and in a half dozen immature retinas. Cells were drawn at various magnifications using a microscope with a camera lucida (drawing tube) attachment.

RESULTS AND DISCUSSION

Four Types of Dorsally-Directed Amacrine Cell in Rabbit Retina

The vertically-oriented, dorsally-directed amacrine cells of rabbit retina may be divided into four types (Figs. 1 and 5), according to particular features of their dendritic field size, branching pattern, and dendritic appendages. All four share the following features : 1) normally placed cell bodies, 2) a unidirectional dendritic tree with all major dendrites ultimately extended in one direction away from the cell body (with some exceptions among type 4 cells), 3) a dendritic tree directed upward (i.e. toward dorsal retina), 4) a dendritic more-or-less vertically aligned with respect to ocular global coordinates, determined with reference to the horizontal 'visual streak' of high ganglion cell density, 5) stratification of the main dendritic branches in the middle of the IPL at the a/b sublaminar border, or in stratum 3.

Type 1 cells (Fig. 1a) have some features in common with the "association amacrine cell" of bird retina (Cajal, '93). The type 1 cell has a fairly stout, single primary dendrites which is covered with long, sometimes branched dendritic spines in its course through sublamina a. Some of the spines emerge directly from the cell soma, and sometimes these are as extensive as small dendritic branches (Fig. 2e). The primary dendrite thins, spines are infrequent, and upon reaching the a/b sublaminar border, a fan of dendritic branches is formed, bearing small boutons en passant and almost no spines. The second type of dorsally-directed amacrine cell resembles the first, except that it lacks the elaborate proximal appendages, bearing only a few short somatic and proximal dendritic spines. The primary dendrites may also be single and thicker (Fig. 1b), or less frequently, dual and thinner (Fig. 1c). Because the exuberance of proximal spines and appendages seems to be variable from animal-to-animal, it is not always easy to distinguish between types 1 and 2 when pooling data, unless examples of both types are present nearby in a given retina.

The third type of dorsally-directed amacrine cell (Fig. 1d) is dramatic in its appearance for two reasons. In contrast to the first two types, its angle of branching is very narrow, no more than 15° with the pivot point at the cell soma. Secondly, in the mid-region of its narrow dendritic field, there is a localized profusion of pedunculated spines and a few short branches. A few dendritic branches, bearing boutons en passant and few dendritic spines pass a significant distance beyond this region to terminate in the same substratum of the IPL. Amacrine cells of somewhat similar morphology, but with a more diffuse localization of spines, have been described in bird retina (Mariani, '83). Some examples of this type have a second primary dendrite which emerges at the ventral side of the cell body (retinal topographic coordinates). This dendrite turns within 50 um and doubles back in the opposite direction to join the dorsally-extended dendritic field, contributing to the spinous plexus, but not extending beyond it.

The fourth type of dorsally-directed amacrine cell (Fig.

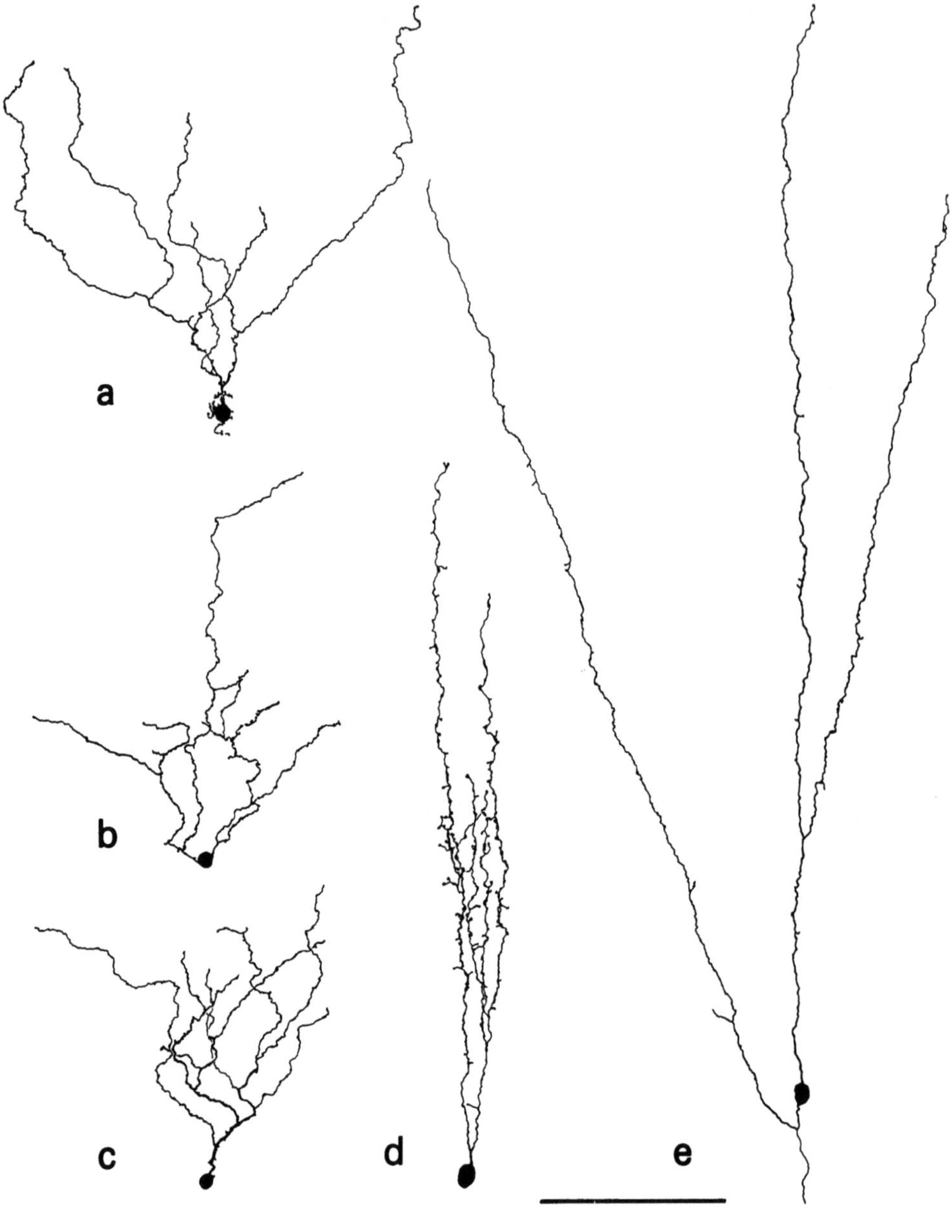

Figure 1. Four types of dorsally-directed, rabbit amacrine cells, in flat view. Camera lucida drawings of Golgi-impregnations. a. type 1, b. and c. variants of type 2, d. type 3, e. type 4. All four stratify in the middle of the IPL. See text for details. Dorsal is up. Calibration bar 100 um.

1e) is less commonly seen in our Golgi preparations. It has few, long, rectilinear dendrites bearing very few spines and few boutons en passant. These dendrites are unbranched in their distal course. In dendritic morphology and wide extent, they resemble wide-field amacrine cells branching in stratum 2, which are plentiful in our preparations. The latter exhibit a preference for horizontal or vertical dendritic orientation, but these may be dorsally or ventrally directed, and one cell typically possesses dendrites directed at right angles. It may be common for the type 4 cell to have a ventrally-directed dendrite, as does the cell of Figure 1e. In this example of a cell from the visual streak, the ventrally directed dendrite turns and becomes dorsally-directed, extending as far as the two branches originating from the dorsally-directed primary dendrite. A short, thin, rudimentary branch remains ventrally-directed.

Directed Dendritic Growth in Immature Retinas: Primordial or "Pruned"?

In the assessment of the possible functional significance of dorsally-directed amacrine cells in rabbit retina, it was thought important to examine such cells in developing retina. The principal question in mind was: does this dendritic field asymmetry arise primordially, or is it a consequence of dendritic "pruning" in response to dendro-dendritic interaction during the main phase of maturation of synaptic physiology, from 8-10 days postnatal (Masland, '77)? It has not yet been possible for us to obtain these amacrine cells at the earliest stages of dendritic outgrowth, but at 5 days of age, when the process of dendritic branching is still at an early stage, many unidirectional, dorsally-directed amacrine cells can be found (Fig. 2a, b, g, and h). In some instances, it is only possible to guess which of the four types these immature amacrine cells will become (e.g. Fig. 2g, h). In the case of type 1 cells, however, it has been possible to follow the developmental sequence from 5 days postnatal to adulthood (Fig. 2a-f). It is clear from these results that the dorsal-directedness of the dendritic trees is primordial, and that the "pruning" which evidently occurs here in the process of dendritic maturation, together with dendritic elongation, shapes the details and the size of dendritic branching, rather than the overall plan.

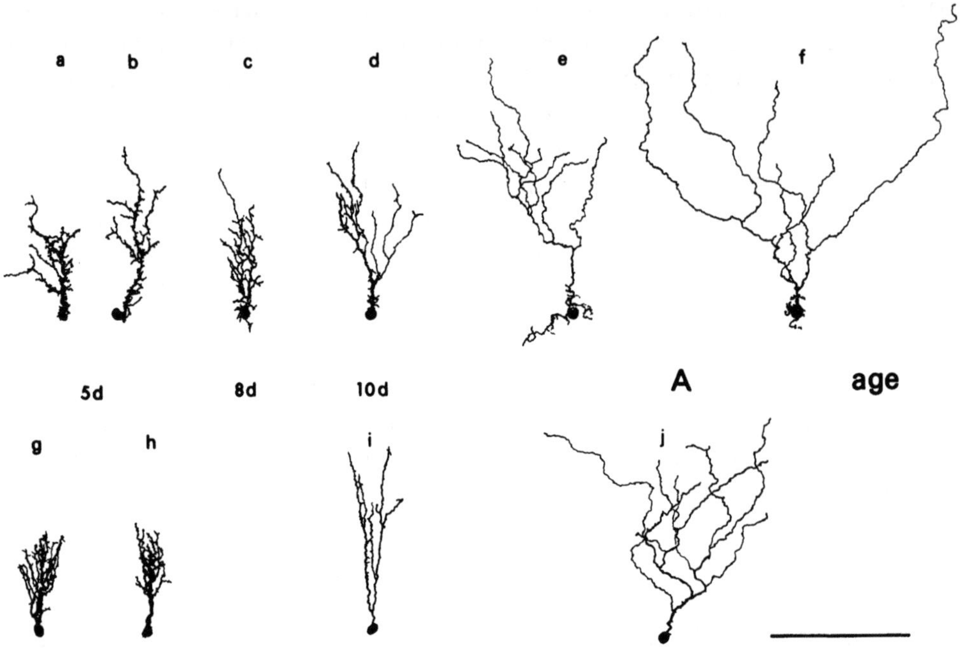

Figure 2. Postnatal development of dorsally-directed amacrine cells. A developmental sequence can be followed for type 1 cells (a-f), from day 5, through days 8 and 10 to adulthood (A). This involves proliferation and then reduction in long spines, and also dendritic extension. Other cells are also dorsally- directed at day 5 (g,h) with no sign of ventrally directed processes; one of these may correspond to type 2 (j) or type 3, and an unidentified late-developing cell is found at day 10 (i). Calibration 100 um.

Are There "Organizers" for Directed Dendritic Growth of Amacrine Cells?

In amphibia, it has been demonstrated that one factor determining the direction and destination of the axons of ganglion cells, as they proceed to map themselves onto the optic tectum, is the topographic location of the parent cell body in the retina (Jacobson, '68). The question arises in respect to dorsally directed amacrine cells, whether or not they may respond to simple cues, such as an axially specific or positional gradient in the retina or optic globe. It seems obvious that something of this nature affects dorsally-directed amacrine cells, but it is difficult to approach this issue directly without knowledge of the molecular cues. In a less direct approach one may consider the extent to which topographic location may affect the specificity and precision of dorsal-directedness.

The precision in the vertical orientation of dorsally-directed amacrine cells required examination in a single retina, where they could be assessed with respect to the best geometrical reference in rabbit retina, the horizontal visual streak. Type 1 and type 2 cells, the most abundant in our sample, were chosen for this study (Fig. 3). Arrows pointing dorsally, were aligned with lines bisecting the angle subtending the whole dendritic field, swept from the cell body. While the "total angle" for type 1 and type 2 cells ranges from 40° to 80°, in this region, just below the visual streak, the bisecting lines are nearly perpendicular to the streak. They vary in general less than 15°, and by no more than 25°.

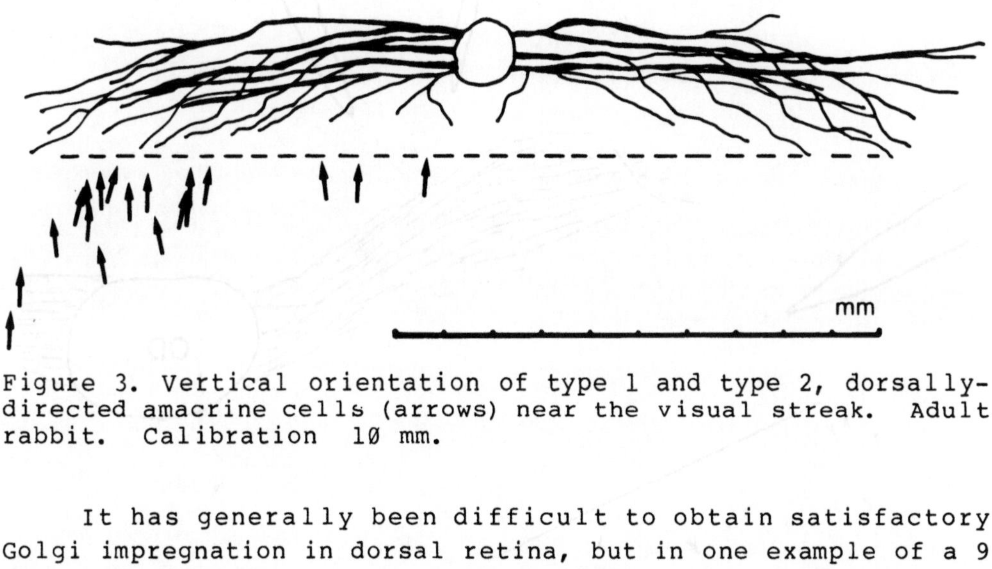

Figure 3. Vertical orientation of type 1 and type 2, dorsally-directed amacrine cells (arrows) near the visual streak. Adult rabbit. Calibration 10 mm.

It has generally been difficult to obtain satisfactory Golgi impregnation in dorsal retina, but in one example of a 9 day old retina, dorsally-directed amacrine cells were labelled both near to the visual streak and in dorsal retina (Fig. 4). Near the streak, type 1 and type 2 cells exhibit the same degree of precision in their alignment as do adult cells (cf. Fig. 3). In dorsal retina, however, this is not the case. None of the four types is ventrally directed, or even reaches a horizontal orientation, but cells in close proximity to one another are directed along axes at 90° to one another. Even type 3 cells with recurving dendrites appear to be rotated as a whole. It is interesting that a type 4 cell crossing the

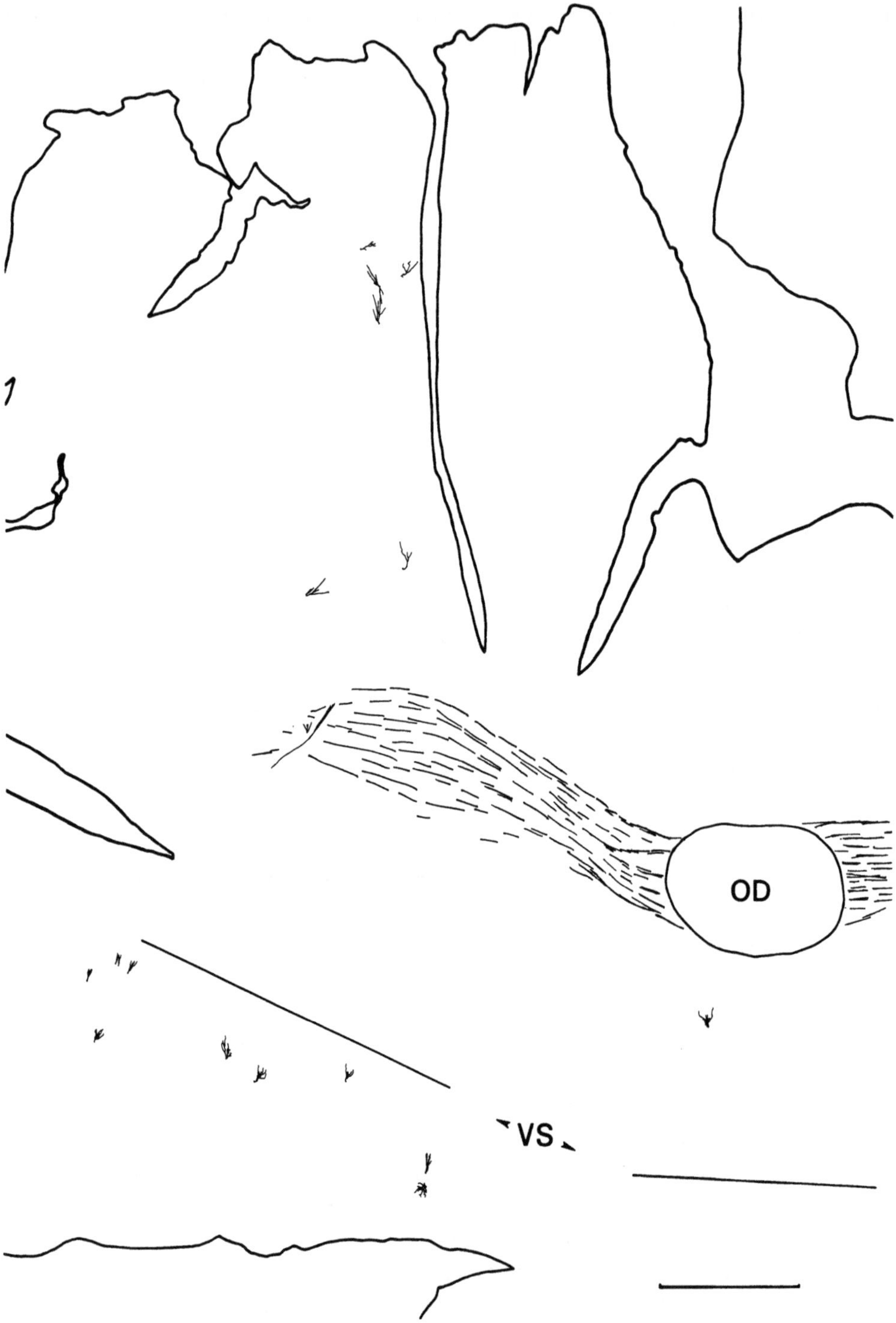

Figure 4. Orientation of dorsally-directed amacrine cells near the visual streak (VS) and in dorsal retina, above streak and optic disc (OD). 9 day rabbit. Cells near the streak are similar to adult (Fig. 3), but dorsal cells are misoriented. Cal. 1 um.

medullary rays in Figure 4 has one thin ventrally-directed dendrite which fails to turn dorsalward, as in the example of Figure 1e, but instead continues ventralward toward the visual streak.

There are perhaps two possible general explanations pertaining to "organizers" that are suggested by these results. The first is that there may be a single organizer affecting dorsally-directed dendritic growth, with its focus in the visual streak; the effect of this "alignment factor" decreases with dorsal distance from the streak. The second possibility is that dendritic direction and alignment are determined by two (or more) competing factors, one of which is focussed in the visual streak, where it exerts the strongest effect on proximate cells. For instance, there may be a second factor operating to orient dendritic trees along radial lines emanating from the center of the retina, as seems to be the case for some ganglion cells in cat retina (Levick and thibos, '82; Levinthal and Schall, '83; Famiglietti, unpublished). The simplicity of the first possibility has its attractions, particularly when coupled with the idea that the centrally-located visual streak may be earlier to mature than more peripheral regions of the retina.

Functional Implications for Ganglion Cells

Dorsally-directed amacrine cells of rabbit retina are unlikely to be involved in the formation of DS responses by virtue of their dendritic asymmetry, because cells of similar morphology are not found pointing in the five other "preferred" directions represented among DS ganglion cells. Similarly, they are not good candidates for inputs to OS ganglion cells, because they do not include horizontally oriented cells (at least near the visual streak, where horizontal-preferring ganglion cells can be found).

Other types of amacrine cells have been found with morphological asymmetried less dramatic than the four types described here, and which are represented in the four cardinal directions. Some of these and related amacrine cells possess dendritic stratification which is more consistent with the dendritic stratification of DS ganglion cells, and will be the subject of a separate communication.

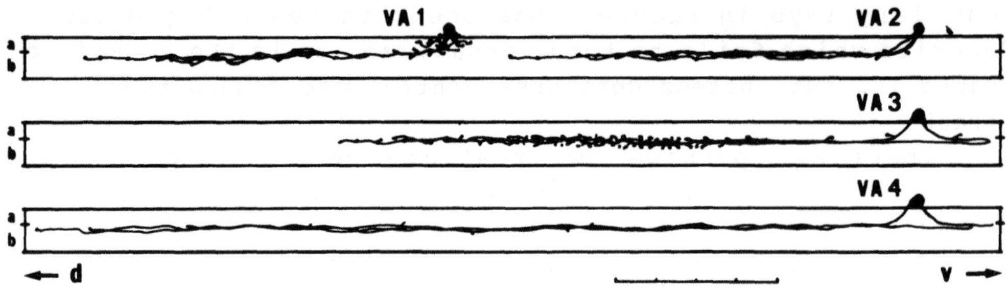

Figure 5. Scale diagram in transverse section of four types of dorsally-directed amacrine cells in rabbit retina. All four types branch at or near the a/b sublaminar border of the inner plexiform layer. The dendritic trees are unidirectional, pointing toward dorsal retina (d) and away from ventral retina (v). Calibration 100 um.

At present it is unclear what possible function dorsally-directed amacrine cells have for ganglion cell function, in respect to their marked dendritic field asymmetry. It is even possible that their unidirectional asymmetry has no bearing on their function. This may be so, in particular, if these amacrine cells operate according to the model of amacrine cell function which proposes that at least some amacrine cells are mosaics of pre- and postsynaptic sites operating strictly in synaptic complexes with no somatic integration of signals. The type 1 cell exhibits a different type of asymmetry from the others, however, a marked dendritic differentiation in separate zones of the dendritic tree, reminiscent of the rod amacrine cell (Famiglietti and Kolb, '75). It may thus function in a manner very different from the other three types of dorsally-directed amacrine cell in rabbit retina.

REFERENCES

Barlow HB, Levick WR (1965) The mechanism of directionally selective units in rabbit's retina. J Physiol (Lond) 178: 477-504

Boycott BB, Dowling JE (1969) Organisation of the primate retina: light microscopy. Phil Trans Roy Soc Lond B 225: 109-184

Brecha NC (1983) Retinal neuro-transmitters: histochemical and biochemical studies. Enson PC (ed) Chemical neuroanatomy Raven Press New York

Cajal SR y (1893) La rètine des vertèbrès. La Cèllule 9: 17-257

Caldwell JH, Daw NW, (1978) New properties of rabbit retinal ganglion cells. J Physiol (Lond) 276: 257-276

Cleland BG and Levick WR (1974) Properties of rarely encountered types of ganglion cells in the cat's retina and lateral geniculate nucleus. J Physiol (Lond) 240: 457-492

Daw NW, Brunken WJ, Jensen RJ (1988) The function of monoamines in the retina. Weiler R and Osborne N (eds). The neurobiology of the inner retina. Springer Heidelberg in press

Daw, NW, Brunken WJ, Parkinson D (1989) Function of synaptic transmitters in the retina. Ann Rev Neurosci in press

Famiglietti EV (1983) ON and OFF pathways through amacrine cells in mammalian retina: the synaptic connections of "starburst" amacrine cells. Vision Res 23: 1265-1279

Famiglietti EV (1987) Morphological classification of ganglion cells in rabbit retina. Abstr Soc Neurosci 13: 380

Famiglietti EV Jr, Kolb H (1975) A bistratified amacrine cell and synaptic circuitry in the inner plexiform layer of cat retina. Brain Res 84: 293-300

Famiglietti EV Jr, Kolb H (1976) Structural basis for ON- and OFF-center responses in retinal ganglion cells. Science 194: 193-195

Famiglietti EV Jr, Siegfried EC (1979) Quantitative analysis of ganglion cells in rabbit retina. Invest Ophthalmol 18 (Suppl): 84

Famiglietti EV Jr, Siegfried EC (1980) The amacrine cells of rabbit retina. Invest Ophthalmol 19 (Suppl): 70-71.

Jacobson M (1968) Development of neuronal specificity in retinal ganglion cells of *Xenopus*. Develop Biol 17: 202-218

Jensen RJ, Daw NW (1984) Effects of dopamine antagonists on receptive fields of brisk cells and directionally selective cells in the rabbit retina. J Neurosci 4: 2972-2985

Kuffler SW (1953) Discharge patterns and functional organization of mammalian retina. J Neurophysiol 16: 37-68

Levick WR (1967) Receptive fields and trigger features of ganglion cells in the visual streak of the rabbit's retina. J Physiol (Lond) 188: 285-307

Levick WR, Thibos LN (1982) Analysis of orientation bias in cat retina. J Physiol (Lond) 329: 243-261

Levinthal AG, Schall JD (1983) Structural basis of orientation sensitivity of cat retinal ganglion cells. J Comp Neur 220: 465-475

Mariani AP (1983) A morphological basis for verticality detectors in the pigeon retina: asymmetric amacrine cells. Naturwissensch 70: 5368

Masland RH (1977) Maturation of function in the developing rabbit retina. J Comp Neur 175: 275-286

AMACRINE CELLS AND CONTROL OF RETINAL SENSITIVITY

M.B.A. Djamgoz, A.J. Capp, J.C. Low and J.E.G. Downing
Imperial College (University of London),
Department of Pure & Applied Biology,
Neurobiology Group,
London SW7 2BB, U.K.

A remarkable ability of the visual systems of terrestrial vertebrates and fish living near the water surface is to remain optimally responsive to light intensities that may vary by as much as 10 orders of magnitude (Rushton, 1965). Signal transmission through the vertebrate retina involves a high degree of convergence with many distal photoreceptors (rods and/or cones) feeding via bipolar cells to relatively few proximal amacrine and ganglion cells. Such a process is inherently one of high gain and this gain needs to be a controlled variable so that the median output level can be modulated over a wide range of possible input levels. It is evident that significant changes can occur in the light sensitivity of retinal neurones independently of any change in the photopigment content of the retina (Witkovsky, 1980; Green, 1986). This mechanism, generally known as network adaptation, involves synaptic processing of signals within the retinal network. Witkovsky (1967) showed originally in carp that neurones in inner and outer retina can undergo sensitivity changes independently. Thus, ganglion cells were found to be sensitive to dimmer backgrounds compared with horizontal cells, whilst in the presence of a strong background that saturated horizontal cell responses (i.e. S-potentials), ganglion cells continued to fire (Witkovsky, 1967). From a stage-by-stage analysis in the mudpuppy retina, Werblin & Copenhagen (1974) concluded that the third and final stage of network adaptation occurs within the inner plexiform layer. The main aim of this article is to review the evidence concerning the role of amacrine cells in this process. Amacrine cell activity has been studied either by intracellular recording of membrane potentials or by extracellular recording of the proximal

NATO ASI Series, Vol. H31
Neurobiology of the Inner Retina
Edited by R. Weiler and N.N. Osborne
© Springer-Verlag Berlin Heidelberg 1989

negative response (PNR), a field potential thought to reflect mainly transient AC activity (Burkhardt, 1970).

Materials and Methods

Our work has been carried out on retinae of cyprinid fish, mostly roach (Rutilus rutilus) and more recently mirror carp (Cyprinus carpio) (Djamgoz et al.,1987). Retinae used in these experiments were isolated from fish dark-adapted for 20-40 minutes. The isolated retina was maintained in moist O_2. Test light stimuli (1 mm^2 spots) were derived from a single-beam optical stimulator and focussed on the surface of the retina from underneath. Micro-electrodes (~100 MOhm tip resistance)

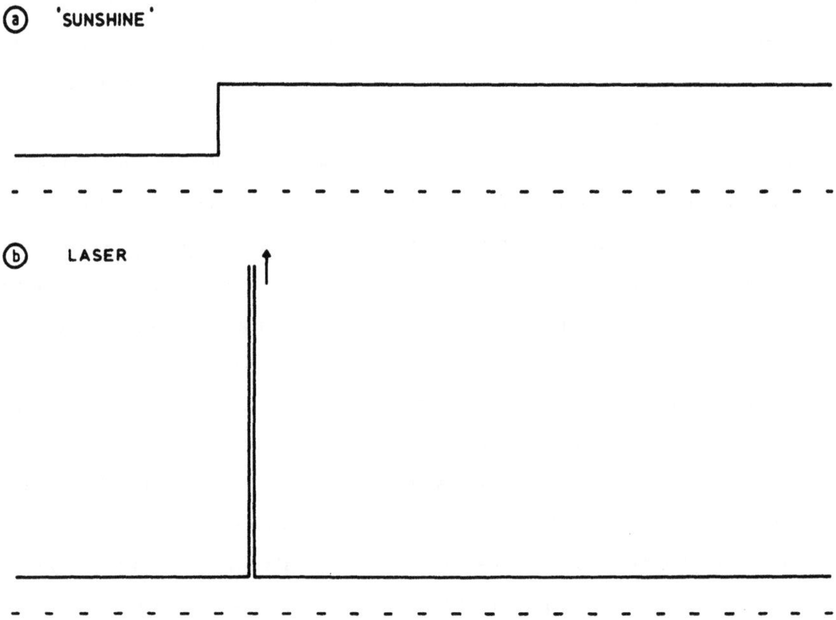

Figure 1. Schematic representation of the two protocols used to investigate sensitivity control by retinal neurones. (a) 'Sunshine'; background illumination of the entire retina; power level, 10^{-2} W/m^2. (b) Laser; flash of 0.1 second duration, giving a power level of some 5 x 10^4 W/m^2. In either case, the cell's responses to test light stimuli (presentations of which are denoted by short bars beneath each trace) were measured and then monitored following the onset of the desensitizing influence.

were advanced from above. The isolated retina was 'desensitized' in two different ways, as illustrated in Fig.1. (1) Background illumination. This was provided by a white-light source placed above the praparation. It could be switched on and off using a manual shutter and provided full-field illumination of the retina lasting many seconds; the power level was 10^{-2} W/m^2. (2) Laser irradiation. For this kind of experiment, an Ar-Kr laser with major emission lines at 647.1 nm and 488 nm was used. The power of the laser beam was some 5 x 10^4 W/m^2, sufficient to bleach a substantial proportion of spectrally-matching cone photopigment(s). Thus, a brief (0.1s) laser flash was used to irradiate a 1 mm^2 area of the retina concentric with the test stimuli (Fig.1b). In either situation, the cells' light-evoked response characteristics were initially determined in the control state of the retina; then the retina was presented with the adaptating light, and the consequences of this on AC functioning were investigated. Furthermore, since one might expect neural events in adaptation to vary according to the ambient condition, parallel experiments have also been performed on HCs in a comparative approach to evalute the relative roles of the two neural systems.

Results

Response-Intensity Profiles

An insight into sensitivity characteristics of ACs was obtained initially by examining the cells' input-ouput (i.e. stimulus intensity - response) relationships. Typical sets of normalized V-logI curves for 'on' and 'off' components of transient amacrine cell (T-AC) responses and their comparison with S-potentials in the roach retina are shown in Fig.2. T-AC responses on the whole are highly dynamic, showing a variety of response profiles. These data reveal a number of characteristics likely to be related to the cells' sensitivity control mechanisms. 1) Highly variable absolute thresholds. 2) Sharply rising response-intensity profiles with operating

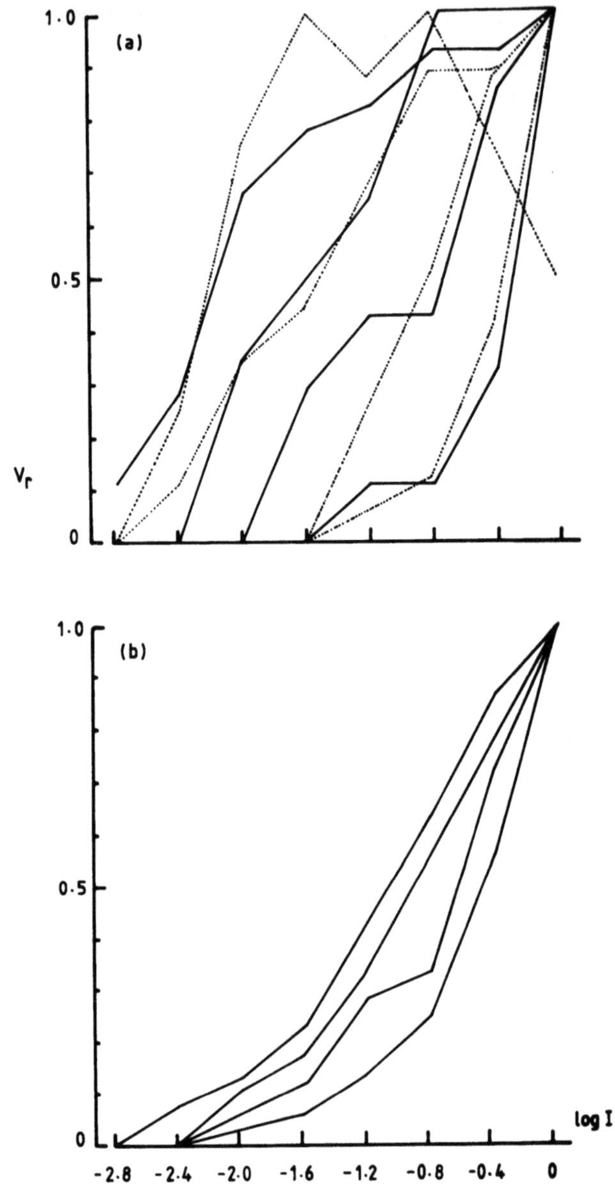

Figure 2. Representative examples of response-stimulus intensity relationships for transient amacrine and horizontal cells of the roach retina (4 units each). (a) 'on' and 'off' components of transient amacrine cells (dotted and solid lines, respectively). (b) horizontal cells. V_r, response amplitude normalised relative to the corresponding maximum. log I, logarithm of light intensity; brightest stimulus used was arbitrarily fixed as 0. The stimuli comprised 1 mm spots of 617 nm light; the brightest stimulus was some 50 W/m^2.

ranges restricted mostly to less than 2 log units. Thus, T-AC responses would appear to involve 'self-facilitation'. 3) Sub-maximal response amplitudes elicited by stimuli brighter than that initially needed to evoke a saturating response (Fig.2a), implying that T-ACs may also adapt rapidly to intense stimuli. In contrast, V-logI curves for S-potentials were much more consistent and did not possess such characteristics (Fig.2b). AC activity in retinae of other vertebrates have also been noted to be associated with narrow operating ranges (e.g. Kaneko & Hashimoto, 1969 - carp; Burkhardt, 1970 - frog; Werblin, 1971; Proenza & Burkhardt, 1973 - mudpuppy; Marchiafava & Torre, 1978 - turtle).

Effects of Steady Background Illumination of Moderate Intensity

An important aspect of the effect of background illumination on T-ACs is its time dependency. This is illustrated in Fig.3a. Following the onset of the background, the 'on' and the 'off' responses of the cell to superimposed mid-range test light flashes were initially suppressed. However, in continued presence of the background field, both response components gradually recovered (Fig.3a & b). On average, immediately following background illumination, responses were reduced to 30-40% of controls, but recovered about one-half of their lost amplitudes to become 60-70% of controls over 60 seconds. Under identical conditions, L_1/H1 type S-potentials were steadily suppressed, remaining at the level attained immediately after application of the background (Fig.3b). We have observed two important characteristics of this slow response recovery effect in T-ACs. Firstly, it was cumulative i.e. the effect got faster during successive presentations of the background. Secondly, some sensitization of responses was seen immediately after turning off the background i.e. response amplitudes became larger compared with control values. Interestingly, this sensitization phenomenon appeared to be more pronounced for the 'off' than the 'on' component of the T-AC response. In contrast, no sensitization of S-potentials was seen under identical experimental conditions.

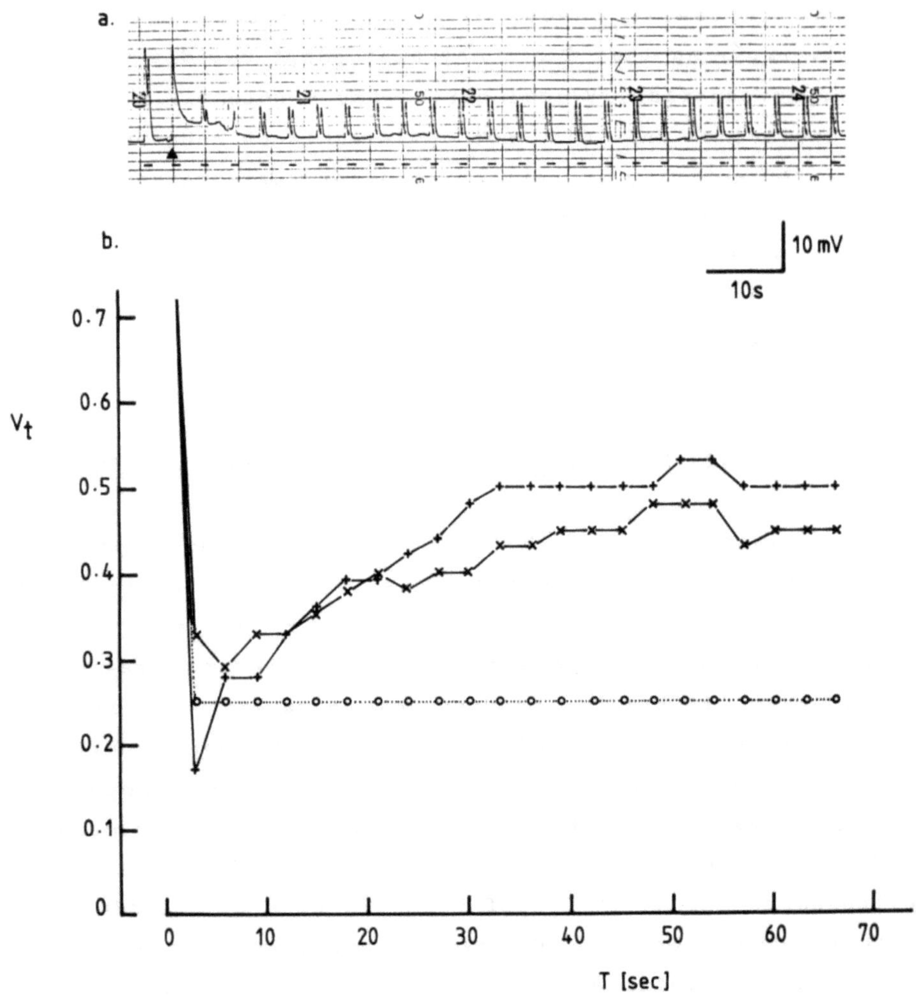

Figure 3. (a) A typical recording showing the recovery of transient amacrine cell responses during steady background illumination of the retina. The arrow-head denotes presentation of the background which was then maintained. Short bars beneath the response trace denote presentations of test light stimuli. (b) Graphical representation of the kind of data illustrated in (a). V_t, fraction of response recovered (V_t = 1.0 corresponds to control value of mid-range response elicited by the test stimulus prior to background illumination). (x) and (+), 'on' and 'off' components of the T-AC. The dotted line represents typical data from a H1 type HC recorded under identical conditions. T, time in seconds.

Effects of Brief, Localized but Intense Irradiation

Further insight into the contribution of ACs to network adaptation has been obtained by studying effects of brief, but intense irradiation of the isolated retina on light-evoked AC responses. Typical recordings are shown in Fig.4. A striking effect apparent in these recordings was the ability of the 'on' and the 'off' transient depolarizations to recover from laser flashes more or less fully. Thus, following laser irradiation of either wavelength, the membrane potentials of the T-ACs were initially clamped in a state of light adaptation. However, within a few seconds, the transient responses began to recover and rapidly increased in amplitude and within 10-15 seconds response amplitudes were comparable to respective controls. Average data for the recovery of the 'on' and the 'off' transients are given in Fig.4c & d, respectively. Clearly both response components recovered more than 80% of the original response amplitudes within some 10 seconds following laser irradiation of either wavelength. In contrast, the effects of laser irradiation on L_1/H1 type S-potentials were maintained for several minutes. Thus, spectrally-matching 647.1 nm laser flashes suppressed the S-potentials to levels less than 10% of controls (Djamgoz, 1988).

Discussion

The main conclusion to emerge from the present work is that following either type of desensitizing influence applied to the retina, T-ACs gradually recover their initially lost 'sensitivity' to test light stimuli, whilst S-potentials remain suppressed. These findings further support the view that a part of the 'network adaptation' process is located in the inner retina and involves <u>sensitization</u> of responses.

A concept of thinking about retinal sensitivity and response properties has evolved based on different levels or stages. A primary level of adaptation is at the photoreceptors and this can in part involve changes in pigment concentration; the second level is in the outer plexiform layer, where the HCs are concerned with modification of photoreceptor-bipolar cell transmission, and is associated with steady-state surround backgrounds; and a third level is at the inner plexiform layer where ACs are involved in suppression of GC responses triggered by changes in the pattern of surround illumination. These

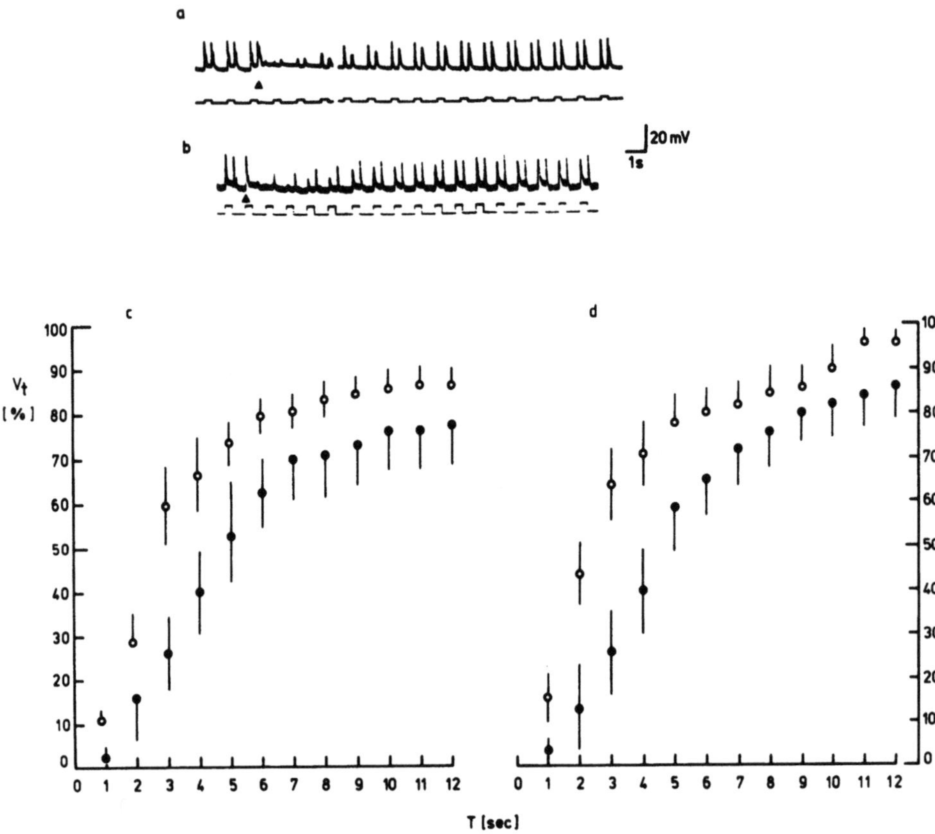

Figure 4. Typical effects of 647.1 nm (a) and 488 nm (b) laser irradiation on T-AC responses. The laser was flashed at the point marked by the arrow-head (further details given in text and Fig. 3 legend). c & d: Average time (T) course of recovery of the 'on' (c) and the 'off' components, following irradiation of retinae with 647.1 nm (full circles) and 488 nm (open circles) laser flashes. V_t, fraction of response recovered (as in Fig. 3, but expressed as a percentage). Each data point is given as the mean ± S.E.M. of 7 measurements on different cells (modified from Djamgoz, 1988).

ideas were developed largely by and from the work of Werblin and co-workers during the 1970's (Werblin & Dowling, 1969; Werblin 1971,1972,1974; Normann & Werblin, 1974; Werblin & Copenhagen, 1974; Copenhagen 1975; Thibos & Werblin, 1978a,b). However, there exists relatively little information on the relative contribution of these stages in modifying the output from ganglion cells (GCs).

Several studies have shown stimulus scaling adaptation of reponses from the inner plexiform layer. Normann & Werblin (1974) showed a shift of V-log I curves for ACs in _Necturus_ retinae by about 1 log unit along the intensity axis in the presence of backgrounds separated in intensity by 1 log unit. Green et al. (1975) have shown that backgrounds too weak to affect the V-log I curves for rods or HCs in skate retinae caused a significant (~1 log unit) shift in PNR (as well as b-wave) curve and reduced their V_{max} values. Green et al. (1975) and Dowling & Ripps (1977) found that to affect the flash threshold for skate rods and HCs a background 2 log units more intense than was required for proximal responses and GCs was needed. They therefore suggested that network adaptation in the proximal retina was important at low light levels. At higher mean levels adaptation was determined by rod behaviour and all units followed a Weber type relationship i.e. it is a fast responding process detecting small perturbations and behaves like a high-pass filter.

Green et al. (1975) and Dowling & Ripps (1977) considered the kinetics of network adaptation and found it could take many (typically 10-15) minutes to recover maximum sensitivity in terms of threshold of reponse when a background was applied. Dowling & Ripps (1976) and Dowling (1977) went on to show that altered extracellular K^+ levels could mimic the effects of network adaptation on retinal sensitivity. In the presence of raised K^+ levels, the b-wave increment threshold increased without any effect on the photoreceptors.

The effects that we have described for T-ACs in cyprinid fish retinae occur on a somewhat faster time scale (seconds to at most few minutes), and involves _sensitization_ of responses. Proceeding proximally through the retina from photoreceptors to ganglion cells, thresholds for light-evoked responses are steadily lowered, whilst operating ranges become narrower. Thus, high-gain, presumably synaptic mechanisms(s) would appear to be present in the inner retina. T-ACs receive synaptic inputs from bipolar cells (BCs) in the inner plexiform layer (Dowling & Werblin, 1969; Witkovsky & Dowling, 1969). The on-BC input generates the transient 'on' of the T-AC response, whilst the off-BC generates the transient 'off' (Kujiraoka et al.,1986). Since the intensity-response curves of BCs are generally steeper than those of HCs, photoreceptor to BC transmission has a higher gain compared with photoreceptor input(s) to HCs (Werblin, 1978; Ashmore & Falk, 1980). A part of the high-gain mechanism apparent in T-AC activity may, therefore, originate in the outer plexiform layer. However, the

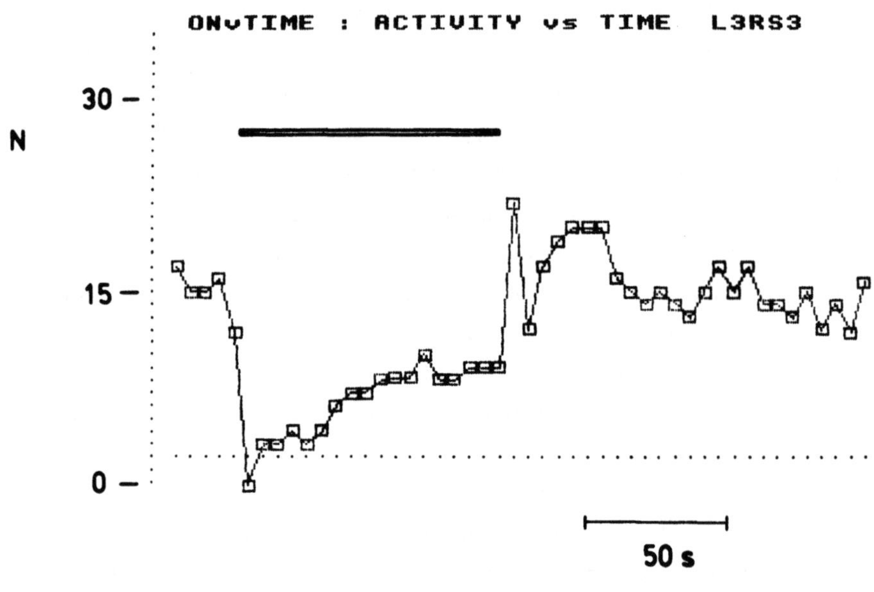

Figure 5. An experiment on a ganglion cell carried out similarly to the one illustrated in Fig. 3, but involving extracellular recording of spike activity. N, total number of spikes in a 200 ms bin counted immediately following the onset of the test stimulus. The horizontal bar denotes the duration of background illumination. The data illustrate the gradual recovery of light-evoked GC responses during the maintained illumination, as in the case of T-ACs (cf. Fig. 3a).

intensity-response functions of the T-ACs are even narrower than those of BCs, covering only some 1 log unit (Fig.2; also, Kaneko & Hashimoto, 1969; Werblin, 1971). It is likely, therefore, that there is further gain in the synaptic transmission from BCs to T-ACs. The neurochemical basis of the sensitization process in T-ACs is not known at present. It would seem possible, however, that the mechanism of response recovery in T-ACs during prolonged background illumination may involve the gradual release of a 'sensitizing agent', accumulation of which within the inner plexiform layer may selectively potentiate synaptic input(s) to T-ACs.

Is the Sensitization Effect in Amacrine Cells Transmitted to Ganglion Cells?

It seems clear that following either steady illumination with moderate intensity or brief, intense irradiation of the isolated retina, T-AC responses rapidly recover whilst S-potentials are suppressed for much longer. It would be interesting to consider, therefore, whether this property of T-ACs is also apparent in GC activity. Evidence suggets that T-ACs and GCs have similar absolute thresholds and operating ranges (Werblin, 1971; Green et al.,1975; Green & Powers, 1982). The effect of background illumination (similar to Fig.1a) on light-evoked responses of a GC in the roach retina is shown in Fig.5. Thus, following the onset of the background, the GC response diminished. However, during continued illumination, the cell recovered some one-half of the lost frequency of response. Clearly, the gradual response recovery effect described for T-ACs is also apparent in GC activity. Copenhagen (1975) has also shown in the mudpuppy retina that upon presentation of a background field of adapting light, GC thresholds first increase but then steadily decrease. ACs are thought to provide at least one-half of the synaptic inputs to GCs in carp retina (Witkovsky & Dowling, 1969), and recent evidence from carp and catfish has suggested that T-AC --> GC transmission could be excitatory (Sugawara, 1985; Sakai & Naka,

1987). Although not conclusive therefore, the available evidence would suggest that the dramatic adaptational properties of T-ACs could be transmitted to GCs, thus influencing the final stage of network adaptation and retinal output.

Acknowledgement

Our work on amacrine cells was supported by The Wellcome Trust.

References

Burkhardt DA (1970) Proximal negative response of frog retina. J Neurophysiol 33: 405-420

Copenhagen DR (1975) Time course of threshold elevation in on-off ganglion cells of Necturus retina: effects of lateral interactions. Vision Res 15: 573-581

Djamgoz MBA (1988) Differential recovery rates of horizontal and amacrine cell responses from intense irradiation in the isolated retina of cyprinid fish. Neurosci Lett 88: 263-270

Djamgoz MBA, Downing JEG, Wagner H-J (1987) Retinal neurones of cyprinid fish : intracellular marking with horseradish peroxidase and correlative morphological analysis. Exp Biol 46: 203-216

Dowling JE (1977) Receptoral and network mechanisms of visual adaptation. Neurosci Res Prog Bull 15: 1-12

Dowling JE, Ripps H (1976) Potassium and retinal sensitivity. Brain Res 107: 617-622

Dowling JE, Ripps H (1977) The proximal negative response and visual adaptation in the skate retina. J Gen Physiol 69: 57-74

Green DG (1986) The search for the site of visual adaptation. Vision Res 26: 1417-1429

Green DG, Dowling JE, Siegel IM, Ripps H (1975) Retinal mechanisms of visual adaptation in tha skate. J Gen Physiol 65: 483-502

Green DG, Powers MK (1982) Mechanisms of light adaptation in rat retina. Vision Res 22: 209-216

Kaneko A, Hashimoto H (1969) Electrophysiological study of single neurones in the inner nuclear layer of the carp retina. Vision Res 9: 37-55

Marchiafava PL, Torre V (1978) The responses of amacrine cells to light and intracellularly applied currents. J Physiol 276: 83-102

Normann RA, Werblin FS (1974) Control of retinal sensitivity. I. Light and dark adaptation of vertebrate rods and cones. J Gen Physiol 63: 37-61

Proenza LM, Burkhardt DA (1973) Proximal negative response and retinal sensitivity in the mudpuppy, Necturus maculosus. J Neurophysiol 36: 502-518

Rushton WHA (1965) The Ferrier lecture: Visual adaptation. Proc R Soc Lond B 162: 20-46

Sakai HM, Naka K-I (1987) Signal transmission in the catfish retina. V. Sensitivity and circuit. J Neurophysiol 58: 1329-1350

Sugawara K (1985) Lateral actions at the inner plexiform layer of the carp retina : effects of turning windmill pattern stimulus. Vision Res 25: 1179-1186

Thibos LN, Werblin FS (1978a) The response properties of the steady antagonistic surround in the mudpuppy retina. J Physiol 278: 79-99

Thibos LN, Werblin FS (1978b) The properties of surround antagonism elicited by spinning windmill patterns in the mudpuppy retina. J Physiol 278: 101-116

Werblin FS (1971) Adaptation in a vertebrate retina : Intracellular recording in Necturus. J Neurophysiol 34: 228-241

Werblin FS (1972) Lateral interactions at the inner plexiform layer of the vertebrate retina: Antagonistic responses to change. Science 175: 1008-1010

Werblin FS (1974) Control of retinal sensitivity. II. Lateral interactions at the outer plexiform layer. J Gen Physiol 63: 62-87

Werblin FS (1978) Synaptic interactions mediating bipolar responses in the retina of the tiger salamander. In 'Vertebrate Photoreception' (Barlow HB, Fatt P, Eds). pp.205-230. Academic Press London

Werblin FS, Copenhagen DR (1974) Control of retinal sensitivity. III. Lateral interactions at the inner plexiform layer. J Gen Physiol 63: 88-110

Werblin FS, Dowling JE (1969) Organization of the retina of the mudpuppy, Necturus maculosus. II. Intracellular recording. J Neurophysiol 32: 339-355

Witkovsky P (1967) A comparison of ganglion cell and S-potential response properties in carp retina. J Neurophysiol 30: 546-561

Witkovsky P (1980) Excitation and adaptation in the vertebrate retina. Curr Top Eye Res 2: 1-66

Witkovsky P, Dowling JE (1969) Synaptic relationships in the plexiform layers of carp retina. Z Zellforschung Mikrosk Anat 100: 60-82

STRUCTURE-FUNCTION CORRELATION: AMACRINE CELLS OF FISH AND AMPHIBIAN RETINAE

Mustafa B.A. Djamgoz and Silvana Vallerga
Imperial College
Department of Pure and Applied Biology, Neurobiology Group
London SW7 2BB, U.K.

Istituto di Cibernetica e Biofisica del Consiglio Nazionale delle Ricerche
16146 Genova, Italy.

Introduction

Morphological diversity of amacrine cells (ACs) is clearly a general characteristic of vertebrate retinae. Beginning with the comparative work of Cajal (1893), much detailed information concerning the cells' morphological features in both the plane as well as the radial extent of the retina has been obtained by Golgi impregnation (Boycott and Dowling 1969; Naka and Carraway, 1975, Perry and Walker, 1980; Ammermuller and Weiler, 1981; Famiglietti and Vaughn, 1981; Kolb et al., 1981; Vallerga and Deplano, 1984; Wagner and Wagner, 1988). The larger the number of parameters considered, the more numerous the types of AC that became morphologically differentiated. Thus, in the retina of the turtle, Kolb (1982) distinguished 22 types of AC, and in the retina of the cyprinid fish, roach, 43 types of AC were found (Wagner and Wagner, 1988). In contrast, an attempt to rationalize the morphological diversity of ACs using dendritic architecture as a qualitative parameter, as opposed to quantitative parameters, such as extent of dendritic field, size, etc. resulted in the definition of five classes of ACs in the sparid fish, bogue (Vallerga and Deplano, 1984). As regards their light-evoked responses, ACs have been divided

NATO ASI Series, Vol. H31
Neurobiology of the Inner Retina
Edited by R. Weiler and N. N. Osborne
© Springer-Verlag Berlin Heidelberg 1989

most commonly into two broad categories: sustained and transient, in turn, the former could be depolarizing or hyperpolarizing (Kaneko and Hashimoto, 1969; Kaneko, 1973, Toyoda et al., 1973; Chan and Naka, 1975, Famiglietti et al., 1977, Murakami and Shimoda, 1977; Mitarai et al., 1978, Djamgoz et al.,1988a,b). AC activity appears to be involved in several aspects of visual processing, including sensitivity control, spatial integration, colour processing and movement detection (Djamgoz et al., 1985). Retinae of fish and amphibia have been used extensively to correlate electrophysiological and structural properties of ACs. This article reviews our current knowledge of such correlations deduced largely from intracellular staining experiments.

Amacrine Cells of Fish Retinae

Early intracellular stainings of ACs with Procion yellow served mainly to identify cellular sites of recordings (Kaneko, 1970; Naka and Ohtsuka, 1975, Chan and Naka, 1976). Detailed morphological characterization of electrophysiologically identified ACs was initiated in carp retinae by Murakami and Shimoda (1977) and Famiglietti et al. (1977), concentrating mainly on the ramification pattern of AC dendrites within the inner plexiform layer (IPL). Murakami and Shimoda (1977) showed that sustained ACs (SAs) had monostratified dendrites, whilst dendrites of on-off transient ACs (TAs) ramified in two or more sublayers of the IPL. Famiglietti et al. (1977) showed further that dendrites of hyperpolarizing SAs were confined in the distal half of the IPL (sublamina a), whilst those of depolarizing ACs were restricted to the proximal half (sublamina b); TA dendrites were found in both sublaminae. Tangential views of electrophysiologically identified ACs were examined by Djamgoz et al. (1984, 1985, 1988a) following intracellular staining with horseradish peroxidase in roach retinae. Initially, the labelled ACs, irrespective of their light-evoked response patterns, appeared to fall broadly into to groups: Types I and II, roughly corresponding to 'giant' and 'stellate' ACs of Cajal (1983). Subsequent studies showed that several other AC types known to exist from Golgi impregnation could be stained; however, again no strict correlation could be found between tangential morphology and electrophysiology (Djamgoz,1988a).So far,we have found that the only AC type with a unique and predictable tangential view is the colour-coded SA of the goldfish retina (see later; Djamgoz et al. 1988b).

Table 1. Correlations between photoresponses and morphological features in amacrine cells of the carp retina (modified from Teranishi et al., 1987).

	Morphological Characteristics				
	Cell body shape	DF size	IPL sub-layer(s)	Peripheral process	Dye coupling
Response Type					
Fast on-off	F	N	d	+	+
Slow on-off	P	W	d	-	-
Centre-on	F	N	b	+	-
	P	M	b	-	+
	P	W	b	-	-
Centre-off	F	N	a	+	-
	P	M	a	-	+
	F	W	a	+	-

A summary of the correlation between photoresponse and morphological features found in ACs of the carp retina is given in Table 1 (Teranishi et al., 1987). Thus, 'fast' TAs had fusiform (F) perikarya and narrow 'n' (<0.3 mm^2) dendritic fields (DFs), whilst 'slow' TAs had pyriform (P) perikarya and wide (>0.8 mm^2) DFs. For either type of SA, there was no direct correlation with size, although centre-on SAs tended to have pyriform cell bodies whilst centre-off cells were more fusiform, presumably in accordance with the ramification patterns of their dendrites in the IPL.

Distal (d-type) dendrites. Axon-like, peripheral processes were first observed arising from main dendrites of all three electrophysiological classes of ACs in the roach retina, and were termed "distal" or "d-type" dendrites to indicate the fact that these processes could extend well over a half of the retina (Djamgoz et al., 1984). Their function(s) is unknown but a number of possibilities based on experimental observations may be considered.

(i) d-type dendrites may be involved in transmission of AC signals over long distances in the retina, although they do not appear to be post-synaptic (Djamgoz et al., 1988a). However, d-type dendrites are usually less than 0.5 μm in thickness, so it is not clear whether non-regenerative activity (e.g. sustained responses) could be propagated effectively. On the other hand, if d-type dendrites have an exceptionally high membrane resistance, as in horizontal cell axons of goldfish, then effective signal propagation might occur (see Yagi and Kaneko, 1988).

(ii) There is a tendency for d-type dendrites to be oriented perpendicularly with respect to optical radii in the roach retina (Djamgoz, et al., 1988a). d-type dendrites may, therefore, be involved in some orientation selective receptive field function in the retina.

(iii) Djamgoz et al. (1988b) have found that in depolarizing SAs with d-type dendrites, the membrane potential recovers from the light-evoked responses steadily more rapidly as the stimulus intensity is increased, and this is the opposite in hyperpolarizing SAs. These findings would suggest that distal dendrites may have a role in temporal tuning of AC responses.

(iv) Finally, in the cat retina, d-type dendrites occur in dopamine-accumulating ACs (Dacey, 1988), but not in serotonin-accumulating ACs (Vaney, 1986). Thus, distal dendrites may be related to neurotransmitter specific function(s) in the retina.

Amacrine Cells of Amphibian Retinae

The simple structured retinae of amphibians have been used successfully for morphofunctional studies, since the first intracellular recordings and stainings of retinal neurones in the mudpuppy (Dowling and Werblin, 1969). The tiger salamander retina, while difficult to study morphologically by Golgi impregnation (Wong-Riley,1974), contains large neurone distributed into thin layers favouring intracellular recordings. The salamander retina has also been used extensively in slice preparations (Werblin 1978), including studies on functional properties of amacrine cells (Werblin et al 1988). Three main morphological types of ACs have been described in the salamander retina, and associated with transient depolarizing responses with variable levels of steady state polarity (Vallerga, 1981). It was difficult to classify responses as purely transient or sustained, expecially by the observed antagonistic organization of the receptive fields, and with given ACs generating mainly transient or mainly sustained responses according to the pattern shape, size and to the intensity of the light stimulus (Fig 1A,C; Vallerga, 1981).

Large field cells with long dendrites monostratified in sublamina b (Fig.1B), produced predominantly sustained responses with opposite polarity for full field (depolarizing) and central (hyperpolarizing) illumination, whilst peripheral stimulation elicited transient responses (Fig.1A). 'Pure' transient responses with a hyperpolarizing plateau (Fig.1C), were recorded from multistratified amacrine cells whose main dendrites gave rise to long fine fibres, ending as small knoblets at the IPL-INL border after extending for several hundreds microns in the distal portion of sublamina a (Fig.1D). Such dendritic extensions appear to be similar to the d-type dendrites described in the cyprinid retina (Djamgoz et al 1988a,b). Small-field broadly stratified amacrine cells, with asymmetric dendritic arborizations (Fig.1F) generated transient on-off responses with a sustained depolarization, the latter being the dominant feature of the response to peripheral illumination (Fig.1E). The morphological substratum for the

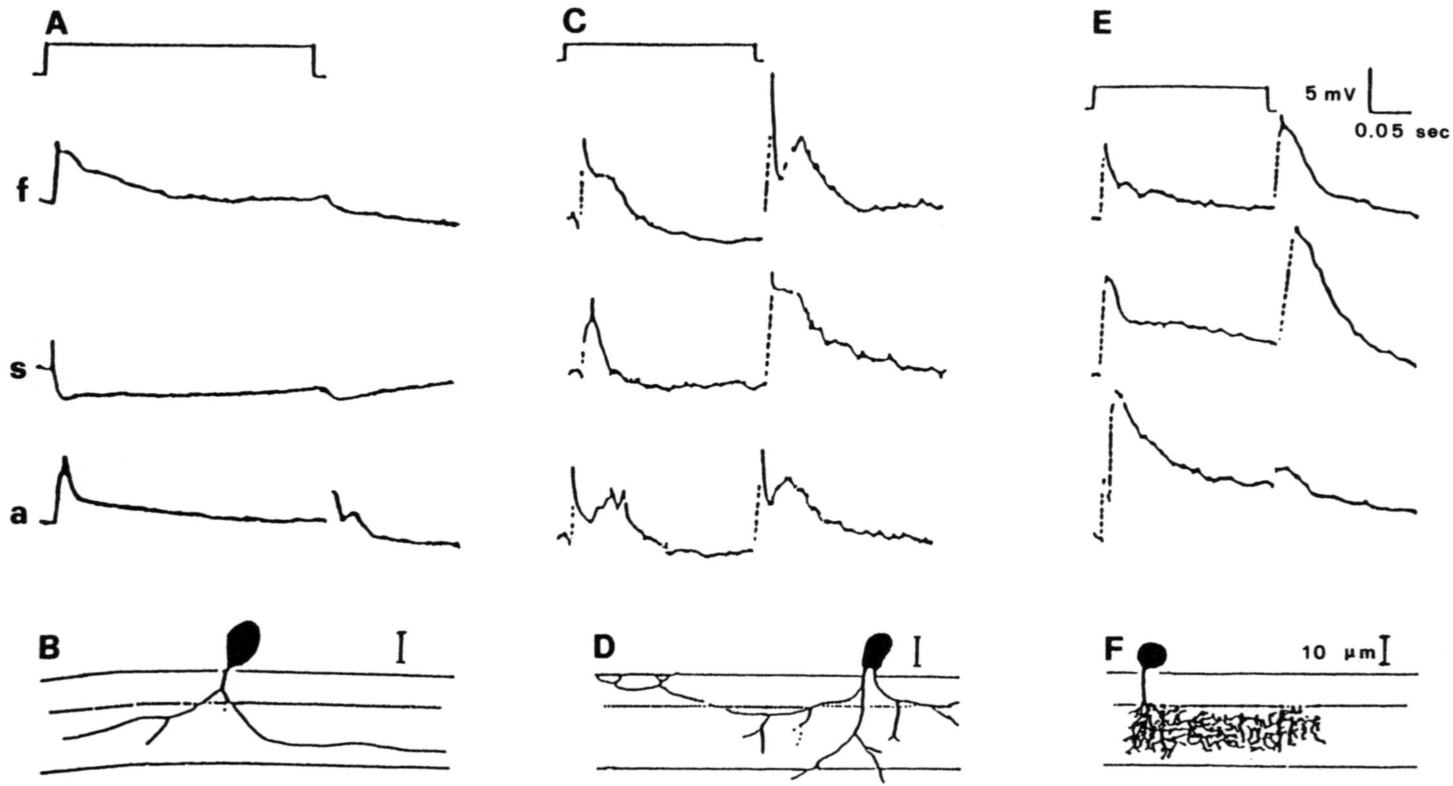

Fig. 1. AC responses to 1250 µm spot (f), 250 µm (s), annulus (a) of moderate white light intensity, and associated morphologies in the larval tiger salamander retina. Presentations of light pulses are depicted by the uppermost steps. Modified from Vallerga, 1981.

depolarizing plateau could be found in the preferential level for dendritic stratification into sublamina b; such ACs, characterized by the asymmetric dendritic arborization have been reported to show substance P-like immunoreactivity (SP-IR) by Yang and Yazulla(1986).

Processing of Visual Information by Morphologically Identified Amacrine Cells.

As noted in the Introduction, ACs have an extensive role in processing of several aspects of visual information. This section summarizes the knowledge that has emerged from morphologically identified ACs of fish and amphibian retinae.

Colour coding. Kaneko (1973), Mitarai et al. (1978) and Djamgoz and Ruddock (1983) have shown that ACs are involved in colour coding. In the goldfish retina, Djamgoz et al. (1988b) stained a particular type of SA which gave depolarizing responses to long and hyperpolarizing responses to short wavelength stimuli(Fig.2A). These cells had a unique morphology with wide, highly elliptical dendritic trees and with no distal dendrites (Fig.2C). The dendrites were broadly stratified in the IPL but were confined to sublamina b (Fig.2B). It was not clear if DF ellipticity was related to the cells' colour coded responses.

Spatial integration. An important aspect of ACs is their spatial organization, whereby they can interact laterally, independently of the horizontal cell network (Werblin, 1972). The potential role of d-type dendrites in this connection has already been mentioned. The data of Teranishi et al. (1987), obtained by intracellular injection of Lucifer yellow (Table 1) suggest that fast TAs, and SAs of either polarity having fusiform cell bodies and narrow DFs are dye-coupled, presumably by gap junctions, forming separate networks (Naka and Christensen, 1981; Marc et al., 1988). The nature of the

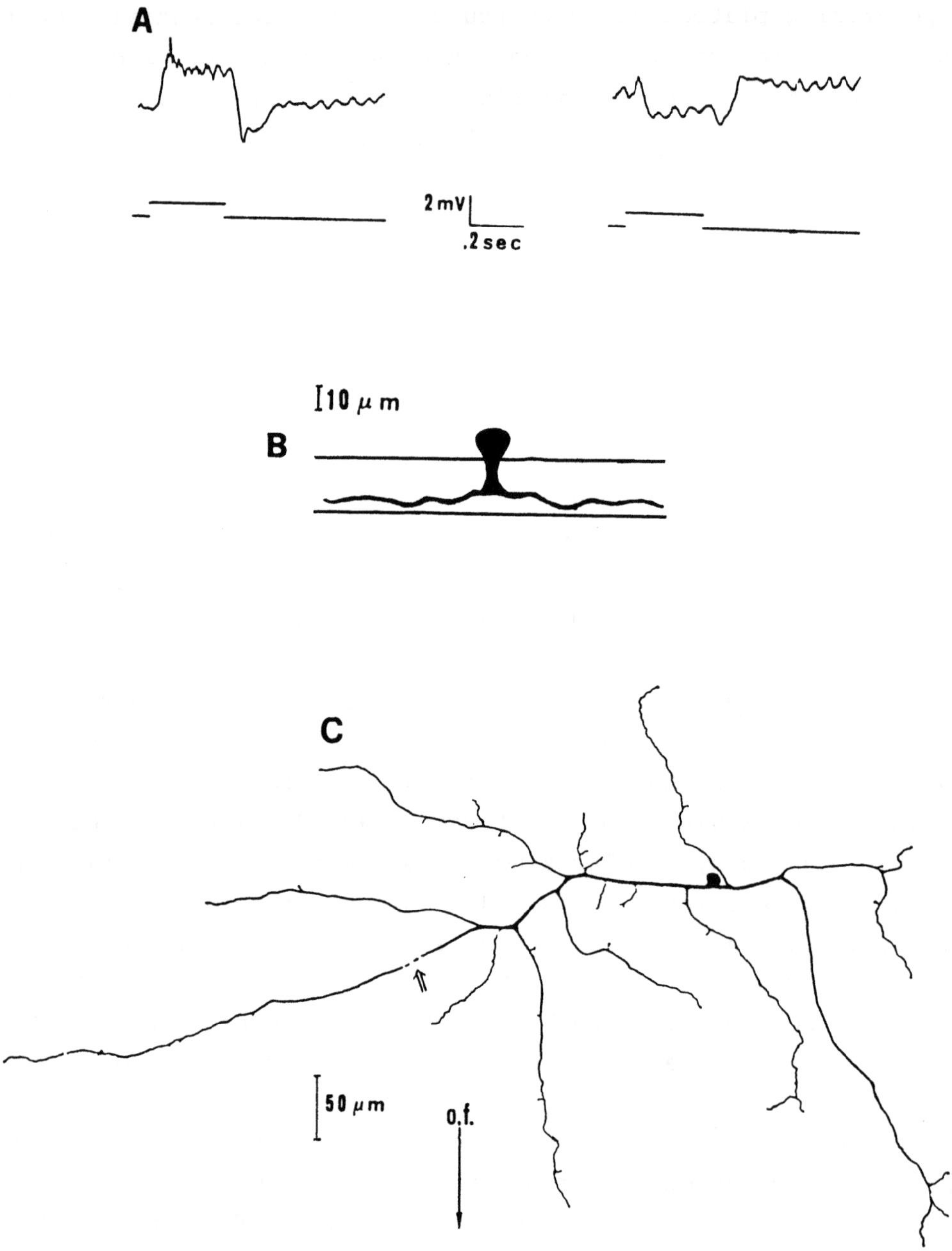

Fig. 2. A. Responses to monochromatic lights (600nm left and 500nm right) of a sustained colour-coded AC in the goldfish retina. The cell depolarizes to long and hyperpolarizes to short wavelength light. B. Radial view of the cell in retinal cross section. The dendrites of the cell ramify in sublamina b. C. Tangential view of the same cell. Direction of optic fibers is given by the single arrow. The double arrow points to the site of HRP injection (modified from Djamgoz et al 1988b).

coupling of other AC types is not known at present, but two possibilities remain: either the other ACs are still electrotonically- (but not dye-) coupled and/or the coupling mechanism is chemical (Djamgoz and Ruddock, 1978).

Movement detection. A movement sensitive AC was successfully stained in the catfish retina by Naka (1980). Type NB ACs responded effectively to a bar of light moving in a direction corresponding to the vertical in the fish's field of view; the response to movement in the horizontal direction was much weaker (Fig.3B). The directional response comprised a sequence of depolarizing - hyperpolarizing - depolarizing membrane potentials. Interestingly, the dendritic fields of the type NB ACs were also asymmetrical (Fig.3A), possible, therefore, that DF asymmetry was related to movement detection,although the long axis of the DF was at right angles to the preferred axis for movement detection.

Movement sensitivity of mudpuppy retinal neurones was originally studied by Werblin (1970). Recently, a circuitry model of neural interaction subserving local change detection and its morphological and pharmacological bases have been proposed for the salamander retina by Werblin et al (1988). The proposed subunit consisted of two bipolar cells and a narrow field broadly stratified AC, which were suggested form a change detecting subunit, with a wide field AC providing the lateral inhibition (Fig. 3C), according to the model of Barlow and Levick (1965). The model has been tested by path clamp recordings in retinal slices; the synaptic and membrane currents of ACs were determined by puffing putative neuro-transmitters at their synaptic input, and subsequently cells were identified morphologically by intracellular staining with Lucifer yellow. The small field AC participating in the subunit had a rather similar dendritic morphology to the small field cell described by Vallerga (1981), differing in the dendritic arborization, which was symmetrical (Werblin et al 1988). They have been found to be GABAergic (Werblin et al., 1988), while the asymmetric small field cells described by Vallerga (1981) have been reported as SP-IR (Yang and Yazulla, 1986).

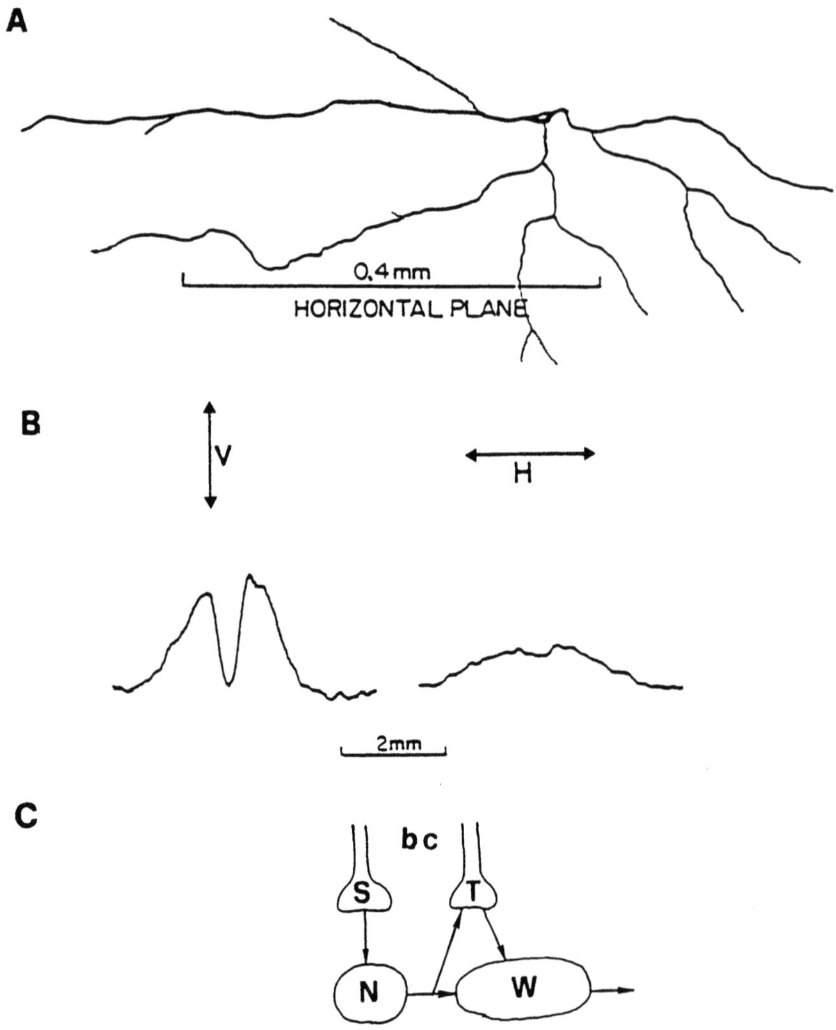

Fig. 3. Involvement of ACs in movement detection in catfish (A and B), and tiger salamander (C) retina. A. Tangential morphology of an AC stained with Procion yellow. B. First order kernels (corresponding roughly to light-evoked responses) of the movement-sensitive AC shown in A. The cell responded much better to stimuli moving in the vertical (V) direction than horizontal (H); modified from Naka (1980). C. The neural circuitry in the 'off' system underlying change detection proposed by Werblin et al (1988) in the salamander retina. Sustained (S) and transient (T) bipolar terminals (BC) excite narrow- (N) and wide- (W) field amacrine cells, respectively. Type-N ACs are GABA-ergic and inhibit transient bipolar terminals and type-W ACs, acting at GABA-B and GABA-A receptors respectively. Type-W ACs are glycinergic and their outputs are propagated laterally (modified from Werblin et al., 1988).

Conclusions

Several conclusions can be made from the available structure - function correlations in ACs. In fish retinae there is, on the whole, no strict correlation between light microscopical tangential morphologies of the cells and their light-evoked response patterns. However, strong correlations exist between electrophysiology and dendritic ramification patterns in the IPL.Importantly,colour-coded SAs generating responses of either polarity depending on the wavelength of the light stimulus are nevertheless monostratified in sublamina b, which is normally associated with depolarizing activity(Famiglietti et al.,1979). It follows tentatively therefore that (i) the functional sub-division of the IPL applies to responses driven by the long-wavelength sensitive cone mechanism, and/or (ii) another important electrophysiological parameter in this connection may be the time course (i.e. sustained vs. transient nature) of the response.

Data from the salamander retina suggest that the functional bisublamination of the IPL holds in the amphibian retina even if the absence of pure sustained responses (Vallerga, 1981). Large field multistratified cells with long dendrites into sublamina a have a distinct hyperpolarizing plateau on the transient responses and cells broadly stratified in sublamina b show a depolarizing steady state level. Amphibian ACs seems to preferentiate sublamina b for dendritic layering (Vallerga, 1981), which is in agreement with the smaller thickness of sublamina a compared to sublamina b. Interestingly in the sparid fish bogue retinae the density of dendritic distribution in sublamina a is about one half of dendritic density in sublamina b (Deplano and Vallerga, 1984).

There are clearly many more morphological varieties of ACs in fish retinae than electrophysiological response patterns and visual functions so far studied. It would appear, therefore, that either the morphological diversity of ACs is of limited physiological significance, or, what might seem more attractive, additional electrophysiological parameters need to be considered in future studies to establish broader correlations of structure and function in ACs.

Acknowledgments

This study was supported by a project grant from The Wellcome Trust (M.B.A.D.) and the Strategic Research Programme of the C.N.R. (S.V.). We are grateful to Dr. Cesare Usai for many useful discussions.

LITERATURE CITED

Ammermuller J, Weiler R (1981) The ramification pattern of amacrine cells within the inner plexiform layer of the carp retina. Cell Tissue Res 220:699-723

Barlow HB and Levick WR (1965) The mechanisms of directionally selective units in rabbit's retina: J Physiol 178:477-504

Boycott BB, Dowling JE (1969) Organization of the primate retina: Light microscopy, Phil Trans R Soc Lond 255:109-184

Cajal, S. R. (1893) La retine des vertebres. La Cellule 9:121-225.

Chan RY, Naka K-I (1976) The amacrine cell. Vision Res 16:1119-1129

Dacey DM (1988) Dopamine-accumulating retinal neurones revealed by in vitro fluorescence display a unique morphology. Science 240:1196-1198

Deplano S, Vallerga S (1984) Quantitative morphology of amacrine cells in teleost retina. In: Borsellino A, Cervetto L (eds) Photoreceptors, Plenum Publ Co, p 295-317

Djamgoz MBA (1984) Electrophysiological characterization of the spectral sensitivities of the horizontal cells in cyprinid fish retina. Vision Res 24: 1677-1687

Djamgoz MBA, Ruddock KH (1978) Properties of amacrine cell responses recorded from isolated fish retinae. J Physiol Lond 7:89-93

Djamgoz MBA, Ruddock KH (1983) Spectral characteristic of transient amacrine cells in a cyprinid fish (roach) retina in vitro. J. Physiol., Lond. 339:19P.

Djamgoz MBA, Downing JEG, Wagner E, Wagner H-J, Zeutzius I (1985) Functional organization of amacrine cells in the teleost fish retina. In: Gallego A, Gouras P (eds) Neurocircuitry of the Retina: A Cajal Memorial, Elsevier New York.

Djamgoz MBA, Downing JEG, Wagner H-J (1987) Retinal neurones of cyprinid fish: Intracellular marking with horseradish peroxidase and correlative morphological analysis. Exp. Biol. 46:203-216.

Djamgoz MBA, Downing JEG, Wagner H-J (1988a) An intracellular horseradish peroxidase study of amacrine cells in the retina of a cyprinid fish. (submitted)

Djamgoz MBA, Spadavecchia L, Usai C, Vallerga S (1988b) Light-evoked response types and morphological identification of amacrine cells in goldfish retina. (submitted)
Dowling JE, Werblin FS (1969) Organization of the retina of the mudpuppy, Necturus maculosus. II Intracellular recordings. J Neurophysiol 32:339-355
Famiglietti EV, Kaneko A, Tachibana M (1977). Neuronal architecture of 'on' and 'off' pathways to ganglion cells in carp retina. Science 198:1267-1269.
Famiglietti EV, Kolb H (1976) Structural basis of 'ON'and 'OFF'-centre responses in retinal ganglion cells. Science 194:193-195
Famiglietti EV, Vaughn JE (1981) Golgi-impregnated amacrine cells and GABAergic retinal neurons: immunocytochemical and histochemical stratification in the inner plexiform layer of of rat retina. J Comp Neurol 197:129-139
Kaneko A (1970) Physiological and morphological identification of horizontal, bipolars and amacrine cells in goldfish retina. J Physiol Lond 207:623-633.
Kaneko A (1973) Receptive field organization of bipolar and amacrine cells in the goldfish retina. J Physiol Lond 235:133-153.
Kaneko A, Hashimoto H (1969) Electrophysiological study of single neurones in the inner nuclear layer of the carp retina. Vision Res 9:37-55.
Kolb H (1982) The morphology of bipolar cells, amacrine cells and ganglion cells in the retina of the turtle, Pseudemys scripta elegans. Phil Trans R Soc Lond B 298:355-393
Kolb H, Nelson R (1981) Amacrine cells of the catfish retina. Vision Res 21:1625-1633
Marc RE, Liu WS, Muller JF (1988) Gap junctions in the inner plexiform layer of the goldfish retina. Vision Res 28:9-24
Marchiafava PL and Torre V (1980) The responses of amacrine cells to light and intracellularly applied currents. J Physiol Lond 251:83-102
Mitarai G, Goto T, Takagi S (1978) Receptive field arrangement of colour-opponent bipolar and amacrine cells in carp retina. Sensory Processes 2:375-382
Murakami M, Shimoda Y (1977) Identification of amacrine and ganglion cells in the carp retina. J Physiol Lond 265:801-818.
Naka K-I (1980) A class of catfish amacrine cells respond preferentially to objects which move vertically. Vision Res 20:961-965
Naka K-I, Carraway NRG (1975) Morphological and functional identification of catfish retinal neurons: I Classical morphology. J Neurophysiol 38:53-71
Naka K-I, Christensen BN (1981) Direct electrical connections between transient amacrine cells in the catfish retina. Science 214:462-464
Naka K-I, Ohtsuka T (1975) Morphological and functional identifications of catfish retinal neurons. II Morphological identification. J Neurophysiol 38:72-91
Perry VH , Walker M (1980) Amacrine cells, displaced amacrine cells and interplexiform cells in the retina of the rat. Proc R Soc Lond B 208:413-431
Teranishi T, Negishi K, Kato S (1987) Functional and

morphological correlates of amacrine cells in carp retina. Neurosci 20:935-950
Toyoda J, Hashimoto H, Ohtsu K (1973) Bipolar-amacrine transmission in the carp retina. Vision Res 13:295-307.
Vallerga S (1981) Physiological and morphological identification of amacrine cells in the retina of the larval tiger salamander. Vision Res 21:1307-1317
Vallerga S, Deplano S (1984) Differentiation, extent and layering of amacrine cell dendrites in the retina of a sparid fish. Proc R Soc Lond B 221:465-477
Vaney D (1986) Morphological identification of serotonin accumulating neurons in the living retina. Science 233:444-446
Yang CY, Yazulla S (1996) Neuropeptide-like immunoreactive cells in the retina of the larval tiger salamander: attention to the symmetry of dendritic projections. J Comp Neurol 248: 105-118
Yagi T, Kaneko A (1988) The axon terminal of goldfish retinal horizontal cells: a low membrane conductance measured in solitary preparations and its implications to signal conductance from the soma. J Neurophysiol 59:482-494
Wagner H-J, Wagner E (1988) Amacrine cells in the retina of a teleost fish. The roach (Rutilus rutilus). Phil Trans R Soc Lond B (in press)
Werblin FS (1970) Responses of retinal cells to moving spots:Intracellular recordings in Necturus maculosus. J Neurophysiol 33: 342-350
Werblin FS (1972) Lateral interactions at the inner plexiform layer of vertebrate retina. Antagonistic response to change. Science 175:1008-1010
Werblin FS (1978) Transmission along and between rods in the retina of the tiger salamander. J Physiol Lond 294:613-626.
Werblin FS, McGuire G, Lukasiewicz P, Eliasof S, Wu SM (1988) Neural interaction mediating the detection on motion in the retina of the tiger salamander. J Visual Neurosci 1
Wong-Riley MTT (1974) Synaptic organization of the inner plexiform layer inn the retina of the larval tiger salamander. J Neurocytol 3:1-33

INTEGRATION OF SYNAPTIC INPUT FROM ON AND OFF PATHWAYS IN MUDPUPPY RETINAL GANGLION CELLS

John S. McReynolds and Peter D. Lukasiewicz[1]
Department of Physiology
The University of Michigan
Ann Arbor, MI 48109-0622 U.S.A.

Based on their responses to steady illumination of the receptive field center, retinal ganglion cells in cold-blooded vertebrates are often classified into three main types, On-center, Off-center and On-Off, which in some cases may be further subdivided according to other criteria such as directional selectivity. In mudpuppy and tiger salamander, electrophysiological studies have revealed that each type of ganglion cell can receive several physiologically different types of synaptic input (for example, sustained and transient excitation, and sustained and transient inhibition) which are integrated to produce complex responses. This article will first summarize briefly the different types of synaptic inputs to ganglion cells in mudpuppy and tiger salamander, and then concentrate on the convergence of inputs from the On and Off pathways.

FUNCTIONALLY DISTINCT SYNAPTIC INPUTS TO GANGLION CELLS

<u>Sustained excitatory inputs</u>. It is generally believed that On- and Off-center ganglion cells receive sustained excitatory inputs, most likely from the corresponding types of bipolar cells (Werblin and Dowling, 1969). On-Off ganglion cells also receive sustained excitatory input (Belgum *et al*. 1983), possibly from both types of bipolar cells. These inputs are most likely mediated by glutamate or a closely related substance, acting at more than one type of receptor (Slaughter and Miller, 1983; Lukasiewicz and McReynolds, 1985). Mudpuppy ganglion cells also receive a tonic, cholinergic, excitatory input in

---

[1]Present address: Graduate Group in Neurobiology, University of California, Berkeley, CA 94720

NATO ASI Series, Vol. H31
Neurobiology of the Inner Retina
Edited by R. Weiler and N.N. Osborne
© Springer-Verlag Berlin Heidelberg 1989

darkness, but this input does not appear to mediate any light-evoked responses (McReynolds and Miyachi, 1986).

Transient excitatory inputs. Transient excitatory responses at the onset and termination of a steady light stimulus are the characteristic responses of On-Off ganglion cells. In cat and rabbit at least some of these responses are due to separate, transient inputs from cholinergic On and Off amacrine cells (Masland et al. 1984), but in lower vertebrates it is not certain whether the transient excitatory inputs are from amacrine or bipolar cells. Possible mechanisms by which transient responses are generated will be discussed later. Transient excitatory responses at both light on and light off can also be seen in some On-center ganglion cells (Lukasiewicz and McReynolds, 1985). In Off-center ganglion cells, transient depolarizing off-responses are often seen at the termination of center illumination; these are probably due at least in part to a transient increase in the sustained excitatory input present in darkness, since hyperpolarizing (Off) bipolar cells often show similar responses.

Transient inhibitory inputs. A wide-field, transient (change-sensitive) inhibitory input has been shown in all three types of ganglion cells (Werblin, 1972; Werblin and Copenhagen, 1974; Frumkes et al. 1981; Wunk and Werblin, 1979; Belgum et al. 1984). This input produces transient inhibitory responses at the onset and offset of steady illumination, and may produce a sustained hyperpolarization in response to appropriate moving stimuli. Although the transient inhibitory responses are often obscured by other components of the ganglion cell response, they can usually be revealed by depolarizing the cell with extrinsic current (Wunk and Werblin, 1979; Belgum et al. 1983, 1984), or by blocking the sustained responses with pharmacological agents (Lukasiewicz and McReynolds, 1985). The transient inhibitory inputs at both light on and light off are selectively blocked by strychnine and thus probably mediated by glycine (Belgum et al. 1984).

Sustained inhibitory inputs. In On-center ganglion cells a tonic inhibitory input is present in darkness, and in Off-center ganglion cells sustained inhibitory input is activated by center illumination (Belgum et al. 1982). Light also activates sustained inhibitory input in many On-Off ganglion cells (Belgum et al. 1983). The sustained inhibitory responses appear to be mediated by GABA (Belgum et al. 1984; Werblin et al. 1988).

CROSSOVER BETWEEN ON AND OFF PATHWAYS IN THE INNER RETINA

It is often implied that separate On and Off pathways are established at the bipolar cell level, with mixing of signals from these pathways only in On-Off cells. However, the nature of the multiple synaptic inputs described above suggests that some On- and Off-center ganglion cells also receive input from both pathways. Here we present pharmacological evidence for extensive signal crossover between the On and Off pathways in the inner retina. The glutamate analogue 2-amino-4-phosphonobutyrate (APB) selectively eliminates the responses of On-center (depolarizing) bipolar cells (Slaughter and Miller, 1981), and thus should block in more proximal cells those responses that are driven, either directly or through interneurons, by On bipolar cells. In the results described here, intracellular responses were recorded from ganglion cells in the superfused mudpuppy eyecup preparation; the recording, superfusion and light stimulation are described in detail elsewhere (Lukasiewicz and McReynolds, 1985).

On-center ganglion cells receive input from the Off pathway. In all of the 8 On-center ganglion cells tested APB blocked the sustained depolarizing response to center illumination; in 3 of these cells it also revealed a transient EPSP at light off, and in 2 of the cells it caused a hyperpolarization in darkness. All of these effects were observed in the cell shown in Fig. 1. The hyperpolarization in darkness was associated with a decrease in input conductance (not shown), and was probably due

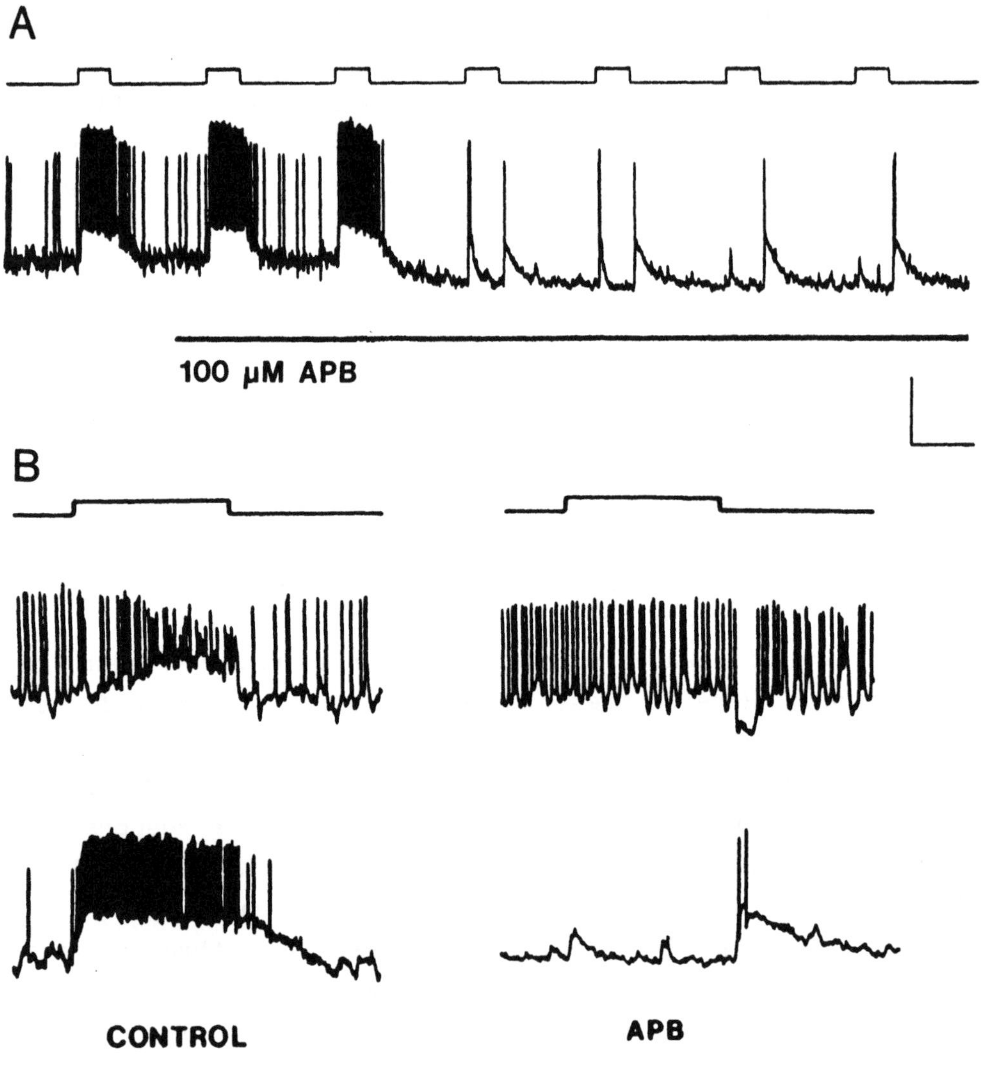

Fig. 1. Effect of APB on On-center ganglion cell response to center illumination. Light stimulus (indicated by upward deflection of line above responses) was a 250 µm dia spot of white light in the center of the receptive field. A. Horizontal line below response trace indicates when superfusate contained APB. Calibration 20 mV, 10 sec. B. Individual responses to same light stimulus as in A in the absence and presence of 100 µM APB. In each column the lower response was recorded at resting potential and the upper response was recorded when the membrane was depolarized (12 mV in control, 15 mV in APB) by a 0.04 nA step of extrinsic current. In the presence of APB a transient EPSP at light off is seen at resting potential and a transient IPSP at light off was revealed when the membrane was depolarized by the current step. The transient IPSP, but not the transient EPSP, was selectively blocked by 10 µM strychnine (not shown). Calibration 10 mV, 2 sec.

to the elimination of background (dark) transmitter release from depolarizing bipolar cells. APB never caused a depolarization in darkness, which suggests that the tonic inhibition active in darkness is driven by the Off pathway, probably via sustained hyperpolarizing (Off) amacrine cells, as shown in the left-hand diagram in Fig. 4. Transient inhibitory responses at light off were revealed when the cell was depolarized with extrinsic current. Since both the excitatory and inhibitory transient responses at light off were resistant to APB, these inputs are probably driven by the Off pathway.

In some On-center ganglion cells illumination of the receptive field surround causes a sustained hyperpolarization even in the absence of center illumination. Concentrations of APB which blocked the depolarizing center response did not block this response to surround illumination in any of the 4 cells tested, indicating that it was due to input from the Off pathway. In 3 cells APB (10-100 μM) actually enhanced the surround response (Fig. 2); this may have been due to the fact that APB can increase coupling between horizontal cells, possibly by interfering with depolarizing bipolar cell input to a modulatory neuron (Dong and McReynolds, 1988).

<u>Off-center ganglion cells receive input from the On pathway</u>. Figure 3 shows that APB blocked the sustained hyperpolarizing response of Off-center ganglion cells to center illumination, revealing a transient inhibitory response at the termination of the light stimulus. This was seen in all of the 7 cells tested. The block of the sustained hyperpolarizing response could not have been due to interference with the sustained excitatory input present in darkness, since APB had no effect on the dark potential. Thus the sustained inhibitory input to Off-center ganglion cells must come from the On pathway, probably via depolarizing sustained (On) amacrine cells, as shown in the left-hand diagram in Fig. 4. Off-center ganglion cells in rabbit and cat also appear to be inhibited by an input from the On pathway, since APB caused an increase in their firing rate (Massey <u>et al</u>. 1983; Wässle <u>et al</u>. 1986); in cat

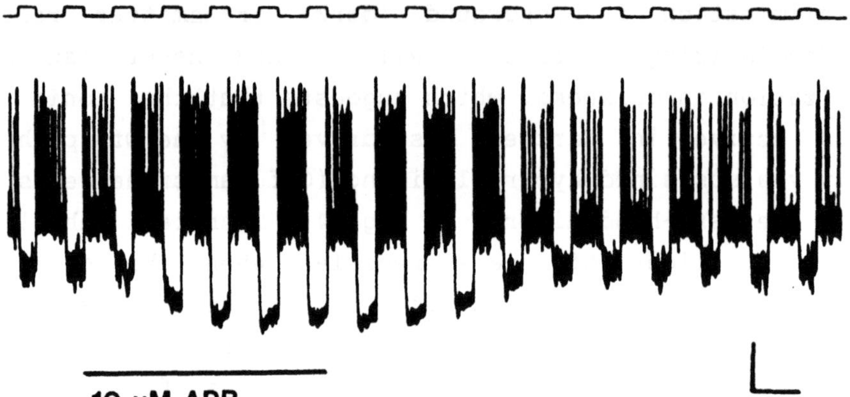

Fig. 2. Effect of APB on On-center ganglion cell responses to surround illumination. Light stimulus was an annulus (i.d. 800 µm, o.d. 2000 µm) concentric with the receptive field center. The response of this cell to illumination of the receptive field center was a sustained depolarization similar to that shown in Fig. 1. Calibration 10 mV, 20 sec.

the inhibitory input appears to be mediated by glycine, since the effect of APB was blocked by strychnine.

Push-pull interaction of sustained excitatory and inhibitory inputs. The demonstration that On- and Off-center ganglion cells receive both sustained excitatory and sustained inhibitory input suggests the possibility that their sustained responses to light could be produced by a simultaneous increase in one input and a decrease in the other. Based on anatomical studies, this kind of 'push-pull' interaction has also been postulated in cat ganglion cells (McGuire et al. 1986), and some electrophysiological results in mudpuppy are consistent with this idea. For example, the sustained depolarizing response of On-center ganglion cells to center illumination is usually associated with an increase in conductance, indicating an increase in excitatory input, but in some cells it was associated with a decrease in conductance, suggesting that it was caused by disinhibition (Belgum et al. 1982). Also, in Off-center ganglion cells the sustained hyperpolarizing response to center illumination is usually associated with an increase in conductance (Belgum et al. 1982), but a decrease in conduc-

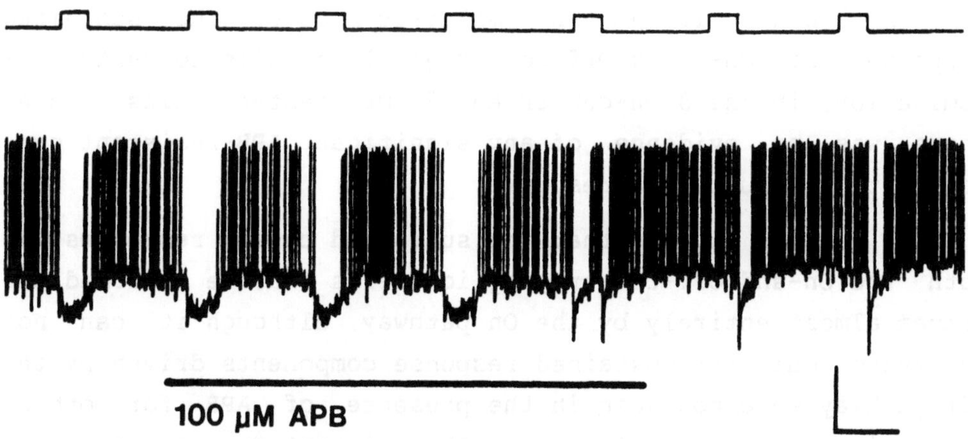

Fig. 3. Effect of APB on OFF-center ganglion cell responses to center illumination. Details as in Fig. 1. Calibration 20 mV, 10 sec.

tance, suggesting disfacilitation, has also been reported (Dacheux <u>et al</u>. 1979). These variations in the sign of the net conductance change associated with a given polarity of sustained response are consistent with a push-pull interaction, but they could also be observed if the sustained response was due entirely to excitation in some On-center ganglion cells and to disinhibition in others, and by inhibition in some Off-center ganglion cells and disfacilitation in others. The evidence for a push-pull interaction is stronger in studies on the antagonistic response to surround illumination; in a single ganglion cell an antagonistic surround response of a given polarity can be associated with either an increase or a decrease in conductance, depending on the amount of center illumination (Belgum <u>et al</u>. 1987).

The results of pharmacological experiments, however, were not consistent with a significant push-pull interaction in the ganglion cells we recorded from. Since APB blocks input from the On pathway to both On- and Off-center ganglion cells, as shown in the preceding section, it should reveal any component of the ganglion cell response that comes from the Off pathway (a disinhibitory response in On-center cells and a disfacilitatory response in Off-center cells). However, as seen in Figs. 1 and 3, 100 μM APB completely blocked the sustained responses of On- and Off-center ganglion cells to center illumination; in the 8 On-center and 7 Off-center cells tested there was no evidence of any significant APB-resistant component to the sustained response.

These results suggest that the sustained center responses of both the On-and Off-center ganglion cells that we recorded are driven almost entirely by the On pathway, although it can not be ruled out that sustained response components driven by the Off pathway were not seen in the presence of APB for other, unexplained reasons. In fact, APB-resistant On responses have recently been reported in goldfish optic nerve discharges (Powers et al. 1988); these responses may have been produced by a decrease in inhibitory input from the Off pathway.

Other investigators have shown that in addition to the type of Off-center cells that we recorded from, some Off-center ganglion cells have sustained center hyperpolarizing responses that are primarily due to disfacilitation and are not blocked by APB, and some have inhibitory responses that become disfacilitatory in the presence of APB (Arkin and Miller, 1988). Thus, there are probably several subclasses of Off-center ganglion cells in the mudpuppy retina. Although center illumination produces similar responses in all of these cells, the mechanisms by which these responses are generated may be quite different; in some cells they appear to be driven mainly by the On pathway, in others mainly by the Off pathway, and in some by both pathways in a push-pull manner. The same variations may exist among On-center ganglion cells. Therefore, the relative strengths of the sustained excitatory and inhibitory inputs shown in the diagram of Fig. 4 may vary considerably among ganglion cells.

ORIGIN OF TRANSIENT INPUTS TO GANGLION CELLS

Although the wide-field transient inhibitory input to ganglion cells could theoretically come from separate On-transient and Off-transient cells, the lack of evidence for

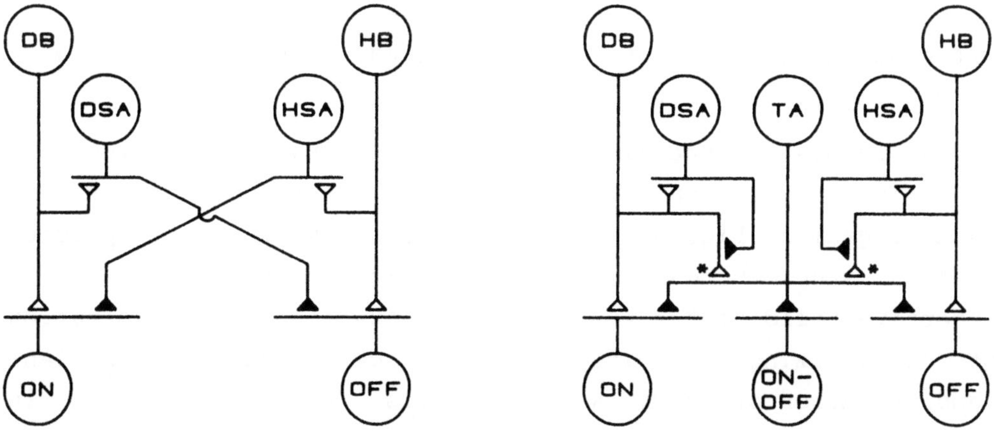

Fig. 4. Proposed synaptic connections underlying sustained and transient inhibitory inputs to mudpuppy ganglion cells. Open triangles indicate excitatory synapses; filled triangles indicate inhibitory synapses. An asterisk beside a synaptic terminal indicates that transmitter release is made transient by feedback from sustained amacrine cells (see text). Diagrams do not indicate actual distribution of processes within IPL. DB, depolarizing bipolar; HB, hyperpolarizing bipolar; DSA, depolarizing sustained amacrine; HSA, hyperpolarizing sustained amacrine; TA, transient (On-Off) amacrine; ON, On-center ganglion; OFF, Off-center ganglion; ON-OFF, On-Off ganglion. Although not shown, DB and HB cells receive sustained input from photoreceptors via sign-inverting and sign-preserving synapses, respectively. Sustained inputs to ON-OFF cells are not shown. Left-hand diagram shows sustained inhibitory input to ON and OFF ganglion cells. Right-hand diagram shows transient inhibitory input to ON, OFF, and ON-OFF ganglion cells. Both diagrams also show sustained excitatory input to ON- and OFF-center ganglion cells. Postulated transmitter substances are: DB and HB, glutamate; DSA and HSA, GABA; TA, glycine.

such cells in mudpuppy suggests that the source of this input is probably a population of glycinergic On-Off amacrine cells. It has been suggested that the transient light-evoked responses in On-Off amacrine cells are due to transient release of transmitter from bipolar terminals (Marchiafava and Torre, 1978; Toyoda _et al_. 1984; Kujiraoka _et al_. 1986), but the mechanism by which this occurs has only recently been elucidated. Studies on tiger salamander suggest that sustained amacrine cells receive excitatory input from bipolar cells and release GABA onto other bipolar terminals, where it acts at

$GABA_B$ receptors to suppress the release of transmitter onto On-Off amacrine cells (Maguire et al. 1988; Werblin et al. 1988; Werblin et al., this volume).

The proposed synaptic pathways underlying the transient inhibitory inputs to ganglion cells are illustrated in the right-hand diagram of Fig. 4. The connections shown here predict that a step increase in illumination of the ganglion cell receptive field center would cause a sustained depolarization of depolarizing sustained amacrine cells and a transient depolarization of On-Off amacrine cells. A step decrease in illumination would cause a sustained depolarization in hyperpolarizing sustained amacrine cells and a transient depolarization in On-Off amacrine cells. On-Off amacrine cells would then provide transient inhibitory input to all three types of ganglion cells. Stimuli which produce sustained responses in On-Off amacrine cells (e.g. certain types of moving stimuli) would cause sustained transmitter release from these cells (Werblin, 1972), and directional selectivity could result from projection of transient inhibitory inputs onto ganglion cells in a specific lateral direction (Maguire et al. 1988; Werblin et al. 1988; Werblin et al., this volume). Transient excitatory inputs to ganglion cells are less well understood; they could be provided by another population of (excitatory) On-Off amacrine cells, by transient increases in a sustained excitatory input, or possibly by a feedback mechanism similar to that described above but acting at synapses from bipolar terminals onto ganglion cells.

In summary, mudpuppy ganglion cells receive multiple synaptic inputs, from both the On and Off pathways, and some of these inputs may interact in a push-pull manner. The variations in response characteristics among ganglion cells of the same class may be in part due to differences in the densities of the different types of input or to physiological modulation of the relative contributions of these inputs.

This work was supported by NIH Grant EY01653.

REFERENCES

Arkin MS, Miller RF (1988) Bipolar origin of synaptic inputs to sustained Off ganglion cells in the mudpuppy retina. J Neurophysiol (in press)

Belgum JH, Dvorak DR, McReynolds JS (1982) Sustained synaptic input to ganglion cells of mudpuppy retina. J Physiol 326: 91-108

Belgum JH, Dvorak DR, McReynolds JS (1983) Sustained and transient inhibitory inputs to on-off ganglion cells in the mudpuppy retina. J Physiol 340: 599-610

Belgum JH, Dvorak DR, McReynolds JS (1984) Strychnine blocks transient but not sustained inhibition in mudpuppy retinal ganglion cells. J Physiol 354: 273-286

Belgum JH, Dvorak DR, McReynolds JS, Miyachi E-I (1987) Push-pull effect of surround illumination on excitatory and inhibitory inputs to mudpuppy retinal ganglion cells. J Physiol 388: 233-243

Dacheux RF, Frumkes TE, Miller RF (1979) Pathways and polarities of synaptic interactions in the inner retina of the mudpuppy. I. Synaptic blocking studies. Brain Res 161: 1-12

Dong C-J, McReynolds JS (1988) APB increases electrical coupling between horizontal cells in mudpuppy retina. Vision Res (in press)

Frumkes TE, Miller RF, Slaughter M, Dacheux RF (1981) Physiological and pharmacological basis of GABA and glycine action on neurons of mudpuppy retina. III. Amacrine-mediated inhibitory influences on ganglion cell receptive-field organization: a model. J Neurophysiol 45: 783-804.

Kujiraoka T, Saito T, Toyoda J-I (1986) Bipolar-amacrine synaptic transmission: effect of polarization of bipolar cells on amacrine cells in carp retina. Neuroscience Res 4 (suppl): 111-119

Lukasiewicz PD, McReynolds JS (1985) Synaptic transmission at N-methyl-D-aspartate receptors in the proximal retina of the mudpuppy. J Physiol 367: 99-115

Maguire G, Lukasiewicz P, Werblin F (1988) Neural interactions underlying the response to change in the tiger salamander retina. J Neuroscience (in press)

Marchiafava PL, Torre V (1978) The responses of amacrine cells to light and intracellularly applied currents. J Physiol 276: 83-102

Masland RH, Mills JW, Cassidy C (1984) The functions of acetylcholine in the rabbit retina. Proc R Soc Lond B 223: 121-139

Massey SC, Redburn DA, Crawford MLJ (1983) The effects of 2-amino-4-phosphono-butyric acid (APB) on the ERG and ganglion cell discharge of rabbit retina. Vision Res 23: 1607-1613

McGuire BA, Stevens JK, Sterling P (1986) Microcircuitry of beta ganglion cells in cat retina. J Neuroscience 6: 907-918

McReynolds JS, Miyachi E-I (1986) The effect of cholinergic agonists and antagonists on ganglion cells in the mudpuppy retina. Neuroscience Res 4 (suppl): 153-161

Powers M, DeMarco P, Bilotta J (1988) APB eliminates ERG b-wave but not optic nerve "ON" response in goldfish. Invest Ophthalmol Vis Sci 29 (suppl): 104

Slaughter MM, Miller RF (1981) 2-Amino-4-phosphonobutyric acid: a new pharmacological tool for retina research. Science 211: 182-185

Slaughter MM, Miller RF (1983) Bipolar cells in the mudpuppy retina use an excitatory amino acid neurotransmitter. Nature 303: 537-538

Toyoda J-I, Fujimoto M (1984) Application of transretinal current stimulation for the study of bipolar-amacrine transmission. J Gen Physiol 84: 915-925

Wässle H, Schaefer-Trenkler I, Voigt T (1986) Analysis of a glycinergic inhibitory pathway in the cat retina. J Neuroscience 6: 594-604

Werblin FS (1972) Lateral interactions at inner plexiform layer of a vertebrate retina: antagonistic response to change. Science 175: 1008-1010

Werblin FS, Copenhagen DR (1974) Control of retinal sensitivity. III. Lateral interactions at the inner plexiform layer. J Gen Physiol 63: 88-110

Werblin FS, Dowling JE (1969) Organization of the retina of the mudpuppy, Necturus maculosus. II. Intracellular recording. J Neurophysiol 32: 331-355

Werblin FS, Maguire G, Lukasiewicz P, Eliasof S, Wu S (1988) Neural interactions mediating the detection of motion in the retina of the tiger salamander. Visual Neuroscience (in press)

Wunk DF, Werblin FS (1979) Synaptic inputs to the ganglion cells in the tiger salamander retina. J Gen Physiol 73: 265-286

STRUCTURE FUNCTION RELATIONSHIPS OF SUSTAINED ON GANGLION CELLS OF THE MUDPUPPY RETINA

Robert F. Miller, Paul Coleman and Marty Arkin
Department of Physiology
University of Minnesota
6-255 Millard Hall
Minneapolis, MN 55455 USA

The technique of whole cell recording (WCR) with patch clamp micropipettes has provided new insights into the properties of nerve cells (Hamill et. al., 1981). This method has received its widest application in dissociated cells, but more recently the "gigaseals" required for success with this technique have been achieved in brain and retinal slice preparations (See Werblin this volume). We have succeeded in obtaining WCR from cells of the intact retina (Coleman and Miller, in Press). Although the major focus of this research has been directed to the mudpuppy, we have also obtained gigaseals from neurons of the rabbit retina, so the overall method may have broad application.

Figure 1 illustrates the general approach we use to obtain WCR in the amphibian retina. For this method, large diameter, low resistance micropipettes are fabricated, filled with a Ringer solution compatible with the intracellular ionic constituents and advanced to the retinal surface. The isolated, perfused retina-eyecup preparation is used for these experiments, with the added treatment of a 3 minute exposure to an enzymatic mixture of collagenase and hyaluronidase to further remove remnants of the vitreous. Blind electrode advancement, combined with intermittent suction to the pipette, often leads to the formation of "gigaseals" of >1 Gohm. Further suction or application of a brief current pulse to the electrode is used to rupture the membrane patch and form continuity between the intracellular environment and the interior of the electrode. If the formation of the gigaseal is not followed by a successful WCR, the electrode can be purged and utilized again; a single electrode can be used for as many as a dozen attempts before it must be replaced.

The advantage of WCR over conventional intracellular recording is that a very stable recording condition is obtained,

NATO ASI Series, Vol. H31
Neurobiology of the Inner Retina
Edited by R. Weiler and N. N. Osborne
© Springer-Verlag Berlin Heidelberg 1989

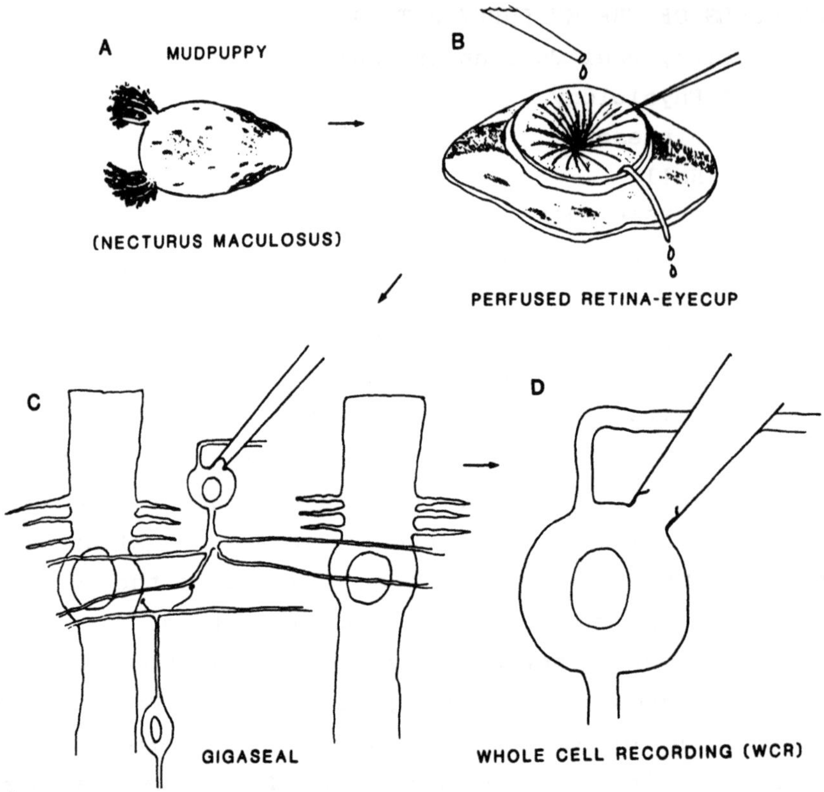

Figure 1

using a microelectrode with a much lower resistance than those normally required for conventional intracellular recording. Typically, a WCR electrode is about 5-10 Mohms in resistance, considerably lower than the several hundreds of megohms in resistance usually required for successful intracellular recordings from retinal neurons.

We studied 8 sustained On ganglion cells using the WCR technique. For each recording, current pulses were routinely applied to evaluate the time constant, input resistance (Rn) and the nature of the charging curve. The average input resistance was 1265 Mohms (S.E. 465 Mohms), with an average time constant of 66 ms (S.E. 7 ms). Figure 2 shows an example of a series of current pulses injected into a mudpuppy sustained ON ganglion cell. One of these curves (lower trace) was analyzed and was quite accurately fit by a single exponential. Additional analysis has been carried out on the charging curves using a second technique which involves Fourier analysis. Both methods show that the charging curves of sustained ON ganglion cells are adequately described by a single exponential. Rall (1969) has pointed out that analysis of the charging curve can be useful in predicting the average electrotonic length of equivalent cylinder dendritic trees. One implication of a single exponential charging curve, is that the dendritic cylinders are electrotonically short. A second implication of this "single compartment" charging curve is that voltage clamp experiments might be rationalized, since voltage decay from the soma to the periphery should be minimal (see Figure 5). It is important to emphasize however, that non-uniformities in membrane resistance and the large conductance

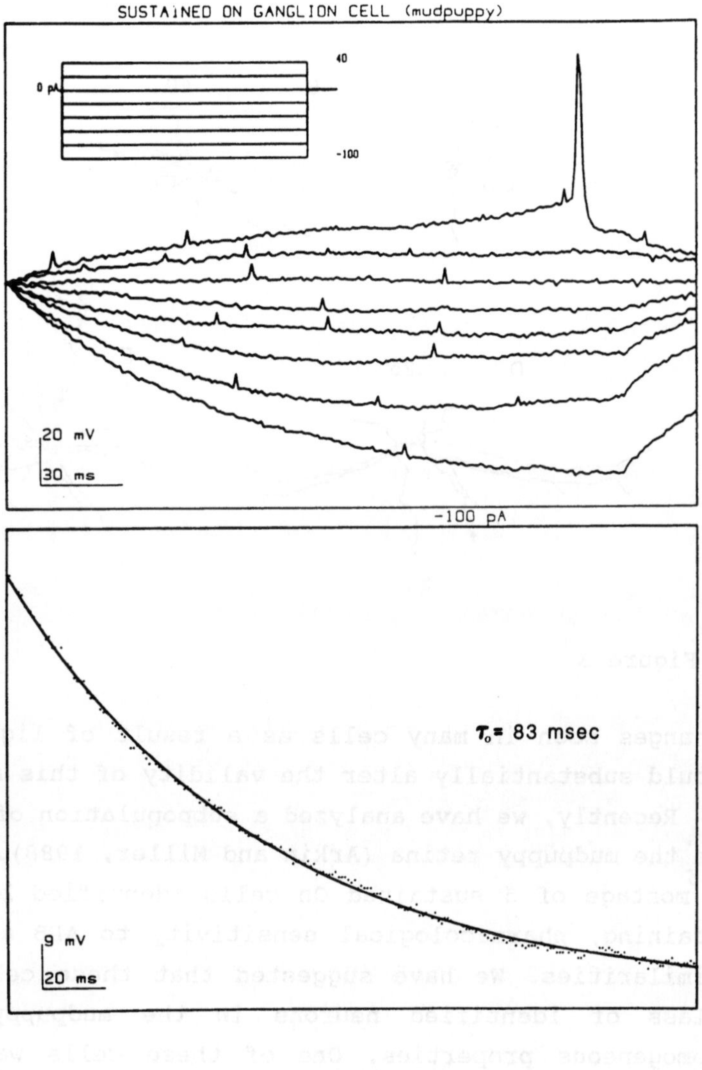

Figure 2

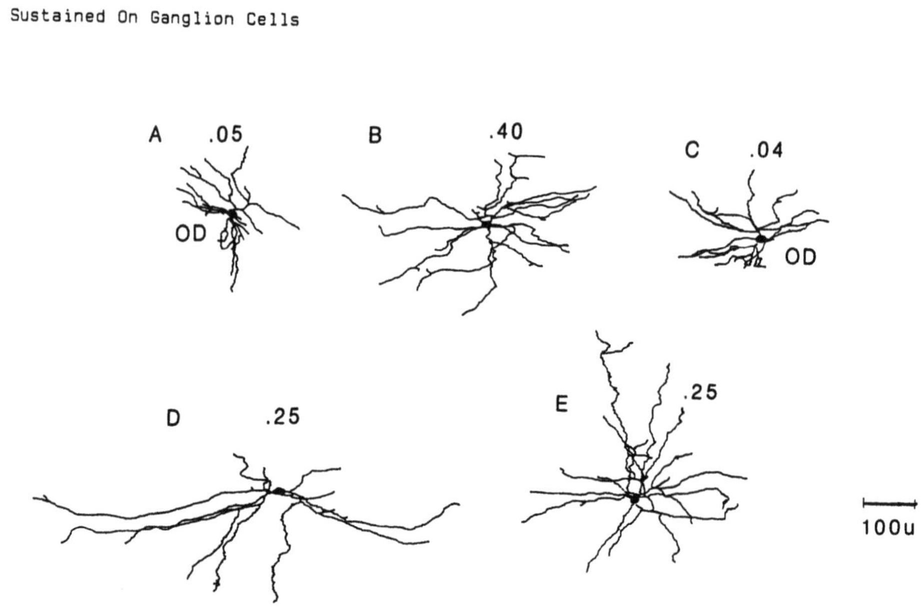

Figure 3

changes seen in many cells as a result of light stimulation, could substantially alter the validity of this assumption.

Recently, we have analyzed a subpopulation of ganglion cells in the mudpuppy retina (Arkin and Miller, 1988). Figure 3 shows a montage of 5 sustained On cells identified by intracellular staining, pharmacological sensitivity to APB and morphometric similarities. We have suggested that these cells represent a class of identified neurons in the mudpuppy, with fairly homogeneous properties. One of these cells was selected for computer simulation analysis. Using an iterative computer program, the structure of the cell was analyzed according to methods established by Rall (1959) and more recently outlined by Miller and Bloomfield (1983). This program uses different values of membrane resistance (Rm) to calculate the input resistance (Rn) at the soma. For the cell we selected for this analysis, the value of Rm needed to match the average Rn (approximately 1200 Mohms) of On ganglion cells was 70,000 ohms cm^2. This number is interesting, because from the time constant measurements,

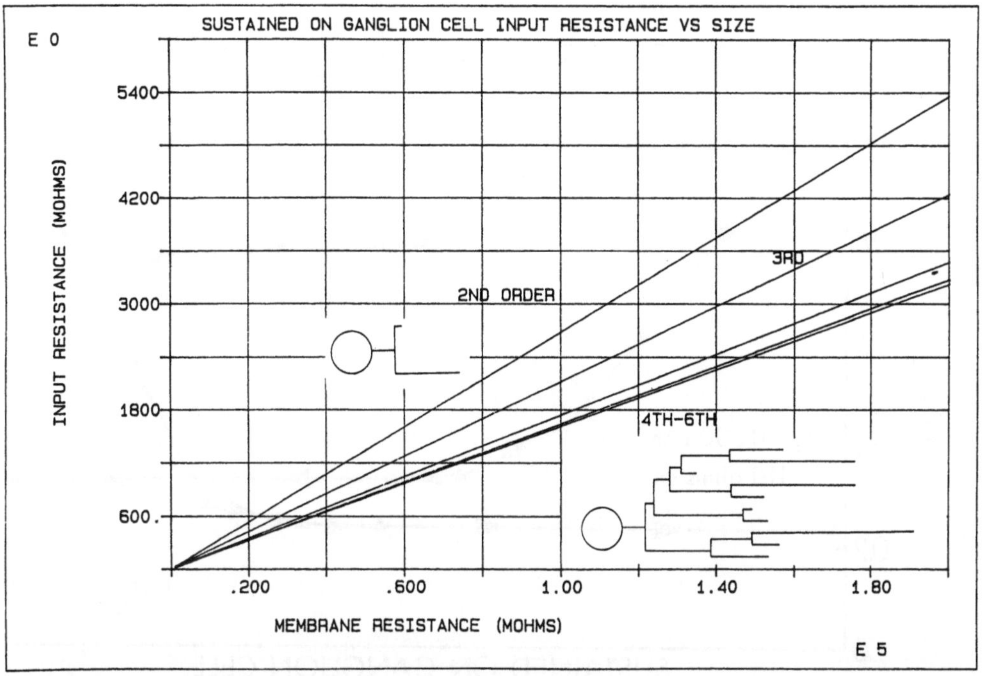

Figure 4

assuming a value of 1 uF/cm^2 for the membrane capacitance, the specific membrane resistance is predicted to be 68,000 ohms cm^2. Thus two independent techniques provide a very similar value for Rm.

Figure 4 illustrates calculations from a single sustained On cell of the mudpuppy retina. In this example, we determined the Rn of the cell for different fractions of its total dendritic tree. The top line shows the relationship between different Rm values and the Rn for the soma together with the first and second-order dendrites. The other lines show the decline in Rn as additional branches are connected to the dendritic tree. Note that the input resistance is reduced noticeably when the 3rd and 4th order dendrites are added, but a much smaller change in Rn is observed when the 5th and 6th order branches are joined. These calculations are consistent with a concept established by Rall (1967), that the Rn of a cell is usually determined by the more proximal portions of the soma-dendritic tree. Figure 4 demonstrates that this general principle applies to the

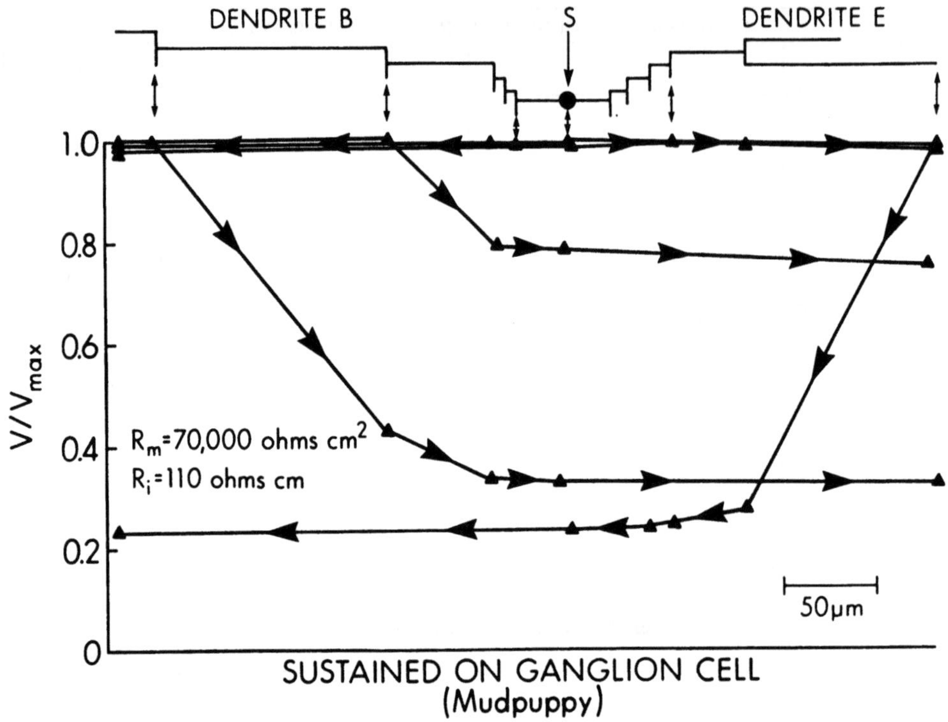

Figure 5

sustained On ganglion cell, independent of Rm. Thus, the distal dendrites contribute little to the Rn of the cell and conductance changes in the periphery will probably be more difficult to detect than those generated closer to the soma, even when Rm is high.

Figure 5 illustrates how steady state current will distribute in the sustained On cell selected for analysis. In this figure current was injected at several sites and the voltage decay calculated by the methodology of Rall (1959). For these calculations, no assumptions were made about equivalent cylinders, but rather the "real" diameter and length values were used to make these calculations. The methods for doing this are well established, though seldom used. For illustrative purposes, the voltage decay along two dendrites is shown, but the calculations were done with all of the dendrites taken into consideration. Note that when current is injected into the soma,

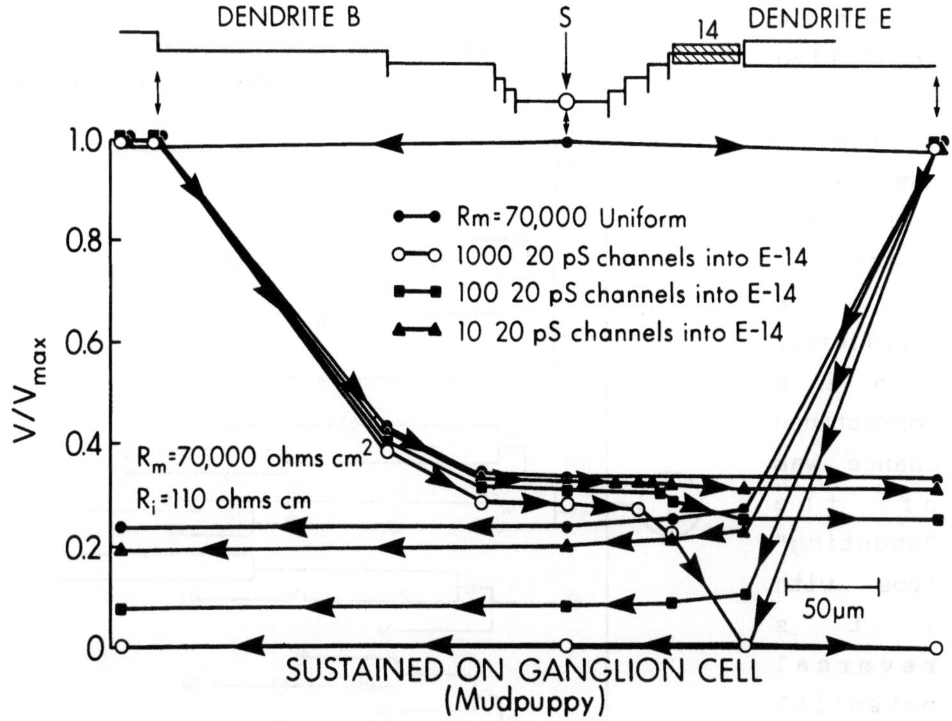

Figure 6

the voltage decay towards the periphery is minimal. This generality also applies when current is injected into any of the dendrites within about 100 uM of the soma. However, peripheral dendrites, because of their relatively small size, have a poor impedance match with more proximal structures and synaptic currents generated on these branches will lead to substantial voltage decay towards the soma. Thus, the concept of isopotentiality relates only to voltages applied to the soma.

One of the next questions we addressed was whether sustained On ganglion cells with high Rm values behaved differently than the same cell with a lower Rm for such features as dendritic inhibition. Figure 6 illustrates one attempt designed to provide some answers to this question. Using the same sustained On cell as in previous calculations, we computed the decay of voltage along the dendritic tree, when current was injected into peripheral branches. On one dendrite we imposed a conductance change consisting of different numbers of 20 pS channels. For simplicity, this conductance change was of the "shunting" type with its reversal potential the same as the membrane potential. Note that for current injected into the same dendrite as the inhibitory conductance change, the steady state decay is much more attenuated than is the voltage decay profile for the other dendrite. In other words, despite the high value of Rm, local dendritic processing can persist in the sense that inhibition localized to a single dendrite is more effective in that dendrite compared to its influence on neighboring dendrites of the same cell. For synaptic mechanisms such as directional selectivity which may depend on local dendritic processing, this feature is generally preserved even when Rm is very large. It is worth pointing out that the magnitude of the conductance change for this example represents a relatively small number of ionic channels. One feature of such a large Rm is that these cells (in the dark-adapted state) are positioned to respond when only a small

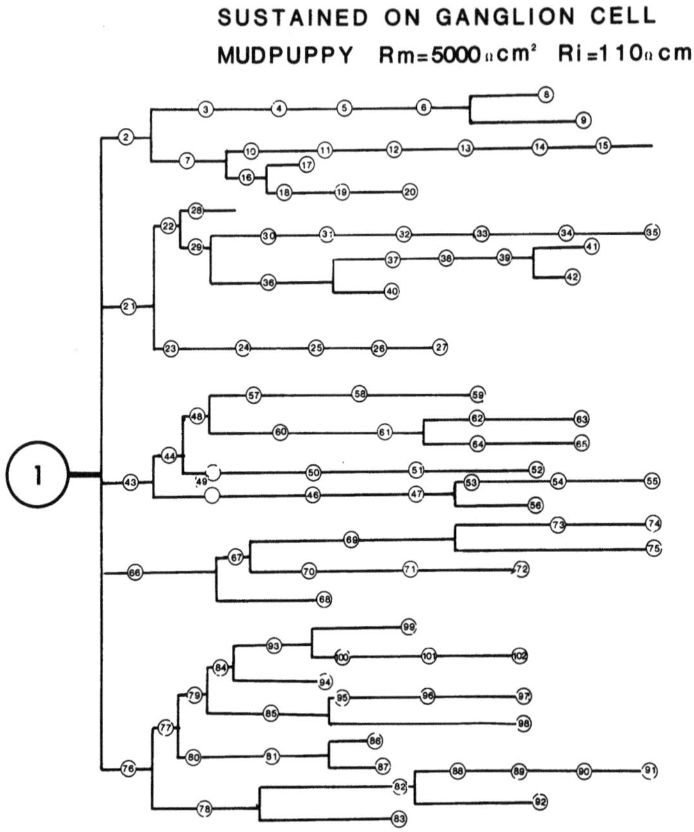

Figure 7

number of ionic channels are activated. However, under light-adapted conditions, sustained On ganglion cells show a large drop in Rn, from more than 1 Gohm to less than 100 Mohms (not illustrated), indicating that in

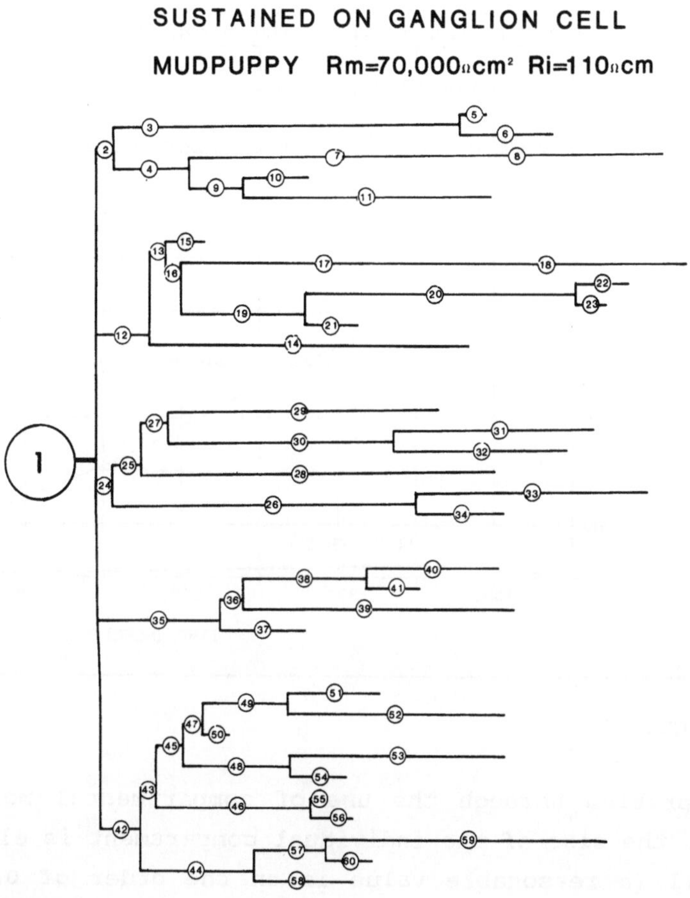

Figure 8

this operative mode, the cell will require an order magnitude larger number of ionic channels to modulate the excitability of the cell.

The results illustrated in figure 5 and 6 show the decay of steady state current injected into various regions of a sustained On ganglion cell with a high, mostly uniform Rm. Light-evoked synaptic potentials have transient components that are generated as a result of a conductance change, not as a result of current injected from an infinite impedance source. Thus, to more realistically simulate synaptic potentials, transients generated by a conductance change need to be studied. For this purpose, calculations of continuous spread of current along the dendritic tree become computationaly cumbersome. As usual, Rall (1967) has provided a general solution for this type

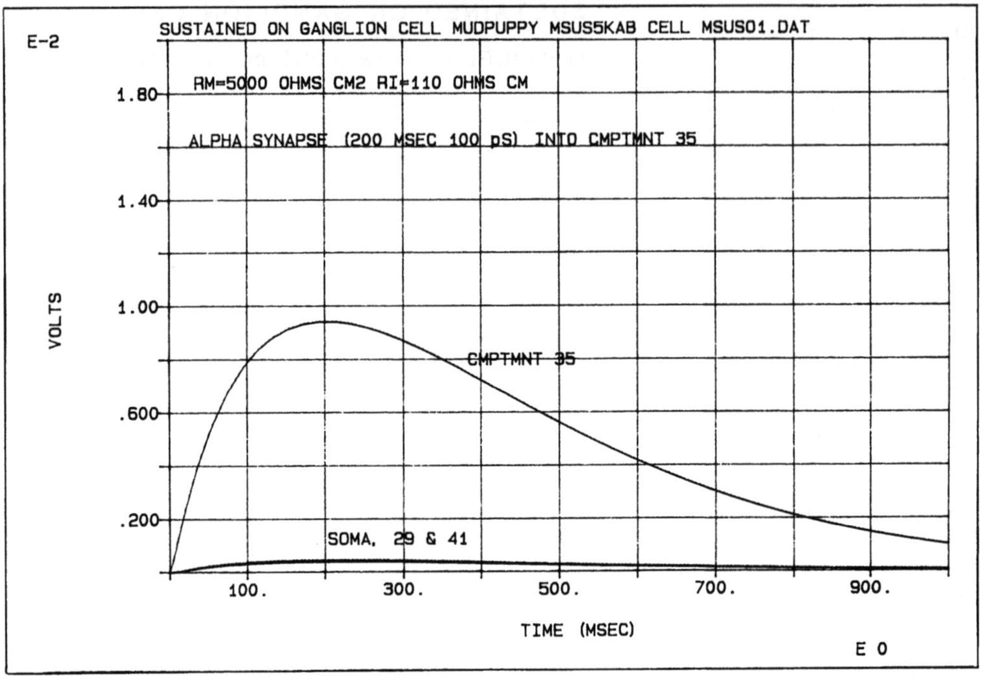

Figure 9

of problem through the use of compartmental models. Provided that the size of the individual compartment is electrotonically small (a reasonable value is on the order of 0.2 lambda) the compartmental model can adequately substitute for a model based on continuous dendrites. Figure 7 illustrates how a sustained On ganglion cell was transformed into a compartmental model, for a cell with an Rm of 5000 ohms cm^2 (top) while figure 8 illustrates a compartment model for the same cell with an Rm of 70,000 ohms cm^2. For these compartmental models, we let each branch consist of at least 1 compartment, irrespective of its electrotonic length and thereafter each compartment was 0.2 lambda or less. Note that the 5000 ohm cm^2 model required 102 compartments whereas the 70,000 ohm cm^2 model could be represented by 60 compartments.

For simulation and spread of "synaptic currents" in the compartmental model, we used a version of the Spice simulator, run on a Vax 1170 computer. Representation of a synaptic conductance change was achieved by connecting a variable

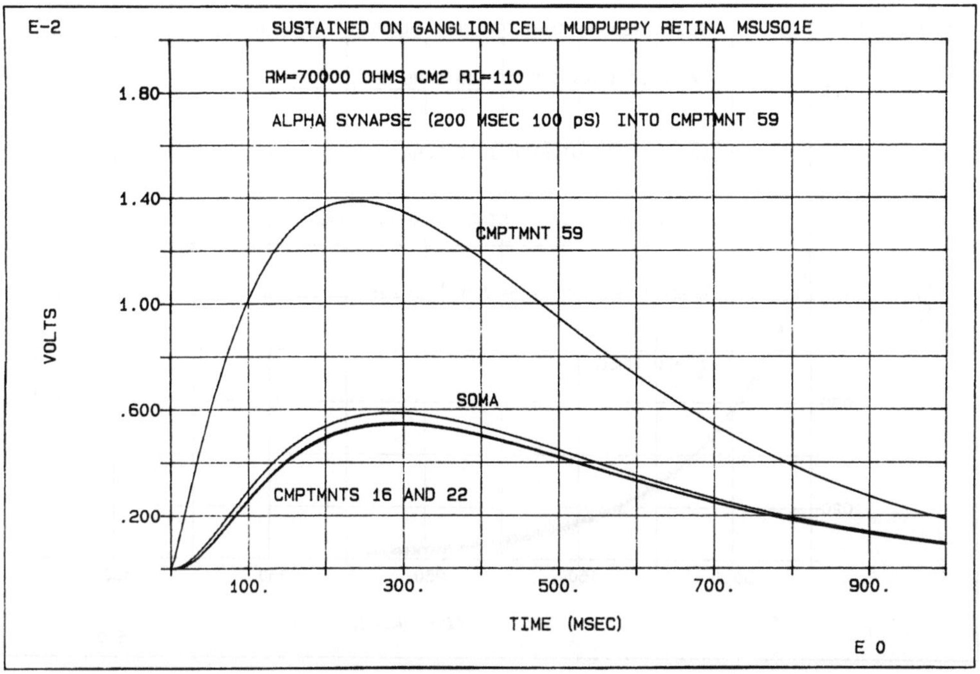

Figure 10

resistor to a 50 mV battery at the site of synaptic injection. The value of the variable resistor was determined by an alpha function (V=t*EXP(-t/alpha)), as illustrated in Figure 7. We selected 200 msec as the value for alpha, since this value was a reasonable match the time variant behavior of mudpuppy bipolar cells in response to medium to low intensity light stimulation. Figure 9 illustrates the injection of a 200 msec alpha conductance change of 100 pS injected into compartment 35. This produced a PSP of almost 10 mV at the site of injection, with considerable attenuation at the soma and more distal dendrites. Note however, that injection of the same conductance change into the same dendrite (compartment 18) with an Rm=70,000 ohms cm^2 results in a somewhat larger response (by about 50%) at the compartment of injection and comparatively less decay of the response at the level of the soma and other compartments. An increase of only about 50% in the size of the synaptic response at the point of current injection reflects the fact that a conductance change was used as the method of current injection.

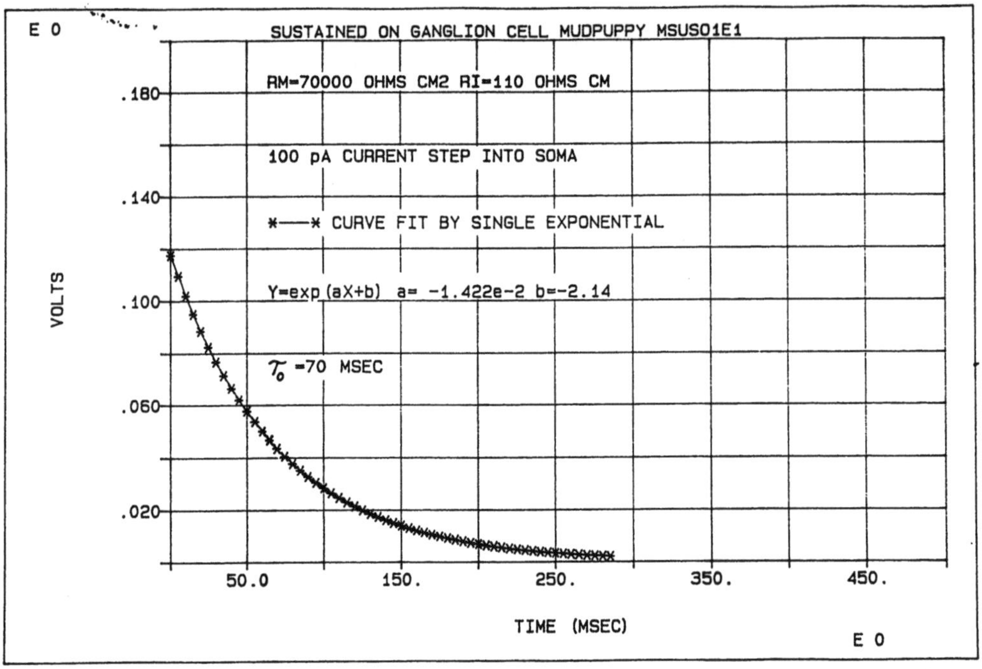

Figure 11

Without the conductance change, a current injection would produce large differences in the resulting voltage for the two models. One benefit of the conductance change is to preserve a more linear response generating mechanism by preventing saturation of the voltage at the emf of the synaptic driving force.

When we have studied ganglion cells in the past, using conventional intracellular recordings, analysis of transients such as that shown in figure 2, consistently demonstrated that more than one exponential contributed to the charging curve of the cell. We wondered therefore whether the shape of the charging curve might be altered by a shunt introduced by the intracellular microelectrode penetration. Figure 11 illustrates the result of a current pulse applied to the soma of the 70,000 ohm cm^2 cell while figure 12 shows the effects of shunting the soma with a 100 Mohm resistor at the soma, in order to simulate the microelectrode shunt. Each charging curve was then plotted and a single exponential was superimposed to determine how well

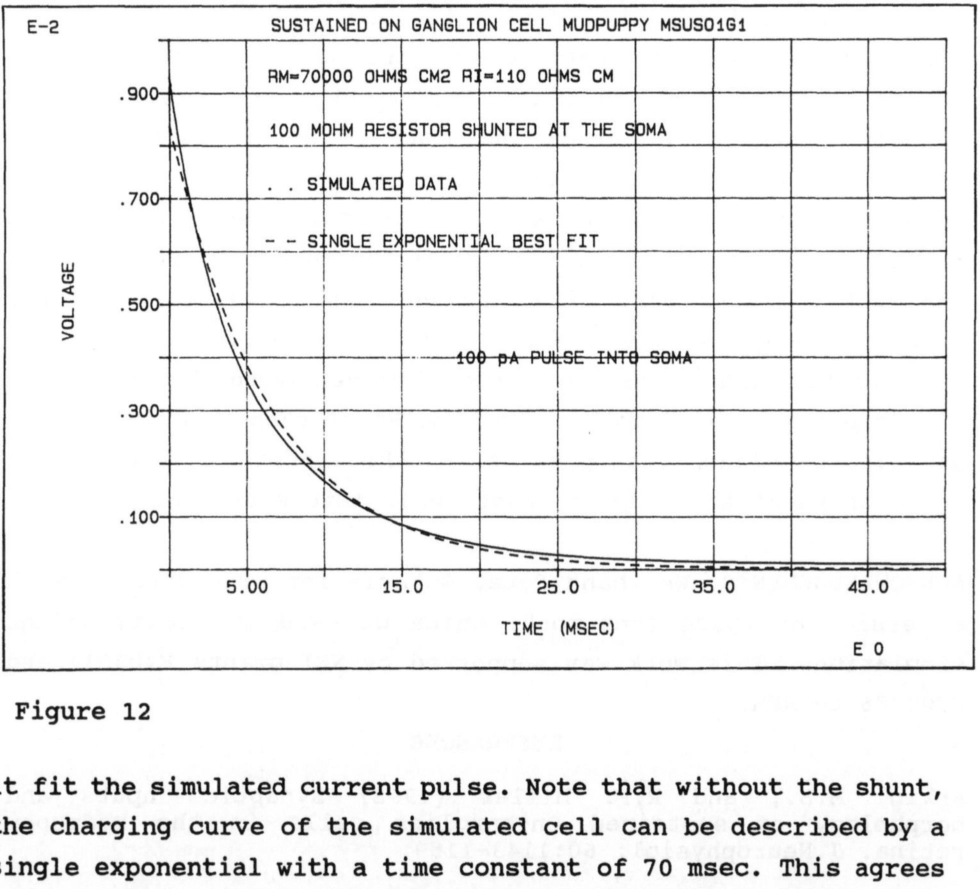

Figure 12

it fit the simulated current pulse. Note that without the shunt, the charging curve of the simulated cell can be described by a single exponential with a time constant of 70 msec. This agrees with the physiological results obtained with WCR. However, with the 100 Mohm shunt, the charging curve is more complex and is no longer fit by a single exponential. Thus, the presence of a shunt introduced into the soma can distort the charging curve and could lead to an erroneous estimate of the electrotonic length of dendritic processes.

In summary, our observations on sustained On ganglion cells in the mudpuppy retina suggest that these neurons have a high membrane resistance and dendrites which are electrotonically short with an average terminal length of about 0.2 Lambda. It is still possible that we are underestimating the magnitude of the membrane resistance. For example, if a 5 Gohm (our best seals) seal is obtained with a 5 Gohm cell, when the seal is ruptured, we could expect to observe an Rn of 2.5 Mohms and hence underestimate the membrane resistance by about half. Even so, we are probably not an order of magnitude away from the value of Rm as we probably have been when derived from studies with intracellular microelectrodes. One might argue that the WCR technique is itself flawed and that it artifactually closes ionic channels. We have recently addressed this issue and suggest that this is unlikely to be the case (Coleman and Miller, in press). We have also attempted to obtain gigaseals in the rabbit retina with mixed success. When gigaseals and WCR were obtained, we have noted the Rn of third-order neurons is much higher than those we have observed with intracellular micropipettes. This implies that the Rm of rabbit retinal cells may also be higher than present. Whether rabbit Rm values will reach those of the mudpuppy must await future studies.

ACKNOWLEDGEMENTS. We thank Peter Guthrie for providing us with a version of spice ("neuros") which we used in our transient simulations. This work was supported by NEI grants EY03014 and EY07376 to RFM.

REFERENCES

Arkin, M.S., and R.F. Miller (1988) Synaptic inputs and morphology of sustained On-ganglion cells in the mudpuppy retina. J.Neurophysiol. 60:1143-1159.

Coleman, P., and R.F. Miller (1988) Measurement of passive membrane parameters with whole cell recording of neurons in the intact amphibian retina. J.Neurophysiol. (In Press)

Hamill, O.P., A. Marty, B. Sakmann, and F.J. Sigworth (1981) Improved patch-clamp techniques for high-resolution currrent recording from cell and cell-free membrane patches. Pflug. Arch. 391:85-100.

Miller, R.F. and S.A. Bloomfield (1983) Electroanatomy of a unique amacrine cell in the rabbit retina. Proc. Natl. Acad. Sci. 80:3069-3073.

Rall, W. (1959) Branching dendritic trees and motoneuron membrane resistivity. Exp.Neurol. 1:491-527.

Rall, W. (1967) Distinguishing theoretical synaptic potentials computed for different soma-dendritic distributions of synaptic input. J.Neurophysiol. 30:1138-1168.

Rall, W. (1969) Time constants and electrotonic length of membrane cylinders and neurons. Biophys. J. 9:1483-1508.

NOT BY GANGLION CELLS ALONE: DIRECTIONAL SELECTIVITY IS WIDESPREAD IN IDENTIFIED CELLS OF THE TURTLE RETINA

R. D. DeVoe, P. L. Carras, M. H. Criswell[1], R. G. Guy[2]
School of Optometry
Indiana University
Bloomington, Indiana 47405
U.S.A.

INTRODUCTION

Directionally selective (DS) ganglion and cortical cells spike more in preferred directions of movement because the underlying postsynaptic potentials (psp's) are large and suprathreshold, compared to smaller (or subthreshold) psp's in antipreferred (or null) directions (Jensen and DeVoe, 1983). What makes the psp's smaller in the antipreferred directions are the earlier arrivals of ipsp's, with negative or positive sign, that add to, or shunt, resp., the later arriving epsp's (Marchiafava, 1979; Watanabe and Murakami, 1984; Douglas et al., 1988). From extracellular recordings, it appears that ACh is an excitatory transmitter and GABA the inhibitory transmitter onto DS ganglion and cortical cells (Sillitoe, 1977; Ariel and Adolph, 1985). It is generally presumed that in the retina the ACh and GABA are released from amacrine cells.

Earlier however, Barlow and Levick, in their seminal 1965 paper on rabbit DS ganglion cells, proposed that horizontal (not amacrine) cells might carry lateral, asymmetric timing information. Subsequently, an heroic EM serial reconstruction of rabbit OPL circuitry was undertaken by Sjöstrand (1974) to find a synaptic basis for directional selectivity in the OPL. The outputs of such circuitry would be DS bipolar cells. However, initial attempts failed to find DS responses in intracellular recordings from preganglionic cells (Werblin, 1970; Marchiafava, 1979), although there were directional changes in transient-OFF responses of an unidentified cell in <u>Necturus</u> (Norton et al, 1970). At best, asymmetric fields of preganglionic cells seem to result only in orientation sensitivities (Norman et al., 1979; Naka 1980; Bloomfield and Miller, 1982).

---

1. Northeastern State University, Tahlequah, Oklahoma 74464, USA
2. Royal Melbourne Institute of Technology, Victoria, Australia

NATO ASI Series, Vol. H31
Neurobiology of the Inner Retina
Edited by R. Weiler and N.N. Osborne
© Springer-Verlag Berlin Heidelberg 1989

The work to be described here began with full confidence that directional selectivity was exclusive to ganglion cells. Our first aim was to identify DS ganglion cells; the second was to study the input-output dynamics of these cells. For a preparation, the isolated eyecup of the turtle Pseudemys scripta elegans was chosen. This was because there was an increased chance of penetrations in a retina with a high percentage of DS ganglion cells, and because much was already known about turtle retinal physiology. The first findings were that the dendritic structures of ganglion cells provided no basis for their directional selectivities (Jensen and DeVoe, 1983; cf. Amthor et al., 1984). That left the timings of the synaptic inputs, which would have to be characterized by recordings from amacrine and bipolar cells. Drifting gratings, imaged on the dorsal retina within a stationary field, were chosen as stimuli in order to elicit sustained responses to movement. With these stimuli, we were quite unprepared to find that about the same percentage of bipolar and amacrine cells as ganglion cells were themselves directionally selective in the turtle retina. Of ganglion cells penetrated (N = 242), 33% were DS, as were 37% of amacrine cells (N = 117) and 42% of bipolar cells (N = 91). In addition, 9 luminosity horizontal cells and 1 accessory cone have been found to be DS so far. The following summarizes these findings.

DIRECTIONALLY SELECTIVE GANGLION CELLS

An example of a (bistratified, ON-OFF) DS ganglion cell's response to a moving grating is shown in Fig. 1 at the left. The obvious point is that spikes occur for only one direction of motion (temporal), where they arise off of larger periodic oscillations of the membrane potential. This cell is thus directionally selective and has a true preferred-null axis. If spikes are lost (as has happened in other cells with injury), the asymmetric amplitudes of periodicities nonetheless persist. Clearly then, the ganglion cell's psp's are still "DS," since the synaptic mechanisms underlying the psp's are unlikely to depend upon the integrity of membrane spiking. By the same token, the directionally dependent amplitudes of periodic responses in nonspiking, preganglionic cells that will be shown later will also be termed directionally selective. Inspection of the responses in Fig. 1 reveals that there is more than just reduction of

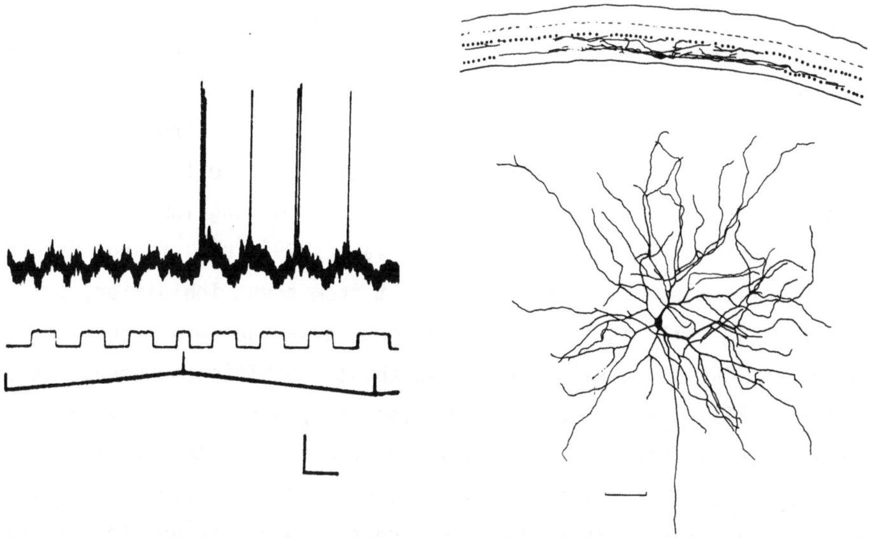

Figure 1 DS bistratified ganglion cell. Right: drawing from Lucifer filled cell; scale = 100 µm. Fine dotted lines in radial view are IPL limits. Nasal is to the right at 90° (see Fig 2). Left: DS responses to 120°/300° movements of a 540 µm grating, 306 µm field, 0.74 Hz contrast frequency. Lowest trace: movement monitor; middle trace: grating stripe monitor. Calibrations = 10 mV, 1 sec. Data of PLC.

periodic amplitudes in the null direction of movement; there is frequency doubling as well (relative to the photocell trace of the stripes of the moving grating). This would be expected from an ON-OFF cell (Jensen and DeVoe, 1983), although frequency doubling often appears reduced or absent in the preferred direction of movement, as shown here (Criswell, 1987). The number of spatial periods per second of a moving grating is the contrast frequency. Thus for Fig. 1, it is the second harmonic of the contrast frequency which is larger (up to 3 times) in this and other null directions. Conversely, the first harmonic (fundamental) was the larger (up to 19 times) in all preferred directions. The second harmonic, even when the larger, was never large enough to elicit spikes in this cell. It seems clear that the changes in harmonic amplitudes with directions of movement must be related to relative arrival times and interactions of ipsp's and epsp's. At its simplest for the first harmonic, it might be supposed that both ipsp's and epsp's a) were sinusoidal, b) varied in relative phases (latencies) as functions of directions, and so c) cancelled or summed in null or preferred directions, resp.. Alternatively, ipsp's might shunt the epsp's and thereby produce higher harmonics too (see below).

This brief description of DS ganglion cell psp's elicited by moving gratings is meant primarily as an introduction to the properties of DS responses in preganglionic cells. Even in DS ganglion cells the analysis of these psp's is a formidable task. We have recorded from and identified at least 12 different DS ganglion cell types in the turtle retina (Jensen and DeVoe, 1983; Criswell, 1987). All but OFF DS ganglion cells have been bi or multi-stratified. Marchiafava (1979) has found unidentified ganglion cells whose DS properties result from shunt inhibition, but it is not known if all turtle DS ganglion cells use this mechanism. In the frog, ON and OFF DS ganglion cells use shunt inhibition, whereas ON-OFF DS ganglion cells have algebraic summation of ipsp's and epsp's (Watanabe and Murakami, 1984). Shunt inhibition would itself generate complex harmonics in psp's, via multiplication of inputs (Torre and Poggio, 1978). It was to characterize such inputs that recordings were begun in bipolar and amacrine cells.

DIRECTIONALLY SELECTIVE BIPOLAR CELLS

In 1983 one of us (MHC) recorded DS psp's in a hyperpolarizing INL cell, and a bipolar cell was found to have been injected with Lucifer yellow. Since then over 38 DS bipolar cells have been penetrated, and 27 have been stained. All tested have been DS only for red lights. Most have been OFF (hyperpolarizing) bipolars; fewer ON (depolarizing) bipolars have been found, in agreement with Weiler (1981). DS OFF bipolars are of two types, quiet and noisy (Criswell, 1987). Quiet DS bipolars could have large, depolarizing off transients, whereas noisy DS bipolars had transient depolarizing potentials (TDP'S) in the dark that were usually reduced during steady illumination. TDP's seemed to be electrically excitable.

Fig. 2 illustrates the results from one DS noisy OFF bipolar. To inferior-superior, back and forth movement, PSP's were larger for 0^{0} (inferior) movement (top left). This difference in amplitudes is the usual finding and is comparable to that for the DS ganglion cell in Fig. 1. This bipolar also exhibited true null psp's, something occasionally found in preganglionic cells. There are no periodic components in the response to the initial, 180^{0} (superior) direction of movement at the left middle of Fig 2, although there are in the opposite direction and during flicker of the same frequency (Fig 2, bottom left). The grating stimuli consisted of 270 µm wide dark and light bars moving within a stationary, 53 µm field. The plateau null response therefore represented a complete failure to respond to periodic darkenings of the visual field. Plateau null responses have also been found in two other DS OFF bipolars. These null responses are not necessarily indicative of inhibition, since inhibition of OFF bipolars generally results in less hyperpolarization (Weiler, 1981). Null responses could reflect synaptic activity in the OPL (see below), or less conceivably might result from shunting by an ion with a hyperpolarizing equilibrium potential near the peak light response.

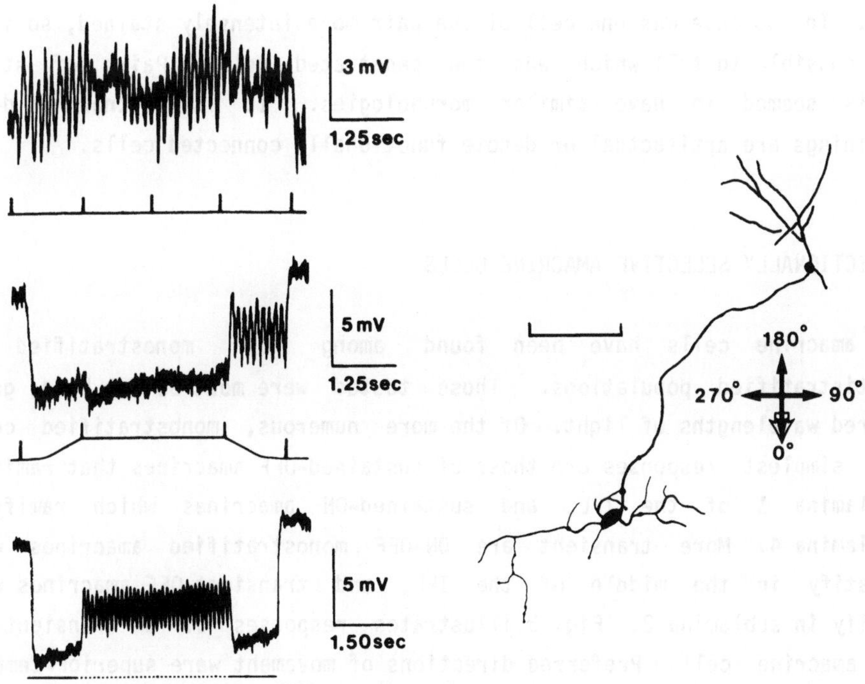

Figure 2 Noisy OFF DS bipolar with inferior (0°) preferred direction. Right scale bar = 50 µm. Responses at left (top to bottom) to 50 µm grating, 53 µm field, 8Hz, 0°/180° movements; to 540 µm grating, 53 µm field, 9 Hz, 180°/0° movement with pause, shutter open between lowest tick marks on movement monitor; to 9.09 Hz flicker, 106 µm field, shutter trace at bottom. From Criswell (1987).

Just as for DS ganglion cells, so too for DS bipolar cells can there be variations in harmonic components with directions of movement. The simplest, seen in quiet OFF cells, is an amplitude variation of nearly sinusoidal responses. More complex are frequency doublings, either in preferred or anti-preferred directions, often accompanied by TDP's in noisy OFF cells (Criswell, 1987). For most OFF DS bipolars, there were nearly constant, baseline hyperpolarizations from which periodic off responses departed, and the DS responses were manifested as diminutions of off responses (with or without TDP's).

Almost all stained DS bipolars, OFF or ON, were bi- or multistratified in the IPL. A few cells had long, lateral processes in the IPL, up to 2 mm for one quiet OFF cell. Additionally, 5 DS OFF cells were stained along with a second, adjacent bipolar cell. Sometimes these cells' axons coursed so closely together that they could scarcely be distinguished as two. In no case was one cell of the pair more intensely stained, so it was not possible to tell which was the penetrated cell. Pairs of stained cells seemed to have similar morphologies. It is not known if double stainings are artifactual or denote functionally connected cells.

DIRECTIONALLY SELECTIVE AMACRINE CELLS

DS amacrine cells have been found among both monostratified and multistratified populations. Those tested were most DS for blue, green, or red wavelengths of light. Of the more numerous, monostratified cells, the simplest responses are those of sustained-OFF amacrines that ramify in sublamina 1 of the IPL, and sustained-ON amacrines which ramify in sublamina 4. More transient are ON-OFF monostratified amacrines which stratify in the middle of the IPL, and transient-OFF amacrines which ramify in sublamina 2. Fig. 3 illustrates responses of a transient-OFF, DS amacrine cell. Preferred directions of movement were superior-temporal on the retina. In these preferred directions, the larger responses appear to be due to TDP's superimposed on the psp's which were also elicited in the nonpreferred directions. Additionally, this cell was noisy in the dark but became less so before and after motion in Fig. 3 (while the shutter was still open). These properties are very much like those of noisy, OFF DS bipolar cells, some of which have terminals in sublamina 2 (Criswell, 1987).

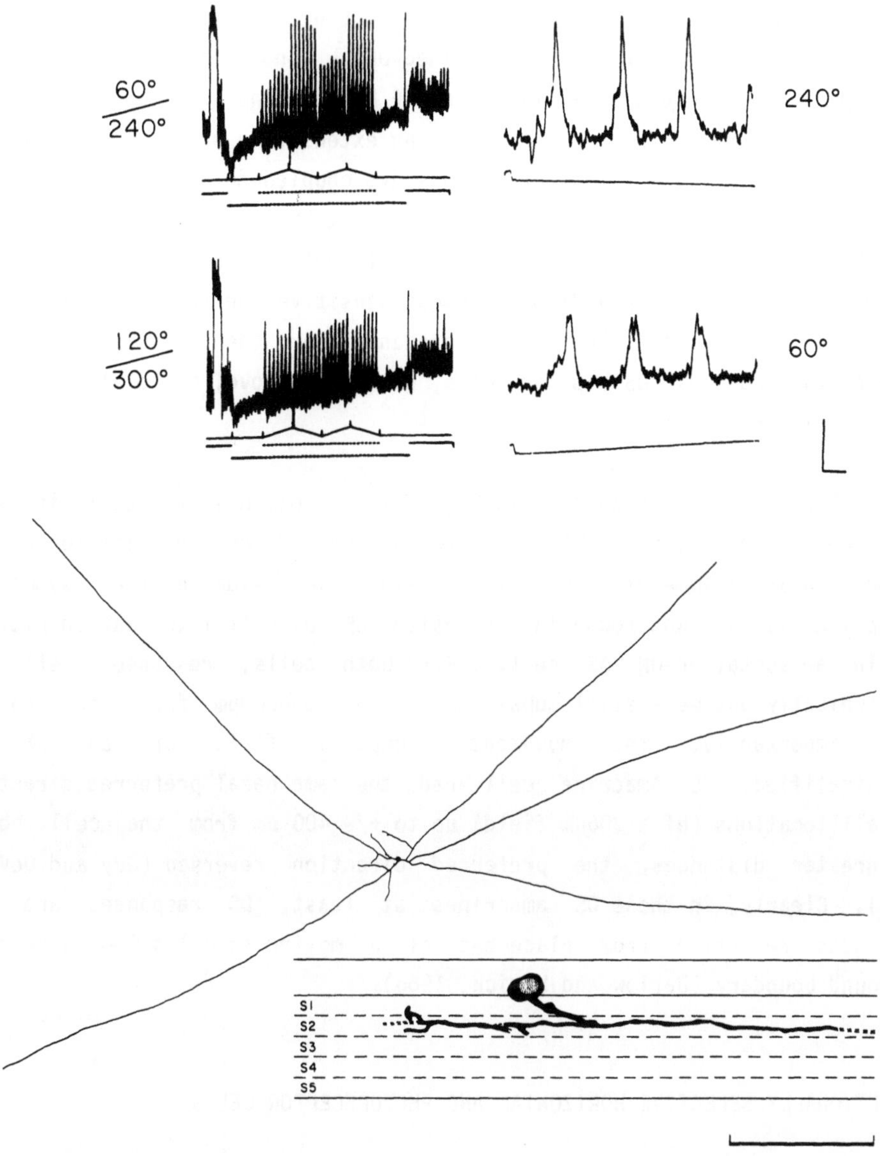

Figure 3 Monostratified DS transient-OFF amacrine cell. Bottom scale bar = 500 μm for flat mount, 50 μm for radial view. Upper left: DS and nonDS responses along two axes, 540 μm grating, 306 μm field, 5.6 Hz contrast frequency. Bottom traces are shutter and stripe monitors. Upper right: expanded views of responses in preferred (240°) and antipreferred (60°) directions. Calibrations = 5 mV, 1 sec (left) or 50 msec (right). Data of RGG.

The DS amacrine cell of Fig 3 had the most extensive dendritic spread of any found, 3.3mm. Another, sustained-OFF, monostratified DS amacrine extended 1.6mm, and a bistratified, ON-OFF DS amacrine extended 1.4mm (Guy and DeVoe, 1988). This last cell had an extensively arborized, asymmetric region, 300 μm in diameter, where it was dye coupled to a ring of 7 other cells of the same type (probably A15 of Kolb (1982)). This raises the possibility that DS amacrines might be functionally connected to like DS amacrines, perhaps providing movement-sensitive networks in the IPL. Results of this sort lend some credance to the possibility that double-stainings of DS bipolar cells, described above, could also indicate functional connections.

With cells as extensive as that in Fig 3, it is possible to test if such amacrines are artifactually DS because of off-center positioning of stimuli in an asymmetric receptive field. No evidence for asymmetric receptive fields was found in a transient-OFF DS cell like that in Fig. 3, or in a sustained-ON DS cell. For both cells, responses fell off monotonically but were still substantial +/- 600-800μm from the center. More importantly, the movement receptive field of an ON-OFF, monostratified, DS amacrine cell had the same nasal preferred direction for all locations (of a 306μm field) up to +/- 400 μm from the cell body. At greater distances, the preferred direction reversed (Guy and DeVoe, 1988). Clearly, in these DS amacrines at least, DS responses are not artifacts resulting from placement of a moving stimulus upon a center-surround boundary (Barlow and Levick, 1965).

DIRECTIONALLY SELECTIVE HORIZONTAL AND PHOTORECEPTOR CELLS

If some photoreceptor (probably cone) terminals in turtles were directionally selective, it would be easy to see how some bipolar cells might be DS too. Since photocurrents of a single photoreceptor can not be DS, there must be synaptic inputs onto terminals to make them DS. DS horizontal cells could provide such inputs. A few DS turtle horizontal cells have now been found (DeVoe et al, 1985; Adolph, 1988), and one such H1AT cell is shown in Fig. 4. For inferior/superior vertical movements, and for temporal horizontal movements, the periodic responses are about the same size, but are only about 55% the size of the responses to nasal

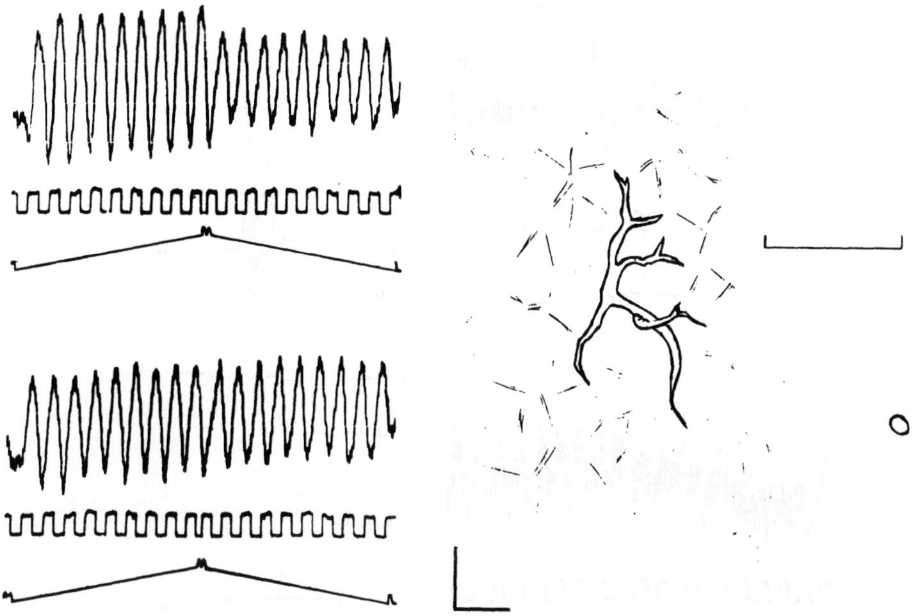

Figure 4 DS H1AT horizontal cell drawn at right, with cell body. Scale = 50 µm. Left: DS response to 90°/270° (top) and 0°/180° (bottom) movements, 906 µm grating, 891 µm field, 5.8 Hz contrast frequency. Calibrations = 2 mV, 0.5 sec. Data of PLC.

horizontal movements. All told, we have recorded from 9 DS horizontal cells (out of a sample of 118). Four were H1AT cells and one was an H1CB cell (the rest were not identified). Two H1AT cells and the H1CB cell were dye-coupled to like cells. Adolph (1988) recorded DS responses from 3 (presumptive but unstained H1CB) horizontal cells (out of a sample of 36), using similar stimuli to those for Fig. 4. Clearly, this population (about 8%) is far smaller than for DS ganglion, bipolar and amacrine cells (35%-42%).

Still smaller is the number of DS photoreceptors (N = 1, and maybe 2), one of which is shown in Fig. 5. This accessory red cone had an superior/ inferior, preferred/antipreferred movement axis. With a grating whose dark and light bars could each just fill the stationary field (Fig. 5, top left), movement in a superior direction gave the larger responses. When a finer grating moved within a larger field (Fig. 5, bottom left), so that there was always illumination somewhere in the field, the cone had a plateau null response to movement in an inferior direction. That is, as for the null responses in Fig. 2, the cone failed to respond to passages

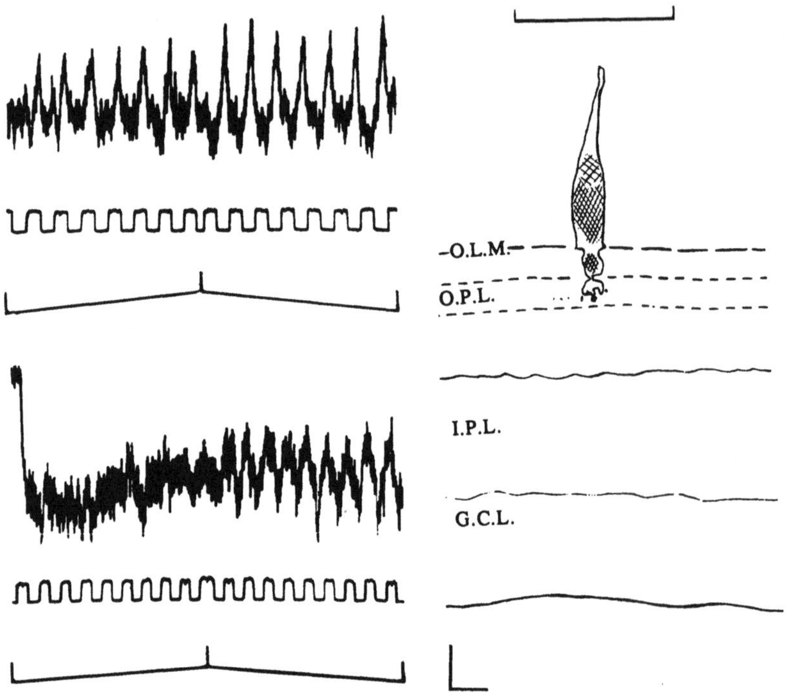

Figure 5 DS accessory cone. Scale bar at right = 50 μm. Left: $0^o/180^o$ movements; Top: 299 μm grating, 146 μm field, 1.47 Hz contrast frequency. Bottom: 171 μm grating, 173 μm field, 1.9 Hz contrast frequency, start of motion at left. Calibrations = 1mV, 1 sec. Data of PLC.

of dark bars. It did hyperpolarize to the onset of the first light bar (Fig. 5, bottom left), but thereafter its response was sustained until superior movement began. The accessory red cone receives inhibitory inputs from H1CB cells (Leeper, 1978), some of which are DS. It generally required larger fields to elicit DS responses from horizontal cells than from this cone, however, which makes problematic the role of DS horizontal cells in the directional selectivity of the cone.

SPECULATIVE CONCLUSIONS

In our hands, moving gratings (also bars and slits) which elicited DS responses from 33% of turtle ganglion cells also elicited DS responses from about the same percentage of bipolar and amacrine cells. We have given evidence that DS responses from amacrine cells are not artifacts.

Also, the percentages of DS bipolars and amacrines seem too high for these cells to be doing nothing of interest. We offer the following speculations, based on all our findings to date.

1. Color-tested bipolars were DS only for red (and white) lights. If rods and red cones share common neural pathways in turtles, and if movement detection also occurs at scotopic illuminations in turtles (carnivores do hunt at night), maybe directional selectivity is encoded for both rods and red cones by OPL cells and is carried by DS bipolar cells.
2. Directional selectivity is not color blind, however; there are ganglion and amacrine cells tested which had the best DS reponses at blue or green (or red) wavelengths. If DS bipolars are only red sensitive, then nonDS bipolars would have to carry blue and green excitation to additional DS synapses in the IPL, where conventional wisdom has long placed such synapses. That the neurochemical machinery for DS transmission exists in both the IPL and OPL (Criswell and Brandon, 1987), can only complicate pharmacological tests of which layer has DS synapses.
3. We find that dendrites of ON, OFF, and ON-OFF ganglion and amacrine cells are stratified as expected in the turtle IPL, but terminals of DS bipolars may be multistratified or terminate in the "wrong" sublaminae. Presumably therefore there are sign-inverting synapses where, for example, an OFF DS bipolar synapses only in the "wrong", ON-sublaminae of the IPL. If so, then might not multistratified DS bipolars have some sign-inverting, and some sign-conserving, synapses onto IPL dendrites, depending upon the sublaminae?
4. Some DS amacrines and bipolars had membrane oscillations (at the stimulus contrast frequencies of 5-8Hz) during illumination following grating movements along the preferred-antipreferred directions, but not following flicker, and not during steady illumination (alone or before movement). We do not know the origins of such oscillations. They could be intrinsic to some DS bipolars that drive DS amacrines. Alternatively, the post-movement oscillations could be reverberations in negative feedback networks (via sign-inverting synapses) between DS bipolars and amacrines.
5. Finally, it is attractive to conclude that some DS ganglion cells surely receive inputs from some preganglionic DS cells. Like most of the above speculations, the proof, if any, lies in the future.

Supported by NIH Grant EY 05163 to R.D.D.

REFERENCES

Amthor FR, Oyster CW, Takahashi ES (1984) Morphology of ON-OFF direction-selective ganglion cells in the rabbit retina. Brain Res 298:187-190
Adolph AR (1988) Center-surround, orientation and directional properties of turtle retinal horizontal cells. Biol Cybern 58:373-385
Ariel M, Adolph AR (1985) Neurotransmitter inputs to directionally sensitive turtle retinal ganglion cells. J Neurophysiol 54:1123-1143
Barlow HB, Levick WR (1965) The mechanism of directionally selective units in rabbit's retina. J Physiol (Lond) 178:477-504
Bloomfield SA, Miller RF (1982) A physiological and morphological study of the horizontal cell types of the rabbit retina. J Comp Neurol 208:288-303
Criswell MH (1987) Cellular mechanisms of movement detection and directionality in the turtle retina. Doct Dissertation, Indiana University, Bloomington
Criswell MH, Brandon CJ (1987) Immunocytochemical evidence that turtle cones are cholinergic. Invest Ophthalmol Vis Sci Suppl 28:278
DeVoe RD, Guy RG, Criswell MH (1985) Directionally selective cells of the inner nuclear layer of the turtle retina. Invest Ophthalmol Vis Sci Suppl 26:311
Douglas JR, Martin KAC, Whitteridge D (1988) Selective responses of visual cortical cells do not depend on shunting inhibition. Nature 332:642-644
Guy RG, DeVoe RD (in preparation) Directional response of turtle retinal amacrine cells to moving stimuli.
Jensen RJ, DeVoe RD (1983) Comparisons of directionally selective with other ganglion cells of the turtle retina: intracellular recording and staining. J Comp Neurol 217:271-287
Kolb H (1982) The morphology of the bipolar cells, amacrine cells and ganglion cells in the retina of the turtle *Pseudemys scripta elegans*. Phil Trans R Soc Lond B 298:355-393
Leeper HF (1978) Horizontal cells of the turtle retina. II. Analysis of interconnections between photoreceptor cells and horizontal cells by light microscopy. J Comp Neurol 182:795-809
Marchiafava PL (1979) The response of retinal ganglion cells to stationary and moving visual stimuli. Vis Res 19:1203-1211
Naka KI (1980) A class of catfish amacrine cells responds preferentially to objects which move vertically. Vis Res 20:961-965
Norman RA, Kolb H, Hanani M, Pasino E, Holub H (1979) Orientation of horizontal cell axon terminals in the streak of the turtle retina. Nature 250:60-62
Norton AL, Spekreijse H, Wagner HG, Wolbarsht ML (1970) Responses to directional stimuli in retinal preganglionic units. J Physiol (Lond) 206:93-107
Sillitoe AM (1977) Inhibitory processes underlying the directional specificity of simple, complex and hypercomplex cells in the cat's visual cortex. J Physiol (Lond) 271:699-720
Sjöstrand FS (1974) A search for the circuitry of directional selectivity and neural adaptation through three-dimensional analysis of the outer plexiform layer of the rabbit retina. J Ultrastruct Res 49:60-156
Torre V, Poggio T (1978) A synaptic mechanism possibly underlying directional selectivity to motion. Proc R Soc Lond B202:409-416
Watanabe SI, Murakami M (1984) Synaptic mechanisms of directional selectivity in ganglion cells of frog retina as revealed by intracellular recordings. Jpn J Physiol 34:497-511
Werblin FS (1970) Responses of retinal cells to moving spots: intracellular recording in *Necturus maculosus*. J Neurophysiol 33:342-350

PHARMACOLOGICAL MODULATION OF THE DARK ADAPTED CAT RETINA

H. Wässle, F. Müller, T. Voigt and M.H. Chun
Department of Neuroanatomy
Max-Planck-Institute for Brain Research
Deutschordenstr. 46
6000 Frankfurt, 71, F.R.G.

In the photopic pathway of the mammalian retina, cones feed into cone bipolar cells, which, in turn, synapse onto ganglion cells. Cone bipolar cells comprise two basic physiological types: ON-bipolar cells that are depolarized and OFF-bipolar cells that are hyperpolarized by a light stimulus (Fig. 1). In the scotopic pathway rods are connected to rod bipolar cells, but instead of these synapsing directly onto ganglion cells (Fig. 1), a distinctive type of narrow-field amacrine cell, the so called AII amacrine cell is interposed (Famiglietti & Kolb, 1975; Kolb, 1979; Kolb & Famiglietti, 1974; Sterling, 1983). Light responses of rod bipolar cells are hyperpolarizing in cat (Kolb & Nelson, 1983; Nelson & Kolb, 1983), and depolarizing in rabbit (Dacheux & Raviola, 1986). AII amacrine cells are also depolarized by light (Dacheux & Raviola, 1986; Nelson, 1982), and in the cat they use the inhibitory transmitter glycine (Pourcho & Goebel, 1985; 1987). They make conventional chemical synapses with OFF-cone bipolar and OFF-ganglion cells (Famiglietti & Kolb, 1975; Sterling, 1983) and contact ON-cone bipolar cells via large gap junctions. In this way (see Fig. 1) they could produce signals of opposite polarity in ON- and OFF-ganglion cells. However it should be remembered that the wiring diagram of the rod pathway is mainly based on ultrastructural observations and the physiological proof is still lacking.

It seems surprising that - given their structural similarity - rod bipolar cells gave depolarizing light

NATO ASI Series, Vol. H31
Neurobiology of the Inner Retina
Edited by R. Weiler and N.N. Osborne
© Springer-Verlag Berlin Heidelberg 1989

responses in rabbit and hyperpolarizing responses in cat. Since intracellular responses of neurones as small as rod bipolar cells are difficult to measure we tried a different approach to settle this question.

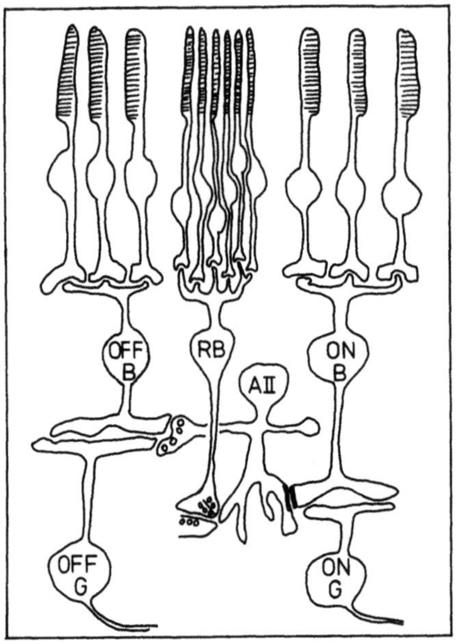

Figure 1: The rod pathway of the mammalian retina. In the centre six rods converge onto a rod bipolar cell (RB), which synapses in the inner plexiform layer with an AII amacrine cell. This bistratified cell makes gap junctions with ON-cone bipolar cells (ON-B), which have excitatory contacts with ON-ganglion cells (ON-G). Conventional inhibitory synapses between AII and OFF-cone bipolar cells (OFF-B) and OFF-ganglion cells (OFF-G) feed the information into the OFF-channel.

In all retinae so far studied, amino-4-phosphonobutyrate (APB) suppresses depolarizing (ON) bipolar cells and blocks their light responses, whereas the light responses of hyperpolarizing (OFF) bipolar cells are hardly affected (Slaughter & Miller, 1985). The present study tested the action of APB application on the scotopic pathway (Fig. 1) in cat. This was based on the assumption that membrane receptors and transmitters in the rod circuitry of the outer plexiform layer have similar properties to those of cones and cone bipolar cells. If the rod bipolar cell were a depolarizing bipolar cell, its light response should be blocked by APB, and neither ON- nor OFF-ganglion cells should respond to light stimuli in the dark adapted retina. The circuitry of Fig. 1 makes a further prediction: since the AII amacrine cell is glycinergic, the glycine antagonist strychnine should block the transmission from AII amacrine cells to OFF-bipolar and OFF-ganglion cells.

The current experiments involved extracellular recordings from ganglion cells in the in vivo cat eye. The animals were anaesthetized and artificially respirated. APB and strychnine were applied iontophoretically from multibarrel electrodes. Individual units were first tested in the light-adapted state. Afterwards the retina was dark adapted and the same units and their drug responses were measured again (for details see Müller et al., 1988).

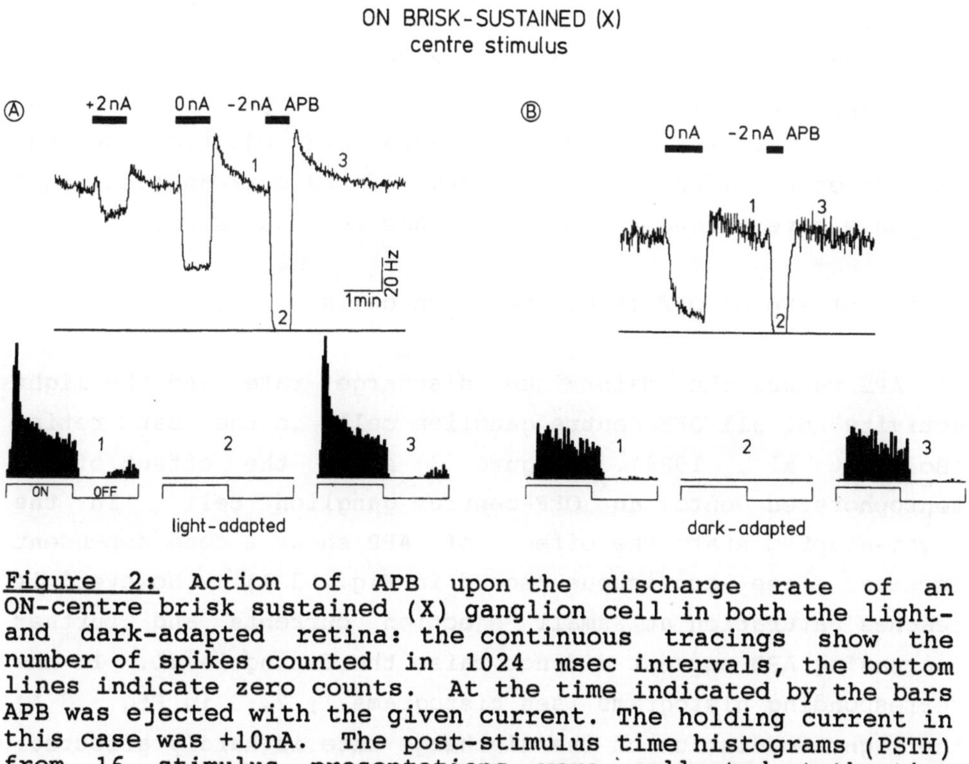

Figure 2: Action of APB upon the discharge rate of an ON-centre brisk sustained (X) ganglion cell in both the light- and dark-adapted retina: the continuous tracings show the number of spikes counted in 1024 msec intervals, the bottom lines indicate zero counts. At the time indicated by the bars APB was ejected with the given current. The holding current in this case was +10nA. The post-stimulus time histograms (PSTH) from 16 stimulus presentations were collected at the time periods indicated by the numbers in the continuous records. The trace under PSTH marks the 512 msec periods when the light was on and off, respectively, and its height indicates 40 spikes per second. A, light-adapted: APB decreases the activity in a dose-dependent manner. After the end of APB application, a remarkable overshoot in activity can be observed from which the cell slowly returns to the predrug level. B, the same cell after 30 min dark-adaptation. The PSTH show a very tonic activity with a latency longer than under mesopic conditions. The continuous tracing shows no rebound after termination of APB application (stimulus diameter 0.8°; A, stimulus increment 4 cd/m² on a background of 1 cd/m²; B, stimulus $4x10^{-3}$ cd/m², no background; eccentricity 5°).

Effects of APB on ON-centre ganglion cells

APB suppresses the maintained as well as the light-driven activity of all ON-centre ganglion cells in the cat retina (Bolz et al., 1984). The effect of APB is dose dependent as illustrated in the continuous record of Fig. 2 A. Even small amounts of APB reduce the firing rate of this cell, and increasing APB ejection completely blocks the light responses (see histograms 1, 2, 3 in Fig. 2 A) and discharge rate (see continous record in Fig. 2 A). When the cell was dark adapted for more than 30 min, the APB action (Fig. 2 B) was similar to the light adapted state: in a dose-dependent fashion the cell was suppressed by APB and both the discharge rate and the light response was blocked by -2nA APB ejection current (Müller et al. 1988). The histograms of the dark-adapted light response have a longer latency and are more sustained.

Effect of APB on OFF-centre ganglion cells

APB raises the maintained discharge rate and the light activity of all OFF-centre ganglion cells in the cat retina (Bolz et al., 1984). Figure 3 shows the effect of APB iontophoresed onto an OFF-centre ganglion cell. In the light-adapted state the effect of APB shows a dose dependent increase (see continuous record in Fig. 3 A). However it reaches saturation at small ejection currents and further increasing APB current did not raise the firing level. In the corresponding histograms (see histograms 1, 2, 3 in Fig. 3 A) the light modulation of the discharge rate is hardly affected. This shows that APB cannot block the light responses of OFF-centre ganglion cells during photopic stimulation. However APB raises the discharge rate of these cells. This might be explained by a network property, where OFF-ganglion cells receive inhibitory input from the ON-channel (push-pull mechanism: Levine & Sheffner, 1977; Sterling, 1983; Wässle et al., 1986) or by direct, non specific actions of the racemic mixture of APB in the OFF channel (Arkin & Miller, 1987).

When the cell is dark-adapted (Fig. 3 B) the results are quite different. The intensity of the light spot stimulating the dark adapted cell was adjusted to yield the same mean firing level as in the light-adapted case (compare continuous

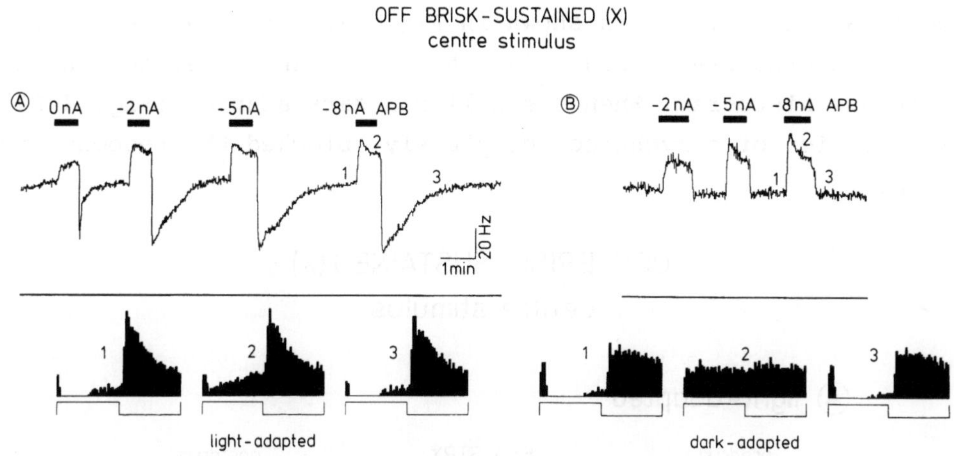

Figure 3: Influence of APB on the activity of an OFF-centre brisk sustained(X) cell. A, light-adapted. APB increases the activity of the cell in a dose-dependent way, with a biphasic action mirror reversed to the action upon ON-ganglion cells. Saturation is reached with low currents. The light-response remains clearly moulated (PSTH 2). B, dark-adapted for 30 min. The APB action is again dose-dependent, but no rebound can be observed. APB blocks the light response of the cell completely (PSTH 2) (centre spot 1°; A, stimulus increment 3 cd/m² on a background of 3 cd/m²; B, stimulus 1.5×10^{-2}, no background; eccentricity 8°; holding current +10nA).

records in Figs. 3 A & B). APB application increases the firing rate, an effect that saturates at -8nA APB ejection current. When -8nA APB are applied, the light modulation of the dark adapted histograms (see histograms 1, 2, 3 in Fig. 3B) is blocked and the cell fires with a rather high maintained discharge rate. Hence in the dark-adapted retina, APB blocks the light responses of both ON-centre (Fig. 2B) and OFF-centre ganglion cells completely.

Effect of strychnine on ganglion cell responses to light stimulation

In ON-centre cells during both photopic and scotopic light adaptation, application of strychnine increased the discharge rate without dramatic effects on the light responses. The effect of strychnine application onto an OFF-centre cell (same cell as in Fig. 3) is shown in Fig. 4. Application in the light adapted state (Fig. 4A) had no dramatic effect on the response histogram. When the cell was dark adapted (Fig. 4 B), application of strychnine completely blocked the response to light.

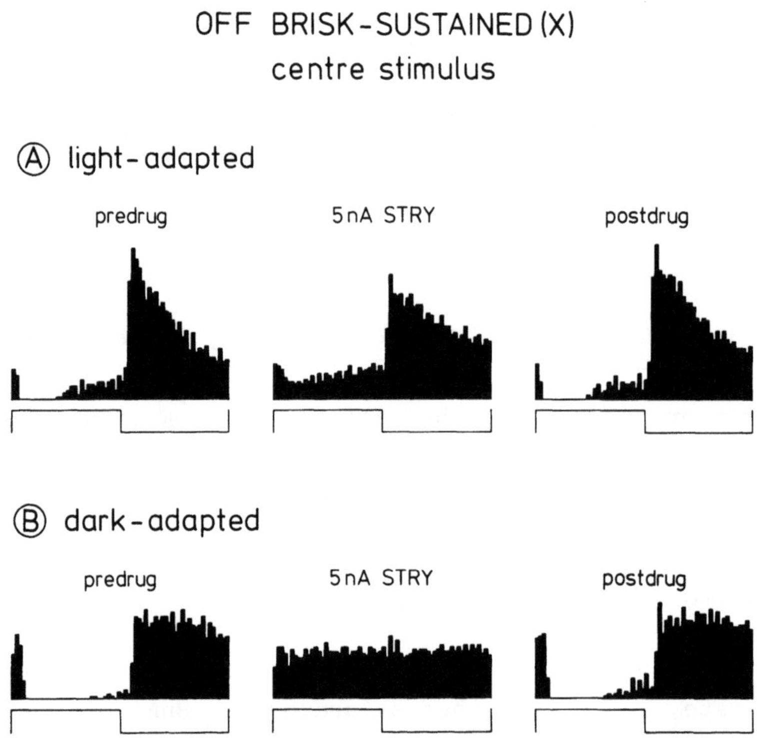

Figure 4: Action of strychnine (STRY) on the activity of an OFF-centre brisk sustained(X) cell (same cell as in fig. 3). A, in the light-adapted state the light response of the cell is not much affected by STRY. B, in the dark-adapted state the same amount of STRY completely blocks the light response (stimulus diameter 1°, A, stimulus increment 3 cd/m² on a background of 3 cd/m²; B, stimulus 1.5×10^{-2} cd/m², no background; holding current -5nA; dark adaptation more than 30 min).

The physiological experiments presented so far are in agreement with the anatomical model presented in Fig. 1: If APB blocks the signal transfer between the rod and the rod bipolar cell, both the light responses of ON- and OFF-centre ganglion cells would be blocked. Strychnine, a glycine antagonist would block the chemical synapse between the AII amacrine cell and the OFF-channel, hence light responses in dark-adapted OFF-ganglion cells would be abolished. Strychnine should not block light responses in ON-ganglion cells, which receive electrical synapses from AII amacrine cells. The ABP results of the present paper together with the known action of APB as a specific blocker of ON (depolarizing) bipolar cells (Shiells et al., 1981; Slaughter & Miller, 1981, 1985) indicate that in the cat rod-bipolar cells give depolarizing light responses like those of the rabbit. However our experiment does not prove this point, only intracellular recordings from rod bipolar cells can give a definite answer.

Modulation of the rod bipolar cell by GABA

Rod bipolar cells invaginate rod spherules in the outer plexiform layer, where they are postsynaptic to the ribbon synapses of the rods with the horizontal cell axon terminals as lateral elements (Kolb, 1974). In the non-mammalian retina several GABA-ergic markers have been described in horizontal cells (Yazulla, 1986). There is less evidence that GABA is a horizontal cell transmitter in the mammalian retina. Using antibodies directed against the GABA-GA-BSA-complex (GABA coupled to bovine serum albumin with glutaraldehyde) we have shown, at both the light-microscopic and ultrastructural level, that cat horizontal cells express GABA-like immunoreactivity (Wässle & Chun, 1988; Chun & Wässle, 1988). Figure 5 A shows a vertical section of the cat retina stained for GABA-like immunoreactivity. Both the outer (OPL) and the inner plexiform (IPL) layers are strongly labelled. Rod spherules are staggered in several rows at the outer nuclear (ONL)/outer plexiform layer (OPL) border and are innervated by the B-type horizontal cell axon terminals (Kolb, 1974).

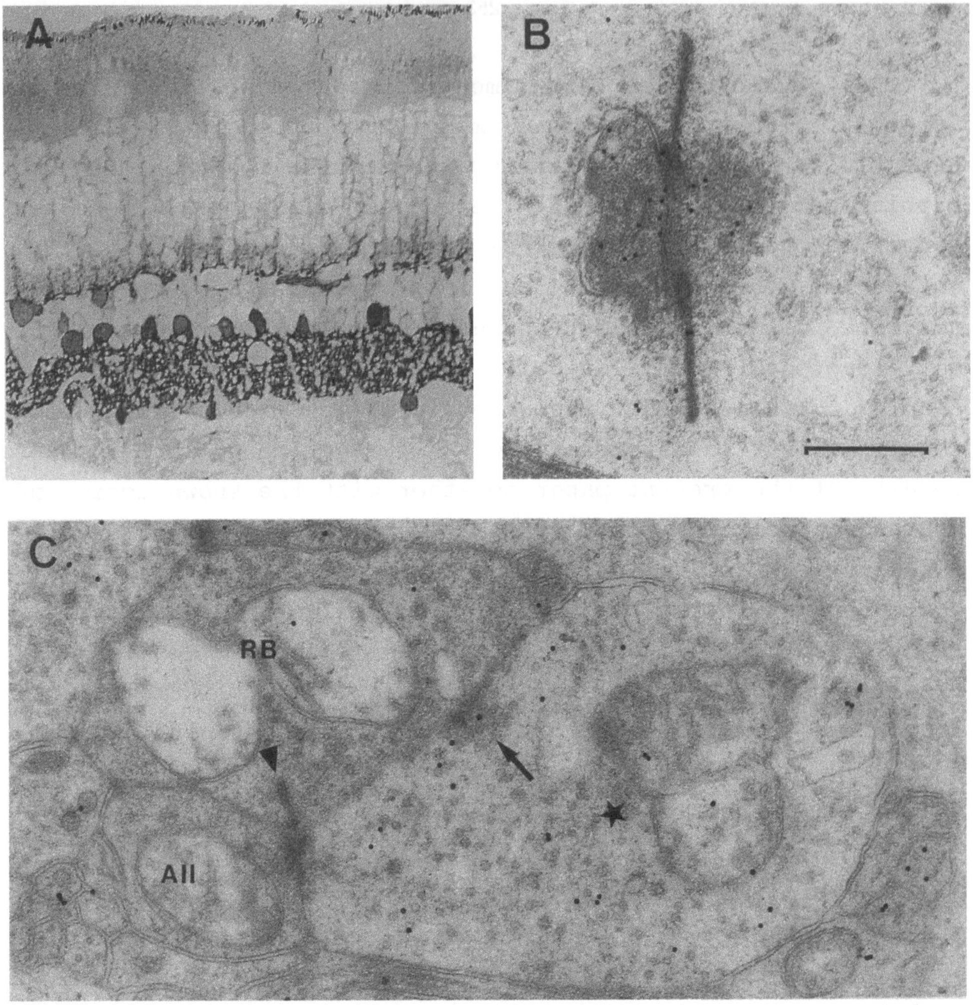

Figure 5: GABA-like immunoreactivity in the cat retina: A, light micrograph of a vertical section stained for GABA-like immunoreactivity using the PAP-method. Horizontal cells, amacrine cells and displaced amacrine cells are labelled. B, electron micrograph of a horizontal section through a rod spherule. On both sides of the vertically oriented ribbon, labelled horizontal cell processes are inserted into the base of the rod spherule. C, electronmicrograph of a rod bipolar cell synapse (dyad) in stratum 5 of the IPL. The rod bipolar terminal (RB) contains a synaptic ribbon (arrowhead). One postsynaptic process is an unlabelled AII amacrine cell (AII). The other postsynaptic process (asterisk) is labelled with gold particles and makes a reciprocal synapse (arrow) (Scale bar 50µm in A and 0.5µm in B, C).

In Fig. 5 A these axon-terminal processes are labelled and can be observed entering the ONL between unstained rod nuclei up to a depth of 10-15 µm. An ultrathin section through a rod spherule stained for GABA-like immunoreactivity is shown in Fig. 5 B. The protein A gold method was used and little gold particles can be detected at the horizontal cell terminals within the ribbon synapse. The demonstration with both the light (Fig. 5 A) and electron microscope (Fig. 5 B) of GABA in the horizontal cell terminals, suggests that the signal transfer between the rod and the rod bipolar cell of the cat retina is modulated by GABA.

Rod bipolar cells have their output synapses in the inner half of the IPL (Kolb & Famiglietti 1974; Famiglietti and Kolb, 1975). As can be seen in Fig. 5 A, strong GABA-like immunoreactivity is present in this part of the retina, suggesting that GABA also modulates the output synapses of rod bipolar cells (Freed et. al, 1987). This is further illustrated in the electronmicrograph of Fig. 5 C. The presynaptic rod bipolar cell (RB) makes a ribbon synapse (arrowhead) with two postsynaptic amacrine cells: one is the AII amacrine cell and the other one is an amacrine cell (asterisk) labelled with gold particles of the GABA-immunoreactivity. This putative GABA-ergic amacrine cell makes a conventional synapse back onto the rod bipolar terminal (Pourcho, 1981; Freed et al., 1987). We also found conventional synapses at rod bipolar axons that were unrelated to ribbons, and expressed GABA-like immunoreactivity. Therefore it is highly probable that GABA is the predominant transmitter modulating the rod bipolar cell axon.

Dopaminergic innervation of AII-amacrine cells

Catecholamine-containing neurones have been found in all vertebrate retinae investigated to date and dopamine is the predominant catecholamine neurotransmitter in the retina. Recently catecholamine-containing retinal neurones have been studied with immunocytochemical methods (Brecha, 1983). In all mammals analysed to date, antibodies against tyrosine hydroxylase (TH) labelled an amacrine cell, which has a rather large cell body, occurs at low density (20 cells/mm²), and has a predominant stratification level close to the amacrine layer in stratum 1 of the IPL (Nguyen-Legros et al., 1983; Brecha et al., 1984; Oyster et al., 1985). In flat mount preparations the dendritic network of these cells became more apparent, and characteristic dopaminergic "ring endings" were described. The rings were interpreted as dopaminergic fibers that surround other amacrine cell bodies (Törk & Stone, 1979). It was suggested from an autoradiographic study that the AII amacrines might be positioned in the dopaminergic rings and it was shown that they receive synapses from ^{3}H-dopamine accumulating amacrine cells (Pourcho, 1982).

In order to test this dopaminergic influence on AII amacrine cells more directly, single AII amacrine cells were injected with Lucifer Yellow (LY) filled microelectrodes (Vaney, 1985). Afterwards the same retina was stained for TH-like immunoreactivity to determine, whether the AII amacrine cells have their cell bodies within the dopaminergic rings (Voigt & wässle, 1987). Figure 6 shows the result of such an experiment in a flat mounted retina (Fig. 6A-C) and in vertical sections (Fig. 6D-F). The TH-immunolabelling in Fig. 6 A shows several dopaminergic rings. Fig. 6B is a double exposure and the LY injected cell body can be detected in the ring in the centre of the micrograph. The micrograph in Fig. 6C was taken with epifluorescence illumination only and it shows the cell body and the characteristic dendrites of the injected AII-cell. The vertical sections (Fig. 6D-F) illustrate that the cell body of AII-cells protrudes into the IPL thus creating a hole in the dopaminergic plexus. In additional experiments, we could show that the AII-amacrine cells receive extended synapses from the TH-immunoreactive fibres surrounding their cell bodies (Voigt & Wässle, 1987). This makes it very likely that AII-amacrine cells are under the modulatory influence of dopamine. However, dopaminergic fibres are also found in the OPL and in the inner part of the IPL (Oyster et al., 1985), hence dopamine can be expected to influence other retinal neurones besides AII amacrine cells.

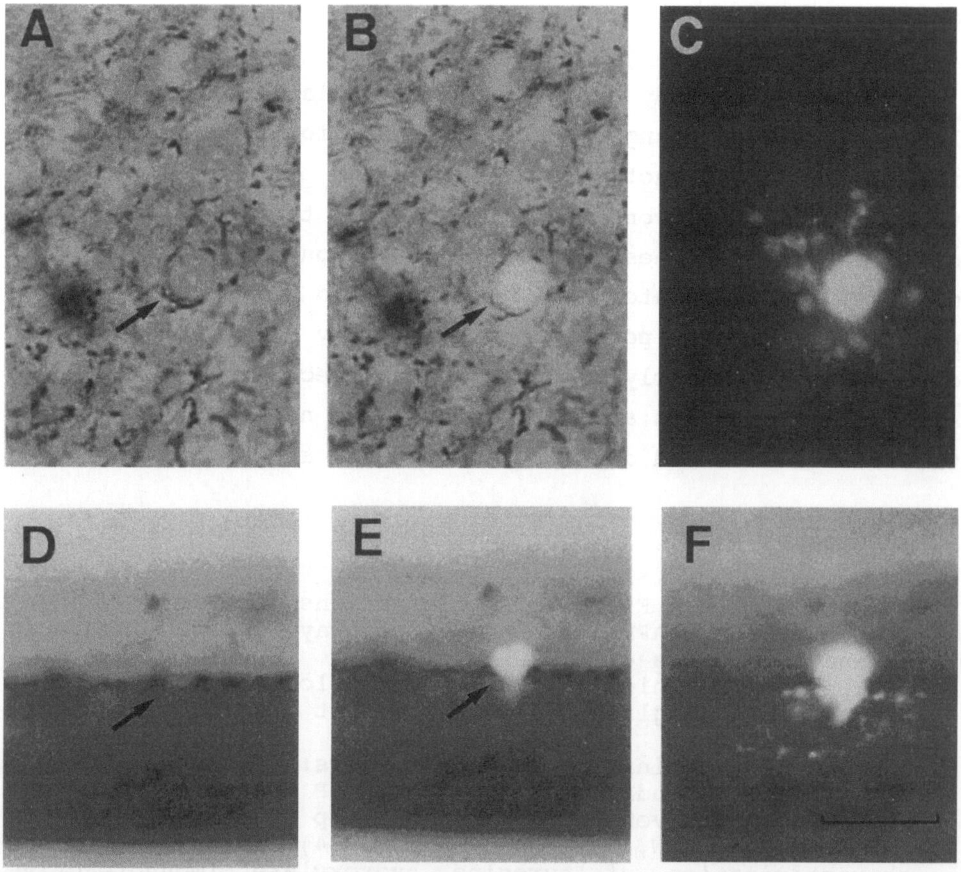

Figure 6: Dopaminergic rings and AII amacrine cells; A, B and C show the same field of a retinal whole mount, where an AII amacrine cell was injected with LY and the retina was stained for TH-like immunoreactivity using the PAP method. A, in bright field illumination the dopaminergic plexus becomes apparent. B, brightfield and epifluorescence illumination show dopaminergic ring (arrow) around AII-cell body. C, epifluorescence shows dendrites and cell body of AII amacrine cell. D, E and F show vertical sections through an LY injected AII amacrine cell after staining the retina for TH-like immunoreactivity. D, in the bright field illumination the gap (arrow) caused by the AII amacrine cell in the TH-band becomes apparent. E and F are micrographs of consecutive sections taken with brightfield and epifluorescence (Scale bar 20μm).

Conclusions

The present study shows that a potentially simple pathway like the rod to ganglion cell connection of the mammalian retina is in fact very complicated if the possible transmitters are considered. However the combination of different techniques enables dissection of this wiring diagram. The modulatory role of GABA and dopamine on the rod pathway has been postulated here only from histological experiments. Clearly, physiological recordings, involving appropriate agonists and antagonists, are necessary to reveal the role of GABA and dopamine in the dark adapted retina.

References

Arkin M S, Miller R F (1987) Subtle actions of 2-amino-4-phosphonobutyrate (APB) on the Off pathway in the mudpuppy retina. Brain Res 426:142-148

Bolz J, Wässle H, Thier P (1984) Pharmacological modulation of On and Off ganglion cells in the cat retina. Neurosci 12:875-885

Brecha N (1983) Retinal neurotransmitters: Histochemical and biochemical studies. In: Emson P C (ed) Chemical Neuroanatomy. Raven Press, New York, pp 85-129

Brecha N, Oyster C W, Takahashi E S (1984) Identification and characterization of tyrosine hydroxylase immunoreactive amacrine cells. Invest Ophthal Vis Sci 25:66-70

Chun M H, Wässle H (1988) GABA-like immunoreactivity in the cat retina: electron microscopy. J comp Neurol in press

Dacheux R F, Raviola E (1986) The rod pathway in the rabbit retina: a depolarizing bipolar and amacrine cell. J Neurosci 6:331-345

Famiglietti E V, Kolb H (1975) A bistratified amacrine cell and synaptic circuitry in the inner plexiform layer of the retina. Brain Res 84:293-300

Freed M A, Smith R G, Sterling P (1987) Rod bipolar array in the cat retina: pattern of input from rods and GABA-accumulating amacrine cells. J comp Neurol 266:445-455

Kolb H (1974) The connectios between horizontal cells and photoreceptors in the retina of the cat: electron microscopy of Golgi preparations. J comp Neurol 155:1-14

Kolb H (1979) The inner plexiform layer in the retina of the cat: electron microscopic observations. J Neurocytol. 8:295-329

Kolb H, Famiglietti E V (1974) Rod and cone pathways in the inner plexiform layer of the cat retina. Science 186:47-49

Kolb H, Nelson R (1983) Rod pathways in the retina of the cat. Vision Res 23:301-312

Levine M W, Shefner J M (1977) Variability in ganglion cell firing patterns; implications for separate "ON" and "OFF" processes. Vision Res 17:765-776

Müller F, Wässle H, Voigt T (1988) Pharmacological modulation of the rod pathway in the cat retina. J Neurophysiol (in press)

Nelson R (1982) AII amacrine cells quicken time course of rod signals in the cat retina. J Neurophysiol 47:928-947

Nelson R, Kolb H (1983) Synaptic patterns and response properties of bipolar and ganglion cells in the cat retina. Vision Res 23:1183-1195

Nguyen-Legros J, Vigny A, Gay M (1983) Post-natal development of TH-like immunoreactivity in the rat retina. Exp Eye Res 37:23-32

Oyster C W, Takahashi E S, Cilluffo M, Brecha N (1985) Morphology and distribution of tyrosine hydroxylase-like immunoreactive neurons in the cat retina. Proc Natl Acad Sci USA 82:6335-6339

Pourcho R G (1981) Autoradiographic localization of (^{3}H)muscimol in the cat retina. Brain Res 215:187-199

Pourcho R G (1982) Dopaminergic amacrine cells in the cat retina. Brain Res 252:101-109

Pourcho R G, Goebel D J (1985) A combined Golgi and autoradiographic study of (^{3}H)glycine-accumulating amacrine cells in the cat retina. J comp Neurol 233:473-480

Pourcho R G, Goebel D J (1987) Visualization of endogenous glycine in cat retina: an immunocytochemical study with Fab fragments. J Neurosci 7:1189-1197

Shiells R A, Falk G, Naghshineh S (1981) Action of glutamate and aspartate analogues on rod horizontal and bipolar cells. Nature 294:592-594

Slaughter M M, Miller R F (1981) 2-amino-4-phosphonobutyric acid: a new pharmacological tool for retina research. Science 211:182-185

Slaughter M M, Miller R F (1985) Characterization of an extended glutamate receptor of the on bipolar neuron in the vertebrate retina. J Neurosci 5:224-233

Sterling P (1983) Microcircuitry of the cat retina. Ann Rev Neurosci 6:148-185

Törk I, Stone J (1979) Morphology of catecholamine-containing amacrine cells in the cat's retina, as seen in retinal whole mounts. Brain Res 169:261-273

Vaney D I (1985) The morphology and topographic distribution of AII amacrine cells in the cat retina. Proc R Soc Lond Biol 224:475-488

Voigt T, Wässle H (1987) Dopaminergic innervation of AII amacrine cells. J Neurosci 7:4115-4128

Wässle H, Schäfer-Trenkler I, Voigt T (1986) Analysis of a glycinergic inhibitory pathway in the cat retina. J Neurosci 6:594-604.

Wässle H, M H Chun (1988) GABA-like immunoreactivity in the cat retina: light microscopy. J comp Neurol (in press)

Yazulla S (1986) GABAergic mechanisms in the retina. Prog Ret Res 5:1-52

EXPRESSION OF TACHYKININ PEPTIDES IN THE MAMMALIAN RETINA

Nicholas Brecha[1-3] and Catia Sternini[3]
Departments of Medicine and Anatomy
UCLA School of Medicine
Los Angeles, California 90024
United States of America

The tachykinins (TKs) are a family of biologically active peptides which are distributed within the vertebrate nervous system, including the retina (Brecha, 1983; Maggio, 1988). These peptides are structurally related and share a common amidated COOH-terminal sequence, Phe-X-Gly-Leu-Met-NH_2, where X is an aromatic or branched aliphatic amino acid (Erspamer, 1981). Although they exhibit similar biological activities in the central and peripheral nervous systems as well as in peripheral tissues, they can be differentiated on the basis of their pharmacological profiles in certain bioassay systems (Henry, 1987). Differences in the NH_2-terminal sequence of these peptides confer differential receptor specificities and at least three receptor classes have been proposed (Henry, 1987; Regoli et al., 1987). The distribution of TK receptor binding sites is consistent with this general classification (Mantyh et al., 1988).

In mammals, three TK peptides, substance P (SP), neurokinin A (NKA; also known as substance K) and neurokinin B (NKB; also known as neuromedin K) have been isolated, sequenced and shown to have a differential distribution in the nervous system (Chang et al., 1971; Kimura et al., 1983; Kangawa et al., 1983; Minamino et al., 1984A; Minamino et al., 1984B; Kanazawa et al., 1984; Ogawa et al., 1985). Recent studies have reported two other TKs, neuropeptide K and neuropeptide gamma which contain the entire NKA amino acid sequence at their COOH-terminus (Tatemoto et al., 1985; Kage et al., 1988). Analysis of cloned

[1]Jules Stein Eye Institute, [2]Brain Research Institute and [3]Center for Ulcer Research and Education, UCLA School of Medicine and VAMC-West Los Angeles, Los Angeles, California 90073

NATO ASI Series, Vol. H31
Neurobiology of the Inner Retina
Edited by R. Weiler and N.N. Osborne
© Springer-Verlag Berlin Heidelberg 1989

cDNAs from mammalian tissues shows that SP and NKA are derived from three distinct mRNAs, preprotachykinin (PPT) A mRNAs, by alternative RNA processing (Nawa et al., 1983; Nawa et al., 1984; Harmar et al., 1986; Krause et al., 1987). Beta and gamma SP/NKA mRNAs encode SP/NKA-containing precursors and account for the majority of PPT A mRNAs. A third PPT A mRNA, alpha-SP mRNA, is a minor occurring SP-encoding transcript. On the other hand, NKB is derived from distinct mRNAs, known as PPT B mRNAs (Kotani et al., 1986; Bonner et al., 1987). SP/NKA and NKB PPT mRNAs are specified by two separate, but structurally related genes (Nawa et al., 1984; Kotani et al., 1986; Krause et al., 1987).

Numerous biologically active peptides, including SP have been described in all vertebrate retinas studied to date (Brecha, 1983; Ehinger and Dowling, 1987). SP-like bioactivity was first described in bovine retina and optic nerve extracts in 1954 by Duner et al., and during the past ten years immunohistochemical and radioimmunoassay studies have reported SP-like immunoreactivity (SP-IR) in representative vertebrate retinas and, in some species, the optic nerve (Brecha, 1983; Eskay et al., 1980). High-performance liquid chromatographic analysis has shown that in mammalian retinas SP-IR elutes in the same position as synthetic SP suggesting the presence of authentic SP (Eskay et al., 1981). The more recent discovery of other TK peptides in mammals has now raised the possibility that other TKs are present in the retina. TK peptides are likely to be detected by many and perhaps all "SP" antibodies which have been employed to date, since most of them have been shown to cross-react with the conserved COOH-terminal region of the TK peptide sequence (Maggio, 1988). Therefore, SP-IR should more properly be referred to as TK-IR.

The general distribution of TK-IR is similar in all vertebrate retinas studied to date, although there are some specific features which are unique for each species (Brecha, 1983).TK-IR cells have been classified as amacrine, displaced amacrine or ganglion cells (Brecha, 1983; Brecha et al, 1987). The majority of TK-IR fibers are distributed to selected laminae of the inner plexiform layer (IPL) and some fibers are distributed to the outer plexiform layer (OPL) and the ganglion cell axon layer (GAL). These general points will be illustrated using the rat and rabbit retinas as examples.

TK-IR in the rat and rabbit retina

Immunohistochemical studies were performed with normal and intraocularly colchicine-treated adult albino rat and New Zealand albino rabbit retinas as well as with rabbit retina and brain after unilateral optic nerve transection. These tissues were fixed in aldehydes and subsequently processed by standard immunohistochemical techniques using a rat monoclonal antibody (NC1/34, from Pel-Freez) directed against the COOH-terminal sequence of SP (Cuello et al., 1979). Specificity of the immunoreaction was assessed by incubating tissues in primary antibody preabsorbed with 10 uM synthetic SP, and in some cases NKA or NKB, which abolished the staining.

In agreement with earlier studies (Fukuda et al., 1981; Osborne et al., 1982; Osborne and Perry, 1985) we have observed TK-IR cell bodies in the proximal inner nuclear layer (INL), IPL and ganglion cell layer (GCL) in the rat retina (Figure 1). Immunoreactive processes are distributed to bands within the IPL at the INL/IPL border, midway between the INL and GCL and at the IPL/GCL border corresponding best to laminae 1, 3 and 5 of the IPL (Ramón y Cajal, 1892). A moderate number of immunoreactive cell bodies are present in the INL and GCL. A few immunoreactive somata are also located in the IPL and these cells are most likely displaced amacrine cells (interstitial cells). TK-IR cell bodies are small in size, and the difficulty of accurate measurements of immunoreactive profiles in cryostat sections precludes their rigorous quantification. An earlier report of SP-IR in rat optic nerve extracts implied the existence of SP/TK-containing ganglion cells (Kanazawa and Jessell, 1976). However, experimental evidence, based on transection of the optic nerve, which results in retrograde ganglion cell degeneration that eliminates most if not all ganglion cells, suggests that the SP-IR cells in the GCL are displaced amacrine rather than ganglion cells (Osborne and Perry, 1985).

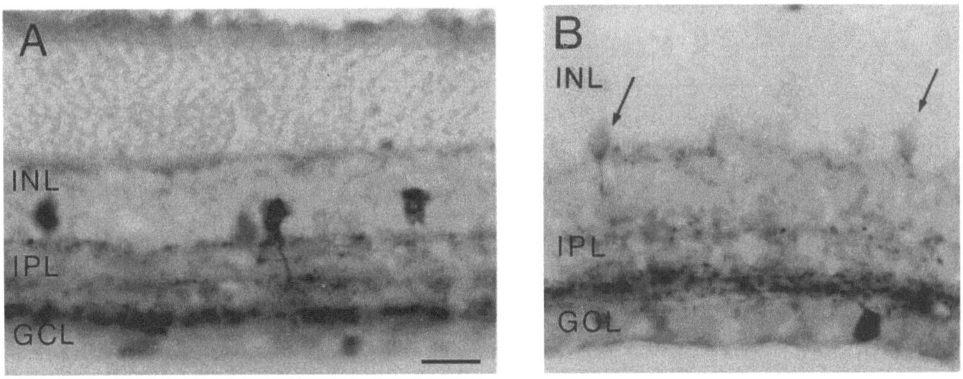

Figure 1: Transverse sections of the rat retina showing TK-IR cells in the INL and GCL and processes in the IPL. A: several somata in the INL. B: one cell in the GCL and two lightly stained cells (arrows) in the INL. Abbreviations are given in the text. Calibration bar, 25 µm.

In the rabbit retina, which has been more extensively analyzed than the rat retina, TK-containing cells are present in the INL and GCL (Brecha et al., 1987). In the INL, two types of cells can be recognized. One is a unistratified amacrine cell which gives rise to a single stout process which passes through the IPL to ramify in lamina 5. The second cell type, that likely corresponds to an amacrine or interplexiform cell, has numerous processes ramifying in several IPL laminae and possibly in the OPL. The majority of TK-IR somata are located in the GCL (Figure 2); they are ovoid in shape and have diameters within the reported range of ganglion and large displaced amacrine cells (Vaney, 1980; Vaney et al., 1981; Oyster et al., 1981; Masland et al., 1984). Immunoreactive somata typically emit 3 or 4 processes which enter and branch extensively in lamina 5 of the IPL. In retinal whole mounts, the primary and secondary processes of these cells can be followed up to 100 µm from their cell bodies before being lost among other stained processes and therefore the absolute dimensions of the dendritic fields of individual cells cannot be determined. Many immunoreactive cells have an axon which can be followed through the GAL towards the optic nerve head providing evidence they are ganglion cells. This suggestion is also supported by radioimmunoassay studies which report SP/TK-IR in optic nerve extracts (Unger et al., 1981). TK-IR cells in the GCL are distributed to all retinal regions and parallel the reported density gradient of ganglion cells across the retina (Oyster et al., 1981; Masland et al., 1984).

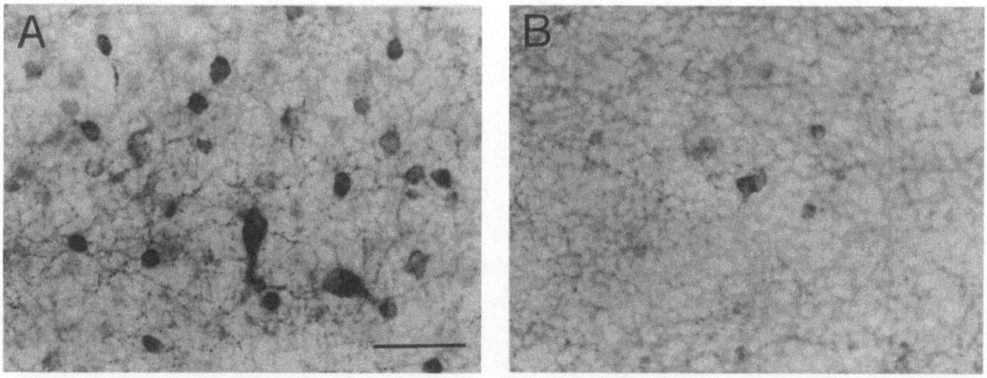

Figure 2: TK-IR somata and processes in the rabbit retina. Whole mount preparations from mid-ventral retinas showing TK-IR somata in the GCL in normal (A) and optic nerve transected (B) retinas. Calibration bar, 50 µm.

Unilateral optic nerve transection resulted in a rapid, marked loss of TK-IR cells in the GCL and processes in the proximal IPL (Figure 2). In central retinal regions, 80-90% of the TK-IR cells in the GCL are eliminated; the remaining immunoreactive cells are small and about the size of the large displaced amacrine cells (Vaney et al., 1981; Masland et al., 1984). On the basis of these experiments, as well as comparison to ganglion cell counts from Nissl stained preparations (Vaney, 1980; Vaney et al., 1981; Oyster et al., 1981; Masland et al., 1984), TK-IR ganglion cells are likely to constitute 25-35% of the total number of ganglion cells. A similar estimate of TK-IR ganglion cells was made in double-label studies in which ganglion cells identified by the retrograde transport of the fluorescent dye, Fast Blue, from the superior colliculus (SC) were also shown to contain TK-IR (Brecha et al., 1987). These studies indicate that the majority of TK-IR cells in the GCL are ganglion cells.

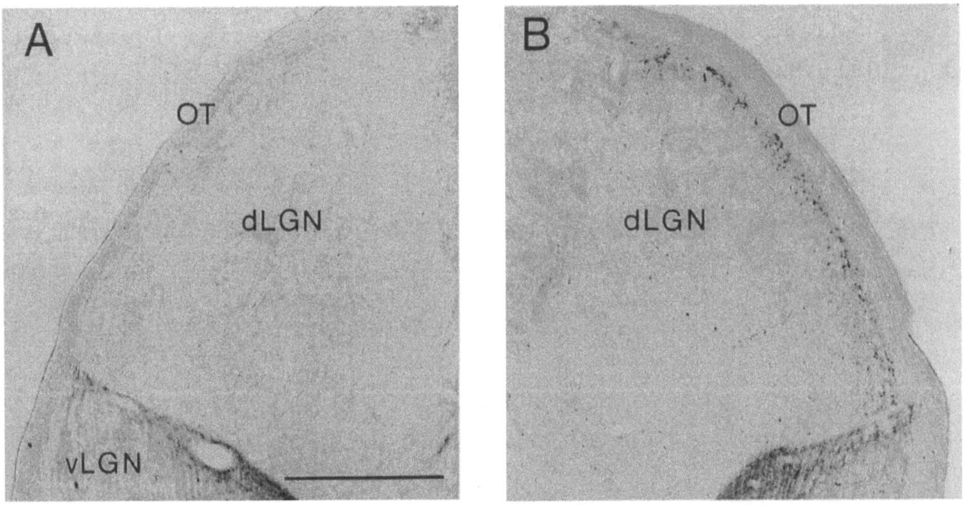

Figure 3: Transverse sections through the rabbit diencephalon after unilateral optic nerve transection showing loss of immunoreactive fibers and terminals in the contralateral dLGN (A) and the normal innervation pattern in the ipsilateral dLGN (B). (From Brecha et al., 1987). OT, optic tract; vLGN, ventral LGN. Calibration bar, 1 mm.

The projection of TK-IR ganglion cells in the adult and neonatal rabbit upon the contralateral dorsal lateral geniculate nucleus (dLGN), SC and accessory optic nuclei (AON) has been extensively analyzed by Donald J. Johnson as part of his doctoral dissertation. In the dLGN there is a well defined plexus of immunoreactive axons and varicosities which forms a band in the alpha sector, adjacent to the optic tract (Figure 3). Immunoreactive plexuses are also present in the stratum griseum superficiale of the SC and in the AON. In all of these structures, the immunoreactive plexuses are less prominent shortly after, and markedly depleted or absent 10 days or more after contralateral optic nerve transection. Optic nerve transection had no detectable effect on the corresponding immunoreactive plexuses in the ipsilateral dLGN, SC or AON.

These studies illustrate that a common feature in both rat and rabbit retina is the presence of TK-IR amacrine and displaced amacrine cells. These generalized features have also been reported for human, monkey, cat and cow retinas (Brecha et al., 1982; Osborne, 1984; Tornqvist and Ehinger, 1988; Pourcho and Goebel, 1988). In rabbit retina, the presence of numerous TK-IR ganglion cells is well documented. To date, similar evidence in other mammalian retinas is lacking. The recent report of TK-IR ganglion cells in frog and chick retina (Kuljis and Karten, 1983; Ehrlich et al., 1987) and the suggestion of several other peptidergic ganglion cell types in the frog (Kuljis and Karten, 1983) raise the possibility that the presence of peptide-containing ganglion cells is a feature common to many vertebrate species.

Cellular localization of SP/NKA and NKB mRNAs

Although there is good circumstantial evidence that multiple TK peptides are present in the retina and they are likely to have a role in retinal processing, there is little information concerning which TK peptides and mRNAs are actually expressed in this structure. We have begun to address these issues by using in situ hybridization histochemistry with the rat retina as a model system.

Aldehyde fixed sections of rat retina were hybridized with ^{35}S-labeled SP/NKA antisense RNA probe (prepared by T7 polymerase transcription from pGlbeta-PPT cDNA) complementary to rat beta mRNAs encoding SP/NKA-containing precursor and recognizing alpha, beta and gamma PPT A mRNAs, or ^{35}S-labeled NKB antisense RNA probes (prepared by SP6 polymerase transcription from pGlNKB17-1 or cRNKp4 cDNAs) complementary to rat NKB-encoding mRNA. SP/NKA and NKB RNA probes in the mRNA orientation (sense

RNA) from pG1beta-PPT and pG1NKB17-1 plasmid DNAs, prepared by SP6 or T7 polymerase transcription, respectively, were used as a control for specific hybridization. For the transcription reactions, 500 µM each of adenosine-, guanidine- and cytidine-triphosphate, 15 µM uridine triphosphate (UTP) and 5 µM of ^{35}S-UTP were used (Wuenschell et al., 1987). The specific activity of each probe was of the order of 4-6 x 10^8 cpm/µg RNA. The labeled RNAs were purified by DNase I digestion, phenol/chloroform extraction and ethanol precipitation. Transcripts were cleaved to 100-150 nucleotides by limited alkaline hydrolysis (Cox et al., 1984). Our in situ hybridization protocol, which will be published elsewhere, is a modification of previously published methods (Chesselet at al., 1987). Slides were processed for autoradiography using Kodak NTB-2 liquid photographic emulsion. Autoradiograms were developed after 7-30 days of exposure and counterstained.

SP/NKA-encoding transcripts were observed in cells distributed to the INL, IPL and GCL (Figure 4). A moderate number of labeled cells were observed in the proximal INL and in the GCL. Occasionally, labeled cells were seen in the IPL. In contrast, NKB-encoding transcripts were primarily found in the GCL (Figure 4) and instances of NKB-encoding transcripts in the INL were extremely rare. Cells expressing SP/NKA and NKB mRNAs have morphological characteristics and distribution similar to the cells containing SP/TK-IR, which have been suggested to be amacrines and displaced amacrines (Osborne and Perry, 1985). However, the possibility that the TK genes are transcribed in a subpopulation of ganglion cells cannot be discounted on the basis of our studies and needs further investigation.

Evidence for the specificity of the labeling includes 1) lack of specific signal above background in sections hybridized with the sense RNA probes, 2) elimination of the labeling in sections pretreated with RNase (100 µg/ml at 37°C for 30 min) before hybridization, 3) absence of labeling in those incubated with the hybridization mixture without the labeled probe and 4) the recognition of SP/NKA- and NKB-encoding transcripts in blot hybridization and protection assays (Brecha et al., 1988) by the same antisense probes used for in situ hybridization histochemistry. The differential distribution of cellular labeling by the SP/NKA and NKB antisense RNAs is an additional evidence for the specificity of the hybridization.

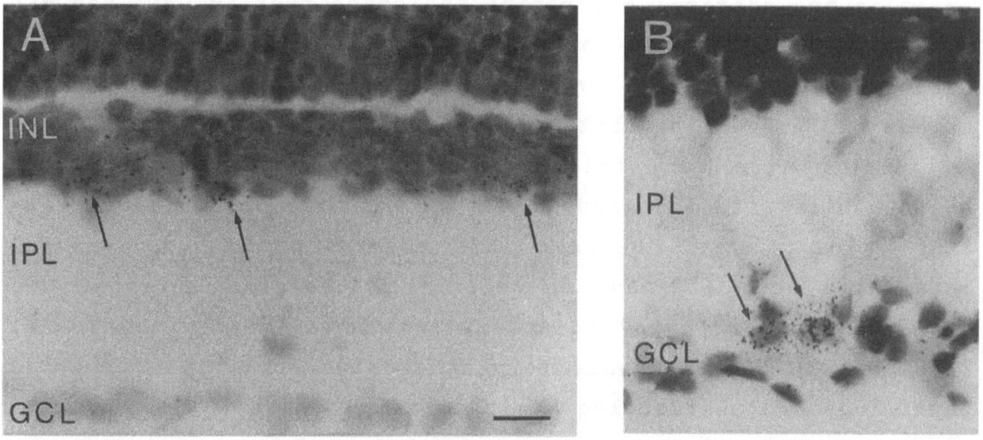

Figure 4: SP/NKA-encoding transcripts in cells located in the proximal INL (A), and of NKB-encoding transcripts in cells located in the GCL (B) of the rat retina. A, focused at the INL. Hybridization appears as clusters of grains over individual cells (arrows). Calibration bar, 25 µm.

In the retina, as has been recently reported in several regions of the brain using in situ hybridization histochemistry with synthetic oligonucleotide probes complementary to rat SP/NKA and NKB mRNAs (Warden and Young, 1988), there is a differential distribution of SP/NKA- and NKB-encoding transcripts. In retina, SP/NKA-encoding transcripts are localized to cells in the INL, IPL and GCL and NKB-encoding transcripts are primarily localized to cells in the GCL. These observations suggest that these transcripts are present in separate cell populations, although the possibility that SP/NKA and NKB mRNAs are also co-expressed in a limited number of cells in the GCL needs further study. A differential tissue expression of the SP/NKA gene due to alternative splicing has also been suggested by protection assays in the bovine nervous system (Nawa et al., 1984). The present studies, however, cannot distinguish between the presence of alpha-SP, beta-SP/NKA and gamma-SP/NKA mRNAs in different retinal cell populations because the antisense RNA probe which we used recognizes all three PPT A mRNA species.

The localization of SP/NKA- and NKB-encoding transcripts to cells in the proximal rat retina indicates that SP, NKA and NKB are synthesized in the retina. PPT A and B RNA probes and TK antibodies label cells having similar appearance, size and distribution, suggesting that both of these experimental approaches identify the same cell types. The *in situ* hybridization histochemical studies argue for a differential cellular expression of the two TK genes. This observation is of general interest since differential cellular expression of these two genes has been reported in other structures (Warden and Young, 1988) and is of importance for the consideration of TK peptide physiology and pharmacology in the retina.

Concluding Remarks

These studies have demonstrated the presence of TK-IR in selected cell types of the rat and rabbit retina. In the rat retina, SP/NKA and NKB mRNAs appear to be expressed in the same cell types which contain TK-IR. These observations, together with previous reports, strengthen the earlier suggestions that TK peptides have a biological role in the proximal retina where they may function as transmitters, modulators or perhaps as growth factors (Maggio, 1988). Moreover, the distinct pharmacological profiles of TK peptides (Henry, 1987) suggest the possibility that each of them has a unique functional role in the retina. Specific high affinity SP, NKA and NKB binding sites have been demonstrated in the mammalian retina (Osborne, 1984; Mantyh et al., 1988) and SP applied to the chick retina evokes phosphatidylinositol hydrolysis (Mantyh et al., 1985) indicating that TK receptors are linked to intracellular messenger systems. To date, only the effects of SP on ganglion cell activity (Dick and Miller, 1981; Glickman et al., 1982) and phosphatidylinositol hydrolysis have been documented and information concerning NKA and NKB action on retinal function is lacking. The differential distribution of SP/NKA and NKB mRNAs in the retina suggests that SP, NKA and NKB have distinctive biological roles in the retina.

Acknowledgements

We wish to thank K. Anderson, M. Cilluffo and J. Gordon for their important contributions to the present studies, D.J. Johnson, whose collaboration and doctoral dissertation have significantly contributed to this subject, Drs. J.E. Krause for the pGlbeta-PPT and pGlNKB17-1 cDNAs, T.I. Bonner for the cRNKp4 cDNA, and L. Kruger for his helpful comments on this manuscript. Supported by NIH grants EY 04067 and DK 38752, VA Merit Review, UCLA Academic Grant #3873 and SKB fellowship.

References

Bonner TI, Affolter H-U, Young AC, Young WS III (1987) A cDNA encoding the precursor of the rat neuropeptide,neurokinin B. Molec Brain Res 2:243-249

Brecha N (1983) Retinal neurotransmitters: histochemical and biochemical studies.In: Emson PC (ed) Chemical Neuroanatomy. Raven Press, New York, p 85

Brecha N, Hendrickson A, Florén I, Karten HJ (1982) Localization of substance P-like immunoreactivity within the monkey retina. Invest Ophthalmol Vis Sci 23:147-153

Brecha N, Johnson D, Bolz J, Sharma S, Parnavelas JG, Lieberman AR (1987) Substance P-immunoreactive retinal ganglion cells and their central axon terminals in the rabbit. Nature (London) 327:155-158

Brecha N, Sternini C, Anderson K, Krause JE (1988) Cellular localization of substance P/neurokinin A and neurokinin B mRNAs in the rat retina. Soc Neurosci Abs (in press)

Chang MM, Leeman SE, Niall HD (1971) Amino-acid sequence of substance P. Nature (London) New Biol 232:86-87

Chesselet M-F, Weiss L, Wuenschell C, Tobin AJ, Affolter H-U (1987) Comparative distribution of mRNAs for glutamic acid decarboxylase, tyrosine hydroxylase, and tachykinins in the basal ganglia: an in situ hybridization study in the rodent brain. J Comp Neurol 262:125-140

Cox KH, DeLeon DV, Angerer LM, Angerer RC (1984) Detection of mRNAs in sea urchin embryos by <u>in situ</u> hybridization using asymmetric RNA probes. Dev Biol 101:485-502

Cuello AC, Galfre G, Milstein C (1979) Detection of substance P in the central nervous system by a monoclonal antibody. Proc Natl Acad Sci USA 76:3532-3536

Dick E, Miller RF (1981) Peptides influence retinal ganglion

cells. Neurosci Lett 26:131-135
Duner H, von Euler US, Pernow B (1954) Catechol amines and substance P in the mammalian eye. Acta Physiol Scand 31:113-118
Ehinger B, Dowling JE (1987) Retinal neurocircuitry and transmission. In: Björklund A, Hökfelt T, Swanson LW (eds) Handbook of chemical neuroanatomy. Vol 5. Integrated Systems of the CNS, Part 1.Elsevier Science Publishers, Amsterdam,p 389
Ehrlich D, Keyser KT, Karten HJ (1987) Distribution of substanceP-like immunoreactive retinal ganglion cells and their pattern of termination in the optic tectum of chick (Gallus gallus). J Comp Neurol 266:220-233
Erspamer V (1981) The tachykinin peptide family. Trends in Neurosci 4:267-269
Eskay RL, Furness JF, Long RT (1981) Substance P activity in the bullfrog retina: localization and identification in several vertebrate species. Science 212:1049-1051
Eskay RL, Long RT, Iuvone PM (1980) Evidence that TRH, somatostatin and substance P are present in neurosecretory elements of the vertebrate retina. Brain Res 196:554-559
Fukuda M, Kuwayama Y, Shiosaka S, Ishimoto I, Shimizu Y, Takagi H, Inagaki S, Sakanaka M, Semba E, Takatsuki K, Tohyama M (1981) Demonstration of a substance P-like immunoreactivity in retinal cells of the rat. Neurosci Lett 23:239-242
Glickman RD, Adolph AR, Dowling JE (1982) Inner plexiform circuits in the carp retina: effects of cholinergic agonists, GABA, and substance P on the ganglion cells. Brain Res 234: 81-99
Harmar AJ, Armstrong A, Pascall JC, Chapman K, Rosie R, Curtis A, Going J, Edwards CRW, Fink G (1986) cDNA sequence of human β-preprotachykinin, the common precursor to substance P and neurokinin A. FEBS Lett 208:67-72
Henry JL (ed) (1987) Substance P and neurokinins. Springer-Verlag, New York
Johnson, DJ (to be published) Observations on the central projections of tachykinin immunoreactive retinal ganglion cells and the ontogeny of the tachykinin retion geniculate system of the albino rabbit. Doctoral dissertation, UCLA
Kage R, McGregor GP, Thim L, Conlon JM (1988) Neuropeptide-γ: a peptide isolated from rabbit intestine that is derived from γ-preprotachykinin. J Neurochem 50:1412-1417
Kanazawa I, Jessell T (1976) Post mortem changes and regional distribution of substance P in the rat and mouse nervous system. Brain Res 117, 362-367.
Kanazawa I, Ogawa T, Kimura S, Munekata E (1984) Regional distribution of substance P, neurokininα and neurokininβ in rat central nervous system. Neurosci Res 2:111-120
Kangawa K, Minamino N, Fukuda A, Matsuo H (1983) Neuromedin K: a novel mammalian tachykinin identified in porcine spinal cord. Biochem Biophys Res Commun 114:533-540
Kimura S, Okada M, Sugita Y, Kanazawa I, Munekata E (1983) Novel neuropeptides, neurokininα and β , isolated from porcine spinal cord. Proc Jpn Acad Ser B 59:101-104
Kotani H, Hoshimaru M, Nawa H, Nakanishi S (1986) Structure and gene organization of bovine neuromedin K precursor. Proc Natl Acad Sci USA 83:7074-7078
Krause JE, Chirgwin JM, Carter MS, Xu ZS, Hershey AD (1987) Three rat preprotachykinin mRNAs encode the neuropeptides

substance P and neurokinin A. Proc Natl Acad Sci USA 84:881-885

Kuljis RO, Karten HJ (1983) Modifications in the laminar organization of peptide-like immunoreactivity in the anuran optic tectum following retinal deafferentation. J Comp Neurol 217: 239-251

Maggio JE (1988) Tachykinins. Ann Rev Neurosci 11:13-28

Mantyh PW, Gates T, Mantyh CR, Maggio JE (1988) Autoradiographic localization and characterization of tachykinin receptor binding sites in the rat brain and peripheral tissues. J Neuroscience (in press)

Mantyh P, Goedert M, Hunt SP, Brecha NC (1985) Functional neurotensin and substance P receptors are present in the retina. Invest Ophthalmol Vis Sci Suppl 26:277

Masland RH, Mills JW, Hayden SA (1984) Acetylcholine-synthesizing amacrine cells: identification and selective staining by using radioautography and fluorescent markers. Proc R Soc Lond B 223:79-100

Minamino N, Kangawa K, Fukuda A, Matsuo H (1984A) Neuromedin L: a novel mammalian tachykinin identified in porcine spinal cord. Neuropeptides 4:157-166

Minamino N, Masuda H, Kangawa K, Matsuo H (1984B) Regional distribution of neuromedin K and neuromedin L in rat brain and spinal cord. Biochem Biophys Res Commun 124:731-738

Nawa H, Hirose T, Takashima H, Inayama S, Nakanishi S (1983) Nucleotide sequences of cloned cDNAs for two types of bovine brain substance P precursor. Nature (London) 306:32-36

Nawa H, Kotani H, Nakanishi S (1984) Tissue-specific generation of two preprotachykinin mRNAs from one gene by alternative RNA splicing. Nature (London) 312:729-734

Ogawa T, Kanazawa I, Kimura S (1985) Regional distribution of substance P, neurokinin α and neurokinin β in rat spinal cord, nerve roots and dorsal root ganglia, and the effects of dorsal root section or spinal transection. Brain Res 359:152-157

Osborne NN (1984) Substance P in the bovine retina: localization, identification, release, uptake and receptor analysis. J Physiol 349:83-93

Osborne NN, Nicholas DA, Dockray GJ, Cuello AC (1982) Cholecystokinin and substance P immunoreactivity in retinas of rats, frogs, lizards and chicks. Exp Eye Res 34:639-649

Osborne NN, Perry VH (1985) Effect of neonatal optic nerve transection on some classes of amacrine cells in the rat retina. Brain Res 343:230-235

Oyster CW, Takahashi ES, Hurst DC (1981) Density, soma size, and regional distribution of rabbit retinal ganglion cells. J Neurosci 1:1331-1346

Pourcho RG, Goebel DJ (1988) Colocalization of substance P and -aminobutyric acid in amacrine cells of the cat retina. Brain Res 447:164-168

Ramón y Cajal SR (1892) La rétine des vertébrés. La Cellule 9: 17-257

Regoli D, Drapeau G, Dion S, D'Orleans-Juste P (1987) Pharmacological receptors for substance P and neurokinins. Life Sci 40:109-117

Tatemoto K, Lundberg JM, Jörnvall H, Mutt V (1985) Neuropeptide K: isolation, structure and biological activities of a novel

brain tachykinin. Biochem Biophys Res Commun 128:947-953
Tornqvist K, Ehinger B (1988) Peptide immunoreactive neurons in the human retina. Invest Ophthalmol Vis Sci 29:680-686
Unger WG, Butler JM, Cole DF, Bloom SR, McGregor GP (1981) Substance P, vasoactive intestinal polypeptide (VIP) and somatostatin levels in ocular tissue of normal and sensorily denervated rabbit eyes. Exp Eye Res 32:797-801
Vaney DI (1980) A quantitative comparison between the ganglion cell populations and axonal outflows of the visual streak and periphery of the rabbit retina.J Comp Neurol 189:215-233
Vaney DI, Peichl L, Wässle H, Illing R-B (1981) Almost all ganglion cells in the rabbit retina project to the superior colliculus. Brain Res 212:447-453
Warden MK, Young WS III (1988) Distribution of cells containing mRNAs encoding substance P and neurokinin B in the rat central nervous system. J Comp Neurol 272:90-113
Wuenschell C, Fisher RS, Kaufman DL, Tobin AJ (1986) In situ hybridization to localize mRNA encoding the neurotransmitter synthetic enzyme glutamic acid decarboxylase (GAD) in the mouse cerebellum. Proc Natl Acad Sci USA 83:6193-6197

THE COEXISTENCE OF MULTIPLE NEUROACTIVE SUBSTANCES IN THE RETINA

C. B. Watt and D. M. K. Lam
Alice McPherson Laboratory of Retina Research
The Center for Biotechnology
Baylor College of Medicine
4000 Research Forest Drive
The Woodlands, Texas 77381

Examples of the coexistence of neuroactive substances in individual neurons have been reported in increasing numbers throughout the central and peripheral nervous systems (Burnstock, 1976; Cuello, 1982; Osborne, 1983; Lundberg and Hokfelt, 1983; Chan-Palay and Palay, 1984; Lam, 1988). However, the obvious questions of whether the coexistence of neuroactive substances reflects co-transmission and/or co-function, as well as the mechanisms and rationale underlying such co-operativities, remain largely unanswered. Over the past three and one-half years, several reports of the co-occurence of neuroactive substances in retinal neurons have emerged, and it is not unanticipated that many additional examples will be revealed in the near future (Lam et al, 1985; Lam et al., 1986). To date, peptide-transmitter (Weiler and Ball, 1984; Watt et al., 1984; Watt et al., 1985b; Watt et al., 1986; Su et al., 1986; Watt et al., 1987; Zucker and Adolph, 1988), peptide-peptide (Li et al., 1985; Katayama-Kumoi, 1985a, Katayama-Kumoi, 1985b; Li et al., 1986), peptide-peptide-transmitter (Watt et al., 1988; Wu and Lam, 1988) and transmitter-transmitter (Osborne and Beaton, 1986; Brecha et al., 1987; Wässle and Chun, 1987; Vaney and Young, 1988; Johnson et al., 1988; O'Malley and Masland, 1988) colocalizations have been demonstrated in the retinas of various submammalian and mammalian species. For this and the following reasons, the vertebrate retina presents itself as an excellent model of the central nervous system for investigating the functional role coexistence plays in the communication of neural information. First, the retina is an easily accessible tissue whose input (light stimuli)

Correspondence: Carl B. Watt, at the address listed above.
Telephone No.: (713) 363-8416

NATO ASI Series, Vol. H31
Neurobiology of the Inner Retina
Edited by R. Weiler and N.N. Osborne
© Springer-Verlag Berlin Heidelberg 1989

and output (e.g., as measured by ganglion cell activities and electroretinogram) can be quantitatively monitored both in intact and isolated retinas. Second, preparations of both isolated retinal cells (Drujan et al., 1972; Lam, 1972; Lam, 1976; Dowling et al., 1983; MacLeish et al., 1984) and living retinal slices (Werblin, 1978; Wu, 1985) are available for multidisciplinary analyses. Third, in terms of its anatomical, neurochemical and functional organization, the retina ranks among the best understood tissues in the central nervous system (Dowling, 1987). This article represents an updated version of two earlier reviews (Lam et al., 1985; Lam, 1988) and seeks to summarize current findings as well as discuss the possible anatomical and functional implications of the coexistence of multiple neuroactive substances in individual retinal cells.

Different Neuropeptides in Putative Glycinergic Neurons

The initial observations of coexistence in the retina revealed the coexistence of a neuropeptide and a neurotransmitter in subpopulations of retinal amacrine cells. Neurotensin was found to colocalize with glycine in turtle amacrine cells (Weiler and Ball, 1984), while the co-occurence of enkephalin and GABA was observed in amacrine cells of the chicken retina (Watt et al., 1984). These findings were extended by recent studies that examined the pattern in which different neuropeptides coexist in putative glycinergic amacrine cells of the chicken retina.

Using immunocytochemistry to visualize enkephalin-, neurotensin- and somatostatin-amacrine cells and autoradiography to visualize cells that exhibit high-affinity uptake of ^{3}H-glycine, it was shown that immunoreactivity for each of these three peptides is present in certain putative glycinergic amacrine cells of the chicken retina (Figure 1, Watt et al., 1985b). This finding was especially interesting because multiple types of morphologically distinct glycine-accumulating neurons have been described in all vertebrate retinas examined to date (Kong et al., 1980; Marc and Lam, 1981; Marc, 1985; Marc and Liu, 1985; Pourcho and Goebel, 1985). Therefore, this finding raised the possibility that morphologically, and perhaps functionally, distinct types of putative glycinergic retinal neurons may in the

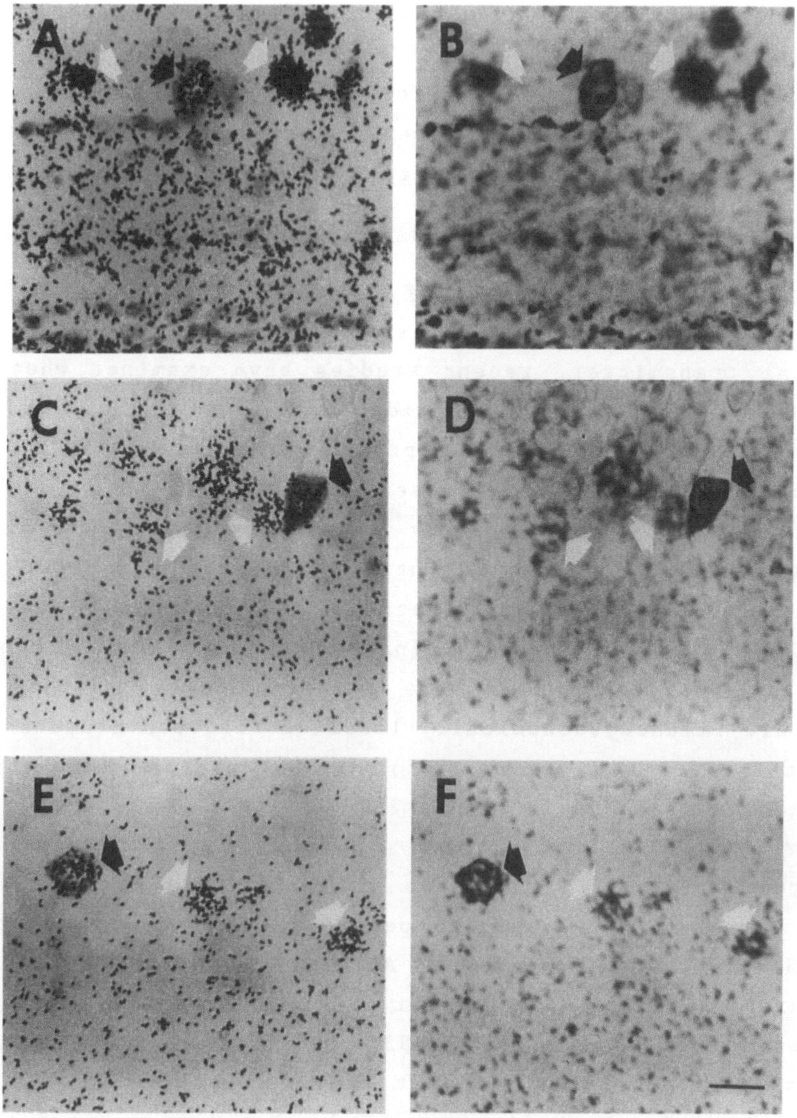

Fig. 1: The simultaneous visualization of either enkephalin, neurotensin or somatostatin (by immunocytochemistry) and [^{3}H] glycine uptake (by autoradiography) in the chicken retina. The paired photomicrographs (A-B, C-D, E-F) are focussed at the level of the emulsion layer (A, C, E) or the level of the underlying immunoperoxidase staining (B, D, F). The black arrows point to cells that are double-labelled for either enkephalin (A-B), neurotensin (C-D) or somatostatin (E-F) immunostaining and [^{3}H] glycine uptake-autoradiography. The white arrows indicate those cells which are only autoradiographically labelled for [^{3}H] glycine uptake. The scale bar equals 10 μm. (From Watt et al., 1985b; with permission).

future be identified and categorized by their co-localization with different neuroactive peptides. If confirmed, such a result could provide a valuable tool for analyses of retinal circuitry, as well as a suitable refinement and confirmation for the correlation between morphology, function and neurochemical specificity.

Coexistence of Two Neuropeptides

In addition to the coexistence of a neuropeptide and a classical transmitter, recent studies have examined whether two chemically distinct neuropeptides coexist in an individual retinal neuron. A direct approach involved visualization by immunocytochemistry of the presence of two peptides in the same retinal section, using primary antibodies raised in different species. With a polyclonal antibody raised in rabbits against neurotensin (Carraway and Bhatnagar, 1980) and a monoclonal antibody raised in mouse against enkephalin (Tavella et al., 1985), it was shown that some amacrine cells in the chicken retina are immunocytochemically labelled for both neurotensin and enkephalin (Li et al., 1985). An example of this co-localization is shown in Figure 2.

The co-localization of enkephalin- and neurotensin-immunoreactivities in the same neuron raised the question of whether these two neuroactive peptides are encoded by the same gene (Scheller et al., 1983). Although the amino acid sequence of neurotensin is not present in any of the known opioid precursors (Boileau et al., 1983; Horikawa et al., 1983; Lewis and Stern, 1983), it is not known whether the mRNAs of neurotensin and enkephalin are co-translated as part of the same gene message (Scheller et al., 1983). From a regulatory point of view, if these two peptides interact with each other to influence chemical transmission in the retina, their simultaneous formation as products of the same gene might ensure that they are co-released upon the proper stimulation of the neuron. Clearly, elucidation of the mechanisms regulating the metabolism and storage of different neuroactive peptides in the same neuron must first await identification of their peptide sequences, mRNAs and genes by molecular genetic approaches.

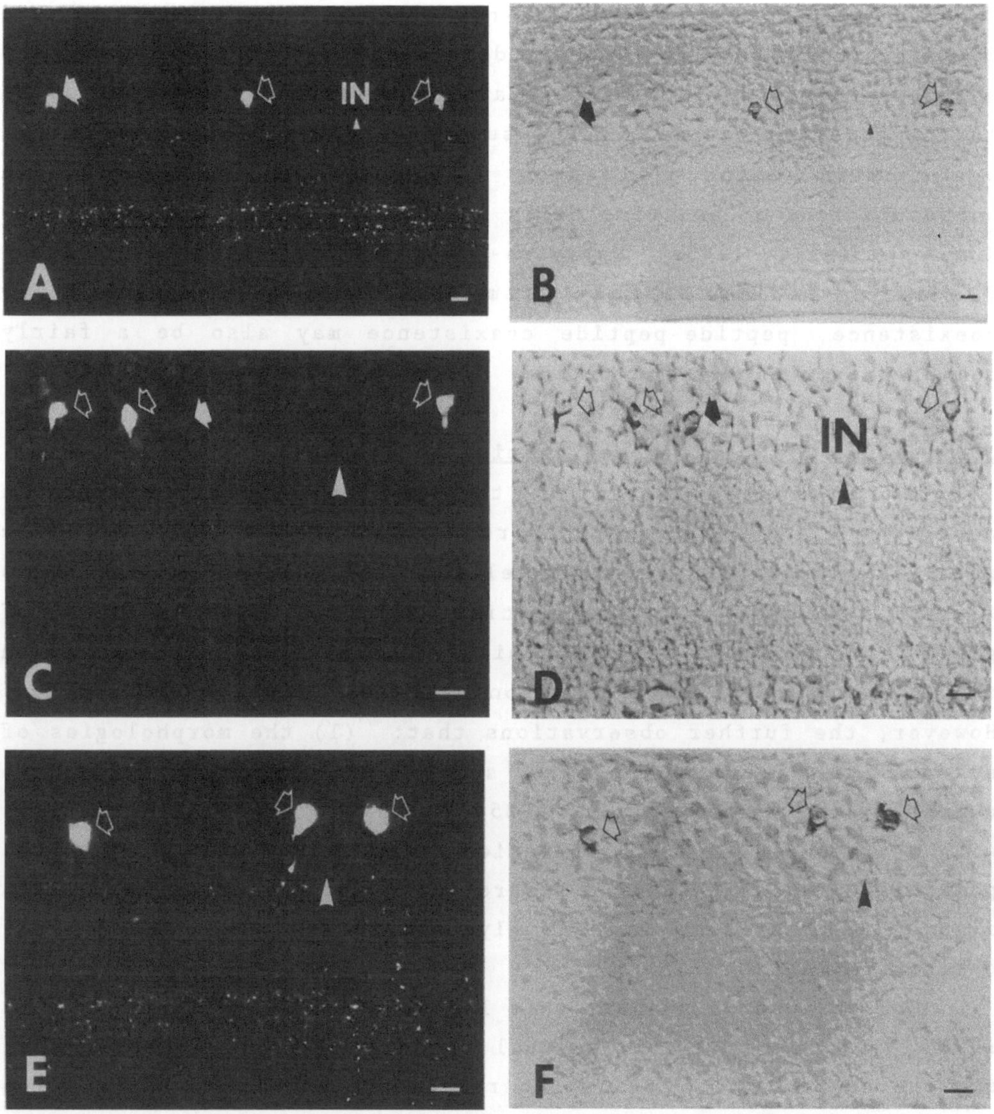

Fig. 2: Paired photomicrographs (A-B, C-D and E-F) demonstrating the same retinal sections processed for immunofluorescence (neurotensin, A, C and E) and immunoperoxidase staining (enkephalin, B, D and F). Amacrine cell somas in the inner nuclear layer (IN) were found to be immunopositive for either neurotensin (compare A and B, solid arrows), enkephalin (compare C and D, solid arrows) or both neurotensin and enkephalin (A-F, hollow arrows). Arrowheads denote border between inner nuclear (IN) and inner plexiform layers. The scale bar equals 10 μm. (From Li et al., 1985; with permission).

In addition to enkephalin-neurotensin coexistence, Li and colleagues (1986) have recently demonstrated that substance P and neurotensin are co-localized in a subpopulation of amacrine cells in the goldfish retina. Also, substance P has been shown to co-occur with either glucagon or pancreatic polypeptide in subpopulations of amacrine cells in the chicken retina (Katayama-Kumoi et al., 1985a; Katayama-Kumoi et al., 1986b). These findings indicate that similar to peptide-transmitter coexistence, peptide-peptide coexistence may also be a fairly common phenomenon in the retina.

Coexistence of Multiple Neuroactive Substances

As presented earlier, the neuropeptides, enkephalin neurotensin and somatostatin were demonstrated to be present in putative glycinergic amacrine cells of the chicken retina. This finding raised the speculation that different types of glycinergic amacrine cells in this retina might be categorized on the basis of their containing one of these three neuropeptides. However, the further observations that: (1) the morphologies of enkephalin-, neurotensin- and somatostatin-amacrine cells are quite similar (Watt et al., 1985a), and (2) enkephalin coexists with neurotensin in a subpopulation of amacrine cells, raised the possibility that enkephalin, neurotensin and perhaps somatostatin might all be present in some glycinergic amacrine cells of the chicken retina.

Recent studies support this latter possibility (Watt et al., 1988). Double- and triple-label paradigms were used to further explore enkephalin's coexisting relationships with GABA, glycine and neurotensin in the chicken retina (Figure 3, Table 1). Double-label studies revealed that 28% of enkephalin-amacrine cells were GABAergic, based on their high-affinity accumulation of the classical neurotransmitter, GABA. Using the same protocol, 53% of enkephalin-amacrine cells were found to be glycinergic. Triple-label experiments (enkephalin-immunocytochemistry combined with both GABA and glycine high-affinity uptake autoradiography) revealed that 81% of enkephalin-amacrine cells exhibit uptake of GABA or glycine and therefore indicate that enkephalin is unlikely to coexist with both GABA

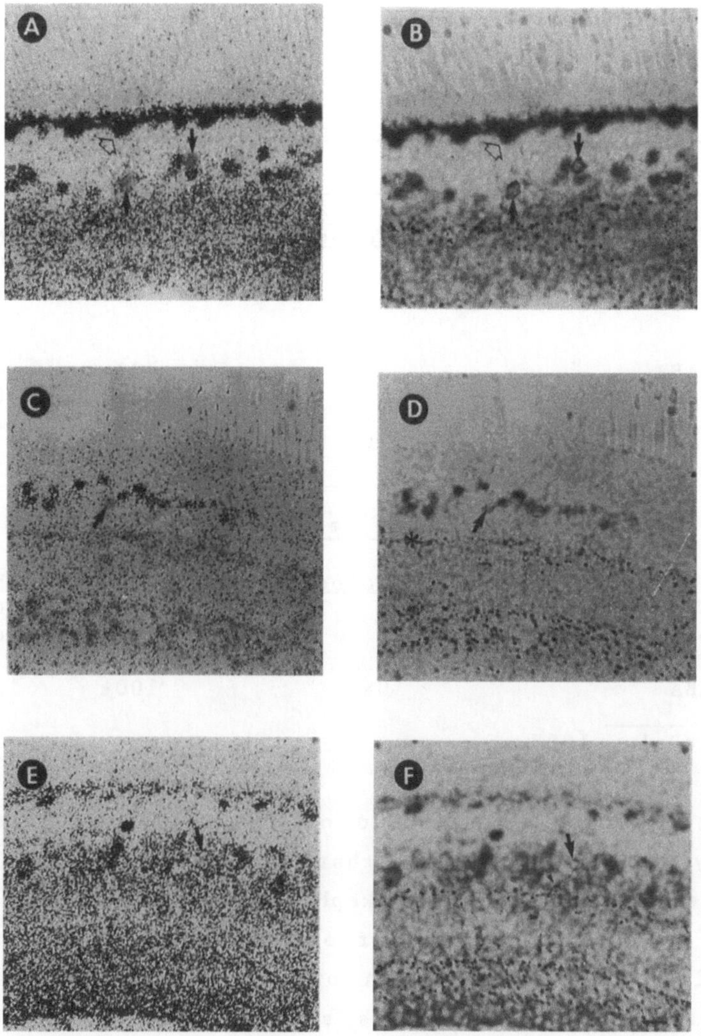

Fig. 3: Photomicrographs demonstrating enkephalin-immunocytochemistry combined with high-affinity uptake autoradiography of: (1) [^{3}H] GABA (A and B); (2) [^{3}H] glycine (C and D); and (3) [^{3}H] GABA and [^{3}H] glycine (E and F). The paired autoradiograms are focussed at the level of the emulsion layer (A, C, E) or the underlying immunoperoxidase staining (B, D, F). The arrows point to enkephalin-immunostained amacrine cell bodies that are not autoradiographically labelled. The open block arrows in A and B point to one of several GABA-labelled horizontal cells. The arrowheads in F point to a process emanating from an enkephalin-immunoreactive cell body. The asterisk in D denotes immunostained puntate deposits in sublayer 1 of the inner plexiform layer (IPL) and demarkates the border between the IPL and the adjacent inner nuclear layer. See Table I for the percentage of double-labelled cells resulting from these experiments. Scale bar equals 10 μm. (From Watt et al., 1988; with permission).

and glycine in chicken amacrine cells.

TABLE I

Quantitative Analysis of Coexisting Relationships

Combinations Tested	% of Total ENK Cells		
	Colocalized	ENK Only	Number of Cells Counted
ENK/GABA	28%	72%	83
ENK/Gly	53%	47%	85
ENK/GABA/Gly	81%	19%	89
ENK/NT*	45%	55%	
	% OF TOTAL NT CELLS		
	Colocalized	NT Only	Number of Cells Counted
NT/GABA	0%	100%	100

*From Li et al., 1985

More importantly and based on the following observations, this study provided evidence that a population of glycinergic amacrine cells contains both enkephalin and neurotensin (Figure 4). First, only 19% of enkephalin-cells were found not to colocalize with either GABA or glycine in the triple-label experiment. Second, a previous study demonstrated that 45% of enkephalin-immunolabelled amacrine cells in the chicken retina are also neurotensin-immunoreactive (Li et al., 1985). Third, coexistence was not observed between neurotensin-immunoreactive amacrine cells and cells that exhibit high-affinity accumulation of ^{3}H-GABA. Therefore, none of the enkephalin-GABA double labelled cells would be expected to contain neurotensin. However, at the least 26% (45% - 19%) of the enkephalin-amacrine cells that accumulate ^{3}H-glycine should also be immunoreactive for neurotensin. Moreover, this prediction was subsequently confirmed by direct triple-label studies which demonstrated that enkephalin and neurotensin immunoreactivities are present in

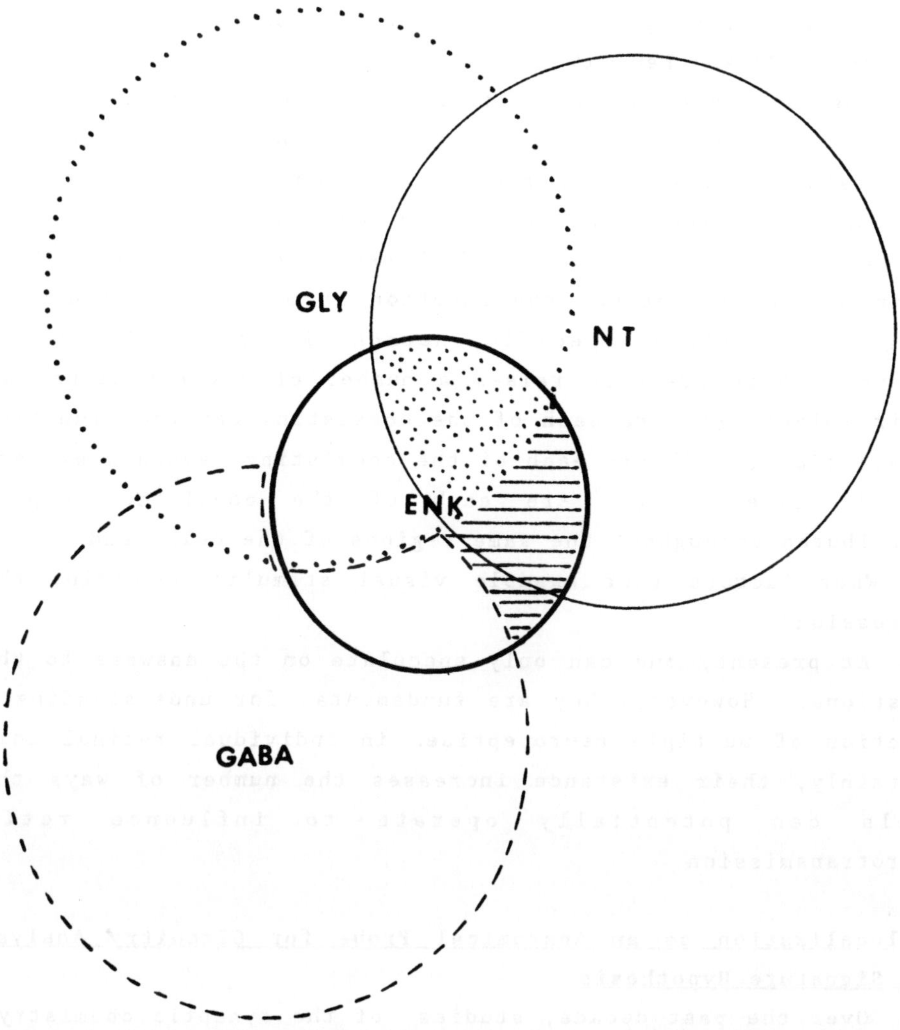

Fig. 4: A set diagram illustrating the relationship between enkephalinergic amacrine cells (ENK) and amacrine cells containing glycine (GLY), GABA and neurotensin (NT). The set boundaries for the ENK, Gly, GABA and NT cell populations are drawn in thick solid line, dotted line, dashed line and thin solid line, respectively. The area shaded with horizontal lines represents the subset of ENKergic cells which do not contain either Gly or GABA. This subset accounts for 19% of the total ENK cells and it is substantially smaller than the overlapped area of the ENK and NT sets (accounts for 45% of the total ENK cells). Therefore, a subset (45%-19%, shaded with small dots) representing amacrine cells containing ENK, NT and Gly must exist. The areas of each set and the overlapping subsets are drawn in approximate scales; the actual percentages of coexistence are listed in Table I. (From Watt et al., 1988; with permission).

glycine-accumulating amacrine cells of the chicken retina (Figure 5, Wu and Lam, 1988).

The observation of multiple peptides in a transmitter-specific retinal neuron is consistent with similar findings in the brain. For example, substance P and thyrotropin-releasing hormone were found in serotonin-containing medullary-spinal projection neurons of the rat (Johansson et al., 1981). In the brain as in the retina, the function of multiple peptides in a single transmitter-specific neuron is presently unknown. However, their presence raises a number of key questions as to their roles: (1) Are each of the coexisting peptides functional at all times? (2) Are each of the coexisting peptides expressed at all times? (3) Are each of the coexisting peptides distributed throughout the same regions of the cell? and (4) What factors (for example visual stimuli) determine their expression?

At present, one can only speculate on the answers to these questions. However, they are fundamental for understanding the function of multiple neuropeptides in individual retinal cells. Certainly, their existence increases the number of ways these cells can potentially operate to influence retinal neurotransmission.

Co-localization as an Anatomical Probe for Circuitry Analyses-The Signature Hypothesis

Over the past decade, studies of the synaptic chemistry of the vertebrate retina have established that even within a single class of neurons, different morphologically and/or functionally distinct subpopulations may use different transmitters (Lam, 1976; Massey and Redburn, 1987). For instance, among the four types of horizontal cells in the goldfish retina, type H1 horizontal cells, which receive input predominantly from red-sensitive cones, are GABAergic, while the other three types are not (Marc et al., 1978; Lam et al., 1979; Lam et al., 1980). Likewise, two different types of goldfish interplexiform cells have been shown to be dopaminergic and glycinergic respectively (Dowling and Ehinger, 1978; Sarthy and Lam, 1979; Chin and Lam, 1980; Marc and Lam, 1981). In other words, these results

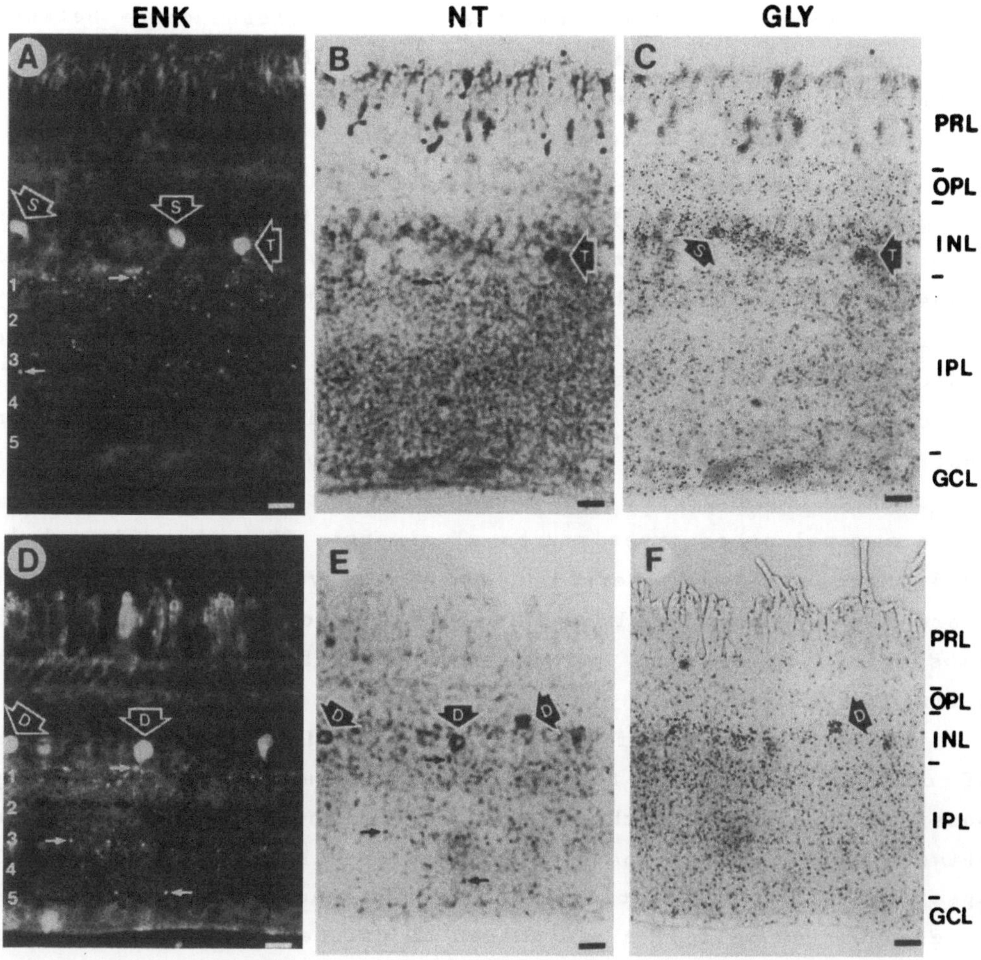

Fig. 5: Simultaneous localization of enkephalin-like (ENK, by immunofluorescence staining, A and D), neurotensin-like (NT, by immunoperoxidase staining, B and E) immunoreactivity and [^{3}H]-glycine (GLY, by autoradiography, C and F) uptake in the chicken retina. The T (triple-label) arrows point to an amacrine cell soma labelled with all three substances. The D (double-label) arrows point to amacrine cell somas labelled with two of the three substances. The S (single-label) arrows point to amacrine cells labelled only with one of the three substances. Small arrows denote fine plexus in sublayers 1, 3 and perhaps 4 and 5 of the inner plexiform layer. [^{3}H]-glycine labels are found in all sublayers PRL: photoreceptor layer; OPL: outer plexiform layer; INL: inner nuclear layer; IPL: inner plexiform layer; GCL: ganglion cell layer; 1-5 denotes distal to proximal sublayers of the IPL. Scale bars equal 10 μm. Retinal sections A-C appear thicker than sections D-F because A-C are more oblique sections. (From Wu and Lam, 1988; with permission).

suggested that there might exist a unique correspondence between the neurochemical specificity and the morphological/physiological specificity of each cell type comprising a single class of retinal neurons. Moreover, it followed that a transmitter-specific label, when used in combination with electron microscopy, might be utilized as an anatomical probe to follow unequivocally the synaptic connectivity of an individual transmitter-specific cell type. While this methodology may still be fruitful for certain situations, recent studies also indicate that this simplistic approach is often inadequate, especially for analyses of the proximal retina. For example, in most vertebrate retinas, a number of morphologically distinct types of amacrine cells are probably GABAergic (Yazulla, 1986), while several other types may be glycinergic (Pourcho and Goebel, 1985; Marc, 1985). Therefore, such observations indicate that there may not necessarily be a one-to-one correspondence between a single transmitter and a morphologically/physiologically-defined cell type.

However, in view of the rapidly expanding number of reports of co-localization in the retina, it should be theoretically possible to preserve the validity of a unique correspondence between neurochemical specificity and morphological/physiological specificity by postulating that within each class of neurons (e.g. amacrine cells), a neuron can be uniquely identified by its characteristic neurochemical signature, i.e. by a combination of its classical transmitters and/or neuroactive peptides. For instance, in the goldfish retina it has been demonstrated that the neurotensin-, substance P- and neurotensin-substance P-immunoreactive amacrine cells each possesses a distinct morphology and pattern of ramification (Li et al., 1986). Likewise, it might be possible to subclassify the different types of glycinergic amacrine cells in the chicken retina with a nomenclature of "glycine-neuropeptide", such as glycine-enkephalin, glycine-neurotensin, glycine-somatostatin, glycine-enkephalin-neurotensin and so forth. It would be of interest to determine whether each of these glycine-peptide amacrine cells correspond to a morphologically and/or functionally distinct cell type. With regard to the retinal GABAergic system, recent studies have revealed that the neuropeptide, enkephalin (Watt et al., 1984; Su et al., 1986; Watt et al., 1986; Watt et al., 1987; Zucker and Adolph et al.; 1988) as well as the classical neurotransmitters, serotonin (Osborne and Beaton, 1986; Wassle and Chun, 1987), acetylcholine (Brecha et al., 1987; Vaney and Young, 1987; Johnson et al., 1988; O'Malley and Masland, 1988) and dopamine (Wässle and Chun, 1987) coexist with GABA in amacrine cells of various submammalian and mammalian species. If the signature hypothesis proves to be generally correct for the various transmitters, peptides and neurons, then the synaptic connectivity of each cell type can be followed unequivocally by simultaneously labelling a cell with the appropriate combination of neuroactive substances. As demonstrated in Figure 6, the simultaneous ultrastructural visualization of ^{3}H-GABA uptake by autoradiography and enkephalin-immunoreactivity by immunocytochemistry provides one of these methods. In addition, with the availability of polyclonal and/or monoclonal antibodies against neuroactive peptides and many of the conventional neurotransmitters as well as their synthesizing enzymes, it is now experimentally feasible to label a neuron simultaneously with at least two antibodies for immunocytochemical studies (Li et al., 1985; Li et al., 1986). Such a double-label immunocytochemical paradigm has recently been used at the electron microscopic level to investigate the coexisting relationship as well as the synaptic connectivity that exists between enkephalinergic and GABAergic cell populations in the turtle retina (Zucker and Adolph, 1988).

The possibility that a neuron may contain two or more neuroactive substances greatly increases the number of cell types that can be uniquely represented neurochemically for a given number of transmitters and neuropeptides. For instance, if the total number of neuropeptides and transmitters present in amacrine cells is η and if each cell type is represented by only two of these substances, then the maximal number of cell types uniquely identified by these substances would be η times (η -1). With respect to the retina, more than a dozen transmitters or neuropeptides have already been localized to amacrine cells, and since there are probably less than one hundred morphologically/

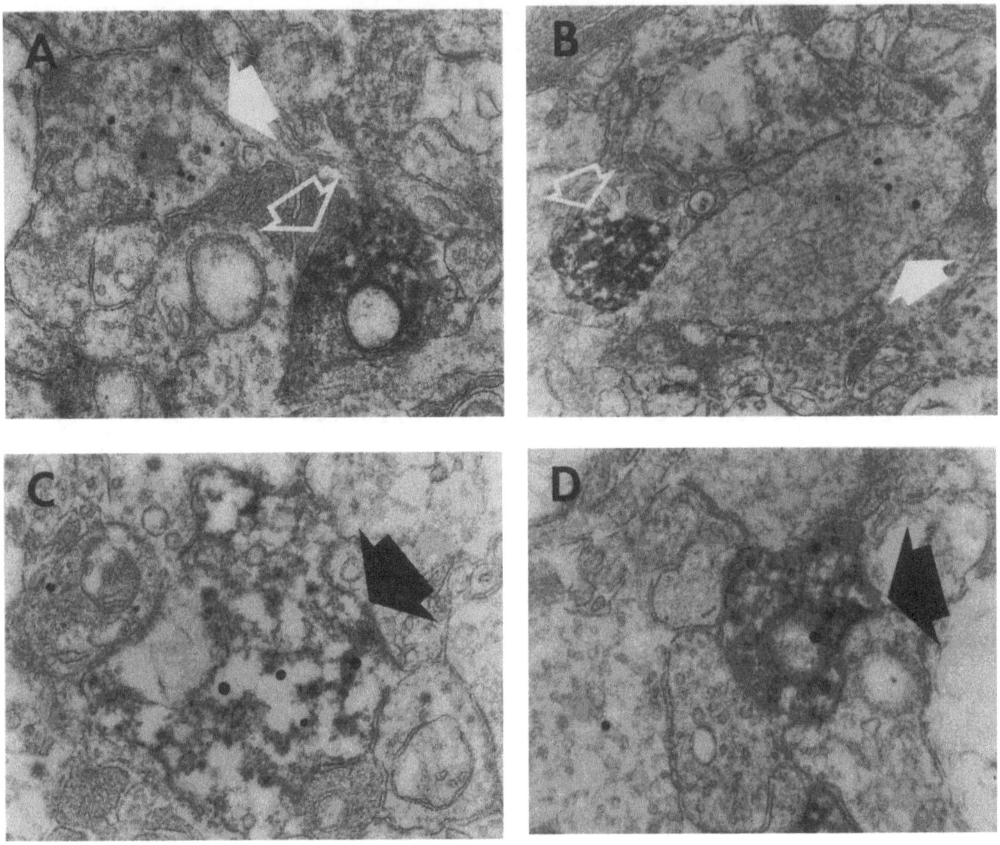

Fig. 6: A montage demonstrating, within the inner plexiform layer of the chicken retina, the simultaneous ultrastructural localizations of enkephalin by immunocytochemistry and [^{3}H]-GABA uptake by autoradiography. In A and B, white solid arrows and white hollow arrows point to processes labelled only for GABA or enkephalin, respectively. In C and D, black solid arrows denote processes that are labelled both for enkephalin-immunostaining and [^{3}H] GABA uptake. Magnification x 37,500. (From Watt and Su, 1988; with permission).

physiologically distinct types of amacrine cells even in the most complex retina, it follows that, if each cell type was represented by at least two neuroactive substances, there would already be more than enough known neuroactive substances to uniquely label every type of amacrine cell. While this certainly does not imply that no additional neuroactive substances will be discovered in amacrine cells, it is also comforting to realize that, at least in theory, the thirty or so morphologically distinct types of amacrine cells thus far described in the retina could each be uniquely categorized neurochemically without the need for identifying an equal number of transmitters or neuropeptides.

Concluding Remarks

Recent demonstrations of the coexistence of multiple neuroactive substances in the nervous system at present reveal only a fairly common morphological phenomenon with as yet little physiological and functional significance. As discussed in the preceding narrative, a number of key questions are raised as to the regulation, expression, distribution and function of multiple neuroactive compounds in an individual neuron. The coexistence of neuroactive substances in a single cell also raises the possibility that in addition to receiving modulatory inputs from surrounding cells, an individual neuron possesses intrinsic mechanisms for regulating closely its own activities and output. This type of self-regulatory mechanism has been postulated as one way to explain the inhibitory effect of enkephalin on the GABA system in the retina (For reviews, see Lam et al., 1986; Lam, 1988; Watt and Su, 1988). An insight into the rules for such autoregulations as well as a general understanding of the functional properties of coexisting neuroactive substances is clearly fundamental to our understanding of neuronal communication. It is likely that such studies may not be totally achieved in intact nervous tissue where neurons are highly interconnected, making it difficult to identify unequivocally the intrinsic properties of an individual neuron. In this regard, the availability of methods for obtaining isolated retinal cells that are viable and readily identified, as well as can be separated into distinct classes, might prove valuable for studying the functional properties of coexistence (e.g. autoregulation) in individual neurons that are isolated from extrinsic cellular interaction.

In the final analysis, the search for retinal neurons that contain more than a single neuroactive substance is clearly only in its infant stage. A systematic continuation of this search and categorization will hopefully provide important insights into the neurochemical basis of retinal organization. It is evident that the quest for a unified relationship between the physiology, morphology and neurochemical specificity of each retinal cell type, if it exists, is not just an academic curiosity. Rather, the availability of a unique neurochemical signature for every cell type within each class of retinal neurons would provide an invaluable probe for understanding, at the cellular and molecular levels, the structural and functional organization of the retina.

Acknowledgements:

We would like to thank Ms. Pat Glazebrook and Ms. Tao Li for expert technical assistance and Ms. Teddy Woodyard for preparing the manuscript. We would like to extend special thanks to Dr. Samuel M. Wu and Dr. Hai-Biao Li for their valuable contributions to the studies presented in this article. This work was supported by NIH grants to Carl B. Watt (EY05622) and Dominic M.-K. Lam (EY02423 and EY02608), and by the Retina Research Foundation (Houston).

References

Boileau G, Barbeau C, Jeannotte L, Chretien, M and Drovin J (1983) Complete structure of the porcine pro-opiomelano cortin m-RNA derived from the nucleotide sequence of cloned c-DNA. Nucleic Acid Res 11:8063-8071

Brecha N, Johnson D, Peichl L and Wässle H (1987) Cholinergic amacrine cells of the rabbit retina express GABA- and GAD-like immunoreactivity. Soc Neurosci Abstr 11:1052

Burnstock G (1976) Do some nerve cells release more than one transmitter? Neurosci 1:239-248

Carraway R and Bhatnagar YM (1980) Immunochemical characterization of neurotensin-like peptides in chicken. Peptides 1:159-165

Chan-Palay V and Palay SL (ed.) (1984) Coexistence of neuroactive substances in neurons. Wiley New York

Chin CA and Lam DMK (1980) The uptake and release of ^{3}H-glycine in the goldfish retina. J Physiol 308:185-195
Cuello AC (ed) (1982) Co-transmission Macmillan Press London
Dowling JE (1987) The Retina. Belknap Press Cambridge, Massachusetts
Dowling JE and Ehinger B (1987) The interplexiform cell system. I. Synapses of the dopaminergic neurons of the goldfish retina. Proc Roy Soc Lond (Biol.) 207:7-26
Dowling JE, Lasater EM, van Buskirk R and Watling KJ (1983) Pharmacological properties of isolated fish horizontal cells. Vision Res 23:421-432
Drujan BD and Svaetichin G (1972) Characterization of different classes of isolated retinal cells. Vision Res 12:1777-1784
Horikawa S, Takai T, Toyosata M, Takahashi H, Noda M, Kakidani H, Kuba T, Hirose T, Inayama S, Hayashida H, Miyata T and Numa S (1983) Isolation and structural organization of the human preproenkephalin B gene. Nature 306:611-614
Johansson O, Hokfelt T, Pernow B, Jeffcoat SL, White N, Steinbusch HWM, Verhofstad AAJ, Emson PC and Spindel E (1981) Immunohistochemical support for three putative transmitters in one neuron: Coexistence of 5-hydroxytryptamine, substance P and thyrotropin releasing hormone-like immunoreactivity in medullary neurons. Neuroscience 6:1857-1881
Johnson D, Brecha N, Peichl L and Wassle H (1988) Localization of GABA- and GAD-like immunoreactivity in cholinergic amacrine cells of the rabbit retina. Soc Neurosci Abstr 29:272
Katayama-Kumoi Y, Kiyama H, Manabe R, Shatani Y and Tohyama M (1985a) Coexistence of glucagon- and substance P-like immunoreactivy in the chicken retina. Neurosci 16:417-424
Katayama-Kumoi Y, Kiyama H, Emson PC, Kimmel JR and Tohyama M (1985b) Coexistence of pancreatic polypeptide and substance P in the chicken retina. Brain Res 361:25-35
Kong YC, Fung SC and Lam DMK (1980) Post-natal development of glycinergic neurons in the rabbit retina. J Comp Neurol 193:1127-1135
Lam DMK (1972) Biosynthesis of acetylcholine in turtle photoreceptors. Proc Nat Acad Sci USA 69:1787-1791
Lam DMK (1976) Synaptic chemistry of identified cells in the vertebrate retina. Cold Spring Harb Symp Quant Biol 40:571-579
Lam DMK (1988) Coexistence and co-function of neuroactive substances in the central nervous system: A view from the vertebrate retina. Proc of the Retina Research Foundation Symp Vol 1 Portfolio Publishing Co Houston 182-197
Lam DMK, Su YYT, Swain L, Marc RE, Brandon C and Wu JY (1979) Immunocytochemical localization of L-glutamic acid decarboxylase in the goldfish retina. Nature 278:565-567
Lam DMK, Su YYT, Chin CA, Brandon C, Wu JY, Marc RE and Lasater EM (1980) GABAergic horizontal cells in the teleost retina. Brain Res Bull 5:137-140
Lam DMK, Li HB, Su YYT and Watt CB (1985) The signature hypothesis: Co-localizations of neuroactive substances as anatomical probes for circuitry analyses. Vision Res 25:1353-1364

Lam DMK, Su YYT and Watt CB (1986) The self-regulating synapse: A functional role for the coexistence of neuroactive substances. Brain Res 11:249-257
Lewis RV and Stern AS (1983) Biosynthesis of the enkephalins and enkephalin-containing polypeptides. Ann Rev Pharmacol Toxicol 23:353-372
Li HB, Watt CB and Lam DMK (1985) The coexistence of two neuroactive peptides in a subpopulation of retinal amacrine cells. Brain Res 345:176-180
Li HB, Marshak DW, Dowling JE and Lam DMK (1986) Co-localiation of immunoreactive substance P and neurotensin in amacrine cells of the goldfish retina. Brain Res 366:307-313
Lundberg JM and Hokfelt T (1983) Coexistence of peptides and classical neurotransmitters. Trends Neurosci 6:325-333
MacLeish, PR, Schwartz EA and Tachibana M (1984) Control of the generator current in solitary rods of the Ambystoma tigrinum retina. J Physiol 348:645-664
Marc RE, Stell WK, Bok D and Lam DMK (1978) GABAergic pathways in the goldfish retina. J Comp Neurol 182:221-246
Marc RE and Lam DMK (1981) Glycinergic pathways in the goldfish retina. J Neurosci 1:152-165
Marc RE (1985) The role of glycine in retinal circuitry. Retinal Transmitters and Modulators: Models for the Brain. Morgan W (ed) CRS Press Boca Raton Fl 120-158
Marc RE and Liu, WLS (1985) ^{3}H-Glycine-accumulating neurons of the human retina. J Comp Neurol 232:241
Massey SC and Redburn DA (1987) Transmitter circuits in the vertebrate retina. Prog in Neurobiol 28:55-96
O'Malley DM and Masland RH (1988) Co-release of acetylcholine and GABA by a retinal neuron. Soc Neurosci Abstr 29:273
Osborne NN and Beaton DW (1986) Direct histochemical localization of 5,7-dihydroxytryptamine and the uptake of serotonin by a subpopulation of GABA neurons in the rabbit retina. Brain Res 382:158-162
Osborne NN (ed) (1983) Dale's principle and communication between neurons. Pergamon Press Oxford
Pourcho RG and Goebel DJ (1985) A combined Golgi and autoradiographic study of (^{3}H)-glycine-accumulating amacrine cells in the cat retina. J Comp Neurol 233:473-480
Sarthy PV and Lam DMK (1979) The uptake and release of ^{3}H-dopamine in the goldfish retina. J Neurochem 32:1269-1277
Scheller RH, Rothman BS and Mayeri E (1983) A single gene encodes multiple peptide-transmitter candidates involved in a stereotyped behavior. Trends in Neurosci 6:340-345
Su YYT, Fry KR, Lam DMK and Watt CB (1986) Enkephalin in the goldfish retina. Cell Mol Neurobiol 6:331-347
Tavella D, Watt CB, Su YYT, Chang KJ, Handlin S, Gaskie V and Lam DMK (1985) The production and partial characterization of monoclonal antibodies against enkephalins. Neurochem Internatl 7:455-466
Vaney DI and Young HM (1988) GABA-like immunoreactivity in cholinergic amacrine cells of the rabbit retina. Brain Res 438:369-373
Wässle H and Chun MH (1987) Dopaminergic and indoleamine accumulating amacrine cells both express GABA-like

immunoreactivity in the cat retina. Soc Neurosci Abstr 13:1051

Watt CB, Su YYT and Lam DMK (1984) Interactions between enkephalin and GABA in an avian retina. Nature 311:761-763

Watt CB, Li HB and Lam DMK (1985a) The presence of enkephalin, neurotensin and somatostatin in glycine-accumulating amacrine cells of the chicken retina. Soc Neurosci Abstr 11:1218

Watt CB, Li HB and Lam DMK (1985b) The presence of three neuroactive peptides in putative glycinergic amacrine cells of an avian retina. Brain Res 348:187-191

Watt CB, Glazebrook PA and Lam DMK (1986) Enkephalins in the teleost retina: localization and coexistence with gamma-aminobutyric acid. Invest Ophthalmol (Suppl) 27:231

Watt CB, LI T, Lam DMK and Wu SM (1987) Interactions between enkephalin and γ-aminobutyric acid in the larval tiger salamander retina. Brain Res 408:258-262

Watt CB, Li T, Lam DMK and Wu SM (1988) Quantitative studies of enkephalin's coexistence with GABA, glycine and neurotensin in amacrine cells of the chicken retina. Brain Res 44:366-370

Watt CB and Su YYT (1988) Enkephalinergic pathways in the retina. In Lam DMK (ed) Proceedings of the Retina Research Foundation Symposium Vol 1 Houston Portfolio Publishing Co 141-161

Weiler R and Ball AK (1984) Co-localization of neurotensin-like immunoreactivity and ^{3}H-glycine uptake system in sustained amacrine cells of turtle retina. Nature 311:759-761

Werblin FS (1978) Transmission along and between rods in the tiger salamander retina. J. Physiol 280:449-470

Wu SM (1985) Synaptic transmission from rods to bipolar cells in the tiger salamander retina. Proc Nat Acad Sci USA 82:3944-3947

Wu SM and Lam DMK (1988) The coexistence of three neuroactive substances in amacrine cells of the chicken retina. Brain Res, in press

Yazulla S (1986) GABA-ergic mechanisms in the retina. In Osborne NN and Chader G (eds) Prog in Retinal Research Pergamon Press Oxford UK 6:1-52

Zucker CL and Adolph AR (1988) Interactions between enkephalin and GABA in the turtle retina. Invest Ophthalmol Vis Sci (Suppl) 29:196

FUNCTIONAL STUDIES ON ENKEPHALIN- AND SOMATOSTATIN-IMMUNOREACTIVE AMACRINE CELLS IN THE CHICKEN RETINA.

I.G. Morgan[1], T.J. Millar[1]*, I. Ishimoto[1]*, M. Boelen[2], M. Dowton[2] and I.W. Chubb[2].

[1]Centre for Visual Sciences, and Visual Sciences Group, Research School of Biological Sciences, Australian National University, Acton, ACT 2601
and
[2]Neuroscience Research Unit, Department of Biology, University of Wollongong, NSW 2500, Australia.

Prominent amongst the neuropeptide-immunoreactive cells of the chicken retina are populations which contain enkephalin-immunoreactive (EI) material (Brecha *et al.* 1979; Tornqvist *et al.* 1981; Fukuda 1982; Ishimoto *et al.* 1983; Millar *et al.* 1984; Watt *et al.* 1984; Watt *et al.* 1985a) or somatostatin-immunoreactive (SI) material (Brecha *et al.* 1981; Buckerfield *et al.* 1981; Tornqvist *et al.* 1981; Fukuda 1982).

The basic morphologies of the two cell types are similar, with pear-shaped cell bodies displaced two to three cell layers from the border of the inner nuclear layer (INL) and the inner plexiform layer (IPL), and dendrites which arborize in strata 1 and 3-5 of the IPL. Displaced EI and SI cells are rarely if ever seen. In the peripheral retina, where the cell

*Present addresses:
Save Sight and Eye Health Institute, Department of Ophthalmology, Sydney Eye Hospital, Sir John Young Crescent, Woolloomooloo, NSW 2011, Australia (TJM): Department of Ophthalmology, Osaka University Medical School, 1-1-56 Fukushima, Fukushimaku, Osaka (553), Japan (II).

NATO ASI Series, Vol. H31
Neurobiology of the Inner Retina
Edited by R. Weiler and N.N. Osborne
© Springer-Verlag Berlin Heidelberg 1989

density is lower, the cells appear to be narrow-field, with dendritic fields of diameter around 50 - 100 um. The EI and SI cells are similarly distributed across the retina, with high densities corresponding to the area of highest ganglion cell density (Ehrlich 1981), and the lowest densities being recorded in the superior and temporal retinas (Kiyama *et al.* 1985; Ishimoto *et al.* 1986, unpublished results). Retinal mosaics of the cells are quite regular, with regularity indices in the 3-5 range.

Despite the many similarities in properties at the light microscopic level, these neuropeptides do not seem to co-exist in the same amacrine cell. As will be detailed below, there are a number of properties in which the EI and SI amacrine cells are distinct. This evidence rules out the total overlap of these cell populations, although a certain percentage of overlap cannot be excluded.

There is no evidence that these EI and SI cell populations are heterogeneous, apart from colocalization studies. In terms of morphology and retinal mosaic, response to neurotoxins (Buckerfield *et al.* 1981; Millar *et al.* 1986), and patterns of storage of peptides under different lighting conditions (see below), the populations of EI and SI cells are homogeneous.

Co-existence of transmitters.

While there is no evidence for the co-existence of EI material and SI material, there does seem to be some co-existence with other putative transmitters. In the chicken retina, the population of EI cells can be broken up into five different groups, those which contain EI material only, those which contain one of GABA, glycine or neurotensin as well, and those which contain both glycine and neurotensin as well (Watt *et al.* 1988). Glycine is found in 53% of the EI cells, neurotensin in 45% and GABA in 28%. The percentage of cells containing glycine and neurotensin as well as EI material was not determined precisely, but must be at least 26%. In the larval tiger salamander retina, almost all EI cells were able to take up GABA (Watt *et al.* 1987). Much less is known about the SI cells, although some co-localization with glycine has been reported (Watt *et al.* 1985b).

The only limitation of these studies concerns the technique used for demonstrating the presence of putative glycinergic and GABAergic neurons, audoradiographic detection of sites of high affinity glycine and GABA uptake. How specific this approach is for glycinergic and GABAergic neurons remains to be established, since uptake sites could also represent mechanisms of inactivation.

The demonstration of considerable heterogeneity in the content of putative transmitters of the EI cells, contrasts with the picture of homogeneity which emerged in the preceding section. Lam and colleagues (Lam *et al.* 1985; Lam *et al.* 1986) have developed hypotheses as to why there might be co-existence. Lam *et al.* (1986) postulate that enkephalins co-released with GABA might inhibit the release of GABA from that terminal via auto-receptors, which would be consistent with the demonstrated ability of met-enkephalin to inhibit GABA release in the retina. The degree of inhibition seems to be large (around 50%), given that only a few percent of the cells which can take up GABA seem to contain EI material, unless the EI material can exert its regulatory effects at a distance.

The second hypothesis is that specific combinations of peptides and non-peptide transmitters may define particular types of cells (Lam *et al.* 1985). For example, in the goldfish retina, three morphologically distinct types of amacrine cells are defined by the presence of substance P-immunoreactivity, neurotensin-immunoreactivity or their co-occurrence (Li *et al.* 1986). Thus the presence of enkephalins in the GABA-labelled or glycine-labelled cells may define specific subpopulations of GABAergic or glycinergic cells. However it should be noted that not all cells of the same morphology and response properties contain the same mix of transmitter. Moreover, the division of the enkephalin cells into five groups does not correspond to any known sub-division, although it is still possible that the particular peptide signature of the cells is related to their target specificity, rather than the essentially pre-synaptic features analysed so far.

Transmitter status.

The characterization of enkephalin as a neurotransmitter in the chicken retina is by far the most complete of any retinal peptide. EI material has been detected by radioimmunoassay, and consists of both met- and leu-enkephalin and a range of precursors (Humbert *et al.* 1979; Watt *et al.* 1984; Millar *et al.* 1984; Su *et al.* 1985; Boelen *et al.* 1989). Radio-labelled methionine has been shown to be incorporated into met-enkephalin, and it has also been shown that such radio-labelled material can be released from the retina *in vitro* by exposure to high K^+ (Su *et al.* 1985). Endogenous leu-enkephalin has also been shown to be released by exposure to high K^+ and it has been further shown that the release of endogenous leu-enkephalin is high in the dark and low in the light (Boelen *et al.* 1989) Enkephalin receptors have been characterized in the chicken retina (Slaughter *et al.* 1985), where they are predominantly of the delta and kappa, rather than the mu type. Evidence for electrophysiological effects of the enkephalin pentapeptides is not available in the chicken retina, but in other species there is some, if sometimes conflicting, evidence. In mud-puppy retina, enkephalins have been reported to inhibit all ganglion cells (Dick and Miller 1981), whereas morphine has been reported to excite about 20% of the ON-OFF ganglion cells in the tiger salamander retina (Watt *et al.* 1987). In the gold-fish retina, evidence was obtained that enkephalins excited ON-ganglion cells and inhibited some OFF-ganglion cells (Djamgoz *et al.* 1981), probably via a poly-synaptic pathway. Another indication of post-synaptic effects comes from the ability of enkephalin to inhibit K^+-stimulated release of GABA (Watt *et al.* 1984), glycine (Su 1986) and dopamine (Su *et al.* 1984) from the chicken retina. However it should be noted that Bauer *et al.* (1985) found that leu-enkephalin stimulated the release of dopamine, but had no effect on the release of GABA and glycine from the chicken retina. Dubocovich and Weiner (1983) found that the enkephalins inhibited dopamine release from the rabbit retina. Some of these effects could be mediated by autoreceptors. Nevertheless, the enkephalins are good candidates for retinal neurotansmitters, although the precise nature of the post-synaptic effects they mediate remains to be established.

In contrast, the status of the SI material in the chicken retina is much less clear. SI material in the chicken retina has been detected by radio-immunoassay (Buckerfield *et al.* 1981; Ishimoto *et al.* 1986), but its molecular form has not yet been established. In other species, somatostatin-14 and somatostatin-28 have been shown to be present (Rorstad *et al.* 1979; Yamada *et al.* 1980; Yamada and Basinger 1982; Marshak *et al.* 1983; Spira *et al.* 1984) and retinal synthesis of somatostatin and precursors has been demonstrated (Yamada and Basinger 1982). The release and physiological effects of somatostatin in the retina have not yet been demonstrated, although somatostatin has been reported to increase the amplitude of the oscillatory potentials of the ERG in the mud-puppy (Wachtmeister 1983), and Bauer *et al.* (1985) reported that somatostatin stimulated serotonin release from the chicken retina.

Synaptology.

One clue as to the functions of amacrine cells may come from the pattern of their synaptic connections. We have previously characterized the synaptic connections of the SI cells in the chicken retina (Ishimoto *et al.* 1986). The SI cells receive only bipolar cell input in stratum 1, but not in strata 3-5, and only amacrine cell input in strata 3-5 but not in 1. They contact other amacrine cells in both strata, but no synapses between SI cells or with ganglion cells have been detected.

There are two conflicting reports on the synaptic connections of the EI cells. Ishimoto *et al.* (1983) reported that the EI cells were purely inter-amacrine, with no input from bipolar cells, and no contact with ganglion cells. However, Watt *et al.* (1985c) reported that EI cells received input from bipolar cells and synapsed onto ganglion cells. To clarify this point we have re-examined the synaptic connections of the EI cells, and have confirmed the results of Watt *et al.* (1985c). Thus we find that the EI cells receive input from both bipolar cells and amacrine cells, and synapse with amacrine cells on ganglion cells. No synapses between EI cells were detected. The frequency of bipolar cell input to the EI cells was much lower than that in stratum 1 for the SI cells, and the frequency of EI cell synapses with ganglion cells was very low. These two factors probably explain the original report of Ishimoto *et al.* (1983).

So far, little is known about the types of bipolar, amacrine and ganglion cells with which the EI and SI cells interact. In addition to the evidence for colocalization, Watt *et al.* (1984) have also reported input from EI processes to processes which have taken up GABA. This is consistent with the model proposed from their physiological studies on the gold-fish retina (Djamgoz *et al.* 1981), that the enkephalins mediate their excitatory effects on ganglion cells, by inhibiting GABAergic inhibitory inputs to the ganglion cells affected.

The distinct distribution of synapses onto the EI and SI cells may point to significant differences in function. If the OFF-ON lamination of the IPL (Famiglietti *et al.* 1977) applies to the avian retina, the EI cells might receive input from both OFF- and ON-bipolar cells, whereas the SI cells might receive input only from OFF-bipolars.

Diurnal variations in peptide levels.

The levels of both EI and SI material vary significantly with lighting conditions (Millar *et al.* 1984; Ishimoto *et al.* 1986). For animals held on a 12:12 light:dark cycle, the levels of EI and SI material rise during the light and fall during the dark period. The changes are not small. In a typical experiment the peptide levels may double from the end of the dark phase to the end of the light, halving them to the end of the dark. The variations in peptide levels are light-driven, not circadian, since prolonging the dark phase leads to a further decline in peptide levels before a plateau is reached, while prolonging the light phase leads to a further increase in peptide levels before a plateau is reached. These variations could also be detected immunohistochemically, except that at the end of the dark phase EI and SI material was virtually undetectable, presumably because of thresholds for detection of immunofluorescence.

In an attempt to explain these changes we postulated that the EI and SI cells were active in the dark, leading to rapid release and breakdown of the peptides which out-stripped synthetic translation and processing rates, thus producing a decrease in the levels of the peptides. In the light the cells could be inactive, leading to decreased release and breakdown of the peptides which would be out-stripped by synthetic rates, allowing the peptides to accumulate. This postulate was based on the simple idea that the potential of visual stimulation to control the rates of transmitter release in visual neurons was obvious, whereas potential controls over translation, processing or degradation were not so obvious. It also implies that degradation of the peptides occurs predominantly after release, while vesicular or cytoplasmic degradation is minimal.

We have now extended these studies to look at the effect of differing light intensities on the levels of EI and SI material in the chicken retina. As shown in figure 1, as soon as the light intensity reaches mesopic levels, i.e. as soon as the cones begin to operate, levels of EI material show the light phase pattern of accumulation. There are no further changes with further increases in light levels. This contrasts with the pattern shown by the SI cells. Peptide levels increase upon moving from scotopic to mesopic conditions, but the level to which the peptides accumulate is roughly proportional to the log of the light intensity up to

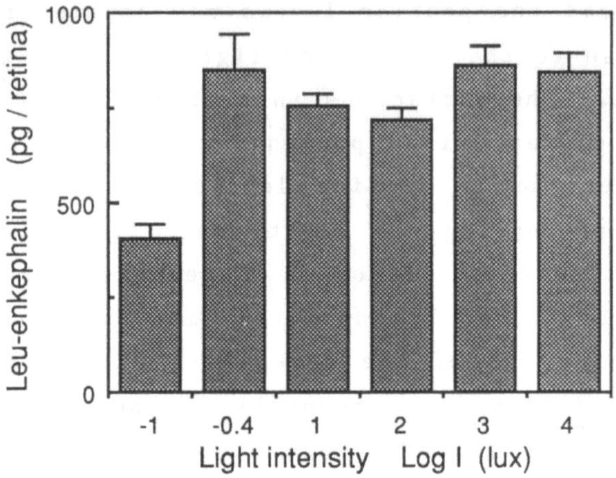

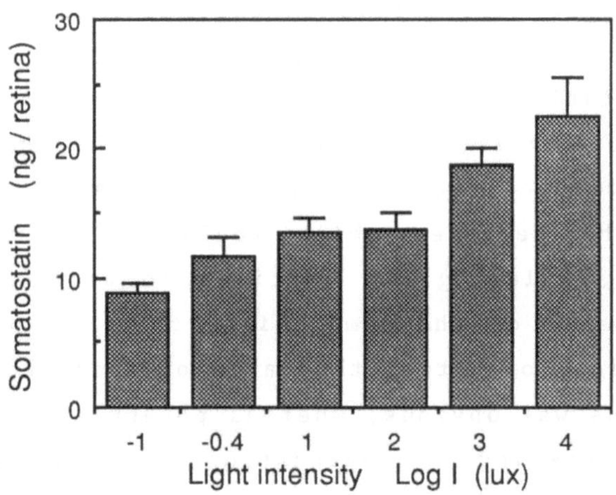

Figure 1: The top figure shows the levels of EI material detected in the retina after 12H of exposure to light of the intensities specified. The bottom figure shows similar data for somatostatin $10^{-0.4}$ lux represents mesopic conditions.

10^4 lux, the highest light intensity we have examined so far.

Thus, the EI and SI cells, although they showed similar variations in peptide levels when held on the same light-dark cycle, differ fundamentally in their response to light. Pursuing our hypothesis that the variations in peptide levels are due to light modulation of the rates of release, the EI cells act rather like a flip-flop device, moving from high levels of activity under scotopic conditions, to a constant low level of activity under mesopic and photopic conditions. In contrast, while the SI cells begin to respond to light with decreased activity under mesopic conditions, they become less and less active with increasing light intensity, acting to some extent as an inverse light meter.

Is modulation of release the crucial factor?

In order to validate our hypothesis, it is first of all essential to demonstrate that light can modulate the rate of release of EI and SI material, and to show that there is little modulation of other potential bottle-necks. Using perfused retinas *in vitro* (Boelen *et al.* 1989), it has now been shown that light decreases the rate of release of EI material, and that the released material is exclusively the pentapeptide, which, if peptidase inhibitors are not added, is then rapidly degraded. The change in rate of release *in vitro* is sufficient on its own, to account for the changes in levels of the EI material *in vivo*, under the different lighting conditions. This is consistent with our original hypothesis, in showing that light can inhibit the release of EI material.

In order to investigate other sites of modulation, the relationships between the levels of precursors and leu-enkephalin were then investigated (Boelen *et al.* 1989). The first point of significance is that the levels of the precursors were high, with the retina containing approximately equal amounts of native leu-enkephalin and leu-enkephalin equivalents in precursors at the end of the dark period. This result implies that the steps between translation to form the original pre-pro-enkephalin A, and formation of native leu-enkephalin are not rapid compared to the rate of synthesis, for if they were, the levels of precursors detectable should be low. Furthermore, while the levels of leu-enkephalin vary with the light-dark cycle, to a first approximation the levels of precursor leu-enkephalin equivalents do not. This result would only be obtained if there were a significant bottle-neck at the final conversion of precursors to leu-enkephalin, for otherwise some precursors should accumulate. This result further implies that there is little modulation of translation or processing rates, unless they are subject to balanced co-regulation to retain constant levels of precursor. The most likely interpretation of these results is therefore that, in the light the rate-limiting step in the metabolism of leu-enkephalin is the rate of release and subsequent degradation. In the dark, the rate of release and subsequent degradation increases markedly exposing rate-limiting steps in the synthesis of native pentapeptide, including a bottleneck at the conversion of pecursors to leu-enkephalin.

Thus, the major, if not the only contributor to the variations in levels of leu-enkephalin during the 12:12 light:dark cycle is the modulation of release rates by light. These results therefore support our original hypothesis, and **allow us to use the rates of accumulation or disappearance of leu-enkephalin as an index of the activity of the EI cells.** Similar experiments have not yet been carried out on SI material, so that our hypothesis for the SI cell is still to be validated, but it will be assumed for the rest of this article.

Control of the state of activity of the EI and SI cells.

The preceding results indicate that in the dark the EI and SI cells are on balance more active, and that in the light the balance shifts so that the cells become less active. This could come about by an increase in inhibition or a decrease in excitation in the light, or by a decrease in inhibition or an increase in excitation in the dark. It is possible to examine whether there are increases in excitation or inhibition by using pharmacological antagonists to block them. Decreases are more difficult to deal with pharmacologically. We have used picrotoxin and strychnine to block possible GABA- and glycine-mediated increases in inhibition, while kynurenic acid, a non-specific excitatory amino acid antagonist (Watkins and Olverman 1987), was used to block excitatory amino acid inputs (fig. 2).

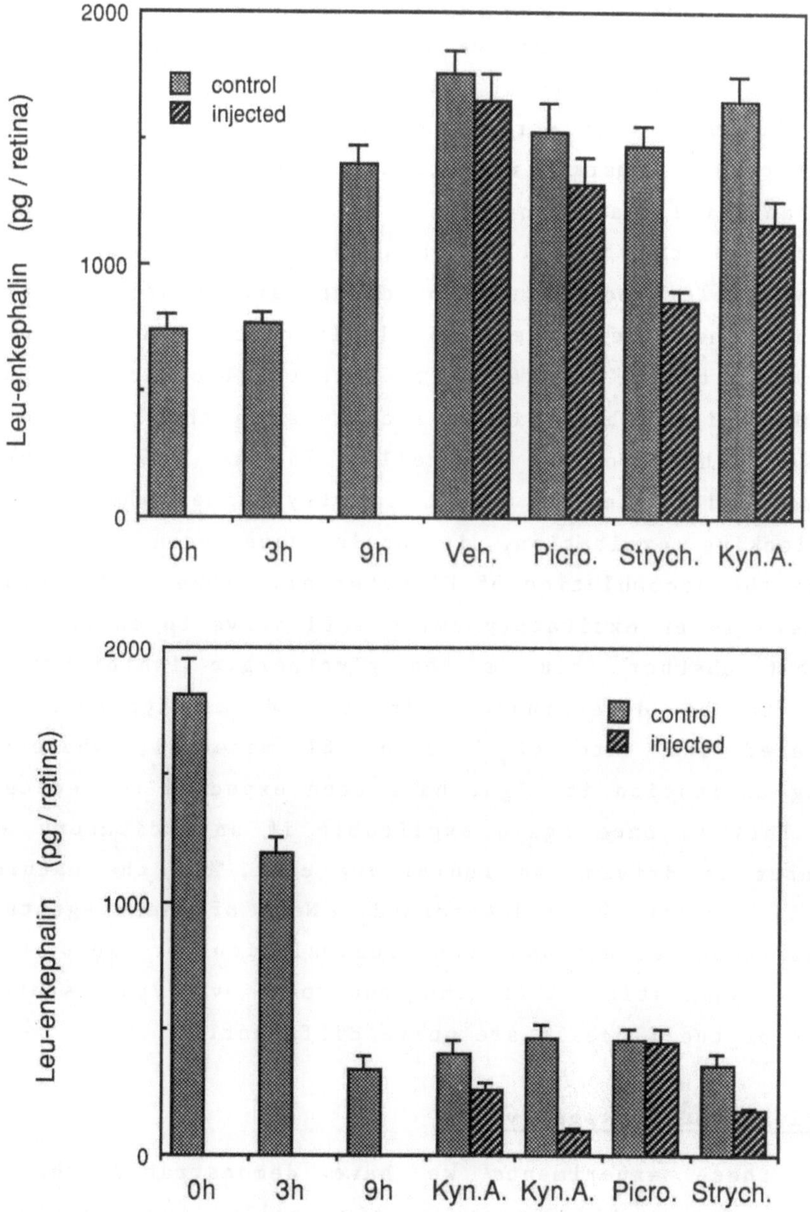

Figure 2: The upper figure shows the effect of intravitreal injection of picrotoxin (10 nmole), strychnine (10 nmole) or kynurenic acid (2 umole) on the accumulation of EI material in the light. The bottom figure shows the effect of intravitreal injection of picrotoxin (10 nmole), strychnine (10 nmole) or kynurenic acid (200 nmole or 2 umole) on the loss of EI material in the dark.

Application of picrotoxin had no effect on the accumulation or disappearance of EI material, indicating that there were no significant GABAergic controls over the EI cells. However, when strychnine was applied during the light phase it caused a marked reduction in the rate of accumulation of EI material, as though it had relieved a glycinergic inhibition of the EI cell. When applied during the dark it slightly accelerated the rate of disappearance of EI material, suggesting that there was also glycinergic inhibition in the dark. Thus, the EI cell seems to receive glycinergic input in the dark, which substantially increases in the light leading to marked inhibition of the cell. In the light, kynurenic acid produced a similar effect to strychnine, whereas if it were blocking excitation, it would have been expected to increase the accumulation of EI material. Thus it is possible that there is an excitatory amino acid drive to an inhibitory cell, but whether this is the glycinergic inhibitory cell remains to be determined. In the dark, kynurenic acid accelerated the rate of loss of EI material, whereas by blocking excitation it might have been expected to reduce the rate. This is once again explicable if an excitatory amino acid input is driving an inhibitory cell, but the nature of that cell remains to be determined. None of these agents had a significant effect on the accumulation or loss of SI material, suggesting that the controls over the state of activity of the SI cells are quite different.

Conclusions and Perspective.

In these experiments we have demonstrated that two populations of peptide-immunoreactive cells, those containing EI and SI material behave in fundamentally different ways in response to light. The EI cells are relatively inactive under mesopic and photopic conditions, but their level of inactivity does not vary with light intensity. In contrast, while the SI cells are less active under mesopic and photopic conditions than under scotopic, their level of inactivity varies with light intensity.

It must be stressed that we are talking about the balance of activity over long periods of time, and that it is not clear how to relate these measures to the operation of ON- and OFF-pathways in the retina, which respond to changes in intensity, rather than to constant light levels. Under scotopic, mesopic and photopic conditions both ON- and OFF-pathways should be operative, the only difference is in whether they are rod- or cone-driven.

The behaviour of the EI cells is consistent with their involvement in the selective processing of rod or cone inputs since they seem to act as a flip-flop device, active in the dark and less active in the light. The SI cells are more likely to be involved in some visual process where absolute light level is important - control of pupil size being one example. Another possibility is that the cells are involved in some way in the processes of dark and light adaptation. Whatever function they serve, it is likely to be locally controlled wihin the retina, since, although there is almost complete retinal coverage, the individual cells are narrow-field, their dendritic fields show little over-lap, and they do not synapse with each other.

The clearest prediction from these results relates to the glycinergic inhibition of the EI but not the SI cells. Although there is no evidence that the input is direct, it will be worthwhile looking for glycine receptors on the EI cells, while they might not be present on the SI cells. Given the lack of effect of picrotoxin, neither might possess GABA receptors. Given the results with kynurenic acid, the source/s of excitation to the EI and SI cells also need clarification, as do the inhibitory inputs to the SI cells, and those other than glycinergic to the EI cells.

These results have implications for peptidergic cells in general. The model by which the major site of modulation of peptide metabolism is release, with litle apparent flexibility in translation and processing even over quite prolonged periods, may be applicable to other peptidergic neurons, both within the retina, and elsewhere in the nervous system. Although other peptidergic cells need not be organized to respond to light-dark cycles in this way, they may still be vulnerable to prolonged periods of stimulation, during which their peptide levels may run down, potentially compromising their functions. This adds further weight to the metabolic inflexibility of peptidergic neurons postulated by Hokfelt *et al.* (1980).

Acknowledgements.

This work has been supported by grants from the National Health and Medical Research Council (IWC and TJM), the Sydney Hospital Research Foundation (TJM) and the OPSM Foundation (TJM).

References.

Bauer B, Ehinger B, Tornqvist K, Waga J (1985) Neurotransmitter release by certain neuropeptides. Acta Ophthalmol 63:581-587

Boelen M, Dawton M, Chubb IW (1989) these proceedings.

Brecha N, Karten HJ, Laverack C (1979) Enkephalin-containing amacrine cells in the avian retina: immunohistochemical localization. Proc nat Acad Sci 76:3010-3014

Brecha N, Karten HJ, Schenter C (1981) Neurotensin-like and somatostatin-like immunoreactivity within amacrine cells of the retina. Neurosci 6:1329-1340

Buckerfield M, Oliver J, Chubb IW, Morgan IG (1981) Somatostatin-like immunoreactivity in amacrine cells of the chicken retina. Neurosci 6:689-695

Dick E, Miller RF (1981) Peptides influence retinal ganglion cells. Neurosci Lett 26:131-135

Djamgoz MBA, Stell WK, Chin CA, Lam DMK (1981) An opiate system in the goldfish retina. Nature 292:620-623

Ehrlich D (1981) Regional specialization of the chick retina: a ganglion cell layer study. J comp Neurol 195:643-657

Famiglietti EV, Kaneko A, Tachibana M (1977) Neuronal architecture of on- and off-pathways to ganglion cells in carp retina. Science 198:1267-1269

Fukuda M (1982) Localization of neuropeptides in the avian retina: an immunohistochemical analysis. Mol Cell Biol 28:275-283

Hokfelt T, Johansson A, Ljungdahl A, Lundberg JM, Schultzberg M (1980) Peptidergic neurones. Nature 284:515-521

Humbert J, Pradelles P, Gross C, Dray F (1979) Enkephalin-like products in embryonic chicken retina. Neurosci Lett 12:259-263

Ishimoto I, Shiosaka S, Shimizu Y, Kuwayama Y, Fukuda M, Inagaki S, Takagi H, Sakanaka M, Saszoka A, Senba E, Sakiyama T, Tohyama M (1983) Leucine-enkephalin-like immunoreactivity in the chicken retina with a special reference to its fine structure. Invest Ophthalmol vis Sci 24:879-885

Ishimoto I, Millar TJ, Chubb IW, Morgan IG (1986) Somatostatin-immunoreactive amacrine cells of chicken retina: retinal mosaic, ultrastructural features, and light-driven variations in peptide metabolism. Neurosci 17:1217-1233

Kiyama H, Katayama-Kumoi Y, Kimmel J, Steinbusch H, Powell JF, Smith AD, Tohyama M (1985) Three dimensional analysis of retinal neuropeptides and amines in the chick. Brain Res Bull 15:155-165

Lam DMK, Li HB, Su YYT, Watt CB (1985) The signature hypothesis: co-localization of neuroactive substances as anatomical probes for circuitry analyses. Vision Res 25:1353-1364

Lam, DMK, Su YYT, Watt CB (1986) The self-regulatory synapse: a functional role for the co-existence of neuroactive substances. Brain Res Rev 11:249-257

Li HB, Marshak DW, Dowling JE, Lam DMK (1986) Co-localization of immunoreactive substance P and neurotensin in amacrine cells of the goldfish retina. Brain Res 366:307-313

Marshak DW, Reeve JR, Shively JE, Hawke D, Takami MS, Yamada T (1983) Structure of somatostatin isolated from bovine retina. J. Neurochem 41:601-606

Millar TJ, Salipan N, Oliver JO, Morgan IG, Chubb IW (1984) The concentration of enkephalin-like material in the chick retina is light dependent. Neurosci 13:221-226

Millar TJ, Boelen M, Morgan IG (1986) Effects of excitotoxins on amacrine cell types in the chicken retina. Proc Aust Biochem Soc 18:73

Rorstad OP, Brownstein MJ, Martin JB (1979) Immunoreactive and biologically active somatostatin-like material in rat retina. Proc Natn Acad Sci USA 76:3019-3023

Spira AW, Shimizu Y, Rorstad OP (1984) Localization, chromatographic characterization, and development of somatostatin-like immunoreactivity on the guinea pig retina. J Neurosci 4:3069-3079

Su YYT (1986) Effect of enkephalins on glycine release in the chicken retina. Soc Neurosci Abstr 12:641

Su YYT, Su S, Lam DMT (1984) An enkephalin in the chicken retina: Receptor binding and physiological studies. ARVO Abstr 292

Su YYT, Watt CB, Lam DMK (1985) Opioid pathways in an avian retina. I. The content, biosynthesis and release of Met^5-enkephalin. J Neurosci 5:851-856

Tornqvist K, Loren I, Kakanson R, Sundler F (1981) Peptide-containing neurons of the chicken retina. Exp Eye Res 33:55-64

Wachtmeister L (1983) The action of peptides on the mudpuppy electroretinogram (ERG). Exp Eye Res 33:429-437

Watkins JC, Olverman HJ (1987) Agonists and antagonists for excitatory amino acid receptors. TINS 10:265-272

Watt CB, Su YYT, Lam DMK (1984) Interactions between enkephalin and GABA in an avian retina. Nature 311:761-763

Watt CB, Su YYT, Lam DMK (1985a) Enkephalins in the vertebrate retina. Prog Ret Res 4:221-240

Watt CB, Li HB, Lam DMK (1985b) The presence of three neuroactive peptides in putative glycinergic amacrine cells of an avian retina. Brain Res 348:187-191

Watt CB, Su YYT, Lam DMK (1985c) Opioid pathways in an avian retina. II. Synaptic organization of enkephalin-immunoreactive amacrine cells. J Neurosci 5:857-865

Watt CB, Li T, Lam DMK, Wu SM (1987) Interactions between enkephalin and gamma-aminobutyric acid in the larval tiger salamander retina. Brain Res 408:258-262

Watt CB, Li T, Lam DMK, Wu SM (1988) Quantitative studies of enkephalin's coexistence with gamma-aminobutyric acid, glycine and neurotensin in amacrine cells of the chicken retina. Brain Res 444:366-370

Yamada T, Basinger S (1982) Biosynthesis of somatostatin-like immunoreactivity by frog retinas *in vitro*. J Neurochem 39:1539-1546

Yamada T, Marshak D, Basinger S, Walsh J, Morley J, Stell W (1980) Somatostatin-like immunoreactivity n the retina. Proc Natn Acad Sci USA 77:1691-1695

PHYSIOLOGY, MORPHOLOGY AND ACTIVITY OF AMACRINE CELLS WITH GLUCAGON-LIKE IMMUNOREACTIVITY IN THE TURTLE RETINA.

Reto Weiler[1], Uwe D. Behrens[1] and Josef Ammermüller[2]
[1]Dept. Neurobiology, Univ. of Oldenburg, Oldenburg, FRG;
[2]Zoological Inst., Univ. of Munich, Munich, FRG

There is ample evidence that the gut peptide glucagon (GLG) serves as a neurotransmitter/neuromodulator in the central nervous system (Saskai et al., 1985, Drucker & Asa, 1988). In the retina, such a role is less clear, although GLG-like immunoreactivity (GLG-IR) has been demonstrated in several vertebrate retinas (Tornqvist et al., 1981; Kuwayama et al., 1982; Tornqvist & Ehinger, 1983; Eldred & Karten, 1983; Brecha, 1983; Bruun et al., 1984; 1985; Brecha & Karten, 1985; Das et al., 1985; Yang & Yazulla, 1986; Gläsener et al., 1988).

In all retinas so far investigated, GLG-IR occurs in amacrine cells having their cell bodies in the inner half of the inner nuclear layer. GLG-IR amacrine cells in all species are multistratified; the level of ramification within the inner plexiform layer (IPL), however, varies from species to species with a predominance in layers 1 and 5. In all retinas the density of GLG-IR amacrine cells is rather low and varies across the retina (Brecha, 1983; Eldred & Karten, 1983). The total content of GLG in different retinas determined by radioimmunoassay is higher (10 - 35 pg/mg protein) in cold blooded vertebrates and birds than in mammals (0.04 - 0.2 pg/mg protein) (Ekman & Tornqvist, 1985). GLG is known to stimulate the formation of cAMP in the chicken retina (Schorderet et al., 1981; Kuwayama et al., 1982) and in cultured glial cells of the embryonic chicken retina (Koh et al., 1984).

NATO ASI Series, Vol. H31
Neurobiology of the Inner Retina
Edited by R. Weiler and N. N. Osborne
© Springer-Verlag Berlin Heidelberg 1989

In the turtle retina there is only one short report demonstrating the existence of amacrine cells with GLG-IR. Their number was estimated to be about 2,500 per retina and they were shown to arborize in layers 1, 3 and 4 (Eldred & Karten, 1983). We have undertaken a study focused on functional aspects of this neuropeptide and a detailed morphological analysis of its elements in this retina.

Morphological elements of a putative glucagonergic system

Retinas of the turtle *Pseudemys scripta* were fixed in 4% paraformaldehyde and either cryosectioned or used as whole mounts. Sections or entire retinas were incubated in a primary antibody directed against porcine or human GLG. The antibodies were either polyclonal (IBL, Hamburg) or monoclonal (Dr. Gregor, FU, Berlin). Standard procedures were used and the concentration of the antibody as well as the incubation time were varied in order to reduce background noise. The specifity of all antibodies was tested by preabsorption experiments. The detection systems to reveal IR were either FITC (Sigma), ABC (Vectastain) or Streptavidin (Amersham).

Amacrine cells with GLG-IR appear to form two subpopulations defined on the basis of their initial ramification patterns (Fig. 1a,b&c). Either one single stout process leaves the soma branching into one thicker and two or three finer processes or several processes emerge directly from the soma. In both types the initial processes branch further and invaginate into the IPL. A remarkable feature of both types is the existence of long processes running along the border between the inner nuclear leayer and the IPL giving rise to several thin processes dropping into the IPL where they branch in layer 3 and layer 5. It seems that there is no significant distinction between the two types regarding the terminating levels within the IPL, although it appears that amacrine cells with a stout principal process contribute to a larger extent to the ramifications at the innermost level of the IPL. In addition to these labeled neurons, GLG-IR cell bodies were also labeled in the

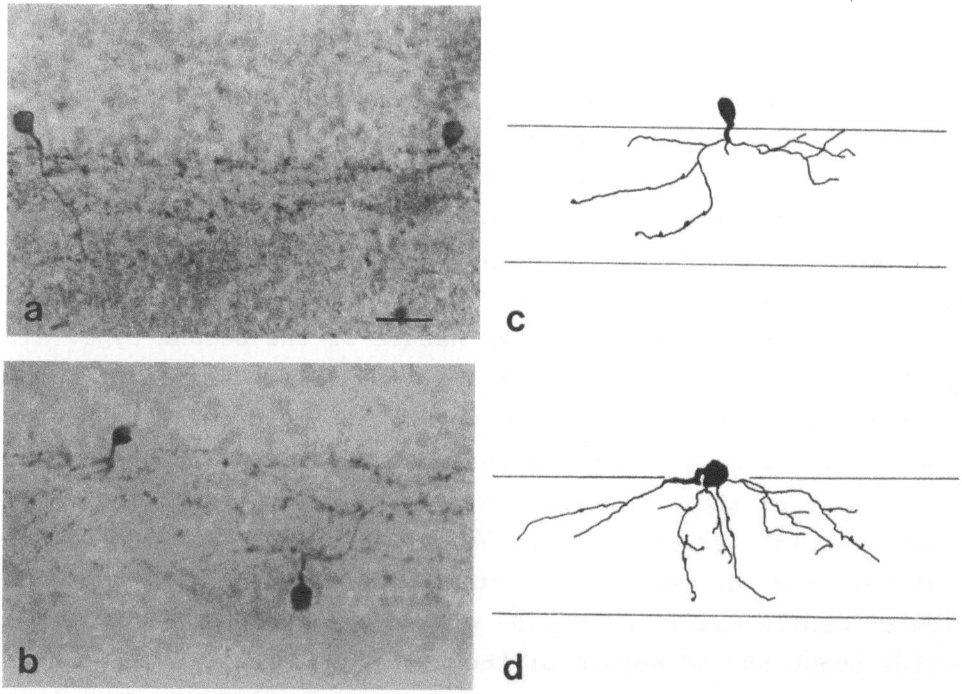

Fig. 1. a,b) Amacrine cells with GLG-IR shown in cross sections using the ABC-Method. Three main levels of arborizations are visible within the IPL. c,d) Camera lucida drawings of the two cell types with GLG-IR. Bar=20 µm.

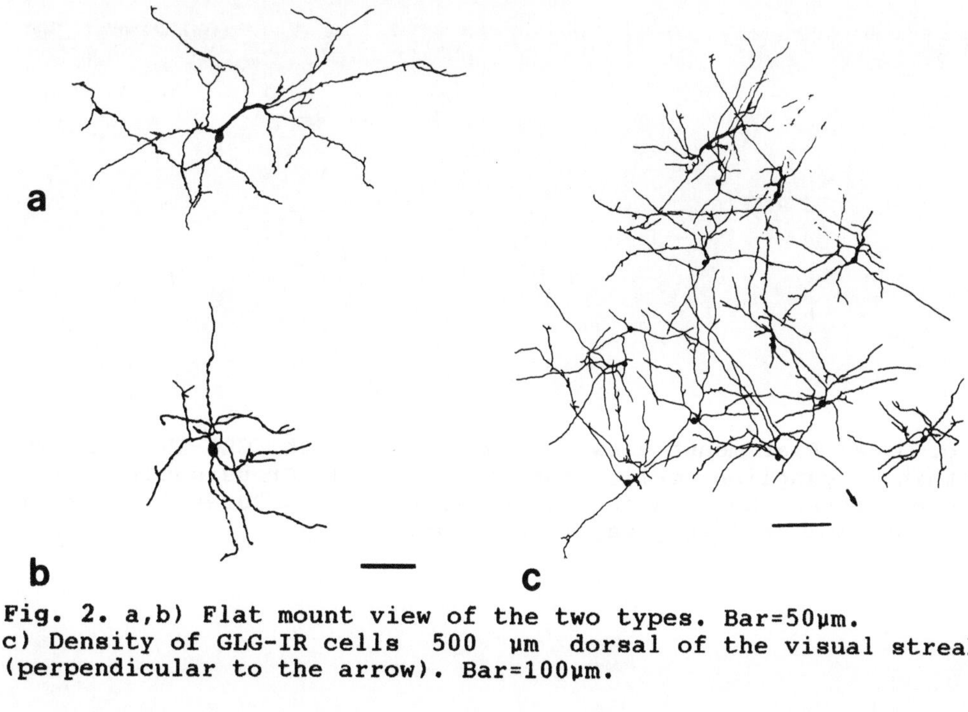

Fig. 2. a,b) Flat mount view of the two types. Bar=50µm.
c) Density of GLG-IR cells 500 µm dorsal of the visual streak (perpendicular to the arrow). Bar=100µm.

ganglion cell body layer (Fig. 1b). Their ramification patterns are essentially identical with the one described. These cells are probably displaced amacrine cells as suggested by Eldred & Karten (1983). In the flat mounted retina (Fig. 2a) the thicker process of the first cell type leaving the stout process is very prominent. Its length is about 70 - 100 µm. Several branches emerse and terminate as bundles of long radiating processes. This gives this type of GLG-IR cell an asymmetric appearance, which contrasts with the nature of the second type. Although it is very difficult to follow all the fine terminal arborizations it seems that the dendritic field of the second type (Fig. 2b) is more circular and smaller than that of the first. Most of the processes of both types have a beaded appearance which seems to be typical of amacrine cells containing neuropeptides or amines. Figure 2c shows the actual distribution of GLG-IR amacrine cells within a region of the central retina about 500 µm dorsal of the visual streak. The main axis of the dendritic field seems not to be oriented. The plexus formed by the processes is not very dense, although that the drawing reflects the plexi at all three levels of the IPL.

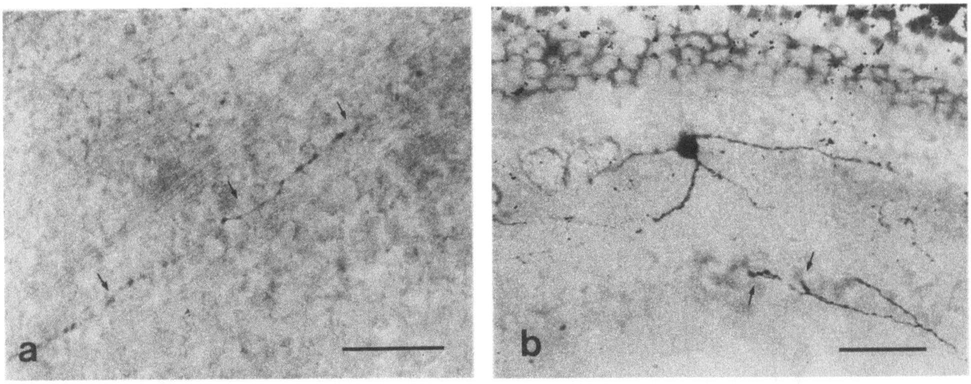

Fig. 3. a) Flat mount view of a fibre with GLG-IR (arrows) within a ganglion cell axon bundle. b) Cross sections show these fibres terminating with knobbey structures (arrows) at the ganglion cell body layer. Bar=50µm

GLG-IR was not only present in processes within the IPL but also in beaded fibres running within the axon bundles (Fig. 3a). These fibres did not contact labeled soma in the ganglion cell body or inner nuclear layer. They enter the retina through the optic nerve and terminate with a knobbly structure (Fig. 3b) within the ganglion cell body layer. The knobbly structure appears to sit on the proximal part of ganglion cell bodies. The total number of fibres within one retina ranged from 0 - 4. These fibres resemble efferent fibres described in the turtle retina (Weiler, 1985; Schütte & Weiler, 1988) and while their origin has not yet been elucidated, we are tempted by the idea that GLG-IR is associated with both efferent projections and intrinsic networks.

Light responses

Attempts were made to localize GLG-IR in amacrine cells which had been intracellularly recorded and labeled with Lucifer Yellow, as previously described for serotoninergic ganglion cells (Weiler & Ammermüller, 1986). This would have provided information about the activity of these neurons in relation to light stimulation. This approach is daunting given that intracellular recordings from amacrine cells are difficult, and the chance to record from one of these sparsely distributed GLG-IR neurons is rare. In addition Lucifer Yellow can reduce the antigenicity of the peptide and is almost vanished by the immunocytochemical procedure thus allowing only one trial. Despite these difficulties we were successful in one instance. Figure 4a-c shows the morphology of the Lucifer Yellow injected amacrine cell in the flat mounted retina. Photographs were taken at three different levels of the IPL. The morphology resembles the second type of GLG-IR amacrine cell as already described. The part of the retina containing the Lucifer Yellow labeled cell was then cryosectioned (25 µm) after fixation with 4 % paraformaldehyde. The section containing the soma (Fig. 4d) was incubated overnight in the primary antibody (polyclonal, IBL Hamburg) and IR was

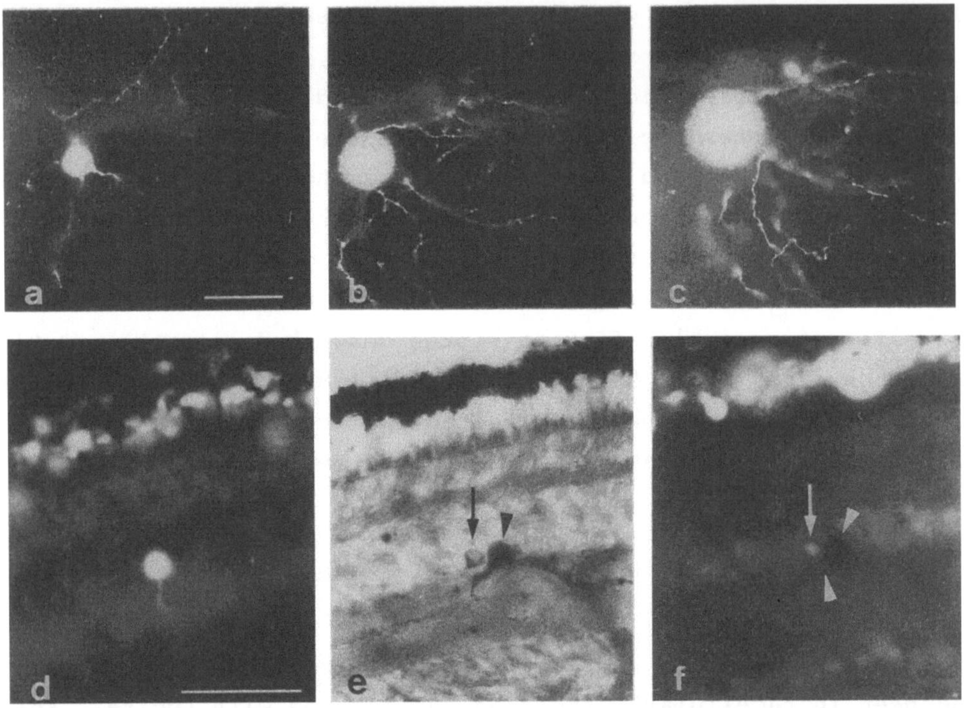

Fig. 4. a-c) Flat mount view of a Lucifer Yellow injected amacrine cell focused at the soma, the middle of the IPL and at the inner border of the IPL. d) The same cell in a cross section. e) The same section after staining for GLG-IR using the PAP method. Arrowheads point to a heavily labeled cell which serves as a landmark. The lighter labeled soma (arrow) corresponds to the soma containing the Lucifer Yellow which is visualized in e) with epifluorescent light. Bar=50µm.

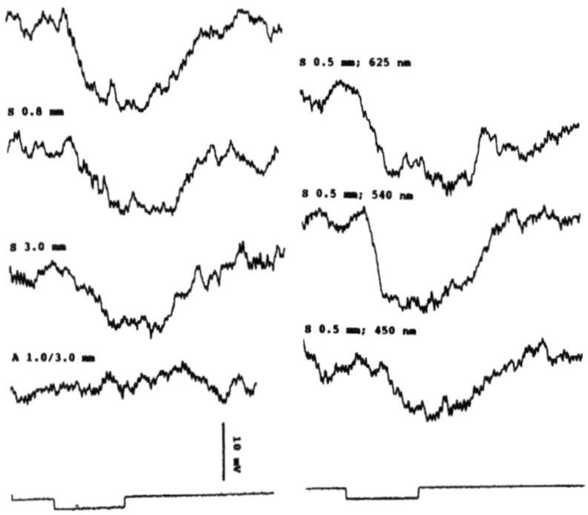

Fig. 5. Light responses intracellularly recorded from the amacrine cell shown in figure 4

revealed using the PAP method with DAB as a chromogen. IR was detectable in amacrine cell bodies and processes within the IPL (Fig. 4e). The weakly labeled soma next to the strongly labeled soma was identical with the one containing the Lucifer Yellow marker seen clearly when the epifluorescent light is used (Fig. 4f). The dark structure seen in the epifluorescent light is the heavily IR-labeled cell seen in figure 4e and serves as orientation marker. The Lucifer Yellow fluorescence in figure 4f is weaker than in figure 4d due to the immunocytochemical procedure. It is primarily restricted to the part of the cell body where the nucleus is located and IR-product is absent (see Fig. 4e). In the part of the soma where the processes emerge, its fluorescence is masked by the IR-product (Fig. 4e&d). The identity of the two cell bodies was further confirmed by projecting the images of the two slides onto a screen with a grid which allowed an accurate alignment of several structures within the section.

The intracellularly recorded responses of this amacrine cell to light stimulation of the retina with white spots and monochromatic light spots are depicted in figure 5. The cell responded to all stimulation except an annulus with a graded hyperpolarization of sustained character. It does not show a prominent center-surround organization, although the peripheral part seems to be slightly inhibitory. It also lacks color opponency. The sustained hyperpolarization during light stimulation indicates that this cell is excited during darkness. Should the GLG-like substance be released from this cell such a release would occur during darkness. A radioimmunoassay system was subsequently used to test this hypothesis.

Release of a glucagon-like substance

Light adapted animals were enucleated and the retina gently dissected from the eye. The left retina was placed in a chamber containing 200 µl turtle Ringer and the right retina in a chamber containing 200 µl turtle Ringer with an elevated potassium concentration (25 mM) which was osmotically balanced by reducing the sodium concentration. The Ringer solutions were continuously bubbled with carbogen and the preparation maintained at room temperature in light. After 10 min the Ringer solutions were gently replaced and a total of 3 fractions were collected and lyophilized for RIA. The retinas were then homogenized in 500 µl 1 M acetic acid. GLG was measured in the Ringer and the tissue fractions using a GLG-RIA-System (IBL-Hamburg). The protein content of retinal fractions was determined by the method of Bradford (1976). The sensitivity of the RIA was 20 pg GLG per mg protein.

Figure 6 shows the data of five experiments. The left column of each pair of columns with the index C represents the data from the control and the right column (K) from the experiment with 25 mM K-Ringer. The height of the columns indicates the total amount of GLG measured which is the sum of GLG measured in the retina homogenates and GLG measured in the Ringer fractions. The lack of a value for the Ringer fractions indicates that GLG was not significantly detectable within our detection limit. Experiments 1 and 2 were made with animals of about 20 cm carapax length, experiments 3 to 5 with animals of about 12 cm carapax length. It therefore seems that the relative amount of GLG depends on the age of the animals. The total amount of GLG measured in the turtle retina which ranged from 50 to 280 pg is higher than the values determined for frog (about 40 pg) or pigeon (about 20 pg) (Ekman & Tornqvist, 1985). The data of experiments 1 and 2 suggest that GLG is released upon incubation in a 25 mM K-Ringer in the order of 10 - 18 %. The amount of GLG in the Ringer of the control was below our detection limit. Since we know from the electrophysiological experiments (see above) that putative GLG amacrine cells respond to light stimulation with a sustained hyperpolarization and the experiments were carried out at constant room light, we did not expect to see any released GLG in the Ringer fractions of the control experiments.

If it is assumed that a release of GLG of the same relative magnitude occurs in experiments 3 - 5, then such a release would not be detected because of the experimental detection limit. This might be a likely explanation of why a release of GLG was not observed in these experiments.

The combined data from the release experiments and electrophysiology suggest that putative GLG-amacrine cells in the turtle retina exhibit their activity during darkness where they continuously release GLG.

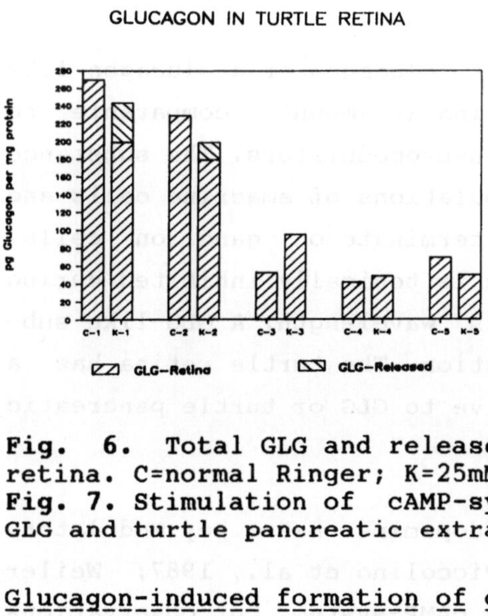

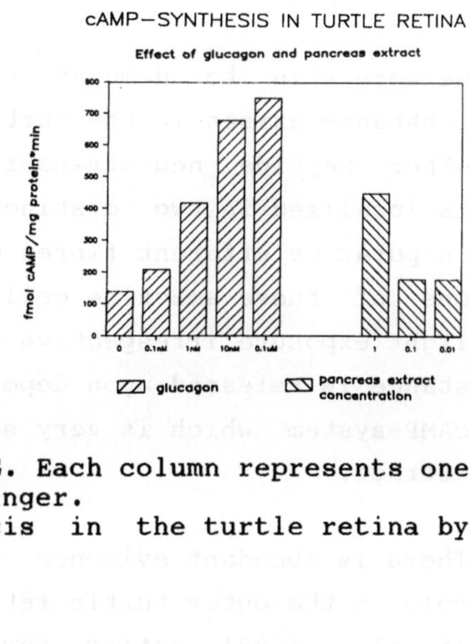

Fig. 6. Total GLG and released GLG. Each column represents one retina. C=normal Ringer; K=25mM K-Ringer.
Fig. 7. Stimulation of cAMP-synthesis in the turtle retina by GLG and turtle pancreatic extract.

<u>Glucagon-induced formation of cAMP</u>

An effect of bovine and porcine GLG and turtle pancreatic extract on the formation of cAMP in the turtle retina was analysed using retinal homogenates. The pancreatic extract was obtained by homogenization turtle pancreas in 3 M acetic acid (1:5, wt/vol); the crude homogenate was centrifuged and the supernatant freeze-dried before suspending in 1 ml of 0.05 M phosphate buffer pH 7.4 containing 500 KIU aprotinin (Trasylol, Bayer). cAMP was assayed using a RIA kit (RIANEN, Du Pont); and the procedure was similar to the one described by Watling & Dowling (1981).

Figure 7 shows that the effect of GLG on the accumulation of cAMP was dose-dependent and as little as 10^{-10} M of GLG was effective. A maximal increase in stimulation of cAMP was obtained for 10^{-7} M. GLG is therefore about 100 times more potent than dopamine which also stimulates cAMP formation in the turtle retina (not shown). Undiluted turtle pancreatic extract also stimulated the formation of cAMP in the retina, though the content of GLG-like material in this crude extract is not known. Whether the nature of the GLG-like material of the extract is identical to bovine or porcine GLG is not known.

Conclusions

We were able to demonstrate that glucagon or a glucagon-like substance exists in the turtle retina in amounts comparable to other peptide neurotransmitters/neuromodulators. The substance is localized in two distinct populations of amacrine cells and in putative efferent fibres which terminate on ganglion cells. One of these amacrine cell types is tonically inhibited during light exposure irrespective of its wavelength. A GLG-like substance is released upon depolarization. The turtle retina has a cAMP-system which is very sensitive to GLG or turtle pancreatic extract.

There is abundant evidence that dopamine has a key modulatory role in the outer turtle retina (Piccolino et al., 1987; Weiler et al., 1988) acting through a cAMP-system. Our observations reported here indicate that there is at least a second system acting on cAMP levels. We do not yet know which are the target neurons of this system. The finding that this system consists of an intrinsic and a most likely retinofugal part makes this search even more important. It also points again to the very important fact that neuronal processing within the retina is much more governed by constant modulatory activity than previously anticipated.

References

Bradford MM (1976) A rapid and sensitive method for the quantitation of microgram quantities of protein utilizing the principle of protein-dye binding. Anal Biochem 72: 248-254.

Brecha N (1983 Retinal neurotransmitters: histochemical and biochemical studies. In: Emson PC (ed) Chemical Neuroanatomy, Raven Press, New York, pp. 85-130.

Brecha N, Karten HG (1985) Localization of biologically active peptides in the retina. In: Morgan WH (ed) Retinal Transmitters and Modulators: Models for the Brain, Vol I, CRC Press Boca Raton, pp. 93-118.

Bruun A, Ehinger B, Tornqvist K (1984) Neurotransmitter candidates in the retina of the mudpuppy Necturus maculosus. Cell Tissue Res 238: 13-22.

Bruun A, Ehinger B, Sytsma V, Tornqvist K (1985) Retinal neuropeptides in the skates, Rana clavata, R. radiata, R. oscekata (Elasmobranchii). Cell Tissue Res 241: 17-24.

Das A, Pansky B, Budd GC (1985) Glucagon-like immunoreactivity in mouse and rat retina. Neurosci Lett 60: 215-218.

Drucker DJ, Asa S (1988) Glucagon gene expression in vertebrate brain. J Biol Chem 263: 13475-13478.

Ekman R, Tornqvist K (1985) Glucagon and VIP in the retina. Invest Ophthalmol Vis Sci 26: 1405-1409.

Eldred WD, Karten HG (1983) Characterization and quantification of peptidergic amacrine cells in the turtle retina: enkephalin, neurotensin and glucagon. J Comp Neurol 221: 371-381.

Gläsener G, Himstedt W, Weiler R, Matute C (1988) Putative neurotransmitters in the retinae of three urodele species (Triturus alpestris, Salamandra salamandra, Pleurodeles waltli). Cell Tissue Res 252: 317-328.

Koh SWM, Kyritsis A, Chader GJ (1984) Interaction of neuropeptides and cultured glial (Müller) cells of the chick retina: elevation of intracellular cyclic AMP by vasoactive intestinal peptide and glucagon. J Neurochem 43: 199-203.

Kuwayama Y, Ishimoto I, Fukuda M, Shimiza Y, Shiosaka S, Inagaki S, Senba E, Sakanaka M, Takagi H, Takatsuki K, Hara Y, Kawai Y, Tokyama M (1982) Overall distribution of glucagon-like immunoreactivity in the chicken retina: an immunohistochemical study with flat-mounts. Invest Ophthalmol Vis Sci 22: 681-686.

Piccolino M, Witkovsky P, Trimarchi C (1987) Dopaminergic mechanisms underlying the reduction of electrical coupling between horizontal cells of the turtle retina induced by d-amphetamine, bicuculline and veratridine. J Neurosci 7: 2273-2284.

Saskai H, Tominaga M, Marubashi S, Katagiri T (1985) Glucagon as a neurotransmitter. Biomed Res Suppl 6: 91-100.

Schorderet M, Sovilla JY, Magistretti PJ (1981) VIP and glucagon-induced formation of cyclic AMP in intact retinae in vitro. Eur J Pharmacol 71: 131-133.

Schütte M, Weiler R (1988) Mesencephalic innervation of the turtle retina by a single serotonin-containing neuron. Neurosci Lett 91: 289-294.

Tornqvist K, Ehinger B (1983) Glucagon immunoreactive neurons in the retina of different species. Graefe's Arch Clin Exp Ophthalmol 220: 1-5.

Tornqvist K, Loren I, Hakanson R, Sundleo F (1981) Peptide-containing neurons in the chicken retina. Exp Eye Res 33: 55-64.

Watling KJ, Dowling JE (1981) Dopaminergic mechanisms in the teleost retina: I. Dopamine-sensitive adenylate cyclase in homogenates of carp retina; effects of agonist, antagonists, and ergots. J Neurochem 36: 559-568.

Weiler R (1985) Mesencephalic pathway to the retina exhibits enkephalin-like immunoreactivity. Neurosci Lett 55: 11-16.

Weiler R, Ammermüller J (1986) Immunocytochemical localization of serotonin in intracellularly analyzed and dye-injected ganglion cells of the turtle retina. Neurosci Lett 72: 147-152.

Weiler R, Kohler K, Kolbinger W, Wolburg H, Kurz-Isler G, Wagner HJ (1988) Dopaminergic neuromodulation in the retinas of lower vertebrates. Neurosci Res Suppl 8: S183-S196.

Yang C-Y, Yazulla S (1986) Neuropeptide-like immunoreactive cells in the retina of the larval tiger salamander: attention to the symmetry of dendritic projections. J Comp Neurol 248: 105-119.

EXCITATORY AMINO ACID RECEPTORS ON FELINE RETINAL GANGLION CELLS

Hisako Ikeda, Jonathan Robbins and Christine Kay,
Vision Research Unit
The Rayne Institute
St. Thomas' Hospital
London SE1 7EH
United Kingdom

INTRODUCTION

There is little doubt that the synaptic transmitter of the retinal bipolar cell is an excitatory amino acid, glutamate or aspartate (Bloomfield & Dowling, 1985), and that at least cone bipolars make synaptic contact with the retinal ganglion cells. However, the properties of the excitatory amino acid receptors on these cells are still obscure. The current theory of excitatory amino acid receptors is that at least three pharmacologically distinct types of glutamate receptors are present in neurons in the central nervous system and these are selectively activated by the glutamate analogues, N-methyl-D-aspartate (NMDA), quisqualate (QA) and kainate (KA) (Watkins, 1981). For many years, the limitation in obtaining clear evidence to support this theory had been the availability of specific antagonists for each of these receptor types.

In 1980, the announcement of a new compound, 2-amino-5-phosphonovalerate (2APV) as a selective NMDA receptor antagonist (Davis et al, 1980) and our receipt of a sample of this initial compound from Dr. Watkins was, therefore, an exciting moment in our laboratory as we were just beginning to study glutamate and aspartate actions on the retinal ganglion cells following a long struggle to establish the intra-retinal microiontophoretic techniques in the optically intact cat eye. We were then delighted to discover that

NATO ASI Series, Vol. H31
Neurobiology of the Inner Retina
Edited by R. Weiler and N. N. Osborne
© Springer-Verlag Berlin Heidelberg 1989

when iontophoretically applied, the sample of ± 2APV powerfully blocked the visually induced response of sustained type retinal ganglion cells, thus suggesting the presence of post-synaptic NMDA receptors on these cells (Ikeda & Sheardown, 1982).

Since our early experiments, many studies of excitatory amino acid receptors on second order retinal neurons, i.e. the horizontal and bipolar cells, which receive glutamate released by the photoreceptors, have emerged (e.g. Slaughter & Miller, 1985; Miller & Slaughter, 1986) as a wider range of receptor antagonists have become available (Watkins & Olverman, 1987). However, in order to characterise excitatory amino acid receptors on single retinal ganglion cells, independently of the modifications in the input caused by the drugs at the outer synapses, microiontophoresis or isolated cell preparation techniques must be applied.

During 1987, we therefore began to re-examine the excitatory amino acid receptors on retinal ganglion cells in the hope that we could characterise the receptor types using the variety of receptor antagonists now commercially available and also the three agonists, NMDA, QA and KA.

METHODS

The procedure for anaesthesia and surgical preparation for intraretinal microiontophoresis in the optically intact refracted eye of the cat have already been described elsewhere (Ikeda & Sheardown, 1982). Our method for classification of retinal ganglion cells into ON and OFF or sustained or transient cells, using bright (120-150 cd/m^2) or dark (20 - 30 cd/m^2) spots or annuli, generated on a TV screen (luminance, 100 cd/m^2), is detailed in Ikeda (1985).

Six-barrelled microelectrodes with a tip total diameter (mean 14.6 ± 0.6 µm) with the resistance of each barrel ranging from 2 to 20 MΩ were advanced close to the soma of a ganglion cell from the vitreous through a tracker tube (inner diameter 1.2 mm) introduced at the upper temporal sclera. (The electrodes used in 1980-81 were four-barrelled with a tip size of 2-4 µm and resistance, 5-20 MΩ advanced through the same tracker tube). Only large, well-isolated spikes (Fussey & Kidd, 1970) were studied. The drugs used in this study are specified in Table 1.

Table 1. Information on the drugs used in this study

	DRUG	Abbreviation	Receptor Potency *	Solvent **	pH	Supplier
ANTAGONISTS	(±) 2-Amino-5-phosphonovalerate	±2APV	NMDA>>QA≥KA	165 mM NaCl	7.0	CRB
	(-) 2-Amino-5-phosphonovalerate	-2APV	NMDA>>QA≥KA	165 mM NaCl	7.0	CRB/TOCRIS
	3((±)-2-Carboxypiperazin-4-yl)propyl-1-phosphonate	±CPP	NMDA>>>QA≥KA	165 mM NaCl	7.0	TOCRIS
	1-(4-Chlorobenzoyl)-piperazine 2,3-dicarboxylate	pCBPzDA	KA>NMDA>QA	165 mM NaCl	7.0	TOCRIS
	L-Glutamate diethyl ester	GDEE	QA>KA≥NMDA	Water	3.5	SIGMA
	γ-D-Glutamylaminomethyl sulphonate	GAMS	KA=QA>NMDA	Water	7.0	TOCRIS
	γ-D-Glutamyl-glycine	γDGG	NMDA>KA≥QA	Water	7.0	SIGMA
	Kynurenate	KYN	NMDA≥KA=QA	Water	8.0	SIGMA
AGONISTS	Kainate	KA	KA	Water	8.0	SIGMA
	N-Methyl-D-aspartate	NMDA	NMDA	Water	8.0	SIGMA
	Quisqualate	QA	QA	Water	8.0	SIGMA

* From Jones et al, 1984; Chapman et al, 1985; Davis et al, 1986; Watkins & Olverman, 1987.
** All drug concentrations 50 mM and negative ejection.

The potency of different antagonists on visually-induced and spontaneous firing of each cell was compared by effects produced by 100 nA of each of the four drugs contained in the electrodes, but initially the ejection current was stepwisely increased from 10 nA to 100 nA in order to obtain dose response curves for each drug. For agonist effects, the current was increased from 0-5 nA to the level which produced a substantial rise in spontaneous firing of approximately equal magnitude from NMDA, QA and KA. When the current level of each drug produced a measurable effect on spontaneous firing, that current level was used to study the effect of the agonist on the visually-driven response of the cell. The visual stimulus used was the optimal sized spot, but with a reduced contrast which caused 50% reduction in the firing compared with the high contrast spot, in order to avoid possible saturation effects. To compare the relative potency of NMDA, QA and KA, the highest magnitude of excitation produced by either one of the three agonists was scaled to 100% and the % effect of others calculated. The values were then divided by the current level used for each drug.

Throughout the experiments, no drug effects which caused spike amplitude or waveform changes or showed no or little recovery were included in the data analysis, as we consider that this discipline is vitally important when considering the post-synaptic effect of iontophoretically applied drugs.

RESULTS

Action of excitatory amino acid receptor antagonists on visually driven firing of retinal ganglion cells

Whilst the original ± 2APV of 1980 could powerfully block the visually-induced response of sustained ganglion cells, the commercially supplied - or ± 2APV produced only a small effect and the new, most potent NMDA receptor antagonist, ±CPP (Davis et al, 1986), caused little effect. In contrast, Kyn, γ-DGG and GAMS which have nearly equal potency to NMDA, QA and KA could block visual excitation of both ON- and OFF-sustained ganglion cells, Kyn being the most powerful of all. GDEE and pCBPzDA (Table 1) were, however, less effective.

The powerful antagonism of Kyn, γ-DGG and GAMS and weak or no antagonism of 2APV and CPP on naturally released excitatory amino acid on visual stimulation are illustrated in Fig. 1. In 11 ON- and 11 OFF-cells, the effects of antagonists on the surround response, i.e. the response to an optimal annulus, were exactly the same as those of the receptive field centre response as shown in Fig. 1. Kyn, γ-DGG and GAMS effectively blocked the surround response.

Figure 2 compares the % inhibition of visually driven excitation of ON- and OFF-sustained cells caused by different antagonists. The potency order appears to be Kyn, γ-DGG, GAMS, 2APV and CPP. Kyn, γ-DGG and GAMS, produced stronger effects upon ON-cells than OFF-cells ($p < 0.025$ for Kyn, γ-DGG and GAMS). Effects of the antagonists on small numbers of transient cells were also studied, but the effects were absent, weaker or variable.

Action of excitatory amino acid receptor antagonists on spontaneous firing of retinal ganglion cells

The effect of antagonists on the spontaneous firing of sustained ganglion cells further produced an intriguing aspect of excitatory amino acid receptors on these cells at resting state. As in the earlier study, - or ± 2APV and ±CPP showed no effects on spontaneous firing, but Kyn, γ-DGG and GAMS reduced spontaneous firing of all ON-sustained cells (n=9), but their effects on OFF-cells were extremely variable, ranging from strongly inhibitory to excitatory.

Action of excitatory amino acid agonists on spontaneous firing of retinal ganglion cells

As illustrated in Fig. 3, NMDA, QA and KA all caused significant increase in spontaneous firing of ON (n=4) and OFF (n=8) type ganglion cells. In an OFF-sustained cell, NMDA caused inhibition, whilst KA and QA caused increase of spontaneous firing. In the majority of cells, the effects of the agonists were monophasic but occasionally the excitatory effect was followed by a period of inhibition before the spontaneous firing returned to the original level. For all cells studied (n =12) KA was the most potent. In most of the cells studied the potency order was KA > QA > NMDA.

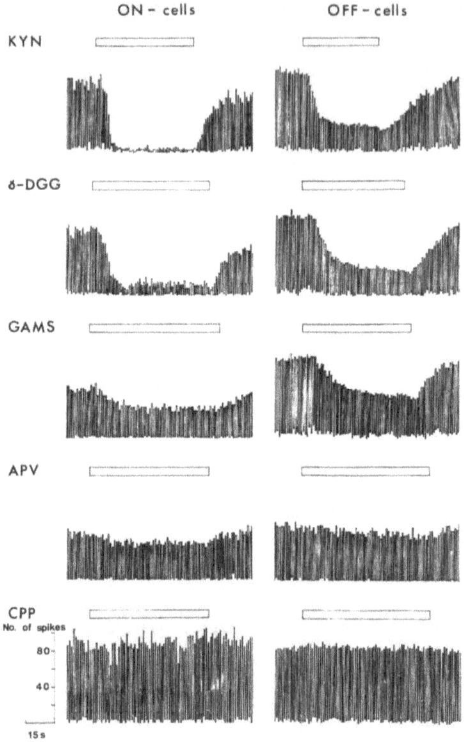

Fig. 1. Effect of 100 nA Kyn, -DGG, GAMS, ± 2APV and ±CPP on response of ON- and OFF-sustained ganglion cells to optimal spot presented at 1 Hz on the receptive field centre. Each upwards deflection is the count of spikes during spot 'on' and downwards deflection, that during spot 'off'. The open bar above each trace indicates the period of iontophoretic application of the drug.

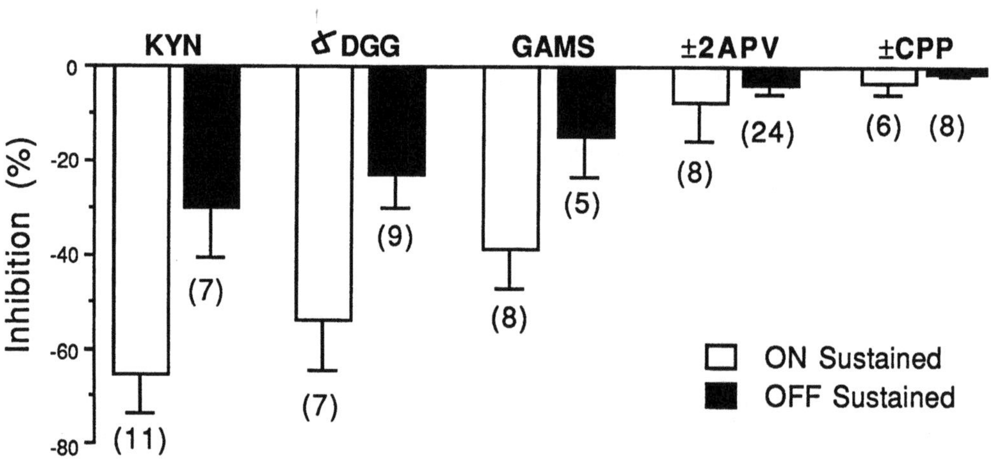

Fig. 2. Potency comparison of different excitatory amino acid antagonists on visually induced response of ON- and OFF-sustained retinal ganglion cells. The potency is expressed by mean and SE of % decrease in the response.

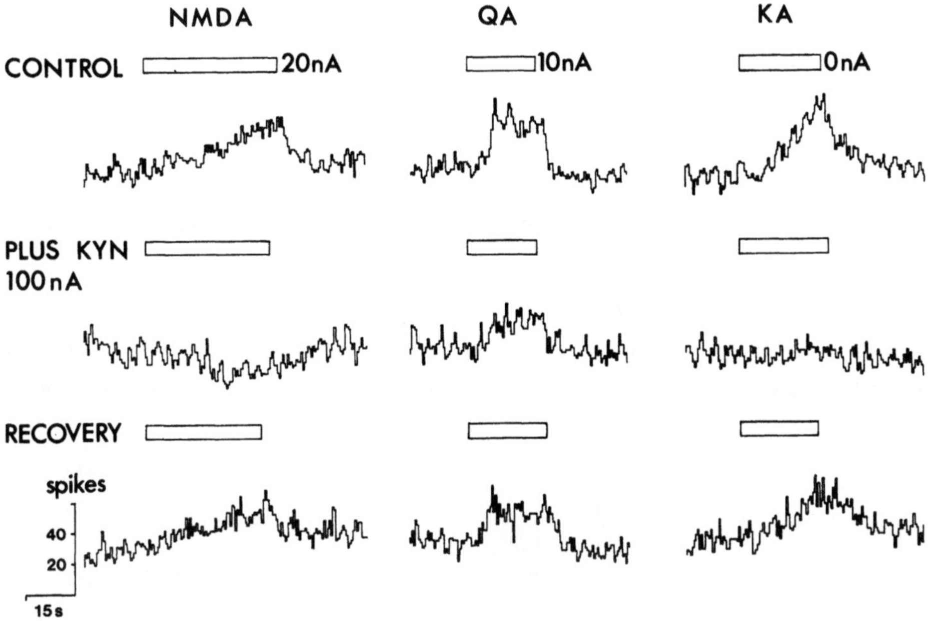

Fig. 3. Action of NMDA, QA and KA on spontaneous firing of ON- and OFF-retinal ganglion cells obtained before, during and after Kyn (100 nA) application. Kyn blocks all the agonist actions, but QA induced response is resistant to Kyn.

The KA and NMDA induced excitation was effectively and QA induced excitation was partially blocked by Kyn, but 2APV or ±CPP blocked only the NMDA effect, although in some cases, ±2APV had weak antagonism on QA effect.

Action of excitatory amino acid agonists on visually driven firing of retinal ganglion cells

As Fig. 4A shows, iontophoretically applied L-glutamtate and L-aspartate increase visually driven firing with very little change in the baseline firing level, although high drug concentration and ejection current have to be used, because of their efficient uptake system. NMDA, QA and KA at low current levels, however, frequently raised the baseline firing level and decreased visually-induced firing, as illustrated by the agonists' actions on responses of an OFF-sustained cell in Fig. 4B, suggesting that these agonists can depolarize ganglion cell membranes to saturation point. We found these cells which showed such saturation effect were those cells which initially had relatively high spontaneous firing rate. Whilst the effects of NMDA, QA and KA are similar to those of L-glutamate or L-aspartate in sustained ganglion cells which had rather low spontaneous firing rate: these agonists increased visually driven firing with very little change in the baseline firing level of these cells.

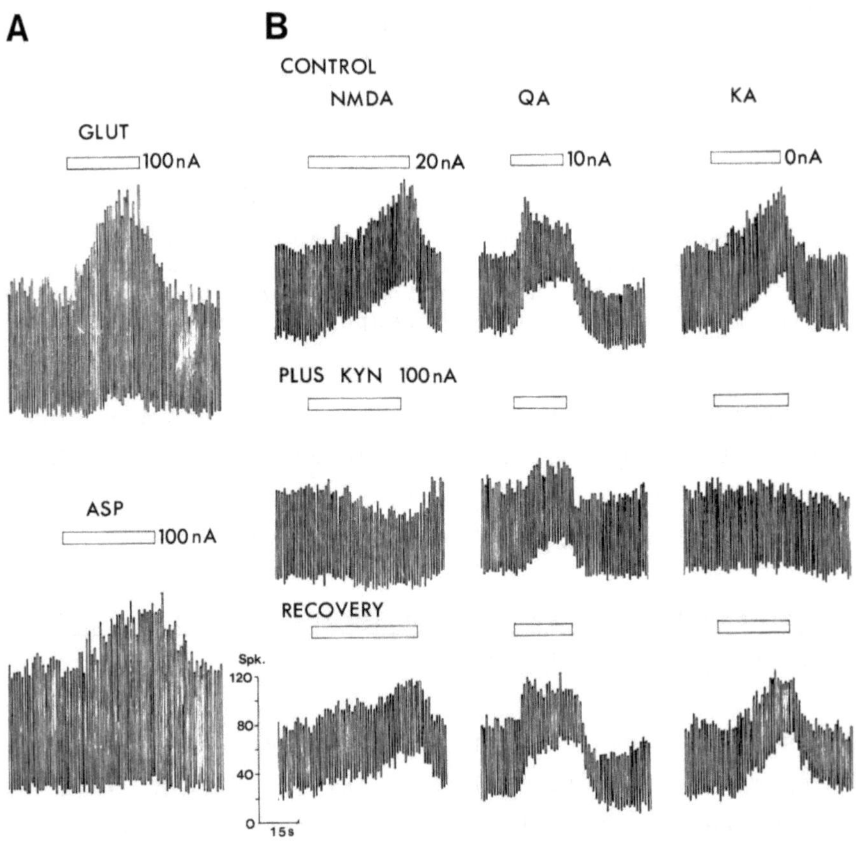

Fig. 4. Actions of L-Glut and L-Asp (A) and NMDA, QA and KA (B) on visually driven response of an OFF-sustained ganglion cell with high spontaneous firing. Note that the visual response amplitude is increased in A, but reduced in B, associated with a significant raise in baseline firing. Kyn blocks the effects induced by the agonists.

Fig. 5. Effects of excitatory amino acid agonists on visually driven response of sustained retinal ganglion cells with low and high spontaneous discharge.

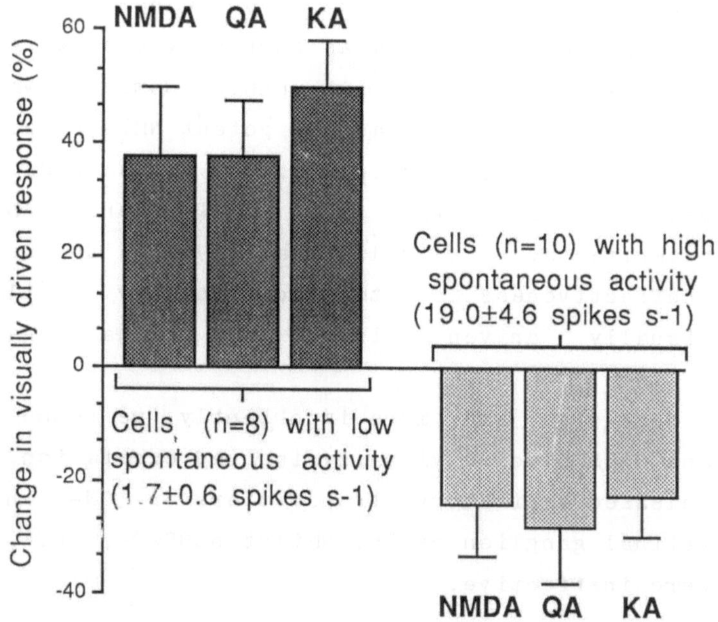

Fig. 5 shows effects of agonists on visually driven response of eight (3 ON, 5 OFF) sustained cells which had low spontaneous firing and ten (4 ON, 6 OFF) cells which had initially high spontaneous fi¨ing. In the low spontaneous firing group, the agonists caused an increase, whilst in the high spontaneous firing group, they caused a decrease of visually driven response associated with significant increase in the baseline firing. Kyn powerfully blocked the NMDA and KA induced changes and less so, the QA induced changes in the response to optimal spot or an annulus of all cells, regardless of their difference in initial spontaneous firing level, whilst CPP blocked the NMDA induced changes only.

DISCUSSION

Firstly, the discrepancy between the effect of the original sample of ± 2APV and that of the recently commercially-available - or ± (purity 99%) on visually driven firing of the retinal ganglion cells remains a puzzle. However, the following two factors lead us to suggest that the original ±

2APV sample which had been produced when rigorous purification techniques such as currently used were not available, may have contained an unknown compound which produced the effect in 1980. 1) The new potent NMDA receptor antagonist, ±CPP was ineffective in blocking visual responses of ganglion cells compared with Kyn, γ-DGG and GAMS in the present experiments. 2) Bolz et al (1984) also found the relative ineffectiveness of the commercially available ± 2APV on visually driven firing of retinal ganglion cells.

Secondly, and more importantly, we found that Kyn, γ-DGG and GAMS blocked visually induced excitation, i.e. naturally released excitatory transmitters on ON- and OFF-sustained retinal ganglion cells, whilst pCBPzDA, GDEE, ± 2APV and ±CPP were ineffective.

This suggests that endogenous excitatory amino acid transmitter action mediating visual excitation of sustained ganglion cells can only be antagonised by the antagonists which have almost equal potency to NMDA, QA and KA as Kyn, γ-DGG and GAMS are (see Table 1), but not by those which have potency to one, more than the other types, regardless of which of the three. All three types of receptors seem to be activated to evoke visual excitation of both ON and OFF-sustained retinal ganglion cells and thus the only antagonists which strongly block all three receptor types can block that excitation. The finding that the antagonist effect was stronger upon ON cells than OFF cells requires further investigation to find out what is the cause for the subtle difference.

The presence of all KA, QA and NMDA receptors are suggested by our agonist studies. Spontaneous firing of both ON- and OFF-sustained ganglion cells were sensitive to all three agonists with the potency order of KA > QA > NMDA and Kyn blocked the KA and NMDA induced response but less so the QA induced response. Interestingly, our *in vivo* results seem to be consistent with those of Aizenman *et al* (1988) on isolated and unclassified retinal ganglion cells in culture. At negative holding current, KA produced inward current in all, whilst QA, in 50% of and NMDA (in absence of Mg) in 68% of isolated retinal ganglion cells: the potency order being KA > QA > NMDA for 50 μM or less concentration, i.e. the same order as in our study. In addition, analogous to our findings, Kyn antagonised KA and NMDA-induced, but not QA induced currents in the isolated ganglion cells.

Furthermore, the effect of NMDA, KA and QA on visually driven response brought out a further aspect of excitatory amino acid transmission at retinal ganglion cells. The ganglion cell's sensitivity to agonists depended on spontaneous firing level of cells. These agonists mimicked the response of L-glutamate or L-aspartate on the visual response (Fig. 4A), i.e. increased the visual response in the ganglion cells which had initially low spontaneous firing. These agonists, on the other hand, caused a reduction of visually driven firing associated with increase of the baseline activity (Fig. 4B), causing saturation in those cells with high rate of spontaneous firing. All the agonist-induced responses were blocked by Kyn while CPP and APV only the NMDA response.

We suggest that glutamate receptors on sustained ganglion cells of the cat are of single molecules, having at least three different binding sites (NMDA, QA and KA); KA binding being most sensitive, and their level of conductance is regulated continuously according to background activity. Such an excitatory amino acid receptor model has also been suggested for cerebellar neurons (Cull-Candy & Usowicz, 1987).

Acknowledgement: The work is supported by the Special Trustees of St. Thomas' Hospital, the University of London Central Funds, and SmithKline Foundation. We thank Dr. Jeff Watkins for many useful discussions and Dr. Gertrude Falk for critical reading of the manuscript.

REFERENCES

Aizenman E, Frosch MP, Lipton SA (1988) Responses mediated by excitatory amino acid receptors in solitary retinal ganglion cells from rat. J. Physiol. 396:75-91

Bloomfield SA, Dowling JE (1985) Roles of aspartate and glutamate in synaptic transmission in rabbit retina. II. Inner plexiform layer. J. Neurophysiol. 53:714-725

Bolz J, Wässle H, Thier P (1984) Pharmacological modulation of ON and OFF ganglion cells in the cat retina. Neuroscience 12:875-885.

Chapman AG, Hart GP, Meldrum BS, Turski, L, Watkins JC (1985) Anticonvulsant activity of two novel piperazine derivatives with potent kainate antagonist activity. Neurosci. Letters 55:325-330

Cull-Candy SG, Usowicz MM (1987) Multiple conductance channels activated by excitatory amino acids in cerebellar neurons. Nature 325:525-528

Davies J, Evans RH, Herrling PL, Jones AW, Olverman HJ, Pook P, Watkins JC (1986) CPP, a new potent and selective NMDA antagonist. Depression of central neuron responses, affinity for [^{3}H]D-AP5 binding sites on brain membranes and anti-convulsant activity. Brain Res. 382:169-173

Davies J, Francis AA, Jones AW, Watkins JC (1980) 2-amino-5-phosphonovalerate (2APV), a highly potent and specific antagonist at spinal NMDA receptors. Br. J. Pharmacol. 70: 52-53

Davies J, Watkins JC (1985) Depressant actions of γ-D-Glutamylaminomethyl sulfonate (GAMS) on amino-acid-induced and synaptic excitation in the cat spinal cord. Brain Res. 327:113-120

Fussey IF, Kidd C, Whitwam JG (1970) The differentiation of axonal and soma-dendritic spike activity. Pflüger's Arch. 321:283-292

Ikeda H (1985) Transmitter action at cat retinal ganglion cells In: Progress in Retinal Research (eds. Osborne NN, Chader GJ) Pergamon Press Oxford New York, pp 1-32

Ikeda H, Sheardown MJ (1982) Aspartate may be an excitatory transmitter mediating visual excitation of 'sustained' but not 'transient' cells in the cat retina: iontophoretic studies in vivo. Neuroscience 72:25-36

Jones AW, Smith DAS, Watkins JC (1984) Structure-activity relations of dipeptide antagonists of excitatory amino acids. Neuroscience 13:573-581

Miller RF, Slaughter MM (1986) Excitatory amino acid receptors of the retina: diversity of subtypes and conductance mechanisms. TINS May 1986, 211-218

Slaughter MM, Miller RF (1985) The role of glutamate receptors in information processing in the distal retina. In: Neuro-circuitry of the Retina, A Cajal Memorial (eds. Gallego A, Gouras, P) Elsevier Science Publishing 51-65

Watkins JC (1981) Pharmacology of excitatory amino acid receptors. In: Glutamate: Transmitter in the Central Nervous System (eds. Roberts PJ, Storm-Mathisen J, Johnston GAR) John Wiley & Sons Ltd 1-24

Watkins JC, Olverman HJ (1987) Agonists and antagonists for excitatory amino acid receptors. TINS July 1987, 265-272

DESEGREGATION: BUSSING OF SIGNALS THROUGH THE RETINAL NETWORK

Malcolm M. Slaughter, Sun-Ho Bai, and Zhuo Hua Pan
Department of Biophysical Sciences
School of Medicine
State University of New York
Buffalo, New York 14214
USA

One of the information processing principles that has been repeatedly discovered in the nervous system is the segregation of signals into separate but parallel pathways. In the visual system, for example, separationist groups such as ocular dominance and orientation columns have been uncovered in the visual cortex (8). This reactionary processing is also prominent in the retina where there is an apparent separation of rod-cone, ON-OFF, and transient-sustained signals (7,9,11, 23). These various modalities decompose images from the outside world into informational components that can be processed by the nervous system. This model of sensory processing proposes that images from the outside world are handled by a multiplicity of parallel pathways, each relaying information about a particular characteristic of the external image, such as its color, orientation, or direction of motion. This theory has found strong support in the discovery of neurons that are preferentially stimulated by one of these trigger features (2,12). Although this appears to be an important mechanism in sensory systems, recent studies in amphibian retina suggest that the segregation of signals is not so absolute, and that in fact there may be a mixing of signals which at face value seems to contradict the principles of decomposition and segregation that enables animals to interpretation the visual world.

In the retina, this issue is most apparent in the photoreceptor system, which is elaborated with specialized detectors for dim or bright light (rods or cones), and for particular wavelengths of color within the cone system. The visible spectrum consists of an enormous number of colors so that if a distinct receptor were required for each color, the retina would be inundated with cones. This dilemma is resolved

NATO ASI Series, Vol. H31
Neurobiology of the Inner Retina
Edited by R. Weiler and N.N. Osborne
© Springer-Verlag Berlin Heidelberg 1989

in primates by using the three cone system to represent the entire spectrum through a relative weighting of primary colors. Each cone type has a response nomogram that permits it to preferentially detect a particular wavelength of light. Thus the spectrum is decomposed into three components, from which the nervous system can reconstruct any color. But this separation is violated by the presence of both electrical and chemical synapses between these cone types, so that each type is activated by a mixture of its own nomogram and those of the other cone types. A similar separation and specialization, followed by a mixing of signals, is seen in the rod/cone pathway. Presumably the rod system is ideally designed to detect a few quanta, a stimulus insufficient to activate the cone system, and this information is conveyed through a specialized network. But the rod and cone system are conjoined anatomically and physiologically at several locations. Rods and cones are connected by gap junctions and the rod amacrine cell is electrically coupled to cone bipolars, and depends on this cone system to carry part of its information (6,10,13,14,15).

Two other parallel pathways, the ON-OFF and the transient-sustained channels, also show a paradoxical convergence. An example of this can be seen in the responses of sustained OFF ganglion cells. In the conventional view, these neurons are exclusively part of the OFF channel, receiving excitatory input from the OFF bipolars in the dark. In the light, when the OFF bipolars are hyperpolarized, the OFF ganglion cells are also hyperpolarized due to decreased excitatory bipolar input (disfacilitation). This hyperpolarizing light response would be associated with a increase in input resistance because of a closing of excitatory channels. Although there are many OFF ganglion cells that are driven in this manner, others show evidence of inhibition during the light stimulus rather than disfacilitation, as shown by Belgum et al.(3). An example is illustrated in figure 1. The light response of this OFF ganglion cell is a sustained hyperpolarization, as shown at the far left of the voltage trace. Brief, negative current pulses

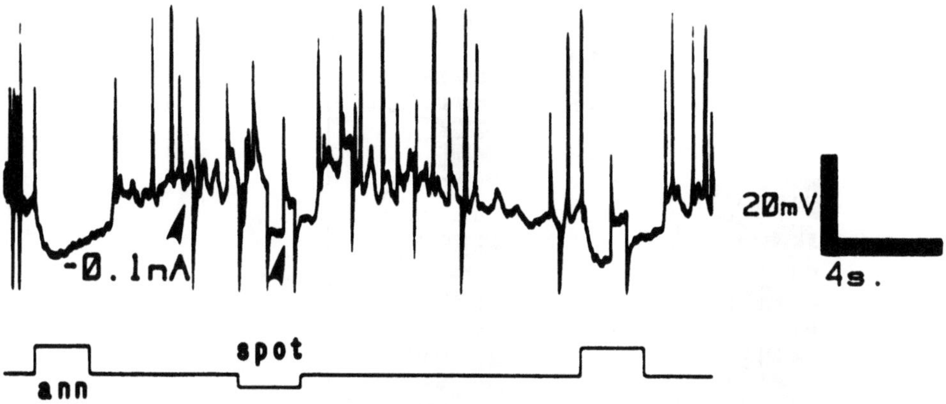

Figure 1 illustrates the conductance changes associated with the light responses of a sustained OFF ganglion cell. Single current pulses (-0.1nA) are shown at arrowheads. Square wave under voltage trace shows timing of spot and annulus stimuli.

were applied to the cell to measure changes in conductance associated with the light responses. A bridge balance was adjusted to almost nullify the voltage responses produced by these current pulses in the dark (first arrow head). During the sustained portion of the light response, the current pulse was reapplied and produced a positive voltage deflection (second arrow head), indicating that the input resistance of the cell was less in the light than in the dark. This indicates that the light response is due to inhibition and that this OFF ganglion cell is not simply following the OFF bipolar. The inhibitory input to these OFF cells cannot come directly from OFF bipolar input, but could come from the OFF channel. However, results such as those illustrated in figure 2 suggest that it originates from the ON pathway. The top two traces in figure 2 show the responses of the same sustained OFF ganglion cell to PDA (cis 2,3 piperidine dicarboxylic acid) and APB (2-amino-4-phosphonobutyrate). PDA suppresses the light responses of OFF bipolars (but not ON bipolars) (22) and also reduces the light responses of third order neurons (21). This effect is due to PDA's action as a glutamate antagonist, the presumptive neurotransmitter of photoreceptors and bipolar cells. As seen

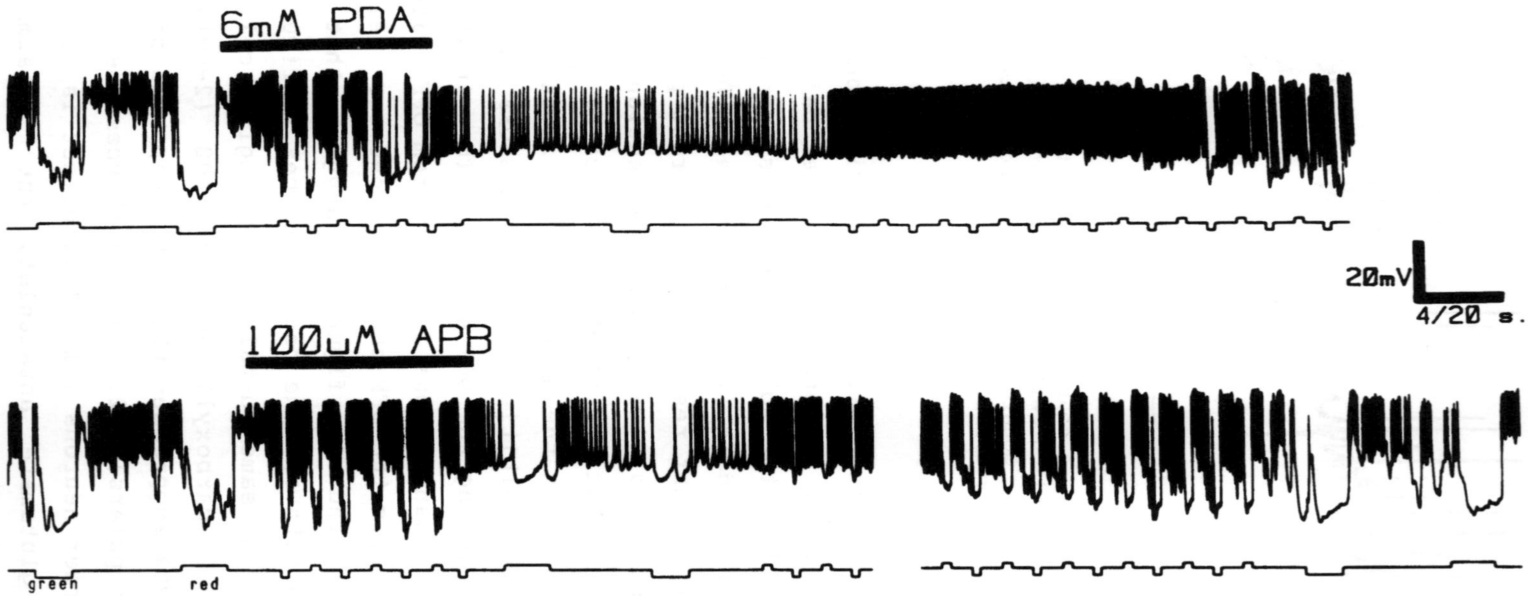

Figure 2 shows the effect of PDA and APB on a sustained OFF ganglion cell. The dark bar above the voltage trace indicates the duration of drug application. Both agents suppress the light response and leave the cell constantly depolarized. The light stimuli are red and green full field illumination. There is a 60 second break in the lower trace during the recovery phase.

in the top trace, PDA suppresses the light response of this cell, but the result of this blocking action is that the cell remains depolarized. If this OFF cell was depolarized by the OFF bipolar, we would expect PDA to hyperpolarize the cell. The middle trace shows the effect of APB, which suppresses the response of ON bipolars (but not OFF bipolars) (20). The effect is very similar to that of PDA. This similarity suggests that the ON bipolar is involved in the light response of this OFF ganglion cell. APB is blocking the input to the ON bipolar, while PDA is blocking the output of the ON bipolar. It appears that most of the light response of this OFF ganglion cell is from the ON pathway. The ganglion cell is spontaneously active in the dark and this activity is inhibited in the light.

In OFF ganglion cells that we have studied in this manner, we have found some that are driven by inhibition and are very sensitive to APB, while others are driven by the more conventional disfacilitatory mechanism and are comparatively insensitive to APB. These experiments indicate that this inhibitory input comes from the ON channel, while the disfacilitatory OFF ganglion cells get their input through the OFF channel. Thus there appears to be a group of OFF ganglion cells that, in terms of their synaptic input, are part of the ON channel. The light evoked hyperpolarization of OFF cells is not always only disfacilitation or inhibition. Sometimes both mechanisms are found in a single cell. In this case, both the ON and the OFF channels are converging. In these OFF ganglion cells, current injections applied to determine conductance changes or reversal potentials give little information under control conditions. But after APB application, a clear disfacilitatory light response can be shown. We interpret this to mean that under control conditions the light response is a product of inhibition, with a conductance increase, and disfacilitation, with a conductance decrease. Both are contributing to the light evoked hyperpolarization, but with opposite conductances and different reversal potentials. But after APB application, the inhibitory input is eliminated, and then current injections can detect the isolated disfacilitatory input. In the amphibian retina it is common to find phasic third order neurons that receive input from both pathways (ON-OFF cells) and it appears that these two pathways may converge on sustained cells as well.

With regard to the separation between transient and sustained signals in the inner retina, we have found that sustained amacrine and ganglion cells have significant transient input (19). This has also been described using white noise analysis in the catfish retina by Sakai and Naka (17,18). Under control conditions this phasic input is masked by the tonic light response, but the relative weighting of phasic and tonic signals can be reversed by activation of a subclass of GABA receptors, termed the GABA/B receptors. This is illustrated in figure 3 which shows the effects of baclofen, a GABA/B receptor agonist (4), on sustained and transient third order neurons. In the sutained ON cell in the top trace, baclofen application suppresses the sustained response and enhances the initial, transient ON response. A transient OFF response is also revealed. In most sustained cells baclofen also hyperpolarizes the cell. This is not true here, suggesting that baclofen's effect is indirect in this case. The middle trace shows a sustained OFF ganglion cell and again baclofen makes the light response of this cell more transient. The cell is hyperpolarized by baclofen, the OFF response becomes larger and more transient while an ON transient response becomes prominent. Thus in both of these sustained cells, activation of GABA/B receptors causes the cells to respond more like transient ON-OFF cells. When recording from transient cells, we have found that baclofen causes them to hyperpolarize and the light evoked EPSPs become enhanced as illustrated in the bottom of figure 3. Before baclofen application, this cell responded with an EPSP and a single spike at light onset and a smaller OFF EPSP. A 0.1nA negative current pulse was applied in the dark (arrowheads) and the bridge was balanced so that under control conditions this current did not produce a voltage

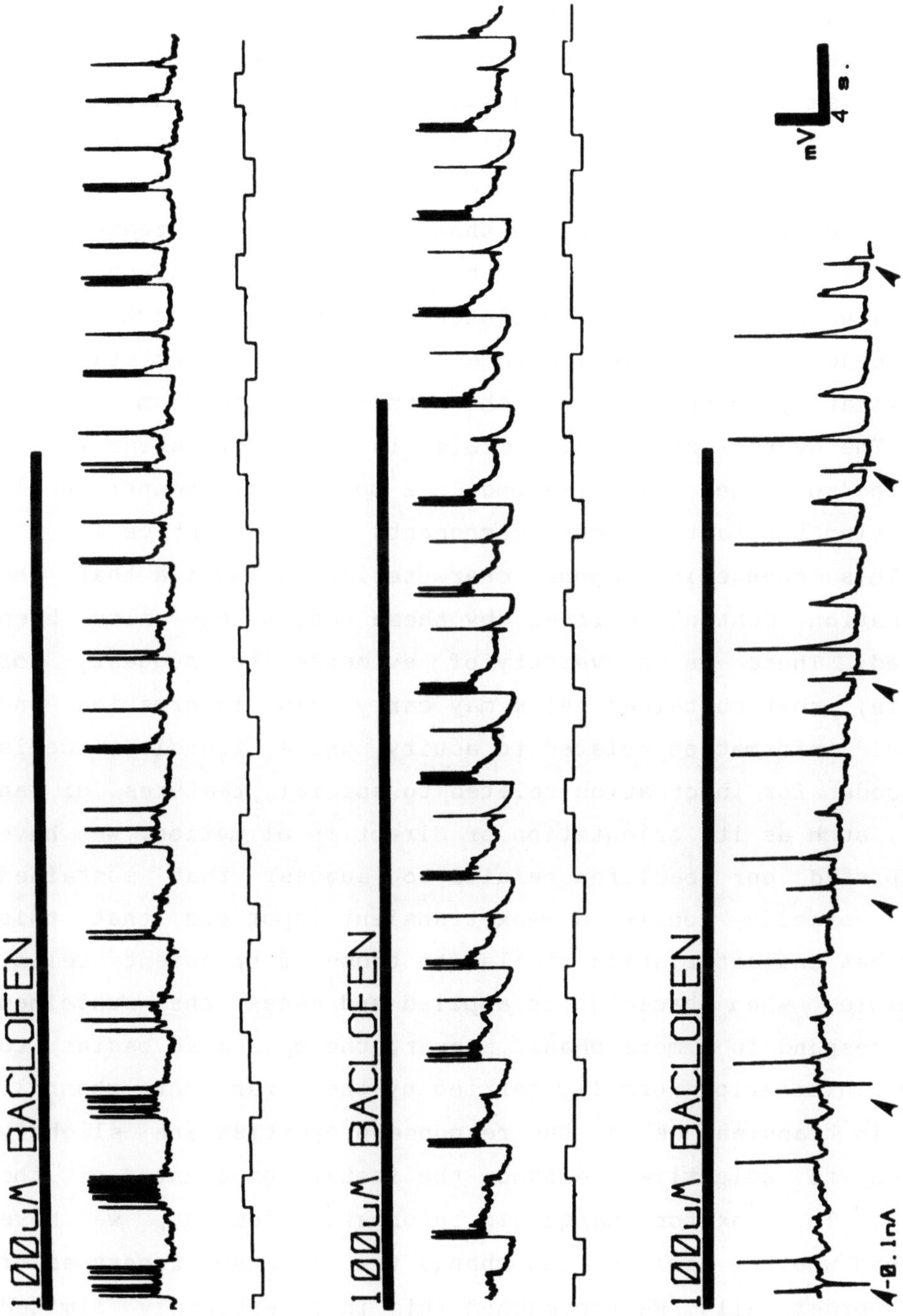

Figure 3 illustrates the effects of baclofen on a sustained ON (top), sustained OFF (Middle), and a transient ON-OFF cell. Baclofen enhances transient responses and suppresses sustained responses. Arrowheads in lowest trace indicate the injection of -0.1nA current pulses. Red and green full field light stimuli were used.

change. Baclofen caused the cell to hyperpolarize and both the ON and the OFF EPSPs became larger.The prolonged timecourse of the EPSPs in this cell is unusual, most often we have found that the EPSP becomes larger but decays faster in the presence of baclofen. The negative current pulse produced a positive voltage deflection indicating that baclofen had decreased the input resistance of the cell by approximately 40Mohms. In general we have found that baclofen has a direct effect on most third order neurons, causing an decrease in input resistance of approximately 50Mohms, due to the opening of potassium channels (19). The overall effect of baclofen in the inner retina is to make the sustained cells respond in a more phasic manner while augmenting the fast response components of transient cells.

This change in response characteristics implies that the information content carried by these cells has also been altered. There is a variety of evidence to suggest, for example, that sustained cells may carry color information and spatial information related to acuity, while transient cells may code for information related to special features of an image, such as its orientation or direction of motion. We have interpreted our baclofen results to suggest that sustained cells normally receive a weak transient input and that this input has characteristics similar to those of transient cells. Therefore, when baclofen is applied and makes the sustained cell respond in a more phasic manner, the cell also begins to relay information normally carried by the transient channel. Even in transient cells, the response properties are slightly altered to selectively enhance the faster components of the signal. To look for changes in information content, we have examined whether baclofen can change the response properties of third order cells. We approached this in an extremely limited way since we were not able to test a variety of stimulus parameters before and after baclofen application and then determine if any of the cell's response properties had changed. We did examine orientation selectivity but we were unable to detect any change in response properties due to baclofen. We

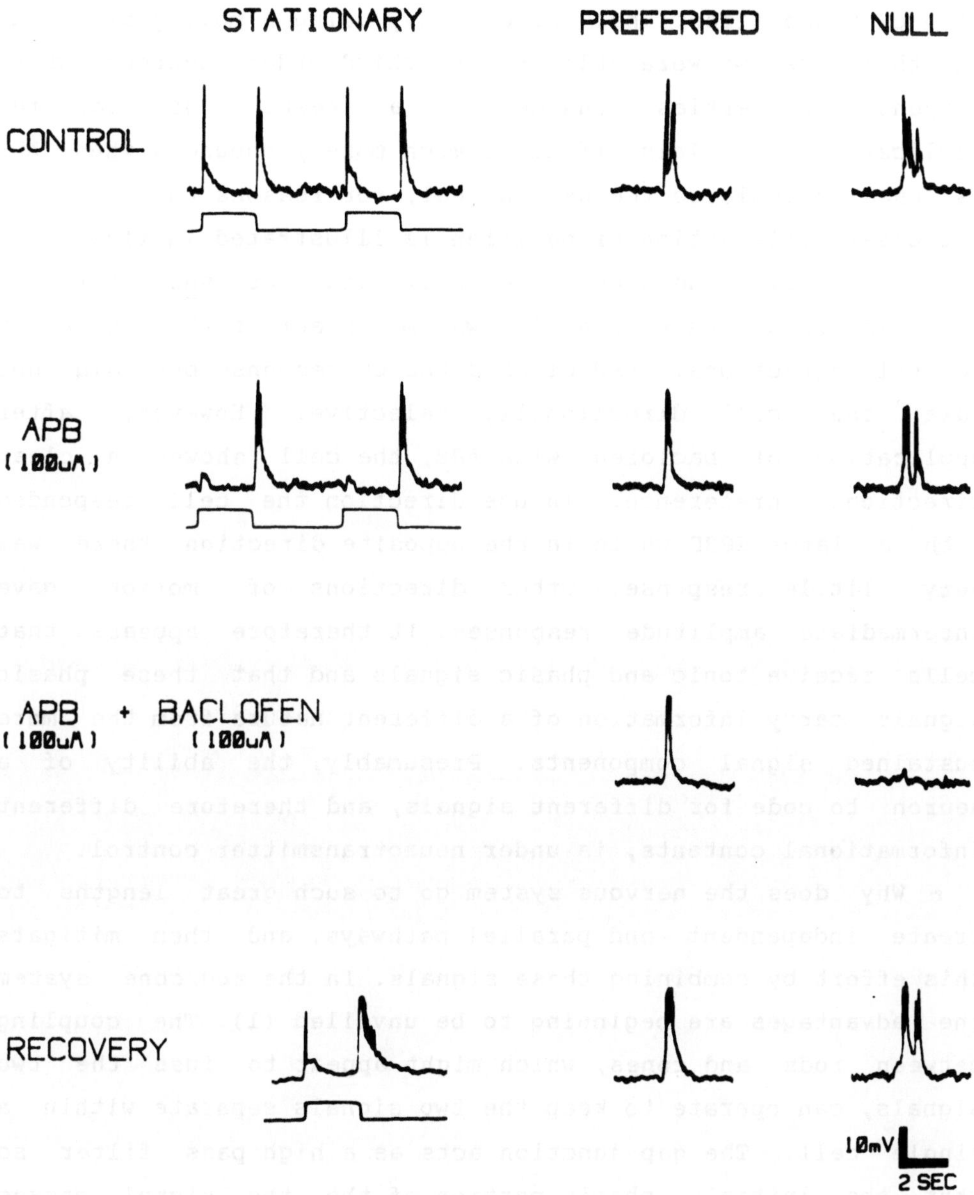

Figure 4 demonstrates that baclofen promotes directional selectivity in some third order neurons. The cell only showed directional selectivity after the application of baclofen and APB. The left column shows the response of the cell to full field white light, the other two columns show the response to a 100um slit moved across the retina.

also examined directional selectivity, using a moving slit, and in this case we were able to find third order neurons whose response properties changed as a result of baclofen application (16). This effect is much more pronounced when APB is used to isolated the OFF channel, for reasons that are not yet clear. This action of baclofen is illustrated in figure 4. Under control conditions the cell did not show markedly differing responses when a slit was moved across the retina in several directions. APB blocked the ON response but did not make the cell directionally selective. However, after application of baclofen with APB, the cell showed a clear directional preference. In one direction the cell responded with a large EPSP while in the opposite direction there was very little response. Other directions of motion gave intermediate amplitude responses. It therefore appears that cells receive tonic and phasic signals and that these phasic signals carry information of a different nature from the more sustained signal components. Presumably, the ability of a neuron to code for different signals, and therefore different informational contents, is under neurotransmitter control.

Why does the nervous system go to such great lengths to create independent and parallel pathways, and then mitigate this effort by combining these signals. In the rod/cone system the advantages are beginning to be unveiled (1). The coupling between rods and cones, which might appear to fuse the two signals, can operate to keep the two signals separate within a single cell. The gap junction acts as a high pass filter so that the initial, phasic portion of the the signal passes through a gap junction, while the tonic component is largely attenuated. Since the dominant part of a photoreceptor's light response is a tonic hyperpolarization, the phasic signal coming across the gap junction is temporally distinct from the cell's own light signal. This temporal separation in photoreceptors is similar to the transient and sustained signals found in the inner retina. Here again, a mixing of these inputs can occur which could still be distinguished based on the frequency components of the responses. Thus the signals can converge into the same cell and yet not lose their capacity to code for different properties.

If these signals can be combined and still remain intact, what advantage does this mixing offer? The answer may be that it permits flexibility to switch network components between a number of functions. In the case of rods and cones, the rod pathway can use the cone system to carry part of its message, rather than simply letting a major portion of the retina lie fallow during scotopic illumination. A similar advantage would exist in the inner retina if the sustained and transient systems could both relay each others signals. We have proposed that the GABA/B receptor may act to control selective attention, not only in the retina but in other regions of the central nervous system (16). If a transmitter can switch the response properties of a cell, it can function to recruit neurons to transmit information based on the requirements of the network.

Supported by National Eye Institute Grant # EY05725.

References

[1] Attwell D, Borges S, Wu SM and Wilson M (1987) Signal clipping by the rod output synapse. Nature 328: 522-524.

[2] Barlow HB and Levick WR (1965) The mechanism of directionally selective units in rabbit's retina. J.Physiol. (Lond.) 178: 477-504.

[3] Belgum JH, Dvorak DR and McReynolds JS (1982) Light-evoked sustained inhibition in mudpuppy retinal ganglion cells. Vision Res. 22: 257-260.

[4] Bowery NG, Hill DR, Hudson AL, Doble A, Middlemiss DN, Shaw J and Turnbull MJ (1980) (-) Baclofen decreases neurotransmitter release in the mammalian CNS by an action at a novel GABA receptor. Nature 283: 92-94.

[5] Famiglietti EV and Kolb H (1975) A bistratified amacrine cell and synaptic circuitry in the inner plexiform layer of the retina. Brain Res. 84: 293-300.

[6] Frumkes TE and Eysteinnson T (1987) Suppressive rod-cone

interaction in distal vertebrate retina: Intracellular records from Xenopus and Necturus. J. Neurophysiol. 57: 1361-1382.

[7] Hartline HK (1938) The response of single optic nerve fibers of the vertebrate eye to illumination of the retina. Amer. J. Physiol. 121: 400-415.

[8] Hubel DH and Wiesel TN (1977) Functional architecture of macaque monkey visual cortex. Proc. R. Soc Lond B 198: 1-59.

[9] Kaneko A (1970) Physiological and morphological identification of horizontal, bipolar, and amacrine cells in goldfish retina. J. Physiol. (Lond.) 207: 623-633.

[10] Kolb H (1977) The organization of the outer plexiform layer in the retina of the cat:Electron microscopic observations. J. Neurocytol. 6: 131-153.

[11] Kuffler SW (1953) Discharge patterns and functional organization of mammalian retina. J. Neurophysiol. 16:37-68.

[12] Lettvin JY, Maturana HR, McCulloch WS and Pitts WH (1959) What the frog's eye tells the frog's brain. Proc. Inst. Rad. Eng. 47: 1940-1951.

[13] Mariani AP and Lasansky A (1984) Chemical synapses between turtle photoreceptors. Brain Res. 310: 351-354.

[14] Nelson R (1977) Cat cones have rod input: A comparison of response properties of cones and horizontal cell bodies in the retina of the cat. J Comp. Neurol. 172: 109-136.

[15] Normann RA, Perlman I, Kolb H, Jones J and Daly SJ (1984) Direct excitatory interactions between cones of different spectral types in the turtle retina. Science 224: 625-627.

[16] Pan ZH and Slaughter MM (1988) A cellular mechanism of selective attention. Science (submitted).

[17] Sakai HM and Naka K-I (1987) Signal transmission in the catfish retina. IV. Transmission to ganglion cells. J. Neurophysiol. 58: 1307-1328.

[18] Sakai HM and Naka K-I (1987) Signal transmission in the catfish retina. V. Sensitivity and circuit. J. Neurophysiol. 58: 1329-1350.

[19] Slaughter MM and Bai S-H (1988) Diffential effects of baclofen on sustained and transient responses of retinal neurons. J Neurophysiol. (in press).

[20] Slaughter MM and Miller RF (1981) 2-Amino-4-phosphonobutyric acid: A new pharmacological tool for retina research. Science 211: 182-185.

[21] Slaughter MM and Miller RF (1983) Bipolar cells in the

mudpuppy retina use an excitatory amino acid neurotransmitter. Nature 303: 537-538.

[22] Slaughter MM and Miller RF (1983) An excitatory amino acid antagonist blocks cone input to sign-conserving second-order retinal neurons. Science 219: 1230-1232.

[23] Werblin FS and Dowling JE (1969) Organization of the retina of the mudpuppy, Necturus maculosus. II. Intracellular recording. J. Neurophysiol. 32: 339-355.

GABA-ACTIVATED CURRENTS IN GANGLION CELLS ISOLATED FROM GOLDFISH RETINA

A. T. Ishida
Department of Animal Physiology
University of California
Davis, California 95616 USA

INTRODUCTION

Since ganglion cells are anatomically and functionally the most proximal neurons of the retina, their physiology must be influenced either directly or indirectly by all of the major types of neuron in the retina. The dominant direct influences appear to be exerted by bipolar and amacrine cells at synapses onto ganglion cells in the inner plexiform layer (see Cajal, 1972; Dowling and Boycott, 1966; Werblin and Dowling, 1969; Miller and Dacheux, 1976; Naka, 1977). Recently, various subtypes of these inputs have been recognized, largely by two experimental approaches.

In one, cell-types and their synapses in the inner plexiform layer have been examined by anatomical methods including Golgi-staining, horseradish peroxidase-injection, thin-sectioning, and serial reconstruction. This work has identified an awesome variety of bipolar, amacrine, and ganglion cells (Cajal, 1972; West and Dowling, 1972; Scholes, 1975; Kolb et al, 1981, 1988), vertically segregated ON- and OFF- pathways in the inner plexiform layer (Famiglietti and Kolb, 1976; Famiglietti et al, 1977), and synaptic profiles along ganglion cell dendrites and on ganglion cell somata (Dowling 1970; Stevens et al, 1980; Kolb and Nelson, 1984; Dacheux and Raviola, 1986; McGuire et al, 1986; Sakai et al, 1986; Marshak et al, 1988).

In the second approach, cells have been marked histochemically (e.g. with a labeled compound which is sequestered by an uptake system), immunocytochemically (e.g. with an antibody directed against a suspected transmitter or an enzyme involved in its synthesis), or by both methods (see Marc, 1982, 1985, 1986; Brecha, 1983; Brecha et al, 1984; Stell, 1985; Yazulla, 1986). Information yielded by this approach include evidence of numerous amacrine cells, e.g. GABA-ergic "pyriform" (Marc et al, 1978), GABA-ergic "polyform" (Yazulla et al, 1986), glycinergic "AII" (e.g. Pourcho and Goebel, 1985), cholinergic "starburst" (see Tauchi and Masland, 1984) and various peptidergic (e.g. Brecha et al, 1984; Marshak et al, 1984; Weiler and Ball, 1984; Eldred and Karten, 1985) and monoaminergic (e.g. Ehinger and Falck, 1971; Ehinger and Floréń, 1976; Vaney, 1986; Dacey, 1988) amacrine cells.

NATO ASI Series, Vol. H31
Neurobiology of the Inner Retina
Edited by R. Weiler and N. N. Osborne
© Springer-Verlag Berlin Heidelberg 1989

The power of this approach is evidenced by its having shown signs of one or more transmitter in each of the five major sublayers of the inner plexiform layer in various species (e.g. goldfish and pigeon; see Karten and Brecha, 1983; Brecha et al, 1984; Yazulla, 1986; Marc, 1986).

Substantial progress has thus been made in delineating structural and neurochemical pathways from the inner nuclear layer into the inner plexiform layer. However, as far as ganglion cells are concerned, two problems persist:

One is that the output of the inner nuclear layer is still not completely charted. To begin with, the transmitters of bipolar cells have yet to be established (though glutamate, serotonin, and glycine have been considered candidates -- see Slaughter and Miller, 1983; Weiler and Schütte, 1985; Sterling, 1983; Marc, 1985; Pourcho and Goebel, 1987), and not more than half of the amacrine cells in any retina have been labeled by all of the neurochemical labels used to date combined (for review, see Marc, 1982). Even if the transmitters used by all bipolar and amacrine cells were known, it is not yet known when and how most of these would be released in response to the wide variety of light stimuli which have been used to identify the numerous functional classes of ganglion cells (see e.g. Maturana et al, 1960; Enroth-Cugell and Robson, 1966; Levick, 1967; Daw, 1968; Davis and Naka, 1980). Finally, even if all of the transmitters and their release patterns were known, it would not be clear which ganglion cells these transmitters would affect, because the types of bipolar and amacrine cells which synapse onto ganglion cells have not yet been completely mapped.

Secondly, the effects of transmitters in the inner plexiform layer in general are not understood in much detail; in some cases, the effects are not yet certain, as they have either not shown a consistent pattern, not yet been tested, or have been inferred from tests of large or unknown doses in situ. Two particular sources of uncertainty in studies in situ are that:

a) Drug effects may be mediated indirectly, particularly if more than one cell-type is sensitive to the chemical at hand[1]. Although transmitters are often tested in the presence of ions (e.g. Mg^{++} or Co^{++}) which suppress transmitter release

---

1: A prime example is GABA, to which some photoreceptor, bipolar, amacrine, and ganglion cells, are sensitive (e.g. Tachibana and Kaneko, 1984, 1987; Miller et al, 1981; Massey and Redburn, 1982; also, see references in Ishida and Cohen, 1988).

from some cells, the usefullness of this strategy hinges on the assumption that these cations act only pre-synaptically. However, since some transmitter-release in the retina appears to be calcium-independent (Yazulla and Kleinschmidt, 1983; Masland and Cassidy, 1987), and since some divalent cations are known to antagonize certain post-synaptic responses (e.g. Evans et al, 1977; Reuter, 1983; Kaneko and Tachibana, 1986b), such protocols are not failsafe.

b) Some drug effects are hardly detectable in situ, unless high concentrations are used, due to diffusion barriers against the penetration of certain chemicals (e.g. Ishida and Fain, 1981). Effects produced by large drug doses, however, may differ from those of lower doses, due to receptor desensitization (Katz and Thesleff, 1957), differential activation of multiple conductance mechanisms (Yasui et al, 1985; see also Takeuchi and Takeuchi, 1971), etc.

To increase our understanding of ganglion cell physiology and pharmacology, (a) the approaches mentioned at the beginning of this introduction should continue to flesh out pre-synaptic inputs to ganglion cells, and (b) to complement this information, some means of directly accessing and investigating the post-synaptic membranes of ganglion cells would clearly be desirable. One straightforward means would be to separate the pre- and post-synaptic cells in question, and proceed by studying the post-synaptic cells alone. With this rationale in mind, we have recently developed methods of identifying individual ganglion cells after isolating them from adult goldfish retinas, and begun to study their response to neurotransmitters using patch-clamp methods (Ishida and Cohen, 1988). Similar approaches have been used to study photoreceptor, horizontal, bipolar, and ganglion cells in a variety of animals (e.g. Bader et al, 1978; Tachibana, 1981, 1983; Lasater et al, 1984; Ishida et al, 1984; Ishida and Neyton, 1985; Shingai and Quandt, 1986; Kaneko and Tachibana, 1986a,b; Attwell et al, 1987; Lipton and Tauck, 1987).

In our initial studies, GABA was used as an agonist because histochemical studies and voltage recordings in situ suggest that it may be used as a neurotransmitter throughout the inner plexiform layer (see Marc et al, 1978; Negishi et al, 1978; Caldwell et al, 1978; Miller et al, 1981; Bolz et al, 1985; Yazulla, 1986).

METHODS

Experiments were performed on ganglion cells isolated from the retinae of adult goldfish (Carassius auratus, ~8 cm body length). Goldfish were used so that the results of our studies in vitro may be compared with anatomical, histochemical, and electrophysiological results obtained from the same animal specie in situ (see Daw, 1968; Raynauld, 1972; Kaneko, 1979; Marc 1982, 1986; Yazulla, 1986; Marshak et al, 1988). Also, we wanted to be able to use cells from adult retinas, rather than embryonic or neonatal tissue, since the in situ results cited above were obtained in adult retinas, since pharmacological and electrophysiological properties in other systems change significantly with age (e.g. Weiss and Horn, 1986; Jaramillo and Schuetze, 1988), and since post-synaptic membranes may be difficult to identify in ganglion cells if synapses have not yet formed in the inner plexiform layer, as in some neonates (e.g. McArdle et al, 1977).

Cells were isolated by trituration of papain-treated retinas, using methods adapted from those described by Tachibana (1981) and by Schwartz and Agranoff (1981). The exposure to enzyme increases the number of viable cells yielded by trituration (cf. Wakshull et al, 1979), and (unless applied intracellularly) does not appear to adversely affect voltage- or transmitter-sensitive conductances in various cell-types (see e.g. Hall and Kelly, 1971; Betz and Sakmann, 1973; Narahashi, 1974). Prior to each dissociation, the optic nerve of each eye used was crushed, as described by Landreth and Agranoff (1976), to stimulate survival of cells in culture.

The electrophysiological methods used in these studies have been described in detail elsewhere (Ishida and Cohen, 1988). Briefly, the responses of single ganglion cells to GABA (γ-aminobutyric acid, Sigma) and related compounds were measured under voltage-clamp in the whole-cell configuration of the patch-clamp method (Hamill et al, 1981). Electrodes were filled with a saline containing an elevated chloride ion concentration (149 mM) to minimize the fade of GABA-activated currents which can occur due to shifts in the chloride equilibrium potential (Huguenard and Alger, 1986; Akaike et al, 1987). In most experiments, the electrodes contained (in mM): 148 CsCl, 17 NaOH, 0.5 $CaCl_2$, 4.7 EGTA, 10 HEPES (pH 7.5), and the control bath contained (in mM) 135 NaCl, 2.5 KCl, 1.0 $CaCl_2$, 10 tetraethylammonium Cl, 10 D-glucose, 21 sucrose, and 3 HEPES, plus 30 nM tetrodotoxin. Bath saline was chilled and superfused at a rate of 1 ml/min over the cells, so that recordings were made at a constant temperature between 9 and 14 ^{o}C. Test compounds were applied to cells either by microperfusion (see Fenwick et al, 1982) or by bath application. Each cell selected for electrophysiological recording was identified as a ganglion cell on the basis of criteria described by Ishida and Cohen (1988), i.e. by the presence of an enlarged nucleolus. This feature is characteristic of ganglion cells in fish and amphibian retinas whose optic nerves have been disrupted (Murray and Grafstein, 1969; Turner et al, 1978), and it is visible in freshly dissociated cells as well as in cells maintained in culture for weeks (Schwartz and Agranoff, 1981). Cells with large neurites were avoided in order to improve the likelihood of an adequate intracellular dialysis and space clamp; some of the cells selected appeared to lack neurites altogether (as in Fig. 1B of Ishida and Cohen, 1988).

RESULTS

We have characterized 4 properties of ganglion cell responses to GABA: receptor type, ion selectivity, elementary conductance, and kinetics.

At negative holding potentials, GABA at ≥ 3 μM elicited inward currents accompanied by substantial increases in current fluctuations. Responses to 3 μM GABA were generally small (10-20 pA), and showed little or no fade during the

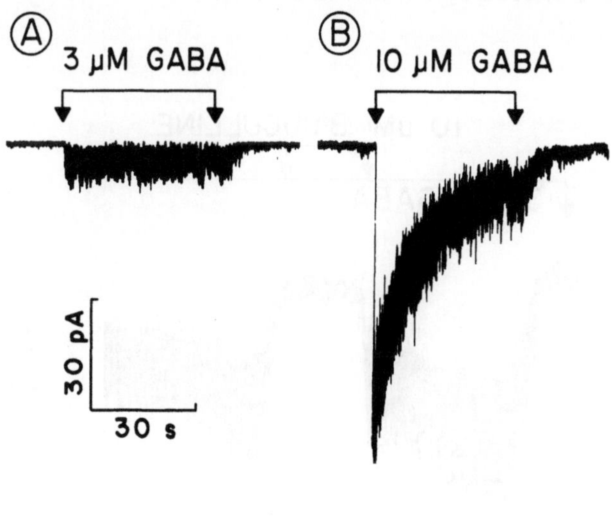

Fig. 1. Whole-cell currents elicited in a single ganglion cell by two concentrations of GABA. GABA applied by microperfusion, first at 3 μM (A), then at 10 μM (B), for the durations indicated by the paired arrows. Holding potential (E_{hold}) = -60 mV. Downward deflection in current traces correspond to inward current (i.e. efflux of chloride ions). (From Ishida and Cohen, 1988.)

entire GABA application (Fig. 1A). Responses to 10 μM GABA initially dwarfed those elicited by 3 μM GABA, but then faded to less than half of their peak amplitude within the first 15 sec of maintained GABA applications (Fig. 1B). Regardless of the mechanism underlying this decay, the fact that it was brought on so strongly by such a small increase in agonist dose obstructed measurement of dose-response curves.

To identify the type of receptor which mediates ganglion cell responses to GABA, we tested for effects of bicuculline, a competitive antagonist of $GABA_A$ receptors (e.g. Enna and Snyder, 1976), and of baclofen, an agonist of $GABA_B$ receptors (e.g. Hill and Bowery, 1981). Ganglion cell responses to 3 μM GABA were inhibited by 10 μM bicuculline (Fig. 2), whereas in cells clamped to negative holding potentials, baclofen alone elicited no response (see Ishida and Cohen, 1988). These results indicate that ganglion cell responses to GABA were mediated by $GABA_A$ receptors under the conditions of our experiments.

$GABA_A$ receptors typically mediate anionic conductance changes (see Bormann, 1988). Thus, to characterize the ionic selectivity of the GABA-activated conductance in ganglion cells, we measured GABA responses at various holding potentials in baths containing different chloride ion concentrations. Cl^- concentrations were altered by equimolar substitution of $NaCH_3SO_3$ for NaCl (cf.

Fig. 2. Antagonism of GABA response by bicuculline. Maintained bath application of 3 μM GABA starts at first arrow; microperfusion of 10 μM bicuculline (with 3 μM GABA) onto cell produces decrease in current amplitude and noise. E_{hold} = -75 mV. (From Ishida and Cohen, 1988.)

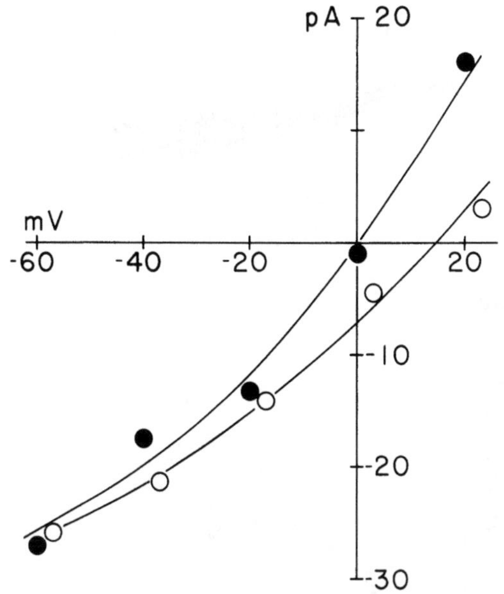

Fig. 3. Current-voltage curves of GABA response in bath saline containing either 100% (●) or 50% (○) of the Cl^- concentration present in the recording electrode (see Methods and text). Peak amplitude of the responses of a single cell to 3 μM GABA, plotted against holding potential. Curves drawn through the points by eye; both curves show slight outward rectification. (From Ishida and Cohen, 1988.)

Ozawa and Tsuda, 1973). In both normal- and low-Cl^- baths, the currents elicited by GABA reversed in polarity at membrane potentials near the Cl^- equilibrium potential (Fig. 3). Thus, GABA appears to increase the permeability of ganglion cells to small anions like Cl^-, but not larger anions like $CH_3SO_3^-$.

At holding potentials on either side of the chloride equilibrium potential, GABA produces a conspicuous increase in noise superimposed on the mean response current (e.g. Fig. 1). The noise in responses to low GABA doses were therefore studied with standard methods of variance and power spectral analysis (Neher and Stevens, 1977) to estimate the amplitude and kinetics of the underlying single-channel currents. Division of the slope of plots of the noise variance increase versus mean current amplitude (e.g. Fig. 4B) by the driving force for the current (i.e., by the difference between the holding and reversal potentials) suggested an elementary conductance of 16 ± 2 pS (mean ± SD, n = 6 cells). Power density

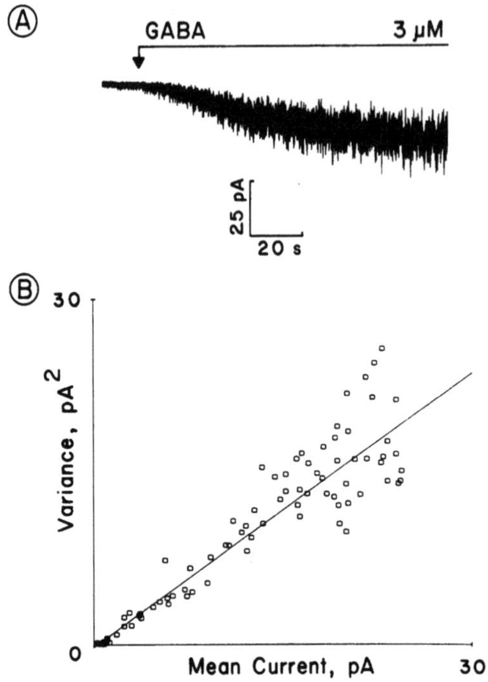

Fig. 4. Estimation of single-channel conductance from whole-cell current. (A) Rise of response to bath-application of GABA (3 µM at steady state); E_{hold} = -60 mV. (B) Increase in current noise variance during the response in A plotted against mean current. Line through data is linear regression, whose slope gives 0.78 pA as estimate of elementary amplitude of GABA-activated current in this cell. (From Ishida and Cohen, 1988.)

spectra were fit largely but in no case completely by single Lorentzian curves, suggesting that the channels activated by GABA may be kinetically complex (cf. Colquhoun and Hawkes, 1977). All power spectra could be fit by sums of 2 Lorentzians, with time constants of ~30 msec and 1-2 msec (e.g. Fig. 5B). Neither these time constants, nor the estimated single-channel conductance, showed an obvious voltage dependence between -20 and -80 mV.

DISCUSSION

Our initial studies have shown that GABA can activate an anionic conductance in goldfish ganglion cells via $GABA_A$ receptors. The elementary properties of this conductance resemble those observed in the response of various other cells to GABA, in their absolute values as well as in their lack of voltage-dependence at

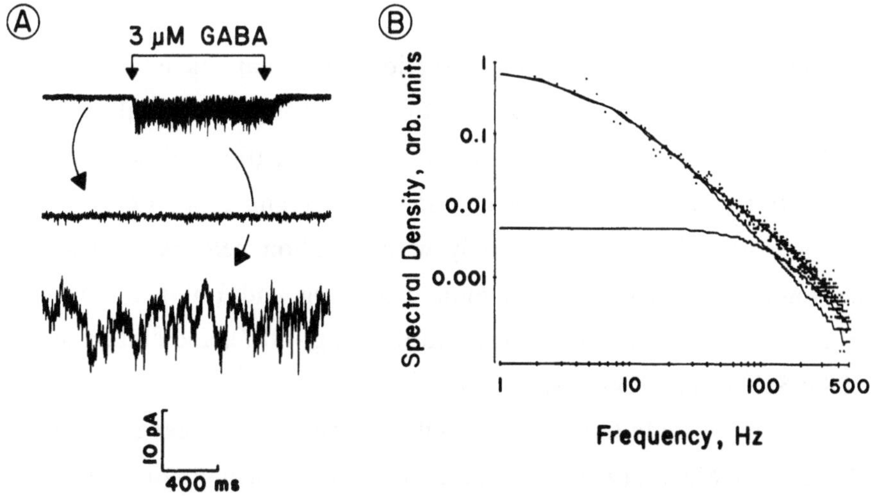

Fig. 5. Power spectrum of GABA-induced current fluctuations. (A) Uppermost trace shows response to 3 μM GABA; duration of application = 45 sec; E_{hold} = -60 mV. Middle and lower traces show portions of current measured before and during GABA application; low-pass filtered at 500 Hz; separated arbitrarily for clarity. Calibration applies to middle and lower traces. (B) Difference of spectra calculated from several control and response current segments taken from recording shown in A. Fit through data points is sum of 2 single Lorentzians plotted under the data points; position of single Lorentzians, selected for best fit by eye of double Lorentzian, correspond to time constants of 32 and 1.5 msec. (From Ishida and Cohen, 1988.)

negative potentials (e.g. McBurney and Barker, 1978; Cull-Candy and Ogden, 1985; Gray and Johnston, 1985; Weiss, 1988; see also the low-conductance states described by Hamill et al, 1983; Bormann and Clapham, 1985; Bormann et al, 1987). In contrast, the conductance activated in salamander ganglion cells by glycine reportedly measures 6 pS (Mittman and Copenhagen, 1985), and thus appears to differ from that activated in goldfish ganglion cells by GABA.

Since several of the cells which we recorded from bore no obvious dendritic or axonal processes, somatic receptors appear to have mediated the response of these cells to GABA. GABA receptors have also been found on somata elsewhere in vertebrate CNS, as have synaptic profiles from glutamic acid decarboxylase (GAD) immunoreactive processes (Ribak, 1978; Alger and Nicoll, 1982; Murakami et al, 1983; Gray and Johnston, 1985). It remains to be seen whether GABAergic synapses onto goldfish ganglion cell somata exist; however, these may come as no real surprise, given that synapses onto ganglion cell somata have been seen in carp, catfish, and cat retinas (Witkovsky and Dowling, 1969; Sakai et al, 1986; Stevens et al, 1980), and that in rat retina, synapses onto ganglion cell somata by GAD-immunoreactive processes have been observed (Vaughn et al, 1981). That these may not be anomalies associated only with ganglion cells is suggested by the synapses found on somata of horizontal (Dowling and Ehinger, 1978), bipolar (Fisher, 1972), interplexiform (Zucker and Dowling, 1987), and amacrine (Dowling et al, 1980; Sakai and Naka, 1985) cells.

This does not preclude GABAergic input to other parts of ganglion cells, and in fact, Muller and Marc (1985) have recently found morphological synapses from GABAergic amacrine cells onto goldfish ganglion cell dendrites. Oddly enough, GABAergic amacrine cell processes are also known to synapse onto the axon terminals of bipolar cells (Marc et al, 1978; Yazulla et al, 1987), and GABA also activates a chloride conductance via $GABA_A$ receptors in these bipolar cells (Tachibana and Kaneko, 1987). Since these cells in turn form excitatory synapses onto ganglion cells (Naka, 1977; Famiglietti et al, 1977), then one might ask if GABAergic input to these bipolar cells has the same effect on ganglion cell activity as direct GABAergic input to the ganglion cell itself. If GABAergic input to a ganglion cell and to its presynaptic bipolar cell are provided by the same amacrine cell, then the two GABA pathways (amacrine→bipolar, amacrine→ganglion) may seem redundant. Alternatively, these pathways may be formed by different amacrine cells, or they could be activated differentially by complex light stimuli. Before resolving this, amacrine cell inputs to bipolar and ganglion cells, as well as the GABA receptor/ionophore complexes in bipolar and ganglion cells, will have to be compared in more detail, and whether or not all bipolar cell inputs to ganglion cells are excitatory should be decided (cf. Sterling, 1983; Marc, 1985).

Nearly all of the cells which we recorded from responded to GABA, and the range in size of the somata which we recover in our dissociations easily encompasses the dimensions of perikarya belonging to various morphological and functional subtypes of ganglion cells (see Ishida and Cohen, 1988). Thus, our results predict that most if not all ganglion cells in goldfish retina may respond to GABA. This is consistent with the observation that GABA can inhibit action potentials in all classes of ganglion cells in carp retina (e.g. Negishi et al, 1978). Assuming that the different ganglion cell sizes which we recover in our dissociations actually reflect different classes, then these cells should respond to various neurotransmitter candidates which have been identified in different sublayers of goldfish inner plexiform layer (see Marc 1982, 1986; Brecha et al, 1984; Yazulla, 1986). For that matter, since convergence of various inputs onto ganglion cells is seen in situ under the electron microscope, it would not be surprising to find single ganglion cells which respond to several transmitters. Ultimately, the results of this work should be helpful in interpreting effects recorded in situ on specific ganglion cell functions (cf. Daw et al, 1982).

REFERENCES

Akaike N, Inomata N, Tokutomi N (1987) J Physiol (London) 391: 219-234
Alger BE, Nicoll RA (1982) J Physiol (London) 328: 125-141
Attwell D, Mobbs P, Tessier-Lavigne M, Wilson M (1987) J Physiol (London) 387: 125-161
Bader CR, MacLeish PR, Schwartz EA (1978) Proc Nat Acad Sci USA 75: 3507-3511
Betz W, Sakmann B (1973) J Physiol (London) 230: 673-688
Bolz J, Frumkes T, Voigt T, Wässle H (1985) J Physiol (London) 362: 369-393
Bormann J (1988) Trends Neurosc 11: 112-116
Bormann J, Clapham DE (1985) Proc Nat Acad Sci USA 82: 2168-2172
Bormann J, Hamill OP, Sakmann B (1987) J Physiol (London) 385: 243-286
Brecha N (1983) In: Emson PC (ed) Chemical Neuroanatomy, Raven Press, NY, p 85
Brecha N, Eldred W, Kuljis RD, Karten HJ (1984) Progr Retinal Res 3: 185-226
Cajal SRy (1972) The Structure of the Retina. CC Thomas, Springfield IL
Caldwell JH, Daw NW, Wyatt HJ (1978) J Physiol (London) 276: 277-298
Colquhoun D, Hawkes AG (1977) Proc R Soc Lond B 199: 231-262
Cull-Candy SG, Ogden DC (1985) Proc R Soc Lond B 224: 367-373
Dacey DM (1988) Science 240: 1196-1198
Dacheux RF, Raviola E (1986) J Neurosc 6: 331-345
Davis GW, Naka K-I (1980) J Neurophysiol 43: 807-831
Daw NW (1968) J Physiol (London) 197: 567-592
Daw NW, Ariel M, Caldwell JH (1982) Retina 2: 322-331
Dowling JE (1970) Invest Ophthalmol 9: 655-680
Dowling JE, Boycott BB (1966) Proc R Soc Lond B 166: 80-111
Dowling JE, Ehinger B (1978) Proc R Soc Lond B 201: 7-26
Dowling JE, Ehinger B, Florén I (1980) J Comp Neurol 192: 665-685
Ehinger B (1966) Z Zellforsch 71: 146-152
Ehinger B, Florén, I (1976) Cell Tiss Res 175: 37-48
Eldred WD, Karten HJ (1985) J Comp Neurol 232: 36-42
Enna SJ, Snyder SH (1976) Brain Res 115: 174-179
Enroth-Cugell C, Robson JG (1966) J Physiol (London) 187: 517-552
Evans RH, Francis AA, Watkins JC (1977) Experientia 33: 489-491

Famiglietti EV Jr, Kaneko A, Tachibana M (1977) Science 198: 1267-1269
Famiglietti EV Jr, Kolb H (1976) Science 194: 193-195
Fenwick EM, Marty A, Neher E (1982) J Physiol (London) 331: 577-597
Fisher SK (1972) Brain Res 43: 587-590
Gray R, Johnston D (1985) J Neurophysiol 54: 134-142
Hall ZW, Kelly RB (1971) Nature New Biol 232: 62-63
Hamill OP, Bormann J, Sakmann B (1983) Nature 305: 805-808
Hamill OP, Marty A, Neher E, Sakmann B, Sigworth FJ (1981) Pflügers Arch 391: 85-100
Hill DR, Bowery NG (1981) Nature 290: 149-152
Huguenard JR, Alger BE (1986) J Neurophysiol 56: 1-18
Ishida AT, Cohen BN (1988) J Neurophysiol 60: 381-396
Ishida AT, Fain GL (1981) Proc Nat Acad Sci USA 78: 5890-5894
Ishida AT, Kaneko A, Tachibana M (1984) J Physiol (London) 348: 255-270
Ishida AT, Neyton J (1985) Proc Nat Acad Sci USA 82: 1837-1841
Jaramillo F, Schuetze SM (1988) J Physiol (London) 396: 267-296
Kaneko A (1979) Ann Rev Neurosc 2: 169-191
Kaneko A, Tachibana M (1986a) Progr Retinal Res 5: 125-146
Kaneko A, Tachibana M (1986b) J Physiol (London) 373:463-479
Karten HJ, Brecha N (1983) Vision Res 23: 1197-1205
Katz B, Thesleff S (1957) J Physiol (London) 138: 63-80
Kolb H, Nelson R (1984) Progr Retinal Res 3: 21-60
Kolb H, Nelson R, Mariani A (1981) Vision Res 21: 1081-1114
Kolb H, Perlman I, Normann RA (1988) Vis Neurosc 1: 47-72
Landreth GE, Agranoff BW (1976) Brain Res 118: 299-303
Lasater EM, Dowling JE, Ripps H (1984) J Neurosc 4: 1966-1975
Levick WR (1967) J Physiol (London) 188: 285-307
Lipton SA, Tauck DL (1987) J Physiol (London) 385: 361-391
Marc RE (1982) Vision Res 22: 589-608
Marc RE (1985) In: Morgan WW (ed), Retinal Transmitters and Modulators: Models for the Brain. CRC Press, Boca Raton, p 119
Marc RE (1986) Vision Res 26: 223-238
Marc RE, Stell WK, Bok D, Lam DM-K (1978) J Comp Neurol 182: 221-246
Marshak DW, Ariel M, Brown E (1988) Exp Eye Res 46: 965-978
Marshak DW, Yamada T, Stell WK (1984) J Comp Neurol 225: 44-52
Masland RH, Cassidy C (1987) Proc R Soc Lond B 232: 227-238
Massey SC, Redburn DA (1987) J Neurosc 2: 1633-1643
Maturana HR, Lettvin JY, McCulloch WS, Pitts WH (1960) J Gen Physiol 43: 129-175
McArdle CB, Dowling JE, Masland RH (1977) J Comp Neurol 175: 253-274
McBurney RN, Barker JL (1978) Nature 274: 596-597
McGuire BA, Stevens JK, Sterling P (1986) J Neurosc 6: 907-918
Miller RF, Dacheux RF (1976) J Gen Physiol 67: 661-678
Miller RF, Frumkes TE, Slaughter MM, Dacheux RF (1981) J Neurophysiol 45: 743-763
Mittman S, Copenhagen DR (1985) Invest Ophthalmol Vis Sci 26: 312s
Muller JF, Marc RE (1985) Invest Ophthalmol Vis Sci 26: 95s
Murakami F, Katsumaru H, Wu J-Y, Matsuda T, Tsukahara N (1983) Brain Res 267: 357-360
Murray M, Grafstein B (1969) Exp Neurol 23: 544-560

Naka K-I (1977) J Neurophysiol 40: 26-43
Narahashi T (1974) Physiol Rev 54: 813-889
Negishi K, Kato S, Teranishi T, Laufer M (1978) Brain Res 148: 67-84
Neher E, Stevens CF (1977) Ann Rev Biophys Bioeng 6: 345-381
Ozawa S, Tsuda K (1973) J Neurophysiol 36: 805-816
Pourcho RG, Goebel DJ (1985) J Comp Neurol 233: 473-480
Pourcho RG, Goebel DJ (1987) J Neurosc 7: 1178-1197
Raynauld J-P (1972) Science 177: 84-85
Reuter H (1983) Nature 301: 569-574
Ribak CE (1978) J Neurocytol 7: 461-478
Sakai HM, Naka K-I (1985) Nature 315: 570-571
Sakai HM, Naka K-I, Dowling JE (1986) Nature 319: 495-497
Scholes, JH (1975) Phil Trans R Soc Lond B 270: 61-118
Schwartz M, Agranoff BW (1981) Brain Res 206: 331-343
Shingai R, Quandt F (1986) Brain Res 206: 331-343
Slaughter MM, Miller RF (1983) Nature 303: 537-538
Stell WK (1985) In: Gallego A, Gouras, P (ed) Neurocircuitry of the Retina. A Cajal Memorial. Elsevier, New York, p 171
Sterling P (1983) Ann Rev Neurosc 6: 149-185
Stevens JK, McGuire BA, Sterling P (1980) Science 207: 317-318
Tachibana M (1981) J Physiol (London) 321: 141-166
Tachibana M (1983) J Physiol (London) 345: 329-351
Tachibana M, Kaneko A (1984) Proc Nat Acad Sci USA 81: 7961-7964
Tachibana M, Kaneko A (1987) Proc Nat Acad Sci USA 84: 3501-3505
Takeuchi A, Takeuchi N (1971) J Physiol (London) 217: 341-358
Tauchi M, Masland RH (1984) Proc R Soc Lond B 223: 101-119
Turner JE, Schwab ME, Thoenen H (1982) Dev Brain Res 4: 59-66
Vaney, DI (1986) Science 233: 444-446
Vaughn JE, Famiglietti EV Jr, Barber RP, Saito K, Roberts E, Ribak CE (1981) J Comp Neurol 197: 113-127
Wakshull E, Johnson MI, Burton H (1979) J Neurophysiol 42: 1410-1425
Weiler R, Ball AK (1984) Nature 311: 759-761
Weiler R, Schütte M (1985) Brain Res. 360: 379-383
Weiss DS (1988) J Neurophysiol 59: 514-527
Weiss RE, Horn R (1986) Ann N Y Acad Sci 479: 152-161
Werblin FS, Dowling JE (1969) J Neurophysiol 32: 339-354
West RW, Dowling JE (1972) Science 178: 510-512
Witkovsky P, Dowling JE (1969) Z Zellforsch 100: 60-82
Yasui S, Ishizuka S, Akaike N (1985) Brain Res 344: 176-180
Yazulla S (1986) Progr Retinal Res 5: 1-52
Yazulla S, Kleinschmidt J (1983) Brain Res 263: 63-75
Yazulla S, Studholme K, Wu J-Y (1986) J Comp Neurol 244: 149-162
Yazulla S, Studholme K, Wu J-Y (1987) Brain Res 411: 400-405
Zucker CL, Dowling JE (1987) Nature 330: 166-168

ACKNOWLEDGEMENTS

The experiments described in this paper were supported in part by NIH grant EY 05234, and performed by the author and B. N. Cohen in the laboratory of G. L. Fain.

THE FUNCTION OF MONOAMINES IN THE RABBIT RETINA

Nigel W. Daw, William J. Brunken* and Ralph J. Jensen#
Department of Cell Biology and Physiology,
Washington University Medical School,
St. Louis, MO, 63110, USA

Over the last several years we have described the action of dopamine and indoleamine drugs on the activity of ganglion cells in the rabbit retina in the mesopic state (Jensen & Daw, 1984, 1986; Brunken & Daw, 1987, 1988a, 1988b). The prime action of dopamine is that it reduces the sensitivity of both ON and OFF responses, while D_1 antagonists reduce or abolish the surround responses in both ON and OFF center cells. The prime action of serotonin is that it increases spontaneous activity and responses in ON cells, while $5HT_2$ antagonists have an opposite effect. For both transmitters there is a second receptor with actions opposite to the first: D_2 receptors with actions opposite to D_1, and $5HT_{1A}$ receptors with actions opposite to $5HT_2$. There are other effects which will be described below, but these are the prime ones.

At the same time, other laboratories have described the connections of the dopamine and indoleamine-accumulating cells to the rod system, and the responses of some of the cells and signs of some of the synapses in the pathway (Dowling & Ehinger, 1978; Ehinger & Holmgren, 1979; Holmgren-Taylor, 1982; Nelson, 1982; Pourcho, 1982; Dacheux & Raviola, 1986; Raviola & Dacheux, 1987; Voigt & Wassle, 1987; Muller et al., 1988). The purpose of this paper is to ask whether the results that we have obtained can be explained in terms of these connections in the inner plexiform layer, what assumptions need to be made, and what predictions result. There is remarkable concordance between the anatomy and the physiology. Generalisations can be made about the function of these monoamines, and a number of interesting conclusions follow.

---

*Address: Department of Biology, Boston College, Boston, MA 02167, USA

#Address: Southern College of Optometry, Memphis, TN, 38104, USA

NATO ASI Series, Vol. H31
Neurobiology of the Inner Retina
Edited by R. Weiler and N.N. Osborne
© Springer-Verlag Berlin Heidelberg 1989

CONNECTIONS OF DOPAMINE AND INDOLEAMINE-ACCUMULATING CELLS WITH THE ROD SYSTEM

Rod bipolar cells in the mammal are connected indirectly to ganglion cells through a rod amacrine cell, called the AII or rod amacrine in the cat (Famiglietti & Kolb, 1975), and the narrow-field bistratified (NFB) amacrine in the rabbit (Dacheux & Raviola, 1986). The rod bipolar cell excites the NFB amacrine cell, and both depolarise in response to light (Dacheux & iola, 1986; Raviola & Dacheux, 1987). The NFB amacrine cell makes sign-conserving connections to the ON pathway through gap junction synapses with depolarising cone bipolar cells, and sign-reversing connections to the OFF pathway through glycine synapses with OFF center ganglion cells (Famiglietti & Kolb, 1975; Pourcho and Goebel, 1985; Muller et al., 1988).

Dopamine cells in the mammal are amacrine cells, with most of their input coming from other amacrine cells, and most of their output going to other amacrine cells (Dowling and Ehinger, 1978). A major target for the output is the NFB amacrine cell (Pourcho, 1982; Voigt & Wassle, 1987). The source of the input is not yet known.

Indoleamine-accumulating cells are also amacrine cells with a variety of morphologies (Ehinger & Floren, 1976; Sandell & Masland, 1986; Vaney, 1986). Their major input comes from rod bipolar cells, they have a minor input from other amacrine cells, and almost all their ouput feeds back onto rod bipolar terminals (Ehinger & Holmgren, 1979; Holmgren-Taylor, 1982; Sandell et al, 1988). They depolarise in response to light (Nelson & Kolb, 1985; Raviola & Dacheux, 1987). All of these connections are summarised in figure 1.

EFFECTS OF DOPAMINE DRUGS ON BRISK CELLS

We will now analyse our results in terms of these anatomical connections. In order to do so, we need to clarify some postulates that have not yet been conclusively proved, although the evidence makes it very likely that they are correct. The first is that the synapse between the dopamine amacrine cell and the NFB amacrine cell is a sign-reversing synapse, with dopamine acting on the postsynaptic membrane through D_1 receptors. The second is that there is a continuous release of dopamine from these cells in the mesopic state, and that the ability of the cell to release dopamine is considerably reduced in the dark-adapted state: this point is supported by a variety of pharmacological experiments (see Parkinson & Rando, 1983;

Godley & Wurtman, 1988). The third is that the dopamine amacrine cells collect input over a wide area, and provide an antagonistic surround to the NFB amacrine cell - a surround that goes through the center in a difference-of-Gaussians manner (Rodieck, 1965).

We have found that dopamine reduces the spontaneous activity in ON center ganglion cells and increases the spontaneous

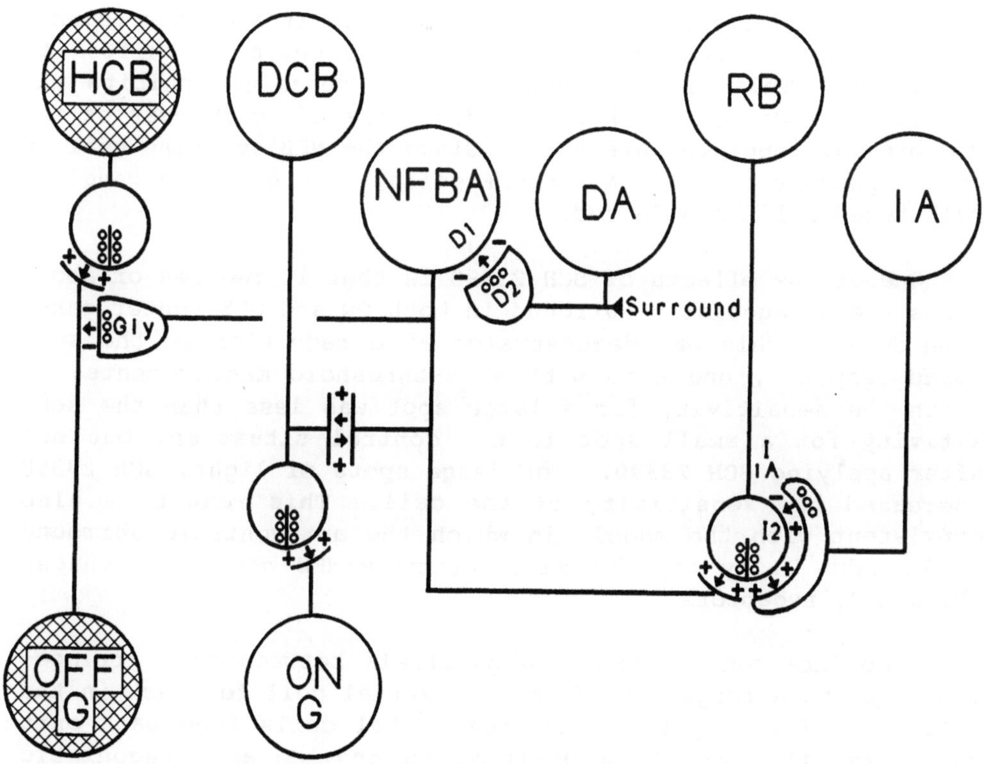

Fig. 1. Diagram of the connections of dopamine and indoleamine-accumulating cells. DA, dopamine amacrine cell; DCB, depolarising cone bipolar; HCB, hyperpolarising cone bipolar; IA, indoleamine-accumulating amacrine; NFBA, narrow-field bistratified amacrine; ON G, ON center brisk ganglion cell; OFF G, OFF center brisk ganglion cell; RB, rod bipolar. + represents a sign-conserving synapse, - a sign-reversing synapse and arrows show the direction of synaptic transmission. D_1, D_2, dopamine receptors. I_{1A}, I_2, indoleamine receptors. Gly, glycine. Cells that depolarise in response to light are open; cells that hyperpolarise are cross-hatched.

activity in OFF center ganglion cells. This is consistent with the model, which shows dopamine acting on ON center ganglion cells with one sign reversal, and on OFF center ganglion cells with two sign reversals. We have also found that SCH 23390, which is a D_1 antagonist, increases spontaneous activity in ON cells and reduces it in OFF cells. This is also to be expected, provided that the postsynaptic receptors are D_1 and there is a continuous release of dopamine from the dopamine amacrine cell.

Dopamine has another effect. It reduces the sensitivity of both ON and OFF center cells. That is, if one measures the response-intensity curve of one of these cells and then remeasures it in the presence of dopamine, the curve is shifted to the right. This is to be expected, if dopamine inhibits the NFB amacrine cell, and thereby reduces the effectiveness of the rod bipolar input to this cell. Since the NFB amacrine cell is on the pathway to both ON and OFF cells, both of these ganglion cell types will be affected.

One of the effects of SCH 23390 is that it reduces or abolishes the antagonistic surround in both ON and OFF center ganglion cells. This was demonstrated as a reduction in the surround response, and also with area-threshold measurements, in which the sensitivity for a large spot was less than the sensitivity for a small spot in the control situation, but not after applying SCH 23390. For large spots of light, SCH 23390 increased the sensitivity of the cell. This result is also consistent with the model, in which the antagonistic surround is brought in by the dopamine amacrine/NFB amacrine synapse through D_1 receptors.

There are some interesting parallels between this arrangement and the arrangement of the horizontal cell feedback in the outer plexiform layer. Since horizontal cells feed back onto receptors, they are in a position to provide an antagonistic surround for both the depolarising and hyperpolarising cone bipolar cells. Similarly, since the dopamine amacrine cell synapes onto the NFB amacrine cell, it is in a position to provide an antagonistic surround for rod influences on both ON and OFF ganglion cells. The input to the dopamine amacrine cell is not yet known, but the model would predict an excitatory input, so that the dopamine amacrine cell should depolarise in response to light.

Many years ago Barlow, Fitzhugh and Kuffler (1957) showed that the surround of ganglion cells in the cat disappears in the completely dark-adapted state. This is also consistent with our model. Pharmacological experiments have shown that the dopamine system in the mammalian retina is less active in the dark than it is in the light (see Parkinson & Rando, 1983; Godley & Wurtman, 1988). The dopamine amacrine/NFB amacrine synapse should therefore be less active in the dark, and the antagonistic surround consequently less effective.

Another set of observations concerns D_2 receptors in the retina. Pharmacological observations suggest that D_2 receptors in the rabbit retina are autoreceptors, acting so that dopamine binding to them reduces the release of dopamine from the presynaptic terminal (Dubocovich & Wiener, 1985). We have found that the D_2 agonist LY 141865 acts like D_1 antagonists, which is to be expected if a D_2 agonist reduces the release of dopamine. The D_2 antagonist sulpiride reduced the spontaneous activity in ON cells, and the sensitivity of OFF cells, which is consistent with potentiation of the release of dopamine. However, it also reduced spontaneous activity in OFF cells: this result is not explained, and it may have been due to the rather high concentrations of sulpiride required to get an effect.

An intriguing result obtained with SCH 23390 was that it turned the sustained ON excitation seen in sustained ON cells and direction selective cells into a sustained ON inhibition. The explanation of this in terms of our model is rather more complicated. The result shows that there has to be both sustained excitatory and sustained inhibitory input to these cells. We know that SCH 23390 increases the spontaneous activity in ON cells, presumably by depolarising them. It could be that this depolarisation shifts the membrane potential of the cell sufficiently that the sustained inhibitory input becomes more powerful than the sustained excitatory input.

EFFECTS OF INDOLEAMINE DRUGS ON BRISK CELLS

In analysing the effects of indoleamine drugs in the rabbit retina, we will assume that they are acting at the synapses of the indoleamine-accumulating cells. Serotonin fulfils all of the requirements for a synaptic transmitter in rabbit retina, except that the presence and release of endogenous serotonin have not been adequately demonstrated (see Osborne, 1984; Brunken & Daw, 1988b). If the transmitter of the indoleamine-accumulating cells is not serotonin, then it is still possible that there is enough cross-reactivity between the transmitter and the indoleamine drugs that we used that the drugs were acting at the indoleamine-accumulating cell synapses.

The indoleamine-accumulating cell receives input from the rod bipolar cell terminal, and feeds back onto this terminal (Ehinger & Holmgren, 1979; Holmgren-Taylor, 1982; Raviola & Dacheux, 1987; Sandell et al., 1988). The rod bipolar cell excites the indoleamine-accumulating cell (Raviola & Dacheux, 1987). We will argue from our results that the synapse from the indoleamine-accumulating amacrine back to the rod bipolar includes a positive feedback.

The indoleamine system includes $5HT_{1A}$ receptors as well as $5HT_2$ receptors. Our results show that these receptors act in reciprocal fashion, just as the D_1 and D_2 receptors act in reciprocal fashion. That is, $5HT_{1A}$ agonists have the same effect as $5HT_2$ antagonists. Positive feedback systems are generally unstable. We will suggest that one reason that the positive feedback synapse from the indoleamine-accumulating cell back to the rod bipolar terminal is not unstable is that the $5HT_{1A}$ receptors act to prevent it getting out of control.

Serotonin increases spontaneous activity in ON center ganglion cells. This is expected with our model, since serotonin acting on the rod bipolar terminal will excite this terminal and consequently the NFB amacrine and ON center cell. Serotonin also increases the ON responses in ON center ganglion cells. This is to be expected, if the feedback synapse onto the rod bipolar terminal is positive.

$5HT_2$ antagonists reduce ON responses and spontaneous activity in ON center cells. This follows if the indoleamine receptors in the rod bipolar terminal are $5HT_2$ receptors. Signals reaching the ON center ganglion cell via the depolarising cone bipolar may also be reduced by shunting into the NFB amacrine network through the gap junctions.

The effect of $5HT_2$ antagonists in OFF center cells is to increase the OFF response or leave it unchanged. The interpretation of this result is more complicated, since nobody really knows how OFF responses are generated in the first place, in terms of the ionic mechanisms that underly the response. Nevertheless, the hypothesis that serotonin is acting on the rod bipolar terminal, and the fact that signals from this terminal excite ON center cells and inhibit OFF center cells, make it reasonable that $5HT_2$ antagonists should increase OFF responses while they decrease ON responses. Essentially, $5HT_2$ antagonists will reduce the NFB amacrine input, and the response of the OFF center ganglion cell will follow the input from the hyperpolarising cone bipolar more closely.

$5HT_2$ antagonists also reduce the surround ON response in OFF center ganglion cells. Our model does not specify how this influence reaches the OFF center ganglion cell. Since the evidence shows that this surround is reduced or abolished by the dopamine antagonist SCH 23390, and the evidence suggests that the indoleamine is acting at the rod bipolar terminal, the model predicts an excitatory pathway between the rod bipolar terminal and the dopamine amacrine cell. However, the route of such a pathway remains to be shown by anatomical methods.

Comparison of the records of responses in the rod bipolar cell and the NFB amacrine cell show two interesting points (Nelson, 1982; Dacheux & Raviola, 1986). First, the response in the NFB amacrine cell rises faster than the response in the rod bipolar cell. This is exactly what one would expect if there were a positive feedback synapse at the rod bipolar terminal. The model therefore predicts that $5HT_2$ antagonists should reduce the rate of rise of the response in the NFB amacrine cell. Second, after the initial fast rise is over, the response in the NFB amacrine cell oscillates for a few milliseconds. One can regard the $5HT_{1A}$ system as a negative feedback system, superimposed on the positive feedback carried by $5HT_2$ receptors. Negative feedback systems tend to oscillate. Consequently the model also predicts that a selective $5HT_{1A}$ antagonist, if one were available, should reduce the oscillations seen in this response. GABA may also be involved in this role, since the indoleamine-accumulating cells also take up GABA (Osborne & Beaton, 1986).

DIRECTION SELECTIVE CELLS

The rabbit retina contains a number of ON-OFF direction selective cells. These cells receive cone input directly from cone bipolar cells, and indirectly from cone bipolar input to acetylcholine amacrine cells (Famiglietti, 1983; Brandon, 1987). There is very little known input from NFB amacrine cells, and no known input from rod bipolar cells (Famiglietti, private communication). The rod input therefore comes almost exclusively through the NFB amacrine/depolarising cone bipolar gap junction synapse.

We have found that $5HT_2$ antagonists reduce ON responses in direction selective cells, just as they do in brisk ganglion cells. In both cases, the pathway for this effect goes through the NFB amacrine/cone depolarising bipolar gap junction. On the other hand, $5HT_2$ antagonists reduce OFF responses in direction selective cells, rather than increase them as happens in brisk ganglion cells. We suggest that this occurs because the NFB amacrine/ganglion cell glycine synapse is present between the NFB amacrine cell and the OFF center ganglion cell, but absent between the NFB amacrine cell and direction selective cell. OFF responses in direction selective ganglion cells may be reduced by $5HT_2$ antagonists because influences that come in to the direction selective ganglion through the NFB amacrine cell/depolarising cone bipolar gap junction affect the body of the direction selective ganglion cell, and therefore influence hyperpolarising cone bipolar input as well as depolarising cone bipolar input to this cell.

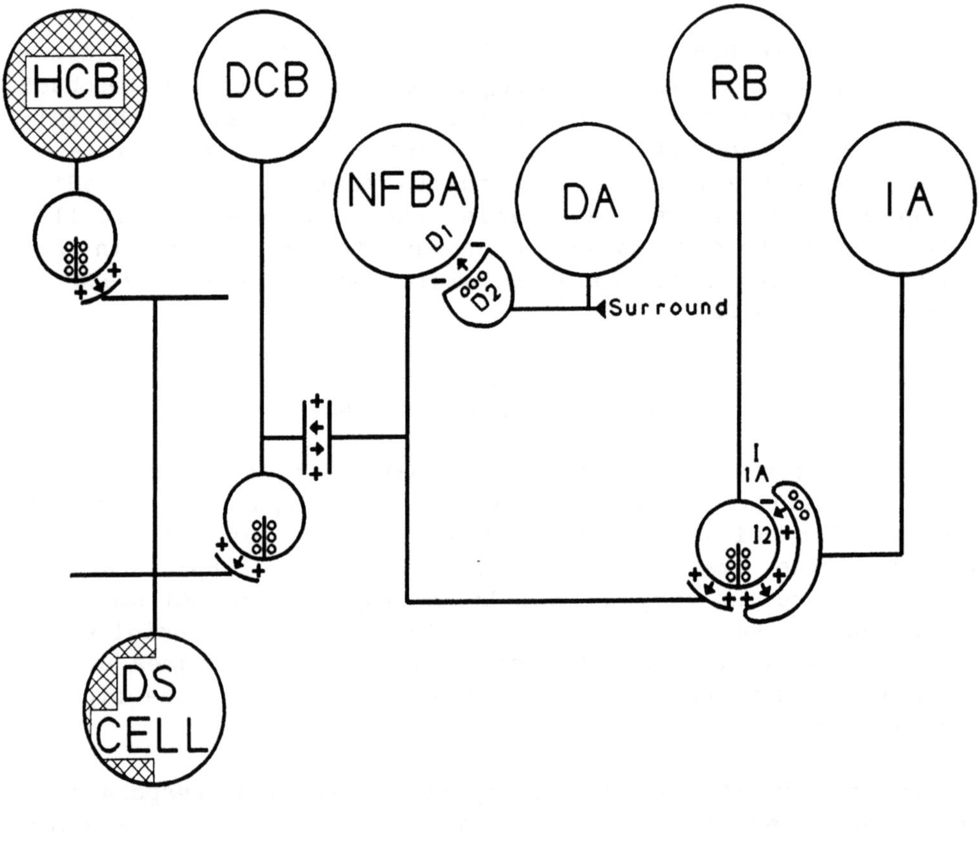

Fig. 2. Diagram of the main connections from the dopamine and indoleamine-accumulating cells to the direction selective ganglion cell. Abbreviations and signs as in Fig. 1.

This line of argument makes a prediction about the response of ON-OFF direction selective ganglion cells in the completely dark-adapted state. This is that they should lose their OFF response, since they have little or no sign-reversing rod input, assuming that rod influences through direct rod/cone connections are not active in this state. They should keep their direction selectivity for the leading edge of a stimulus, and have little or no response to the trailing edge of a stimulus.

CONCLUSIONS

These results and arguments lead to an overall picture of the function of monoamines in the mammalian retina.

Dopamine cells are there to provide the antagonistic surround in the rod system. A single class of dopamine amacrine cells suffices for this purpose, because they act on the NFB amacrine, and through the NFB amacrine on both ON and OFF center cells. The surround is absent in the completely dark-adapted state, because the release of dopamine from the dopamine amacrine cells is reduced in the dark. This serves to increase the sensitivity of the system in the dark.

Indoleamine cells are there to increase the gain of the rod bipolar/NFB amacrine synapse. This will increase signal/noise ratio for all sources of noise occurring after the rod bipolar/NFB amacrine synapse, such as "dark light" (rhodopsin isomerisations resulting from heat rather than light) coming into the ganglion cells from other rod bipolars, and noise in the NFB amacrine and ganglion cell membranes.

Predictions can also be made about what should happen in the completely dark-adapted state. This is best summarised in terms of the response-intensity curve recorded from ganglion cells. Dopamine should shift this curve to the right, just as it does in the mesopic state, because the D_1 receptors are still there. Under these conditions D_1 antagonists should have little effect, because the release of dopamine from these cells is reduced in the dark. $5HT_2$ antagonists should reduce the slope of this curve, since they will abolish the positive feedback at the rod bipolar terminal. The action of serotonin itself will be complicated, since it acts at both $5HT_2$ and $5HT_{1A}$ receptors, and it would be hard to time its application to coincide with the pulse of light, but it would probably increase the slope of the curve if applied at the correct time.

SUMMARY

The paper can best be summarised in terms of the models drawn in Figs. 1 and 2, a list of observations that are consistent with the models, and a list of predictions.

The observations are:

--- Dopamine reduces the sensitivity of both ON and OFF responses.
--- D_1 antagonists reduce or abolish the surround response in both ON and OFF center cells.
--- DA reduces spontaneous activity in ON cells, increases it in OFF cells, and D_1 antagonists do the reverse
--- D_2 agonists act like D_1 antagonists, and D_2 antagonists in some respects have the opposite effect
--- 5HT increases ON responses and spontaneous activity in ON cells
--- $5HT_2$ antagonists reduce ON responses in all types of cell
--- $5HT_2$ antagonists reduce OFF responses in direction selective cells, and have little effect or increase them in brisk cells
--- $5HT_{1A}$ agonists act like $5HT_2$ antagonists

The predictions are:

--- The dopamine amacrine/NFB amacrine synapse is inhibitory
--- D_1 antagonists will abolish the surround response in the NFB amacrine cell
--- The indoleamine amacrine feedback synapse onto the rod bipolar terminal is a positive feedback synapse, such that $5HT_2$ antagonists will reduce the slope of the response-intensity curve in the NFB amacrine cell
--- $5HT_{1A}$ receptors act to curtail the positive feedback, and $5HT_{1A}$ antagonists, when found, will reduce the oscillations seen in the NFB amacrine cell response
--- Direction selective cells in the completely dark-adapted state will have little OFF response or response to the trailing edge of a moving stimulus
--- The response-intensity curve of brisk cells in the completely dark-adapted state (primarily ON cells, since one is concerned with increments of light rather than decrements of light) will be shifted to the right by dopamine, not much affected by D_1 antagonists, and their slope reduced by $5HT_2$ antagonists.

REFERENCES

Barlow HB, Fitzhugh R, Kuffler SW (1957) Change of organisation in the receptive fields of the cat's retina during dark adaptation. J Physiol 137: 338-354

Brandon C (1987) Cholinergic neurons in the rabbit retina: dendritic branching and ultrastructural connectivity. Brain Res 426: 119-130

Brunken WJ, Daw NW (1987) The actions of serotonin agonists and antagonists on the activity of brisk ganglion cells in the rabbit retina. J Neurosci 7: 4054-4065

Brunken WJ, Daw NW (1988a) The effects of serotonin agonists and antagonists on the response properties of complex ganglion cells in the rabbit's retina. Vis Neurosci 1: 181-188

Brunken WJ, Daw NW (1988b) Neuropharmacological analysis of the role of indoleamine-accumulating cells in the rabbit retina. Vis Neurosci 1: in press

Dacheux RF, Raviola E (1986) The rod pathway in the rabbit retina: a depolarizing bipolar and amacrine cell. J Neurosci 6: 331-345

Dowling JE, Ehinger B (1978) Synaptic organisation of the dopaminergic neurons in the rabbit retina. J Comp Neurol 180: 203-220

Dubocovich ML, Weiner N (1985) Pharmacological differences between the D-2 autoreceptor and the D-1 dopamine receptor in rabbit retina. J Pharmacol Exp Therap 233: 747-754

Ehinger B, Floren I (1976) Indoleamine accumulating neurons in the retina of rabbit, cat and goldfish. Cell Tissue Res 175: 37-48

Ehinger B, Holmgren I (1979) Electron microscopy of the indoleamine-accumulating neurons in the retina of the rabbit. Cell Tissue Res 197: 175-194

Famiglietti EV (1983) ON and OFF pathways through amacrine cells in mammalian retina: the synaptic connections of "Starburst" amacrine cells. Vision Res 23: 1265-1279

Famiglietti EV, Kolb H (1975) A bistratified amacrine cell and synaptic circuitry in the inner plexiform layer of the retina. Brain Res 84: 293-300

Godley BF, Wurtman RJ (1988) Release of endogenous dopamine from the superfused rabbit retina in vitro: effect of light stimulation. Brain Res 452: 393-395

Holmgren-Taylor I (1982) Electron-microscopical observations on the indoleamine-accumulating neurons and their synaptic connections in the retina of the cat. J Comp Neurol 208: 144-156

Jensen RJ, Daw NW (1984) Effects of dopamine antagonists on receptive fields of brisk cells and directionally selective cells in the rabbit retina. J Neurosci 4: 2972-2985

Jensen RJ, Daw NW (1986) Effects of dopamine and its agonists and antagonists on the receptive field properties of ganglion cells in the rabbit retina. Neurosci 17: 837-855

Muller F, Wassle H, Voigt T (1988) Pharmacological modulation of the rod pathway in the cat retina. J Neurophysiol 59: 1673-1689

Nelson R (1982) AII amacrine cells quicken time course of rod signals in the cat retina. J Neurophysiol 47: 928-947

Nelson R, Kolb H (1985) A 17: a broad field amacrine cell in the rod system of the cat retina. J Neurophysiol 54: 592-614

Osborne NN (1984) Indoleamines in the eye with special reference to the serotonergic neurones of the retina. Prog Retinal Res 3: 61-104

Osborne NN, Beaton DW (1986) Direct histochemical localisation of 5,7 dihydroxytryptamine and the uptake of serotonin by a subpopulation of GABA neurons in the rabbit retina. Brain Res 382: 158-162

Parkinson D, Rando RR (1983) Effect of light on dopamine turnover and metabolism in rabbit retina. Inv Ophthal 24: 384-388

Pourcho R (1982) Dopaminergic amacrine cells in the cat retina. Brain Res 252: 101-109

Pourcho RG, Goebel DJ (1985) A combined Golgi and autoradiographic study of ^{3}H glycine accumulating amacrine cells in the cat retina. J Comp Neurol 233: 473-380

Raviola E, Dacheux RF (1987) An excitatory dyad synapse in the rabbit retina. Proc Nat Acad Sci 84: 7324-7328

Rodieck RW (1965) Quantitative analysis of cat retinal ganglion cell response to visual stimuli. Vision Res 5: 582-601

Sandell JH, Masland RH (1986) A system of indoleamine-accumulating neurons in the rabbit retina. J Neurosci 6: 3331-3347

Sandell JH, Masland RH, Raviola E, Dacheux RF (1988) Connections of indoleamine-accumulating cells in the rabbit retina. J Comp Neurol : in press

Vaney DI (1986) Morphological identification of serotonin-accumulating neurons in the living retina. Science 233: 444-446

Voigt T, Wassle H (1987) Dopaminergic innervation of A II amacrine cells in mammalian retina. J Neurosci 7: 4115-4128

A ROLE OF THE ANGIOTENSIN-RENIN SYSTEM FOR RETINAL NEUROTRANSMISSION?

E. Zrenner, P. Dahlheim, K.-H. Datum

Laboratories of the Max-Planck-Institute for
Physiological and Clinical Research
in the University Eye Hospital,
Mathildenstr.8, 8000 Munich 2, FRG

Summary

Angiotensin converting enzyme (ACE) catalyzes the formation of angiotensin II (A II) which presumably acts as central neurotransmitter/modulator. We determined the ACE-activity in various ocular tissues and the effect of the ACE inhibitor Enalapril (MK 422, MSD) as well as of the A II-receptor blocker saralasin (Röhm) on the electroretinographical responses of the arterially perfused cat eye.

The highest specific ACE-activity was found in the retina (190 ± 4.5 µmol/g/h; 4 eyes), while ACE-activity in the choroid was only 1/5th and in other ocular tissues 1/40-1/100th of retinal activity. Injecting MK 422 in concentrations ranging from 0.1 - 100 ng/ml/min for 20 min into the ophthalmociliary artery of the perfused cat eye in low concentrations increased b-wave amplitude by 30-40% reversibly and enhanced sensitivity, while higher concentrations decreased b-wave amplitude and sensitivity. In the light adapted cone-ERG b-wave and off-effect , amplitudes were increased by low concentrations by a factor of 2-3 but abolished by high concentrations that produced an irreversible P III component. Flow rate was not altered by these MK 422 concentrations.

The A II-receptor blocker saralasin, injected in similar concentrations for 20 min, transiently increased flow rate moderately, but enhanced b-wave by 300-500% with the enhancement persisting for several hours (halftime of saralasin is 3.2 min). These effects lead to the strong assumption that the angiotensin-renin-system has, besides its vasoactive role, a direct or indirect role as neuromodulator system in retinal neurons.

NATO ASI Series, Vol. H31
Neurobiology of the Inner Retina
Edited by R. Weiler and N. N. Osborne
© Springer-Verlag Berlin Heidelberg 1989

Introduction

Peptides have received increasing attention in retinal neurophysiology. The peptides that are involved in the angiotensin-renin system, however, have been rarely studied in respect to their role in retinal neurotransmission, although it is well established that they occur in the retina (Igic and Kojovich, 1980 and Ferrari-Dileo et al., 1988). There is no doubt that angiotensin has transmitter-like functions in neurons of the central nervous system (Severs and Daniels-Severs, 1973; Felix, 1982; Sessle and Henry, 1987; Phillips, 1987). It therefore seems worthwhile to have a closer look at possible roles of the angiotensin-renin system in retinal physiology.

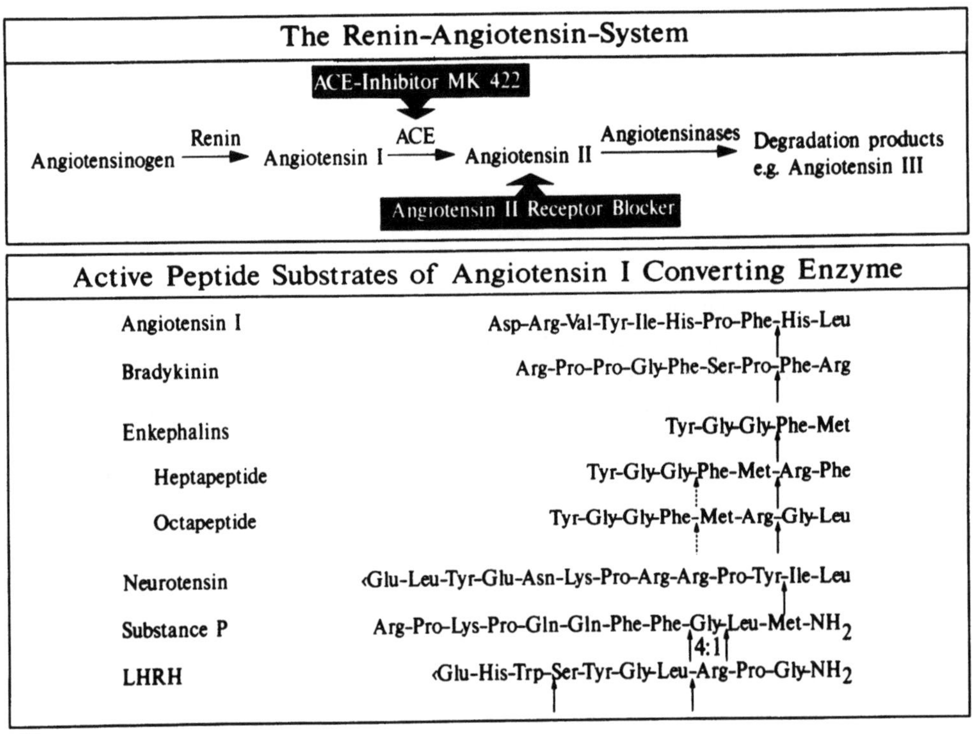

Fig. 1. The angiotensin-renin system (top) and the interactions with other peptides (bottom) found in the retina of various species. Filled arrows denote primary sites of cleavage by human ACE, broken arrows denote secondary ones. (Modified from Skidgel a Erdös, 1987).

As shown in fig. 1 (top), angiotensin I-converting-enzyme (ACE) catalyzes the formation of the octapeptide angiotensin II from the decapeptide angiotensin I. Accordingly, the angiotensin-converting-enzyme acts as a dipetidase. Ferrari-Dileo et al. (1988) found ACE in the plasma, vascular endothelium, corneal endothelium, as well as in the retinal vessels, in the choroid and the neural retina of cat, bovine and human tissue. As also shown in fig. 1 (bottom), ACE has a rather broad substrate specificity (from Skidgel and Erdös, 1987). ACE can metabolize a number of substances that are thought to play a role in retinal neurotransmission. Not only is angiotensin I converted into angiotensin II, but sites of cleavage were found among others in enkephalins, neurotensin, substance P and LHRH. Moreover it is discussed that the methenkephalinase and ACE may act biochemically identical (Mellstrom et al., 1986; Swerts et al., 1979). This strongly points to a possible role of ACE in retinal neurotransmission, since methenkephalinergic neurons are reported in retinae of many species (Watt and Su, 1988; Morgan, 1989, this volume).

On the other hand, it is not excluded that angiotensin II (A II) itself, which acts as a very potent vasoconstrictor might also have a neurotransmitter-like action or at least serve as a modulator of neuronal activity. According to Sirret et al. (1977), specific angiotensin binding in brain neurons is more potent than in any other tissue (for a survey, see Ganten et al., 1982). Mallorga et al. (1988) report binding sites in the rabbit neural retina, in the choroid, in retinal blood vessels, in the ciliary processes, the iris and the ciliary body. The role of angiotensin II for retinal function has not been studied yet, although in the central nervous system it affects neurotransmission: It is well known that A II increases the rate of noradrenalin synthesis (Story and Ziogas, 1987); it prolongs and potentiates the action of acetylcholin (Severs and Daniels-Severs, 1973), it increases the dopamine turn-over and it affects serotonin synthesis in synaptosomal fractions of the hypothalamus and brain stem (Phillips, 1987). These findings make a modulator-like action of A II very probable. There is not only an endogenous brain renin-angiotensin system but angiotensin II is found also in neurons of the CNS by immunocytochemistry (Lind et al., 1985). In addition, A II excites CA1 pyramidal cells of the hippocampus, as shown by intracellular recordings; these effects can be reversed by the A II-antagonist saralasin (Haas et al., 1980). Moreover, A II receptors satisfy the criteria of specificity, saturability, reversibility of binding and high affinity (see Felix, 1982). It seems therefore very worthwhile to look for actions of the angiotensin-renin system on retinal neurotransmission.

The goal of the present studies therefore was to determine ACE distribution in ocular tissues and to study the effects of ACE inhibitors and A II receptor blockers on retinal function by means of electroretinographical recordings in the arterially perfused cat eye. This paper reviews the effects of the ACE inhibitor MK 422 on the electroretinogram, reported by Dahlheim et al. (1988a, b, c) as well as the data on the A II receptor blocker saralasin, where a full paper is in preparation (Dahlheim et al., 1989).

Methods

1. Electrophysiological recordings from the arterially perfused cat eye

The investigations were performed in the arterially perfused eyes of 12 cats, as described by Gouras and Hoff (1970); Niemeyer (1975); Zrenner and Gouras (1979); and Dahlheim et al. (1988a). It not only provides remarkable stability in electrophysiological recordings, but permits to exactly determine the flowrate of the perfusate which is very important in a study applying vasoactive peptides. Moreover, the concentration of the drug can be precisely determined at the place of action, in contrast to experiments in the intact animal. Also metabolites cannot play a role, since there is no systemic passage of the drug.

12 eyes of 9 adult cats were enucleated and perfused with a mixture of medium 199 and calf serum after oxygenation (details see Dahlheim et al., (1988a, 1989). The perfusion medium was transported into the eye via a cannula inserted into the ophthalmociliary artery with the pressure adjusted so that a perfusion rate of approximately 2 ml/ min was obtained. A schematic drawing is shown in fig. 2. Drugs were injected by means of a pump through a very small tubing that is concentric to the perfusion tube and ends shortly before the insertion of the cannula into the ophthalmociliary artery.

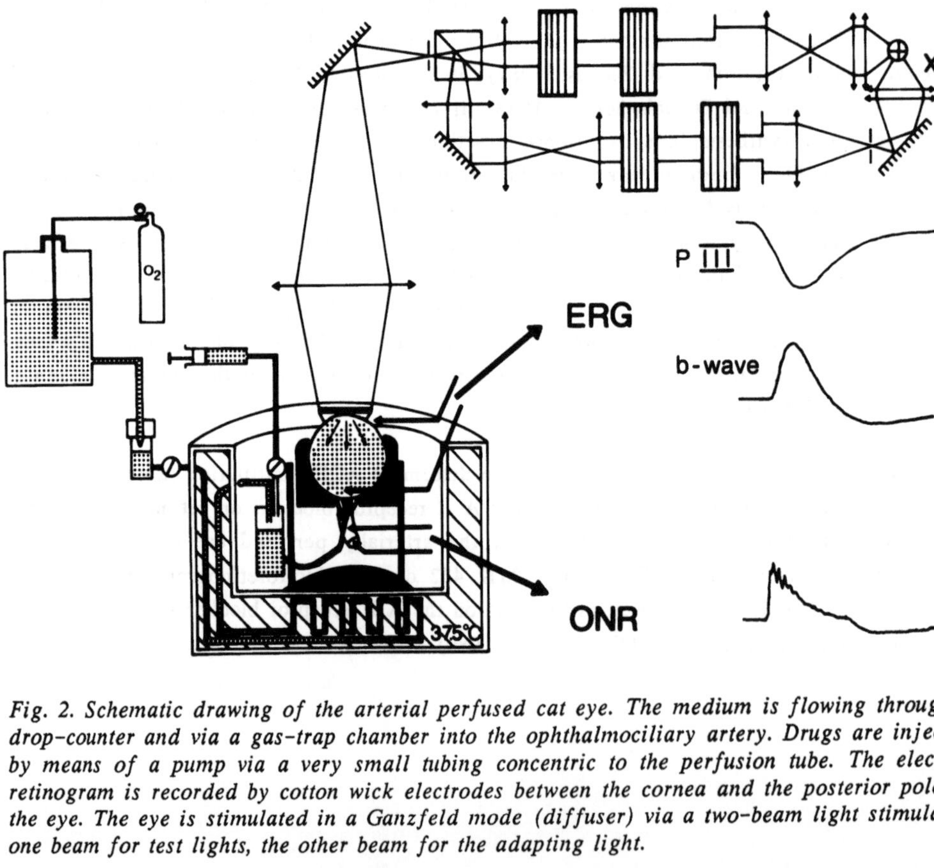

Fig. 2. Schematic drawing of the arterial perfused cat eye. The medium is flowing through a drop-counter and via a gas-trap chamber into the ophthalmociliary artery. Drugs are injected by means of a pump via a very small tubing concentric to the perfusion tube. The electroretinogram is recorded by cotton wick electrodes between the cornea and the posterior pole of the eye. The eye is stimulated in a Ganzfeld mode (diffuser) via a two-beam light stimulator, one beam for test lights, the other beam for the adapting light.

inhibitor MK 422 approximately two hours before enucleation (right column) was significantly higher than that measured in the group perfused with blood. This might point to an increased rate of ACE-synthesis, induced via a feedback mechanism by the passage of the ACE-inhibitor.

	Ocular distribution of ACE-activity		
tissue perfused with:	blood [µmol/g/h]	culture medium [µmol/g/h]	culture medium + MK 422 [µmol/g/h]
Retina	108.96 ± 6.60	164.19	190.52 ± 4.43
Choroid	55.10 ± 4.82	15.41	40.10 ± 20.70
Nervus opticus	5.38 ± 0.99	4.97	5.29 ± 0.51
Ciliary body	2.87 ± 0.22	2.63	3.31 ± 0.72
Iris	2.75 ± 0.36	4.56	4.09 ± 1.78
Sclera	1.59 ± 0.41	2.60	2.85 ± 0.52
Cornea	0.48 ± 0.10	0.64	0.72 ± 0.20
Vitreous	0.34 ± 0,04	---	0.78
Lens	0.09 ± 0.03	0.12	0.14 ± 0.06
	n=4	n=1	n=5

Table 1: The distribution of specific ACE-activity in various ocular tissues. Three groups were formed, group 1 consists of eyes perfused with blood, group 2 of eyes perfused with culture medium, group 3 consists of eyes that were perfused with culture medium and had received the ACE-blocker MK 422 approximately 2 hours after enucleation. (From Dahlheim et al., 1988b)

2. Effects of the ACE-inhibitor on the rod electroretinogram

Fig. 3 shows the effects of the ACE-inhibitor MK 422 (Enalaprilat, Merck, Sharp and Dohme) onto the rod ERG. The inhibitor was applied in concentrations ranging from 0.5 to 50 ng/ml/min (1.3-130 nmol/ min resp .) for 20 minutes. Only the effects obtained with concentrations of 1.0 and 10.0 ng/ml/min culture medium are shown in fig. 3. The amplitudes of the b-waves are plotted against the logarithm of irradiance of the stimulus (508 nm), presented for 200 ms every 5 seconds after dark adaptation for at least 40 minutes. Such V-log I functions were measured before application of the inhibitor (triangles), during application (squares) and approximately one hour after application (circles).

The eye was stimulated in a Ganzfeld mode by means of a two-beam light stimulator, originating in a xenon lamp, one beam for monochromatic test stimuli, the other beam for background light, serving for the isolation of the cone system. The a- and b-wave of the electroretinogram was recorded with 2 cotton wick silver-wire electrodes, one placed near the limbus of the cornea, the other one near the posterior pole of the eye. If isolated by hypoxia, also recordings of the P III could be made, indicative of the electrical activity of the outer retina. The optic nerve could be sucked into a third electrode, so that the summed activity of the optic nerve fibres was recorded in the so-called optic nerve response (ONR). The electrophysiological responses were filtered between 0.3 and 300 Hz and recorded on an averager (Nicolet 1070). The drugs were resolved in identical amounts of the perfusion medium in various concentrations and injected for 20 min at a constant flow rate. Electrophysiological recordings were made several times before, during and after the injection of the drugs.

After the experiment, the eye tissues were dissected, cut and homogenized with ultrasonics, for determination of ACE activity in the various tissues. Further details on methods see Dahlheim et al. (1988a).

2. Determination of angiotensin I-converting-enzyme activity

ACE activity was determined by means of a fluorimetric assay according to Piquilloud et al. (1970). The synthetic substrate Z-PHE-HIS-LEU, (Bachem, USA) is cleaved by ACE. The resulting dipetide HIS-LEU can be transformed into a fluorescent complex by means of o-phthaldialdehyd. The concentration of the fluorescent complex is measured by means of a Zeiss spectral photometer with additional fluorescent facilities. Further details can be found in the paper of Dahlheim (1988a).

Results

1. Ocular distribution of ACE activity

Table 1 shows ACE-activity in the various retinal tissues as obtained in three different groups of eyes: The first group consisted of eyes perfused only with blood (left column), the second group was perfused with culture medium (center column), the third group was perfused with culture medium to which the ACE-inhibitor MK 422 had been added for 20 min and then had been washed out. Surprisingly, the highest ACE-activity was found in the retina and not in the choroid. In the other tissues (optic nerve, ciliary body, iris, sclera, cornea, vitreous, lens) ACE-activity was more than 10 times lower. The differences between the three groups are small; interestingly, the activity in the group that had received the ACE-

Next to each set of curves, original responses are shown, elicited by stimulus irradiances indicated by the appropriate values of log E. The V-log I function obtained before application shows the typical sigmoidal course. During application of MK 422 in low concentrations, (left side of fig. 3), amplitudes grow so that the V-log I function (squares) shifts to the left and becomes steeper, indicating an increase in sensitivity and gain. One hour after application (circles), the V-log I function is compressed and shifts to the right. There is no complete recovery as monitored for several hours after these experiments have been performed. On the right side of fig. 3 similar experiments with an inhibitor concentration 10 times higher than on the left side are shown. Due to a wider range of test lights applied, the V-log I function before application (triangles) shows two phases, the ascending sigmoidal phase and a descending phase at very high test light intensities. Contrary to the results shown on the left, higher blocker concentrations reduce b-wave amplitudes by at least 50 % (squares) immediately during application. One hour after application (circles), the compression of the

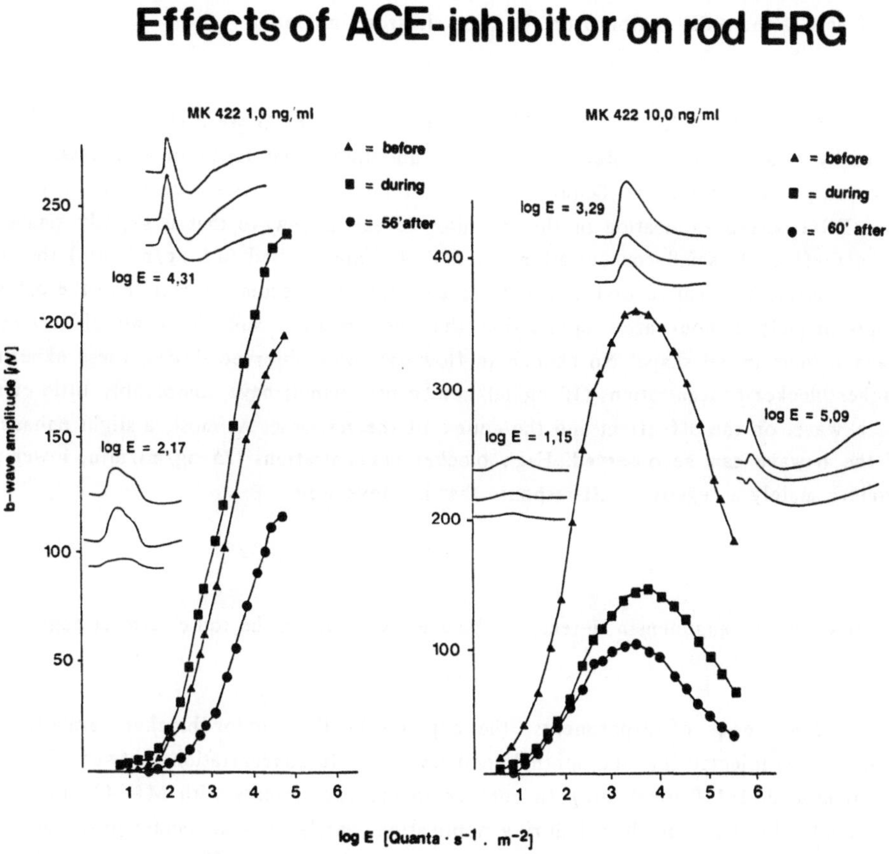

Fig. 3: V-log I functions obtained in the arterial perfused cat eye before (triangles), during (squares) and approximately 1 hour after (circles) application of the ACE-inhibitor MK 422 in 2 different concentrations (1.0 ng/ ml . min. on the left side, 10.0 ng/ ml . min. on the right side). The data were obtained after 30 minutes of dark adaptation with 508 nm flashes of 200 ms duration, presented with an interstimulus interval time of 5 seconds. The inhibitor was applied for 20 minutes. Original recordings before (top), during (center) and after application (bottom of each triplet) are shown besides the curves for 2 different test light irradiances (log E). Modified from Dahlheim et al. (1988a)

V-log I function becomes more pronounced, although sensitivity does not change dramatically. Latencies were prolonged by about 10% during application but shortened by about 15% one hour after application, despite the highly reduced amplitude.

Flowrate was continuously monitored by an electronic drop counter and did not change during the entire course of the experiment. The effects can therefore not be explained simply by alterations of the haemodynamic parameters. Even if the inhibitor would have had an effect on the haemodynamic parameters, the flowrate would have been enhanced since the vasoconstrictor effects would have become weaker which usually causes a b-wave increase but not the observed decrease.

3. Effects of the ACE-inhibitor on the cone electroretinogram

The effects of the ACE inhibitor MK 422 on the cone ERG were studied by isolating cat cones by means of strong blue-green background lights that result in a spectral sensitivity function peaking at 555 nm (Schuurmans und Zrenner, 1983). As shown in fig. 4, the typical cone ERG before application of the inhibitor reveals a b-wave that is slightly smaller than the off-effect. If small concentrations of MK 422 are applied (0.5 ng/ml/min) the b-wave grows considerably for several hundred percent and even becomes larger than the off-effect. Approximately 1 hour after application the cone b-wave and the cone off-effect have reached their initial shape. No change in flow-rate was observed during these experiments. Higher blocker concentrations (10 ng/ml/min, center panel) have comparably little effect on the b-wave, on the off-effect and the course of the response. At most, a slight enhancement of the b-wave can be observed. High blocker concentrations (25 ng/ml/min, lower panel) produce mainly a negative P III response that is irreversible.

4. Effects of the angiotensin II-receptor-blocker saralasin on the rod electroretinogram

In a second series of experiments, the angiotensin II receptor blocker saralasin (Röhm Pharma) was injected into the ophthalmociliary artery in concentrations between 0.1 and 100 ng/ml/min (0.11-110 nmol/min). In contrast to the experiments with MK 422 the synthesis of angiotensin II is not altered during application. Saralasin is an octa-peptide and differs from angiotensin only in respect to 3 amino-acids. It has a high affinity to the angiotensin II receptors. The systemic halftime value lies between 3 and 5 min.

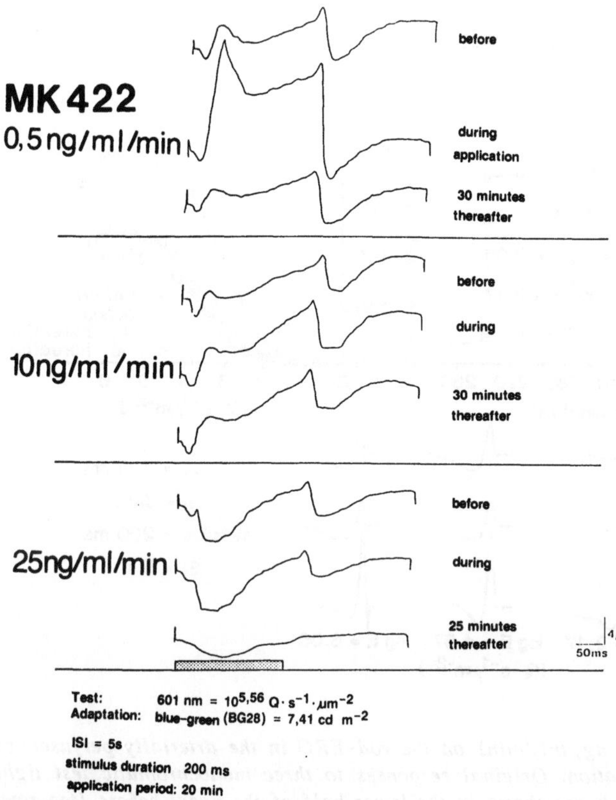

Fig. 4: Effects of the ACE-inhibitor on cone ERG: Cones were isolated by adapting the eye for 10 minutes to a blue-green Ganzfeld background light (Schott BG 28 filter, 7.41 cd.m^{-2} on the diffuser and stimulating the retina with long-wavelength-light (601 nm, $10^{5.6}$ Quanta.s^{-1}µm^{-2}). Stimulus duration was 200 ms and the stimulus interval 5 s; application of the inhibitor lasted 20 min. The effects of 3 different concentrations are shown as indicated besides each block of data that presents the original recordings before, during and after application of MK 422. (From Dahlheim et al. 1988a)

The effects of a very low concentration of saralasin on the the rod ERG of cat is shown in fig. 5. Original responses to 3 different test light irradiances (log E) are shown in the lower half of the figure before (top), during (center) and approximately 1 hour (bottom) after application for 20 min. Stimulus and recording conditions were identical to those mentioned in fig. 3. As can be seen in the original responses the amplitudes grow enormously and reach values above 1000 µV, never obtained in intact or arterial perfused cat eyes in our laboratory. The time course of the amplitude during a period of 3 1/2 hours is shown in the upper left (lefthand ordinate) for the three test light irradiances presented in the lower half. Under all three conditions of stimulation the b-wave grows at least 300% to 500% and maintains these high amplitude values. Even several hours longer than shown here, b-waves were still higher than before application of saralasin. In the upper left of fig. 5 also the change of flow rate is shown (crosses, right hand ordinate). Flow rate was markedly increased after application of saralasin, however, only for a relatively short period of time. While the flowrate approached its initial value after approximately 2 hours, b-waves still continued to rise and never returned to their initial level during the observation time. These facts make it very improbable that the flowrate is responsible for the dramatical enhancement of the b-wave. In the right upper panel the change in the V-log I functions during the course of the experiment is shown. Absolute sensitivity at low testlight levels is changed only little by saralasin. However, the V-log I function becomes steeper which points to a change in gain.

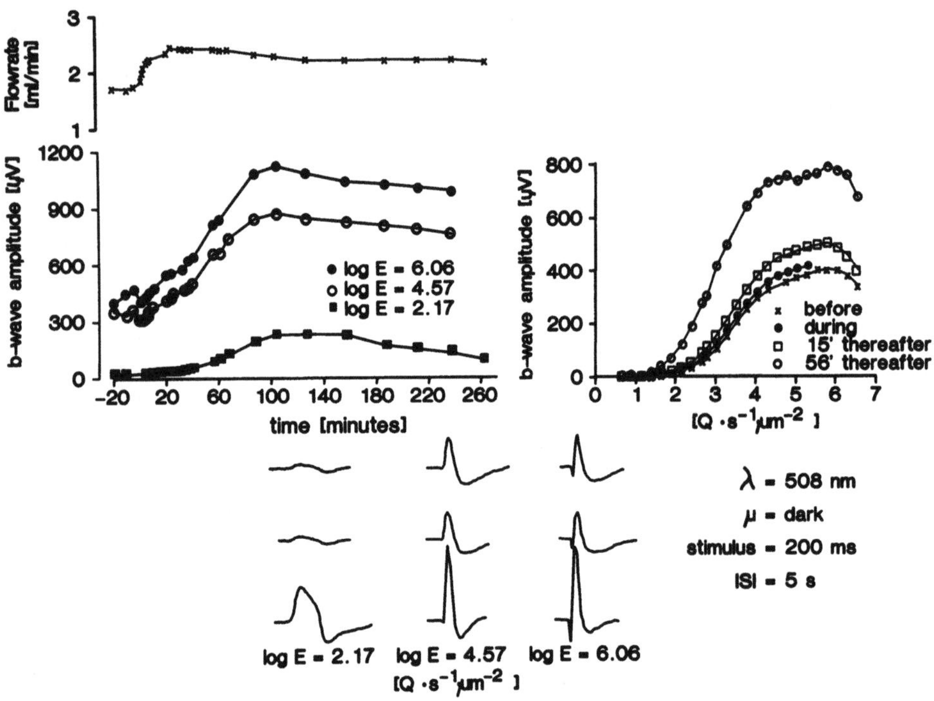

Fig. 5: Effects of saralasin (0.5 ng/ml/min) on the rod-ERG in the arterially perfused cat eye after 30 min of dark adaptation. Original responses to three monochromatic test lights (508 nm) of different irradiance E are shown in the lower half of the panel before (top row), during (center row) and after (bottom row) saralasin application. The time course of the amplitude of these three conditions is depicted in the upper left, together with the flowrate (crosses). The V-log I functions during the several stages of the experiment are shown in the upper right panel. Remaining stimulus and recording conditions are identical to those mentioned in fig. 3 (from Dahlheim et al., 1988c, 1989)

Discussion

A number of points support the evidence that the angiotensin-renin system has a neurophysiological effect outside the vascular system. The highest concentrations of ACE-activity were found in retinal samples and not in the vascular layers of the eye. Additionally, studies of the ACE distribution in the bovine retina (Ferrari-Dileo et al., 1988) have shown that ACE activity in the neural retina is approximately double that found in retinal blood vessels. At low concentrations, the ACE inhibitor MK 422 has effects on retinal responsivity: Enhancement of b-waves during application, suppression of b-wave after application. Rods and cones are approximately equally affected, although they differ in respect to their sensitivity to various concentrations. There seems to be a difference in the effect on cone on- and off-responses, the on-responses being more affected by ACE inhibitors than off-responses. All these effects of the ACE inhibitors are not accompanied by alterations in perfusion rate which makes it quite likely that the observed effects represent actions of modulation within the angiotensin-renin system on retinal neurons.

According to our experiments, it seems feasible to assume that the MK 422 induced reduction of ACE activity lowers the A II concentration which is accompanied by increased b-wave amplitudes. On the other hand, saralasin also -by different means- reduces A II action and enhances the b-wave as well. The common denominator of both effects would be that reduction of A II activity results in increased b-wave amplitudes. This view is supported by the fact that the high ACE in vitro activity measured two hours after ACE inhibitor application (see Table 1) and the thereby induced increase of A II activity was still accompanied by strong reduction in b-wave amplitude (fig. 3, circles).

Several other mechanisms of action of the angiotensin-renin system on retinal neurotransmission can be discussed. The most probable explanation can be found in the broad ACE substrate specifity (see Table 1). Since methenkephaline is a substrate of ACE, ACE could easily alter methenkephaline action present in the retina (Mellstrom et al. 1986) and thereby result in response modulation in the electroretinogram. ACE actions on other peptides modulated by light in the retina are possible as well, such as those shown in Table I. Moreover, angiotensin II could alter the function of cholinergic neurons (Severs and Daniels-Severs, 1973) and/ or dopaminergic neurons (Phillips, 1987) and thereby alter retinal transmission. A cascade of such events cannot be excluded, either. All these mechanisms would be assumed to take place in the inner retina. This suggestion is validated by experiments with the ACE inhibitor in P III preparations, where no effects were observed.

Last but not least an action of the angiotensin-renin system on Müller cell function should be discussed. Since the potassium-metabolism in the Müller cell (glial cell) contributes heavily to the b-wave of the electroretinogram, modulation of Müller cell function would most probably show up in the ERG. Nevertheless, such actions on the Müller cells would alter neurotransmission in the retina as well, as we could show by several simultaneous recordings from the optic nerve sucked into a glass pipette electrode, the amplitudes of which were changing in parallel to the b-wave amplitude. Even if the angiotensin-renin system should not be able to exert direct neurotransmitter-like actions, the effects on retinal glial cells certainly would modulate signals at the retinal output, as observed in the recordings from the optic nerve.

In conclusion: The results presented here point to a role of the angiotensin-renin system in retinal neurotransmission, although the site of action in the retina cannot be determined by the present experiments. Other methods, such as immunohistochemical techniques have to be employed to further elucidate the site of ACE and A II action in the retina. Since there are reports of effects of widely used ACE-inhibitors on the visual performance of hypertensive patients (Coulter, 1988), possible effects of such drugs upon the visual system should be carefully monitored.

Acknowledgements

We are greatly indebted to Prof. Dahlheim for his advice and to Mr. Reiter for help with the fluorimetric assay. The research was supported by the German Research Council (Az. Zr 1/1-2, Zr 1/1-3, DA 93/8, SFB 220).

References

Coulter DM (1988) Eye pain with nifedipine and disturbance of taste with captopril: a mutually controlled study showing a method of postmarketing surveillance. Br Med J 296: 1086-1088

Dahlheim P, Schneider T, Reiter W, Zrenner E (1988a) Angiotensin I-converting enzyme (ACE): Aktivitätsverteilung im Auge und direkte Wirkungen auf die Netzhautfunktion. Fortschr Ophthalmol (in press)

Dahlheim P, Zrenner E, Datum K-H (1988b) Angiotensin I-converting enzyme (ACE): Distribution in ocular tissues and neurophysiological effects in cat retina (in preparation)

Dahlheim P, Reiter W, Datum K-H, Zrenner E: Ocular angiotensin I-converting enzyme: Does it play a role in retinal neurotransmission? Pflügers Archiv, Suppl No 1 to Vol: 411, (1988c)

Dahlheim P, Datum K-H, Reiter E, Zrenner E (1989) (in preparation)

Felix D (1982) Angiotensin, neurohormone and neurotransmitter? TIPS 3: 208-210

Ferrari-Dileo G, Ryan JW, Rockwood EJ, Davis EB and Anderson DR (1988) Angiotensin-Converting enzyme in bovine feline, and human ocular tissues. Invest Ophthalmol Vis Sci 29: 876-881

Ganten D, Printz M, Phillips MI and Schölkens BA (eds), The Renin-Angiotensin system in the brain. Heidelberg, Springer (1982)

Gouras, P and Hoff, M (1970) Retinal function in an isolated, perfused mammalian eye. Invest Ophthalmol Vis Sci 9: 388-399

Haas HL, Felix D, Celio MR and Inagami T (1980) Angiotensin II in the hippocampus. A histochemical and electrophysiological study. Experientia 36: 1394-1395

Igic R and Kojovich V (1980) Angiotensin I Converting Enzyme (Kinase II) in ocular tissues. Exp Eye Res 30: 299-303

Lind RE, Swansson LW, Bruhn TO, Ganten D (1985) The distribution of angiotensin II-immunoreactive cells and fibers in the paraventriculo-hypophysial system of the rat. Brain Res 338: 81-89

Mallorga P, Babilon RW, Sugrue MF (1988) Distribution of Angiotensin-II receptors in the albino rabbit eye. Invest Ophthalmol Vis Sci (ARVO Suppl) 29: 201

Mellstrom B, Iadarola MJ, Yang H-YT and Costa E (1986) Inhibition of MET^5-Enkephalin-ARG^6-PHE^7 degradation by inhibitors of dipetidyl carboxypeptidase. J Pharmacol Exp Ther 239: 174 (1986)

Morgan IG, Millar TJ, Ishimoto I, Boelen M Dowton M, Chubb IW (1989) Somatostatin- and Enkephalin-immunoreactive amacrine cells of the chicken retina. (this volume)

Niemeyer, G (1975) The function of the retina in the perfused eye. Doc Ophthalmol 39: 53-116

Phillips M I (1987) Functions of angiotensin in the central nervous system. Ann Rev Physiol 49: 413-435

Piquilloud Y, Reinharz A, Roth M (1970) Studies on the angiotensin Converting Enzyme with different substrates. Biochim Biophys Acta 206: 136-142

Schuurmans RP, Zrenner E: The arterially perfused eye: Colour vision mechanisms and neurotransmitters. In: Color Vision in Clinical Pharmacology, Nr 4, eds: Rietbrock and Woodcock, Vieweg & Sohn, Braunschweig, Wiesbaden pp 88-104 (1983)

Sessle BJ, Henry JL (1987) Angiotensin II excites neurones in cat solitary tract nuclei which are involved in respiration and related reflex activities. Brain Res 407: 163-167

Severs WB and Daniels-Severs AE (1973) Effects of angiotensin on the central nervous system. Pharmacological Reviews 25: 415-449

Skidgel RA and Erdös EG (1987) The broad substrate specificity of human angiotensin I Converting Enzyme Clinical and Experimental Theory and Practice, A 9 (2 & 3): 243-259

Story DF and Ziogas J (1987) Interaction of angiotensin with noradrenergic neuroeffector transmission. TIPS 8: 269-271

Sirret NE, McLean AS, Bray JJ and Hubbard JI (1977) Distribution of angiotensin II receptors in rat brain. Brain Res 122: 299-312

Swerts JP Perrisot R, Maltroy B, Schwartz JC (1979) Is "enkephalinase" identical with "angiotensin converting enzyme"? Eur J Pharmacol 53: 209

Watt CB, Su Y-Y T (1988) Enkephalinergic pathways in the retina. In: Proceedings of the Retina Research Foundation Symposium, Vol 1: 141-162 Portfoliô Publ Compl, The Woodlands, Texas

Zrenner E, Gouras P: Blue-sensitive cones of the cat produce a rodlike electroretinogram. Invest Ophthalmol Vis Sci 18: 1076-1081 (1979)

PROBLEMS POSED BY PRIMATE GANGLION CELLS FOR FUNCTIONAL ANATOMY AND PSYCHOPHYSICS

Barry B. Lee
Department of Neurobiology,
Max Planck Institute for Biophysical Chemistry,
D-3400 Göttingen,
W. Germany.

There are many features common to the the retinae of different mammals. There are, for example, often two distinct types of horizontal cell (Cajal, 1933) and there is a distinct class of ganglion cell which can be visualised by neurofibrillar staining (Peichl et al., 1987). Yet it has become increasingly evident over the last few years that the primate retina possesses unique features, which distinguish it from retinae of other mammals such as the cat. One feature is the presence of three distinct cone types. Although other mammals may possess a rudimentary ability to distinguish wavelength differences, simian primates are unique in their well-developed colour vision. How far the primate retina has been remodelled to process spectral as well as spatial composition is uncertain, but interpretation of results from primate retina are coloured by the fact that wavelength as well as intensity is being encoded.

The primate retina is of especial interest since the old-world primate possesses trichromatic colour vision similar to that of man, and available evidence suggests animals such as the macaque see much as we do (DeValois et al., 1974). There is psychophysical evidence for the presence in the human visual system of different channels responsible for different aspects of psychophysical performance. Most of this evidence rests on detection experiments of one sort or another, which indicate that there exist an achromatic, luminance channel and also a number of chromatic channels. Identification of these channels with different ganglion cell classes in for example, the macaque retina, has up to now received little unambiguous experimental support.,

The primate retina has attracted the attention of anatomists since Cajal, and many details of its anatomy are well-established. However, following the early recordings of Gouras and his collaborators (de Monasterio and Gouras, 1975; de Monasterio, 1978) recording from monkey ganglion cells has been rare. This is unfortunate, since primate ganglion cells offer an opportunity for comparing physiological results on the one hand with behavioural tests on man, and on the other hand with anatomical findings relevant to function. I shall review here some recent results obtained from monkey ganglion cells,

NATO ASI Series, Vol. H31
Neurobiology of the Inner Retina
Edited by R. Weiler and N.N. Osborne
© Springer-Verlag Berlin Heidelberg 1989

and attempt to show that the ability to correlate results from three levels, behavioural, physiological and anatomical, is a powerful means of understanding retinal function.

Classes of macaque ganglion cells

The two main classes of ganglion cell give rise to two cell systems that maintain their integrity up to striate cortex. Tonic, wavelength-opponent ganglion cells send their axons to the parvocellular layers of the lateral geniculate nucleus, and phasic, non-opponent ganglion cells send axons to the magnocellular layers of the nucleus (de Monasterio and Gouras, 1975; de Monasterio, 1978; Wiesel and Hubel, 1966; Dreher et al., 1976; Creutzfeldt et al., 1979; Perry et al., 1984). Tonic ganglion cells are wavelength-opponent by virtue of antagonistic inputs from different cones. Most tonic cells (ca. 80%) receive input solely from middle-wavelength and long-wavelength (M– and L–) cones. Cells with +M–L–cone input may be of the green on-centre or red off-centre type, whereas cells with +L–M cone input may be red on-centre or green off-centre (de Monasterio and Gouras, 1975). Of cells with short-wavelength (S–) cone input, those with excitatory S–cone input are in the majority, although a distinctive group of cells receives inhibitory S–cone and excitatory M–cone input (Valberg et al., 1986). Phasic ganglion cells receive M– and L–cone input to both centre and surround (de Monasterio, 1978).

Although the anatomical equivalents of these physiological types have not yet been fully established, it is likely that tonic cells near the fovea have the morphology of midget cells (Cajal, 1933; Polyak, 1957) and that phasic ganglion cells have parasol-like morphology (Leventhal et al., 1981; Perry et al., 1984). Although ganglion cells projecting to the parvocellular layers have been subsumed into the Pβ class (Perry et al., 1984) and the ganglion cells projecting to the magnocellular layers into the Pα class, it is uncertain if these classes are homogeneous over the retina. More recently, Shapley and Perry (1986) renamed these cells 'P-' and 'M-'cells after their targets in the geniculate, for there is increasing evidence that tonic and phasic cells of the monkey differ from the sustained (β) and transient (α) ganglion cells of the cat in many important respects.

Ganglion cells and psychophysical channels

The first indication that the tonic and phasic cell systems could have distinct roles was obtained in the geniculate nucleus. When achromatic gratings are used as stimuli, phasic cells have much higher contrast sensitivity than tonic cells (Kaplan and Shapley,

1982; Hicks et al., 1983), and we have confirmed this result in retinal ganglion cells. The sensitivity of phasic cells approximates that of a human observer viewing the same stimulus. This suggested that phasic cells might underlie detection of small luminance differences, and could relate to the so-called luminance channel of psychophysics.

Psychophysical evidence for such a channel originates from several experimental paradigms, the most important of which is heterochromatic flicker photometry. This technique was devised as a way of specifying the spectral sensitivity of a human observer. Results obtained with it were used to define the human luminosity function, V_λ, by the CIE in 1924. A human subject is presented with a field (typically 2°) which alternates between two different wavelengths at, say, 10 Hz. Their relative intensities are then adjusted until the sensation of flicker is minimised or abolished. The two lights are then defined to be of equal luminance.

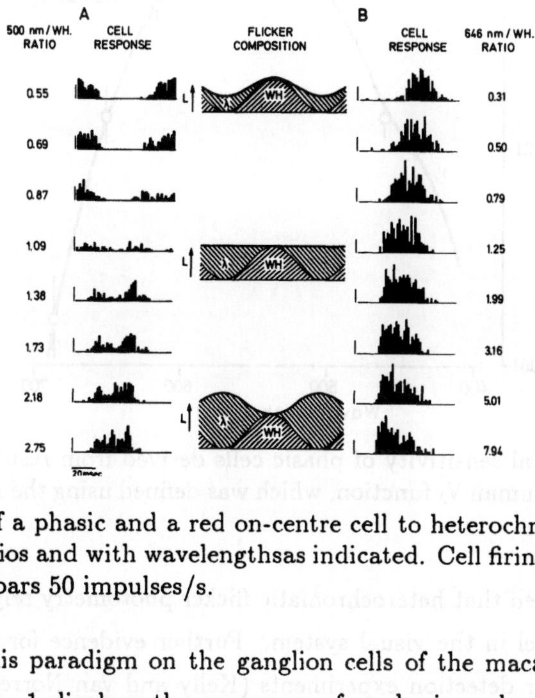

Fig. 1. Responses of a phasic and a red on-centre cell to heterochromatic flicker at 10 Hz, at luminance ratios and with wavelengthsas indicated. Cell firing was averaged over 6 s, binwidth 2 ms, bars 50 impulses/s.

We have used this paradigm on the ganglion cells of the macaque retina (Lee et al., 1988a), and Fig. 1 displays the responses of a phasic and a tonic ganglion cell to heterochromatic flicker. In the middle is shown the composition of the flicker; a monochromatic wavelength was alternated with a white reference light, and the relative intensities of the components systematically varied. The luminance ratio between them (calibrated relative to the human V_λ function) is indicated by each histogram, which contain the response to one cycle of flicker. The response of a phasic cell (*A*) almost disappears close to a ratio of one. Thus, a minimum in cell response closely corresponds to minimisation of flicker by a human subject. For a red on-centre cell (*B*), with red-white flicker, a gradual phase shift occurs without any indication of a response minimum. This gradual phase change can be explained in terms of a centre-surround phase delay (Lee et al., 1988b).

From the relative intensities at which phasic cells' responses were minimised, we determined their spectral sensitivity, and this is shown in Fig. 2. The open symbols represent mean values from phasic cells and the solid line the V_λ function. There is clearly a good correspondence. We demonstrate elsewhere that other psychophysical properties of heterochromatic flicker photometry can also be found in the responses of phasic cells (Lee et al., 1988a), and we conclude that when an observer minimises flicker in this paradigm, minimisation of responses of phasic cells is the substrate of this task.

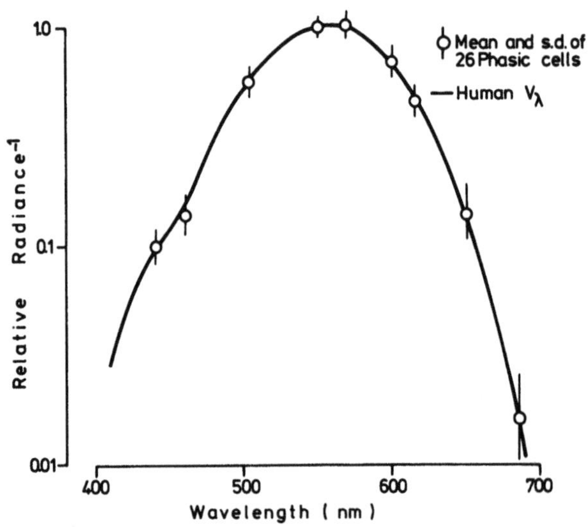

Fig. 2. Mean spectral sensitivity of phasic cells derived from results such as in Fig. 1 compared with the human V_λ function, which was defined using the same psychophysical paradigm.

It is often assumed that heterochromatic flicker photometry relys on an achromatic or luminance channel in the visual system. Further evidence for such a channel was obtained from flicker detection experiments (Kelly and van Norren, 1977). If human subjects are required to detect achromatic, luminance flicker, sensitivity is maximal around 10-20 Hz, with less sensitivity at lower temporal frequencies. If a light varies in colour but not in luminance, sensitivity is maximal around 1 Hz and falls off rapidly above 2 Hz. These results have been explained in terms of luminance and chromatic flicker tapping independent luminance and chromatic channels with different temporal characteristics.

We measured sensitivity of macaque ganglion cells to luminance and chromatic flicker, and compared it with sensitivity of human subjects viewing the same stimulus (Fig. 3; Lee et al., 1988c). With luminance flicker, phasic ganglion cells are more sensitive than tonic ganglion cells, especially at low temporal frequencies. The sensitivity of phasic ganglion cells is similar to that of human observers, with peak sensitivity occurring at a frequency of 10-20 Hz. It is feasible therefore that detection of luminance flicker relies on the phasic cell system.

With red-green or blue-green chromatic flicker at equal luminance, tonic cells are more sensitive than phasic cells, which can, however, detect red-green chromatic flicker because of a non-linearity of M– and L–cone summation (Lee et al., 1988d). With blue-green chromatic flicker, the wavelengths chosen differentially stimulated only the S–cones, and so cells with only M– and L–cone input (e.g., red on-centre, green on-centre and phasic cells) do not respond. The blue on-centre cells shown received S–cone input.

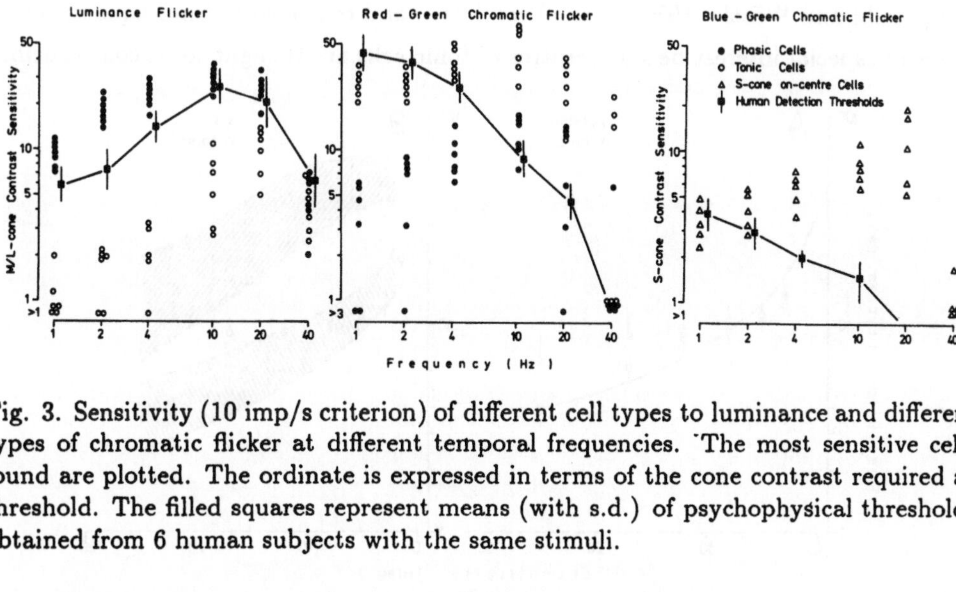

Fig. 3. Sensitivity (10 imp/s criterion) of different cell types to luminance and different types of chromatic flicker at different temporal frequencies. The most sensitive cells found are plotted. The ordinate is expressed in terms of the cone contrast required at threshold. The filled squares represent means (with s.d.) of psychophysical thresholds obtained from 6 human subjects with the same stimuli.

A striking discrepancy between physiological and psychophysical results is apparent with chromatic flicker. At low temporal frequencies, tonic cell sensitivity approaches that of human subjects, and these cells could thus support detection. The psychophysical sensitivity curve, however, drops off sharply above 2 Hz, whereas tonic cell sensitivity remains high. It was striking to see tonic cells responding to chromatic flicker not perceived by human subjects. This result implies that high-frequency components of tonic cells' signals are somehow filtered out at later stages in the visual pathway.

The results in Fig. 3 support the thesis that different cell types support detection of different types of flicker. For luminance flicker, phasic cells are the substrate of a luminance channel mediating detection, and for red-green and blue-green chromatic flicker different groups of tonic cells form the substrate of chromatic channels mediating detection.

With such comparisons of ganglion cell activity and human performance, one can identify the physiological basis of different channels and mechanisms postulated by psychophysics. Such experiments are helpful in interpretation both of psychophysical results and of retinal function.

Visual resolution and ganglion cell centre size

Visual resolution and centre size are parameters of ganglion cell performance which are of interest from the viewpoint of functional anatomy. In the cat retina, there is a direct relationship between centre size and visual resolution for both X and Y ganglion cells (Peichl and Wässle, 1979). Visual resolution and centre size of macaque ganglion cells is of especial interest because centres of tonic cells are thought to be cone specific.

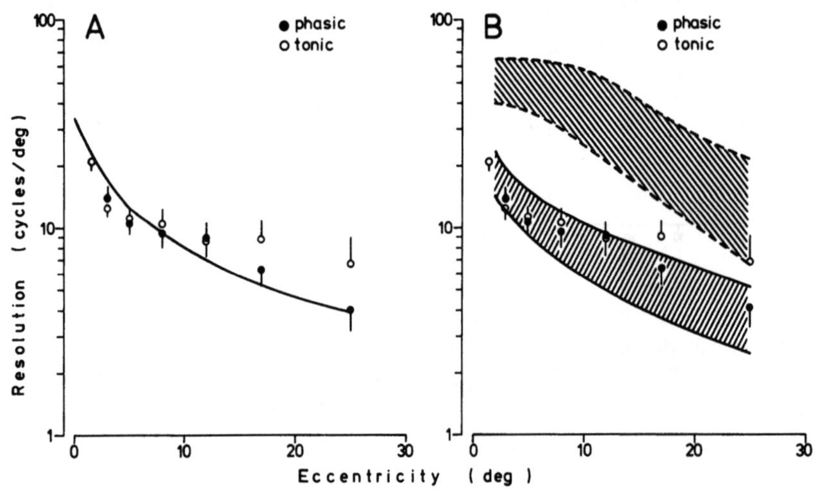

Fig. 4. *A* Visual resolution of macaque ganglion cells plotted against retinal eccentricity to achromatic gratings. Means and s.d. for tonic and phasic ganglion cells are compared with psychophysical measurements from man. Resolution comparable to that of man is attained by both cell types. *B* The same physiological data compared with resolution expected if dendritic tree diameter of Pα and Pβ cells was identical with centre diameter. There is a reasonable correspondence for the former but not for the latter. Reproduced by permission of the Physiological Society.

We have recently carried out a systematic survey of the visual resolution of macaque ganglion cells (Crook et al., 1988) and found that at a given retinal eccentricity the visual resolution of tonic and phasic cells is similar. This is illustrated in Fig. 4. Ganglion cells have been divided into different groups according to eccentricity. The mean visual resolution of tonic cells at a given eccentricity is similar to that of phasic cells over most of the retina. In the central two degrees, it is difficult to record from phasic cells, for they are buried deep in the layers of ganglion cells around the foveal pit (Perry and Silveira, 1987). However, even in this region their resolution is likely to be similar to that of tonic cells, for Blakemore and Vital-Durand (1986), recording from the lateral geniculate nucleus, found visual resolution in the parvocellular and magnocellular layers to be similar at all eccentricities, including the foveal representation. The solid line in Fig. 4*A* shows a comparison with the visual resolution of human observers, scaled slightly to take into account the smaller eye size in the monkey. Remarkably, at eccentricities above 10^0, some tonic ganglion cells had a higher visual resolution than that expected psychophysically.

The results of Fig. 4 are surprising because centre size of tonic, opponent ganglion cells is thought to be small, less than 7 minutes of arc up to 10^0 (de Monasterio and Gouras, 1975). In the cat, Peichl and Wässle (1979) showed that cells could resolve gratings up to limit at which about two grating cycles fitted into the centre. If this were so for tonic cells of the monkey, a 7 minutes of arc centre size should provide a visual resolution of 20 cycles/degree. This is substantially higher than that that found at eccentricities in excess of 5^0, both in physiological and psychophysical data. On the other hand, centre size estimates for phasic ganglion cells (de Monasterio and Gouras, 1975) would be consistent with their visual resolution.

How far might one expect dendritic tree diameter of primate ganglion cells to be related to centre size? In the cat, centre size bears a relation to dendritic tree diameter, though it is larger by a factor of about 1.4 for Y cells and by 1.8-2.0 for X cells (Peichl and Wässle, 1979; Wässle et al. 1983a,b). Perry et al. (1984) made measurements of dendritic tree size of ganglion cells of the macaque and found that for their Pβ, putatively tonic, ganglion cells, dendritic tree diameter remained small up to 10^0, being of the order of a few minutes of arc. For Pα cells, dendritic tree diameter increased monotonically with eccentricity. From Perry's data, we calculated the visual resolution expected on the assumption that dendritic tree diameter was equal to centre size, and the range of resolution expected is shown as the shaded areas in Fig. 4*B*. For tonic cells resolution is much lower than the prediction from dendritic tree diameter. The latter remains approximately constant up to an eccentricity of 10°, but resolution decreases, so that at 10° a substantial difference is present. On the other hand, there is a clear correlation between resolution of phasic cells and dendritic tree diameter of Pα cells, .

In fact, the discrepancy for tonic cells may not be so surprising. Dendritic tree diameter will only be related to centre size if a ganglion cell, in building its centre, summates inputs from subunits with fields smaller than the eventual centre size. Input elements with fields of significant diameter may account for the fact that centres of α and β cells of the cat are larger than their dendritic trees (Peichl and Wässle, 1979; Wässle et al, 1983). For tonic cells, many of which probably belong to the midget system, input from a single midget bipolar cell would result in a ganglion cell's centre size being determined by that of the bipolar cell.

The problem for functional anatomy now becomes clear. If a midget bipolar cell receives input from only one cone, a centre size similar to the cone diameter should be expected. Although cone outer segment diameter increases with eccentricity, at 5-10° it is only a few minutes of arc, and thus on the basis of the calculations of Peichl and Wässle (1979), a cone should be able to resolve gratings of much higher resolution than seen in the ganglion cell to which it is presumed to provide the sole centre input. Optical factors should not play a role with the dimensions involved at 5-10° of eccentricity.

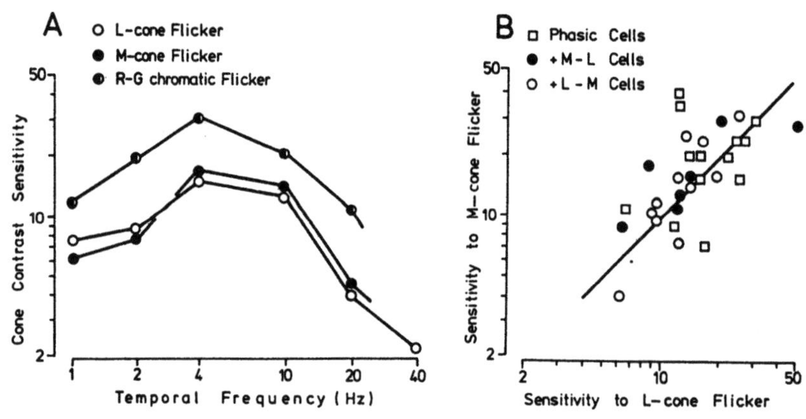

Fig. 5. *A* Sensitivity of a red on-centre cell (10 imp/s criterion) to M–cone or L–cone flicker at different temporal frequencies, and to chromatic flicker. Sensitivity of each cone is similar, indicating centre and surround have cone inputs of similar strength, and chromatic flicker sensitivity is about double that of the individual cones, indicating linear summation. *B* Comparison of sensitivity of red on-centre, green on-centre and phasic cell sensitivity to M– and L–cone flicker at 10 Hz. Sensitivity of all cell types is similar.

One possible explanation of the lower visual resolution of tonic ganglion cells is that, because the centre is small, it is insensitive, so that resolution is poorer than expected on the basis of an analysis such as that of Peichl and Wässle (1979). Though the low contrast sensitivity of tonic cells to achromatic gratings might be interpreted as being due to intrinsically low centre sensitivity, it is also possible that tonic cells have low contrast sensitivity due to the cone-opponent interaction of centre and surround. To distinguish between these alternatives, we have isolated the opponent mechanisms of centre and surround by means of silent substitution, and determined separately their sensitivity. For example, for a red on-centre cell the centre should receive input from the L–cone and the surround from the M–cone.

Fig. 5 show results from such an experiment. We used a red-green flickering field (4^0), and adjusted the intensities of the red and green components so that either the L– or M–cone was stimulated, the other being silent. By adjusting the modulation depth of the flicker, we could determine sensitivity to M– or L–cone flicker. In Fig. 5*A*, we show sensitivity of a red on-centre cell to M– and L–cone flicker at different temporal frequencies. Sensitivity has been expressed in terms of cone modulation necessary to produce a 10 impulses/s modulation of the cell's firing. Clearly, sensitivity to M– and L–cone flicker is similar, and this indicates that centre and surround have similar sensitivity; the centre is not intrinsically insensitive. Also shown is sensitivity to red-green chromatic flicker, which is about twice as great as sensitivity to M– or L–cone flicker. This shows that sensitivity to chromatic flicker is a linear sum of contributions from M– and L–cones.

In Fig. 5*B* is compared sensitivity to M– and L–cone flicker at 10 Hz of a sample of red on-centre (+L–M), green on-centre (+M–L) and phasic cells. The cell populations overlap, so that one can conclude that the cone inputs to all these cell types are of similar sensitivity, and centres of tonic cells are not intrinsically insensitive.

A further possibility which would explain resolution data is that, for technical reasons, earlier studies underestimated centre size of tonic cells. We thus measured centre size of a sample of macaque ganglion cells and compared it to their visual resolution. We constructed area-threshold curves with flashed white spots, for this technique avoids difficulties arising from the different contrast sensitivities of tonic and phasic cells. Area-threshold curves are shown in Fig. 6*A*, and the inflection in these curves give an estimate of equivalent centre size (Peichl and Wässle, 1979; Crook et al., 1988). On theoretical grounds, one would expect that the grating just resolved should have a period length of about half the centre diameter, and Fig. 6*B* shows a sample of tonic and phasic cells in which this comparison was carried out. Because of the difficulty of recording phasic cells close to the fovea, the cells with the smallest centres and best resolution are tonic ones, but clearly for both tonic and phasic cells there is a similar relationship between resolution and centre size. The slope of the regression line was 2.1, and did not differ significantly from the value of 1.82 found by Peichl and Wässle (1979) for cat ganglion cells.

The results of Fig. 6 suggest that centres of tonic ganglion cells may indeed be larger than previously reported. How this result may be made consonant with anatomical data, which suggest the midget bipolar cells synapse with only one cone, remains uncertain. One possibility is a degree of cone-cone coupling, but as yet no anatomical data is available which parallels the steep decline in resolution as one moves away from the fovea.

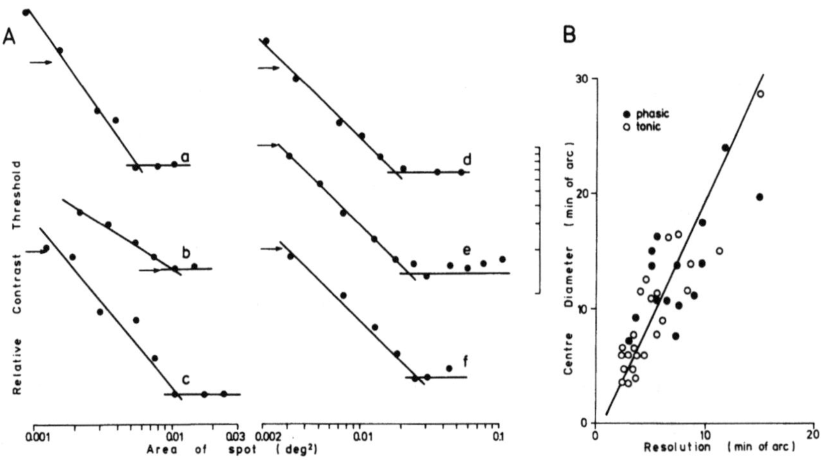

Fig. 6. *A* Centre size of phasic and tonic cells was estimated from area-threshold curves. Spots of different diameters were presented upon a 100 cd/m^2 background and the contrast required to produce a threshold response estimated. With increasing spot size, threshold decreases until a plateau is reached. From the intersection of the two lines equivalent centre size can be calculated. *B* Centre size compared to the period length just resolved for tonic and phasic cells. The regression line through the points has a slope close to two, as expected on a theoretical basis.

Conclusions and perspective

It is encouraging that psychophysical evidence for chromatic and achromatic channels in the visual system can have a neurophysiological substrate in the activity of neurophysiologically distinct cell types. The evidence presented here suggests that hypotheses by which tonic, opponent cells do 'double duty' as both chromatic and luminance channels, depending on stimulus conditions, are untenable. Sensitivity of an observer to change in luminance or chromaticity would seem to be set at a rather peripheral level, and we presume behavioural threshold to be determined by the most sensitive class of cells available to the animal. Further exploration of cell behaviour using other psychophysical paradigms may reveal more of the physiological substrate of visual performance.

Visual resolution of phasic and tonic cells is similar, and both cell types could aid in psychophysical performance. However, resolution of a pattern by an array requires not only that the individual elements have an adequate resolution, but also that these elements be present in sufficient density. Phasic cells probably undersample the retinal image. On the other hand, it is uncertain whether different tonic ganglion cell types form separate sampling systems. This, and the fact that different tonic cell classes may change their resolution limit according to the colour of a grating, tend to complicate arguments about the sampling density available through tonic cells.

One problem posed by the results presented here is that the centres of cone-opponent ganglion cells may be larger than has been supposed. If it were true that a single cone provided input to a midget ganglion cell via a midget bipolar cell, then that ganglion cell's resolution should be related to the diameter of the cone. Tonic cell's resolve gratings less well than would be expected on that basis. Whether centres of tonic ganglion cells could be effectively enlarged due to inter-cone coupling or at more proximal sysnaptic sites can only be the subject of speculation.

REFERENCES

Blakemore C, Vital-Durand F (1986) Organization and post-natal development of the monkey's lateral geniculate nucleus. J. Physiol. 380:453–491

Cajal SR (1933) The structure of the retina. Thomas, Springfield, Illinois

Creutzfeldt OD, Lee BB, Elepfandt A (1979) A quantitative study of chromatic organisation and receptive fields of cells in the lateral geniculate body of the rhesus monkey. Exptl Brain Res 35:527–545

Crook JM, Lange-Malecki B, Lee BB, Valberg A (1988) Visual resolution of macaque retinal ganglion cells. J. Physiol. 396:205–224

de Monasterio FM (1978) Properties of concentrically organised X and Y ganglion cells of macaque retina. J. Neurophysiol. 41:1394–1417

de Monasterio FM, Gouras P (1975) Functional properties of ganglion cells of the rhesus monkey retina. J. Physiol. 251:167–195

DeValois RL, Morgan HC, Polson MC, Mead WR, Hull EM (1974) Psychophysical studies of monkey vision. I. Macaque luminosity and colour vision tests. Vision Res. 14:53–67

Dreher B, Fukuda Y, Rodieck RW (1976) Identification, classification and anatomical segregation of cells with X-like and Y-like properties in the lateral geniculate nucleus of old-world primates. J. Physiol. 258:433–453

Hicks TP, Lee BB, Vidyasagar TR (1983) The responses of cells in macaque lateral geniculate nucleus to sinusoidal gratings. J. Physiol. 337:183–200

Kaplan E, Shapley RM (1982) X and Y cells in the lateral geniculate nucleus of the macaque monkey. J. Physiol. 330:125–144

Kelly DH, van Norren D (1977) Two-band model of heterochromatic flicker. JOSA 67:1081–1091

Lee BB, Martin PR, Valberg A (1988a) The physiological basis of heterochromatic flicker photometry demonstrated in the ganglion cells of the macaque retina. J. Physiol. In Press

Lee BB, Martin PR, Valberg A (1988b) Responses and phase behaviour of macaque retinal ganglion cells to luminance and chromatic flicker. Submitted for publication

Lee, B.B., Martin, P.R. & Valberg, A. (1988c). Sensitivity of macaque retinal ganglion cells to luminance and chromatic flicker. Submitted for publication.

Lee BB, Martin PR, Valberg A (1988d) A non-linearity seen on summation of M– and L–cone inputs to phasic retinal ganglion cells of the macaque. Submitted for publication.

Leventhal AG, Rodieck RW, Dreher B (1981) Retinal ganglion cell classes in old-world monkey: morphology and central projections. Science 213:1139–1142

Peichl L, Ott H, Boycott BB (1987) Alpha ganglion cells in mammalian retinae. Proc Roy Soc B 231:169–197

Peichl L, Wässle H (1979) Size, scatter and coverage of ganglion cell receptive field centres in the cat retina. J. Physiol. 291:117–141

Perry VH, Oehler R, Cowey A (1984) Retinal ganglion cells that project to the dorsal lateral geniculate nucleus in the macaque monkey. Neuroscience 12:1110–1123

Perry VH, Silveira LCL (1988) Functional lamination in the ganglion cell layer of the macaque's retina. Neuroscience In Press

Polyak SL (1957) The vertebrate visual system. Univ. of Chicago Press, Chicago.

Shapley R, Perry VH (1986) Cat and monkey retnal ganglion cells and their visual functional roles. TINS 9:229—235

Valberg A, Lee BB, Tigwell DA (1986) Neurones with strong inhibitory S-cone inputs in the macaque lateral geniculate nucleus. Vision Res. 26:1061–1064

Wässle H, Boycott BB, Illing R-B (1981) Morphology and mosaic of on- and off-beta cells in the cat retina and some functional considerations. Proc Roy Soc B 212:177–195

Wässle H, Peichl L, Boycott BB (1981) Morphology and topography of on- and off-alpha cells in the cat retina. Proc. Roy. Soc. B 212:157–175

Wiesel TN, Hubel DH (1966) Spatial and chromatic interactions in the lateral geniculate body of the rhesus monkey. J. Neurophysiol. 29:1115–1156

AMACRINE-BIPOLAR CELL INTERACTIONS MEDIATE CHANGE DETECTION IN THE RETINA OF THE TIGER SALAMANDER

F.S. Werblin, G. Maguire, Peter Lukasiewicz

Neurobiology Group
University of California at Berkeley
Berkeley, California 94720

INTRODUCTION

Neural interactions that underlie movement-gated inhibition and that generate responses to movement are found throughout the visual system. The retinal ganglion cells in many species including frog (Maturana et al., 1960), salamander (Werblin, 1970), and cat (Cleland & Levick, 1974) show directionally selective movement detection. A general scheme for directionally selective movement detection in rabbit has been proposed (Barlow & Levick, 1965), and numerous theoretical schemes have been described (Torre & Poggio, 1978;1981). The cells, synaptic pathways and interactions mediating change and movement detection have not yet been identified. Only recently has it become possible to probe within the inner plexiform layer to discern the neuronal interactions underlying change-gated activity (Maguire et al., 1988; Werblin et al., 1988).

A brief account of the interactions involved in change detection is presented here, and then interactions are extrapolated to a hypothetical model that can account for directionally selective movement detection as proposed by Barlow and Levick (1965).

METHODS

The techniques used in these studies are described in detail in Werblin et al. (1988) and Maguire et al. (1988). Briefly, synaptic currents in retinal cells were measured using whole cell patch clamp (Hamill et al., 1981). Retinal slices (Werblin, 1978) were used because cells can be patch clamped without enzymatic treatment, so they retain most of their synaptic connections, allowing the measurement of synaptic currents. The patch pipettes were filled with Lucifer yellow so that cells and processes could be identified during the recording session (Stewart, 1978).

NATO ASI Series, Vol. H31
Neurobiology of the Inner Retina
Edited by R. Weiler and N.N. Osborne
© Springer-Verlag Berlin Heidelberg 1989

The retina was stimulated by "puffing" the excitatory analogs kainate and glutamate, at the dendrites of the bipolar cells. This depolarized the OFF bipolars but left the ON bipolars unaffected (Maguire et al., 1988). At the level of amacrine cells we saw responses only in those elements with processes falling within the OFF sublamina; ON amacrine cells were silent, suggesting that there is little crosstalk between ON and OFF systems at this level.

Electrical and pharmacological properties of cells in the change-sensitive subunit.

The key interaction in the change sensitive circuit is a $GABA_B$ modulation of the calcium current involved in transmitter release (Maguire et al., 1988B) at a population of bipolar cell terminals. The interactions are shown schematically in Fig 1.

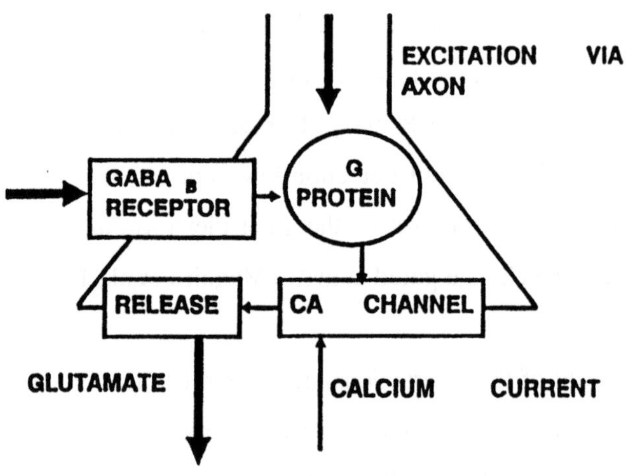

Figure 1. Schematic of the $GABA_B$ modulation of transmission at that class of bipolar cell terminals thought to release transmitter transiently. GABA (or baclofen) activates a class of $GABA_B$ receptors at the terminal. They act (via a G-protein) to reduce the terminal calcium current. This current is thought to be associated with transmitter release. The $GABA_B$-mediated block of transmission is thought to be delayed with respect to excitation, so the transmitter release is truncated after an initial transient burst.

We characterize the membrane events surrounding this $GABA_B$ modulation, and then describe the specific classes of amacrine cells that are thought to provide the

GABAergic input to, and receive the glutamatergic output from the bipolar terminals (Slaughter & MIller, 1983). Finally, we combine the activity of these cellular elements to generate the change-sensitive circuit.

Interactions at the bipolar terminal

In a class of bipolar cell, with axon terminals that ramify near the midline of the inner plexiform layer the voltage gated calcium currents are decreased by baclofen, a $GABA_B$ receptor agonist (Robertson & Taylor, 1986). The baclofen modulation suggests the presence of $GABA_B$ receptors as shown in Fig 1. Sustained calcium currents are recorded only in those bipolar cells with intact synaptic terminals. The effects of baclofen in reducing the voltage-gated calcium current in bipolar terminals are shown in Fig 2. These calcium currents were never well clamped: there was always an inflection in the calcium current, suggesting that the axonal conductance is small relative to the calcium conductance itself.

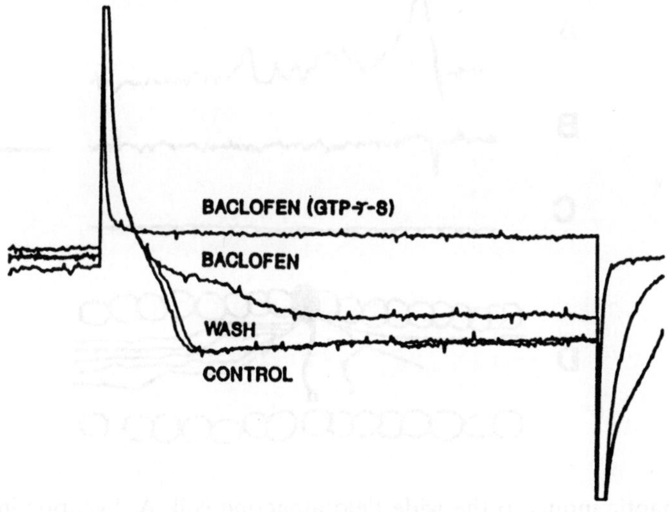

Figure 2. $GABA_B$ modulation of calcium currents in bipolar terminals. Control shows normal voltage-gated inward calcium current of about 100 pA. Baclofen reduces this calcium current by about 50 per cent. In the presence of GTP-g-S the calcium current is completely eliminated. The action of GTP-g-S suggests the presence of a G-protein in the pathway between $GABA_B$ receptor and calcium channel. Control response 100 pA; trace width 1 sec.

When the patch electrode contained GTP-g-S, the blocking action of baclofen was greater. Fig 2 shows that in the presence of GTP-g-S, virtually all of the calcium current was blocked by baclofen. The action of GTP-g-S is consistent with a second messenger system that involves a G-protein (Dolphin & Scott, 1987) as described by Gilman (1987).

Synaptic output of bipolar terminal

An amacrine cell type in which the excitatory synaptic input is also blocked by baclofen (Maguire et al., 1988; Werblin et al., 1988) has characteristic processes that spread widely across the IPL spanning up to 1 mm. Most of the processes ramify in a region near the midline of the IPL. The excitatory input to these amacrine cells is quite transient as shown in Fig 3B, and, like the bipolar calcium current, it is blocked by baclofen. Baclofen has no effect upon the postsynaptic membrane.

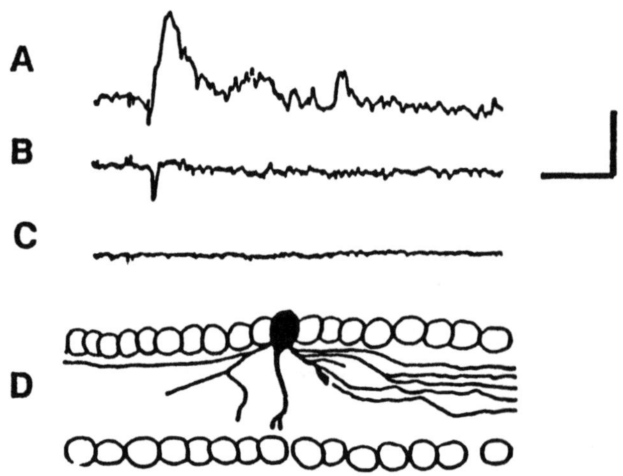

Figure 3. Synaptic inputs to the wide field amacrine cell. A. Synaptic input in response to a kainate puff delivered at the bipolar dendrites. There is an initial transient inward current followed by a slower outward current. B. In the presence of picrotoxin, the outward current is blocked but the inward current remains intact. C. In the presence of baclofen (and picrotoxin) the inward current is eliminated. Baclofen has no direct effect upon the amacrine cell, so baclofen is probably acting presynaptically, possibly at the bipolar terminal. D. Typical wide field morphology of this amacrine cell type. Calibration: 100 pA; 200 msec; in D slice width = 250 um.

The baclofen block is consistent with the notion that the wide field transient amacrine cells are *postsynaptic* to those bipolar terminals that are affected by baclofen, although we cannot rule out the possibility of an interneuron lying between the bipolar terminal and the amacrine cell. It is thought that these wide field amacrine cells release glycine because wide field lateral inhibitory signals to other amacrine cells are blocked by strychnine (Barnes & Werblin, 1987). All ganglion cells receive a wide field transient inhibitory input (Wunk & Werblin, 1978), and strychnine blocked all transient inhibitory inputs to ganglion cells (Belgum et al., 1984). The electrical and morphological properties of the wide field amacrine cell are outlined in Fig 3.

Synaptic input to the bipolar terminal

The presence of $GABA_B$ receptors at the bipolar terminal suggest that a subset of terminals receives a GABAergic input. Anatomical evidence suggests that bipolar terminals receive input from amacrine cells (Dowling & Werblin, 1969; Wong-Riley, 1974). Pharmacological-anatomical studies have suggested that a sustained GABAergic amacrine cell may provide this synaptic input

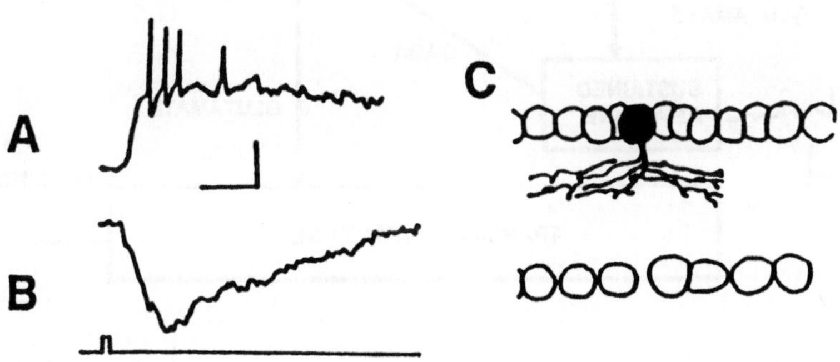

Figure 4. Synaptic inputs and response form of the narrow field sustained amacrine cell. A. Voltage response to a kainate puff at the bipolar dendrites showing a sustained train of spikes. B. Synaptic current measured in response to the puff puff. This current is much more sustained than that of the transient amacrine cell in Fig 3. C. A sketch of a typical OFF amacrine cell with narrow span of processes confined to the outer half of the IPL. Calibration: 50 pA; 20 mV, 200 msec; in D slice width =200 um. Puff shown in lower trace in B.

(Vaughn et al., 1981; Marc et al., 1978) to both ON and OFF bipolar cells (Marc et al., 1988). We have identified a class of amacrine cell with relatively narrow spread of processes, confined to either the ON or OFF sublamina (Famiglietti et al., 1977) of the IPL (Maguire et al., 1988). These cells receive a slowly inactivating excitatory synaptic input, and respond with a concomitant train of spikes as shown in Fig 4.

In a separate study we found that amacrine cells identified as narrow field also reacted with anti-GABA (Werblin et al., 1988), suggesting that they are GABAergic. Therefore it is likely that this class of narrow field sustained amacrine cell provides the GABAergic input to bipolar terminals.

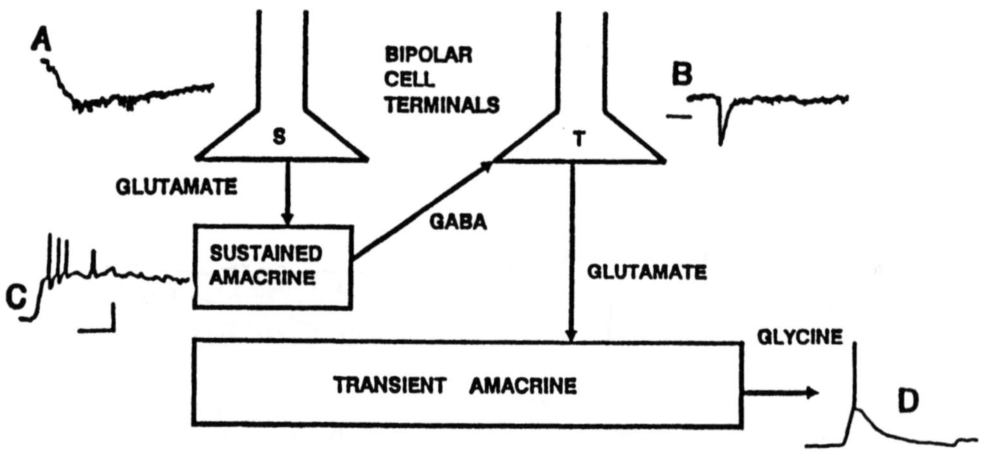

Figure 5. Proposed change-sensitive neural subunit. One class of bipolar terminal (S) drives the narrow field amacrine cells with sustained excitation (Fig 4). Feedback to the transient bipolar terminal (T) is thought to be mediated by a GABAergic, narrow field sustained amacrine cell that acts through a $GABA_B$ receptor to truncate transmission from the terminal. The output of the terminal is received by a wide field transient amacrine cell. A. Sustained input to the narrow field amacrine cell. B. Transient input to the wide field amacrine cell. C. Sustained GABAergic output of the narrow field amacrine cell. D. Single spike output of the wide field amacrine cell. Calibration: 20 mV; 50 pA, 100 msec.

A change-detecting subunit.

A change-detecting circuit is shown in Fig 5. This circuit utilizes the $GABA_B$ modulation at the bipolar terminal (T). Transient release from this (T) terminal is formed because sustained GABAergic input from the narrow field sustained amacrine cell inhibits release from the bipolar terminal. This truncates the otherwise sustained excitatory signal arriving via the (T) bipolar cell axon. The transient, GABA-blocked output drives the wide field transient amacrine cell.

A separate class of bipolar terminal (S) is devoid of this truncating feedback, and therefore its release is sustained. It provides the input to the sustained, narrow field amacrine cell. This scheme requires that the signal in the inhibitory GABAergic pathway be *delayed* by about 100 msec with respect to the excitatory signal in the bipolar axon, setting the interval of the transient excitatory signal as shown in B of Figs 3 and 5.

Implementation of the change detecting subunit in a movement detecting scheme

The interactions described above provide a possible set of neural components with which the movement detecting functions described by Barlow (1953) in frog, or Barlow & Levick 1965) in rabbit could be implemented. They can also account for the movement-sensitive activity described by Werblin (1970) as well as movement-gated inhibition (Werblin, 1972; Enroth Cugell & Jakiela, 1980).

In this scheme, the change sensitive function is mediated by the subunit described above. A population of subunits with these properties could provide the integrated input to the movement-sensitive ganglion cells. The lateral inhibitory arm of the system (W) would be formed by the wide field amacrine cell that receives a movement-gated excitation and delivers a lateral, glycine-mediated inhibition. A scheme incorporating a population of these elements to form a movement-sensitive ganglion cell receptive field is shown in Fig 6.

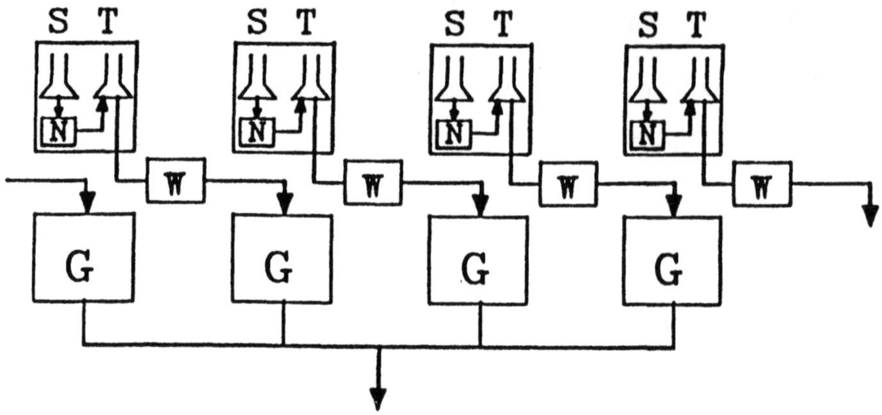

Figure 6. An hypothetical implementation of a movement-detecting scheme using the subunits described earlier. A series of subunits spans the receptive field of the ganglion cell. Each generates a transient signal that drives the element (W) directing a lateral inhibitory signal to the right in anticipation of the moving target, where it acts to veto activity elicited by the right most subunit.

SUMMARY

This paper describes a change-sensitive circuit whose function depends upon the $GABA_B$ decrease of calcium current associated with decrease of synaptic release from a class of bipolar cell terminals. A sustained GABAergic signal acts to truncate the action of a sustained excitatory signal within the bipolar cell, leading to transient release. The $GABA_B$ pathway appears to involve a G-protein, which may be responsible for the necessary delay in turning off the excitatory input to the wide field amacrine cell resulting in transient release.

We have demonstrated, using immunocytochemistry, that a narrow field, sustained responding amacrine cell is GABAergic (Werblin et al., 1988). The anatomy and pharmacology then suggest that the narrow field cell provides the input to the $GABA_B$ receptors at the bipolar terminal.

A wide field, transient responding amacrine cell is thought to receive the excitatory output from the transient bipolar terminal. This amacrine cell's input, like the calcium current in the bipolar terminal, is blocked by baclofen.

The circuit requires a separate class of bipolar terminal that is *not* truncated, but that releases transmitter in a *sustained* manner. The complete subunit would include both an ON and an OFF subunit, with separate bipolar and narrow field amacrine cells feeding their resulting transient excitation to the wide field lateral element. A mosaic of these change sensitive subunits could be iterated to generate a network capable of mediating directionally selective movement detection as shown in Fig 6.

REFERENCES

Barlow, H.B. & Levick, W.R. (1965). The mechanisms of directionally selective units in rabbit's retina. J. Physiol. 178:477-504.

Barnes, S. & Werblin, F.S. (1987). Direct excitatory and lateral inhibitory synaptic inputs to amacrine cells in the tiger salamander retina. Brain Res. 406: 233-237.

Belgum, J.H., Dvorak, D.R. & McReynolds, J.S. (1984) Strychnine blocks transient but not sustained inhibition in mudpuppy retinal ganglion cells. J. Physiol. 354:273-286.

Cleland, B.G & Levick, W.R. (1974) Properties of rarely encountered types of ganglion cells in the cat's retina and an overall classification., J. Physiol. 240: 457-492.

Dolphin, A.C. & Scott, R.H. (1987). Calcium channel currents and their inhibition by (-)Baclofen in rat sensory neurons: modulation by guanine nucleotides. J. Physiol. 386:1-17.

Dowling, J.E. & Werblin, F.S. (1969) Organization of retina of the mudpuppy Necturus maculosus. I. Synaptic structure. J. Neurophysiol. 32315-338.

Enroth Cugell, C. & Jakiela, H.G. (1980) Suppression of cat retinal ganglion cell responses by moving patterns. J. Physiol. 302: 49-72.

Famiglietti, E. V. Jr., Kaneko, A., & Tachibana, M. (1977). Neuronal architecture of ON and OFF pathways to ganglion cells in carp retina. Science 198:1267-1269.

Gilman, A.G. (1987) G-proteins: transducers of receptor-generated signals. A. Review Biochem. 56: 615-650.

Hamill, O.P., Marty, A. Neher, E., Sakmann, B. & Sigworth, F.J. (1981) Improved patch clamp techniques for high resolution current recording from cells and cell-free membrane patches. Pflugers ARch. 391:85-100.

Maguire, G., Maple, B., Lukasiewicz, P. & Werblin F. (1988) Calcium channel currents of bipolar cell axon terminals are modulated via $GABA_B$ receptors. Neuroscience Abstracts.

Maguire, G., Lukasiewicz, P. & Werblin F. (1988) Amacrine cell interactions underlying the response to change in the tiger salamander retina. J. Neurosci. (in the press).

Marc, R.E., Stell, W.K., Bok, D. & Lam, D.M.K. (1978). GABA-ergic pathways in the goldfish retina. J. Comp. Neurol. 182:221-245.

Marc, R.E. Liu, S. & Muller, J.F. (1988) Multiple GABA-mediated surround channels in goldfish retina. Arvo Abstracts 29:272.

Maturana, H.R., Lettvin, J.Y., Pitts, W.H. & McCulloch, W.S. (1960). Anatomy and physiology of vision in the frog (Rana pipiens) J. gen. Physiol. 43 supp. 129-175.

Robertson, B. & Taylor, W.R. (1986). Effects of g-aminobutyric acid and (-) Baclofen on calcium and potassium currents in cat dorsal root ganglion neurones in vitro. Br. J. Pharmac. 89:661-672.

Slaughter, M.M., & Miller, R.F. (1983). Bipolar cells in the mudpuppy retina use an excitatory amino acid neurotransmitter. Nature 303:537-538.

Stewart, W.W. (1978). Functional connections between cells as revealed by a highly fluorescent naphthalimide tracer. Cell 14:741-759.

Torre, V. & Poggio, T. (1978) A synaptic mechanism possibly underlying directional selectivity to motion. Proc. R. Soc. Lond.B 202:409.

Torre, V. & Poggio, T. (1981) An application: A synaptic mechanism possibly underlying motion detection. In Theoretical Approaches in Neurobiology pp., 39-46 MIT Press, Cambridge, MA.

Vaughn, J.E., Famiglietti, E.V., Barber, R.P., Saito, K., Roberts, E. & Ribak, C.E. (1981). GABAergic amacrine cells in rat retina, immunocytochemical identification and synaptic connectivity. J. Comp. Neurol. 197:113-127.

Werblin, F.S. (1978) Transmission along and between rods in the retina of the tiger salamander. J. PHysiol. 294:613-626.

Werblin, F.S. (1977). Regenerative amacrine cell depolarization and formation of on-off ganglion cell response. J. Physiol. 264:767-785.

Werblin, F.S. (1972). Lateral interactions at inner plexiform layer of vertebrate retina: Antagonistic response to change.
Science 175:1008-1010.

Werblin, F.S. (1970). Responses of retinal cells to moving spots:Intracellular recording in Necturus maculosis. J. Neurophysiol. 33:342-350.

Werblin, F.S., & Dowling, J.E. (1969). Organization of the retina of the mudpuppy, II. Intracellular recording. J. Neurophysiol. 32:339-355.

Werblin, F.S., Maguire, G., Lukasiewicz, P., Eliasof, S. & Wu. S.M. (1988) Neural interactions mediating the detection of change in the retina of the tiger salamander. Visual Neuroscience. (in the press).

Wong-Riley, M.T.T. (1974). Synaptic organization of the inner plexiform layer in the retina of the tiger salamander. J. Neurocytology 3:1-33.

Wunk, D.F. & Werblin, F.S. (1978). Synaptic inputs to the ganglion cells in the tiger salamander retina. J. gen. Physiol. 73:265-286.

MEMBRANE CURRENTS OF RETINAL BIPOLAR CELLS: GOLDFISH *vs.* MOUSE.

A. Kaneko, M. Tachibana and L. H. Pinto*
National Institute for Physiological Sciences
Okazaki 444 Japan
and
*Department of Neurobiology and Physiology
Northwestern University
Evanston, IL 60201 USA

The time course of light-evoked responses differs greatly between neurons of the distal retina and the proximal retina. In the distal retina, photoreceptors, horizontal cells and bipolar cells show sustained light responses, while in the proximal retina responses of amacrine and ganglion cells have prominent transient components even in those cells classified as a sustained type. Shaping of signals from sustained to transient may involve various mechanisms; activation of specific membrane conductances of amacrine and ganglion cells, interaction between excitatory and inhibitory signals that arrive to a postsynaptic cell with time delay, or a transient release of transmitter substance from presynaptic (bipolar) cells.

A few years ago Kaneko and Tachibana (1985) studied ionic currents of bipolar cells isolated from the goldfish retina, and identified Ca current and three other types of cationic currents. The Ca current in goldfish bipolar cells was all sustained showing no decay during the 2 s depolarization.

More recently we analyzed membrane currents of bipolar cells of the mouse retina (Kaneko, Pinto & Tachibana, 1989). Despite of a vast amount of work on retinal interneurons in lower vertebrates, reports in mammals are very limited (photoreceptors of monkey, Baylor, Nunn & Schnapf, 1984; mouse horizontal cells, Suzuki & Pinto, 1986; cat retinal neurons, see Kolb & Nelson, 1984). We wanted to fill this gap in our knowledge on the mammalian retina. Furthermore, mouse has a large variety of mutants which affect the visual system without causing gross degeneration (Balkema, Mangini, Pinto & Vanable, 1984). Comparison of the retina of these mutants with those of the normal animal will allow certain inferences to be made about the function of the affected gene products; first requirement for this comparison is a thorough study of the retinal neuron in the wild type animal. To our surprise, the Ca current of mouse

NATO ASI Series, Vol. H31
Neurobiology of the Inner Retina
Edited by R. Weiler and N.N. Osborne
© Springer-Verlag Berlin Heidelberg 1989

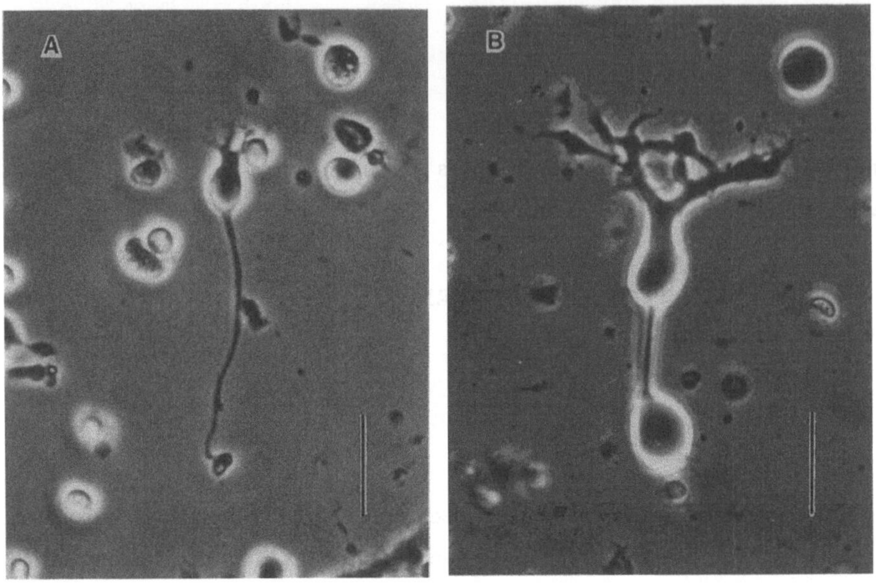

Figure 1: Solitary bipolar cells dissociated from the mouse (A) and from the goldfish (B) retinae. Scale bar 20 μm.

bipolar cells has a character that is different from the Ca current of goldfish bipolar cells. Here, we summarize the properties of Ca current of mouse bipolar cells by contrasting them against those of goldfish bipolar cells.

Membrane currents of solitary bipolar cells of the mouse

We obtained solitary bipolar cells from adult goldfish and mouse (C57BL/6J and C57BL/6crSlc) retinae by enzymatic (papain) dissociation (Fig. 1). Membrane currents of these cells were measured by whole-cell voltage clamp using patch pipette. Voltage-dependent currents carried by monovalent cations were similar in bipolar cells of the mouse and goldfish. As shown in Fig. 2A, hyperpolarization from the holding voltage, V_h, of -46 mV evoked a slowly-activated inward current (identified as an h-current, since this current component was blocked by 10 mM Cs^+ applied to the superfusate), and depolarization evoked a TEA-sensitive, outward current (probably a combination of $I_{K(V)}$ and $I_{K(Ca)}$).

Since a 2 s depolarization to around -10 mV evoked a sustained Ca current in goldfish bipolar cells (Kaneko and Tachibana, 1985; also see Fig. 9 of this article), we used the same command protocol in experiments of mouse bipolar cells initially. However, no Ca current was seen even after the counteracting K currents were suppressed by

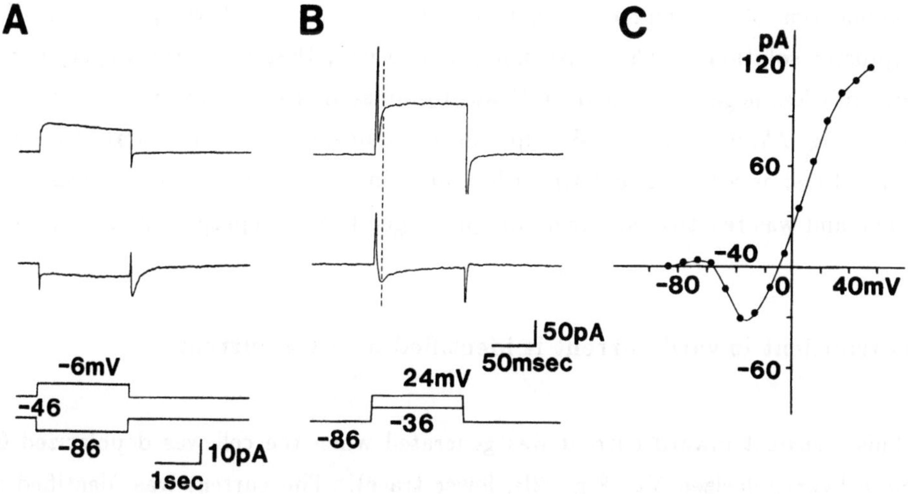

Figure 2: Membrane currents recorded from a bipolar cell of the mouse. A. Currents evoked by 2 s depolarizing (upper trace) and hyperpolarizing (lower trace) command pulses. V_h = -46 mV. Note an inward current tail at the cessation of the hyperpolarizing command. B. Currents evoked by 100 ms depolarization to 24 mV (upper trace) and to -36 mV (lower trace). Records are displayed on a much faster time scale than in A. The vertical broken line indicates the time at which the current amplitude was measured and plotted in the current-voltage relation curve of C. V_h = -86 mV.

an application of Cs^+ and/or TEA to the outside of the cell or by replacing K^+ of the pipette solution by Cs^+. We noticed, however, that when the hyperpolarizing commands less negative than -70 mV was terminated, a transient inward current was evoked (Fig. 2A, lower trace). By experiments to be described, this inward tail current was identified as a Ca current which had been inactivated while the cell was held at -46 mV and was reactivated during the prolonged (>2 s) hyperpolarizing command.

The transient inward current is identified as a Ca current

This transient inward current was generated when the cell was depolarized from a more hyperpolarized V_h (Fig. 2B, lower trace). The current was identified as a Ca current from the following observations. First, the transient inward current was resistant to TTX (5 μM). The same concentration of TTX immediately blocked action potentials of ganglion cells (identified by a large cell size, 4-5 long dendrites radially extending from the perikaryon) in the same dish. Secondly, the transient inward current became undetectable in a medium to which no Ca was added (Fig. 3A), and became larger in high Ca medium (Fig. 3B). Thirdly, the transient inward current was blocked by Co^{2+} (Fig. 4). The difference between the current records obtained under the control condition and in 4 mM Co^{2+} gives the Co^{2+}-sensitive current component, which is essentially I_{Ca}.

I_{Ca} thus isolated was detectable at membrane potentials more positive than -60 mV, reached the peak at about -30 mV, and became undetectable at around +40 mV. Decrease in amplitude with strong depolarization of the membrane is probably due to the reduction of driving force, because at these depolarized potential range, activation of calcium conductance (gCa) is nearly equal to gCa at -30 mV (see later).

I_{Ca} is present both at perikaryon and at axon terminal

Bipolar cells have distal dendrites where they receive synaptic input from photoreceptors and a proximal axon at which they send out signals to amacrine and ganglion cells. We found that I_{Ca} flowed both at the perikaryon and at the axon terminal. Local I_{Ca} was obtained as a fraction of total I_{Ca} blocked by a focal application of 4 mM Co^{2+} from *ca.* 1 μm tip pipette placed as close to the cell surface as possible. I_{Ca} of the axon terminal had a similar time course as I_{Ca} of the perikaryon, though the former was smaller in amplitude (Fig. 5). These observations indicates the possibility that

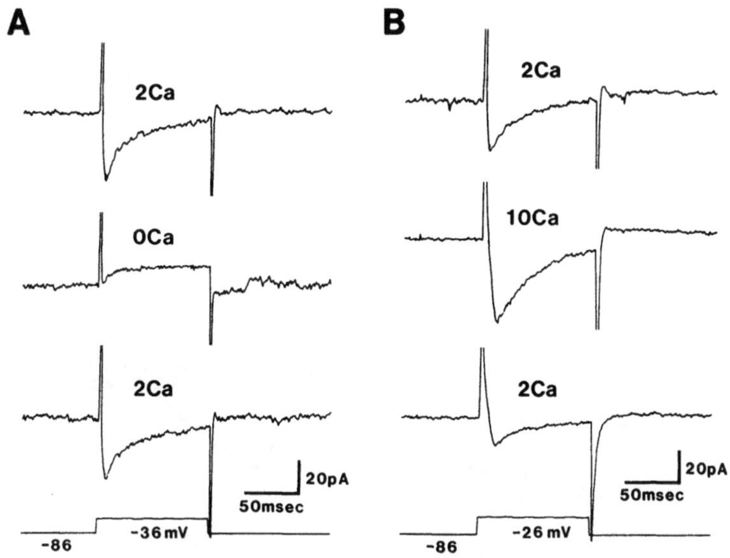

Figure 3: Effects of Ca^{2+}-removal (A, middle trace) and of elevation of $[Ca^{2+}]_o$ (B, middle trace) on the transient inward current of mouse bipolar cells. The inward current was activated by depolarization to -36 (A) or to -26 (B) mV from the V_h of -86 mV. Top traces, control; bottom traces, recovery. Pipette solution contained 120 mM-CsCl. (Reproduced by permission from Kaneko *et al.*, 1989)

this current may participate in transmitter release from the axon terminal.

I_{Ca} can be modified by membrane potential within the physiological range

Activation As shown in Fig. 5, I_{Ca} was detected as an inward current at membrane potentials between -60 mV and +40 mV. Since the current amplitude is the product of the driving force (the potential difference between the membrane potential and the equilibrium potential of Ca^{2+}, E_{Ca}) and the conductance (gCa), it is important to know the relation between gCa and the membrane potential. Since accurate estimation of E_{Ca} is difficult, we directly measured gCa by using tail current. Tail current analysis was performed in the following way. First, gCa was activated by membrane depolarization to a specific potential level. At the timing of full activation of gCa, depolarizing command is terminated by driving the membrane potential back to V_h. Then, Ca (tail) current flowed through gCa that declined with time. As seen in Fig. 6, the amplitude of the tail current was proportional to the amplitude of I_{Ca} that flowed during the depolarizing command. Since the driving force during the tail current (potential difference between E_{Ca} and the holding potential) remained constant independent of

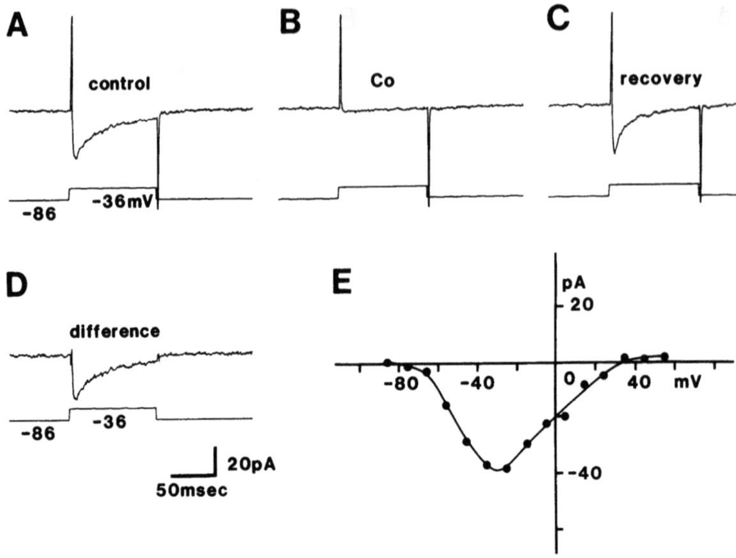

Figure 4: Effect of Co^{2+}. The transient inward current of mouse bipolar cell (A, control) is completely blocked by an application of 4 mM Co^{2+} in the superfusate (B). D, the Co^{2+}-sensitive current obtained by subtracting the trace B from trace A. E, current-voltage relation of the Co^{2+}-sensitive current. Pipette solution contained 120 mM-CsCl. (Reproduced by permission from Kaneko *et al.*, 1989)

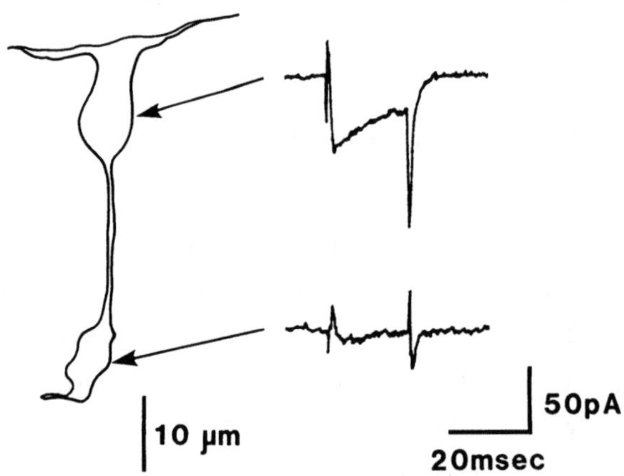

Figure 5: Spatial distribution of Co^{2+}-sensitive current. The upper trace illustrates the difference between the two current records obtained (1) under the control condition and (2) when a solution containing 4 mM Co^{2+} was applied to the soma. The lower trace illustrates the difference between the two current records obtained under the control condition and when Co^{2+} was applied to the axon terminal. Currents were evoked by a 20 ms depolarizing command to -36 mV from a holding voltage of -96 mV. Pipette solution contained 120 mM-CsCl. (Reproduced by permission from Kaneko *et al.*, 1989)

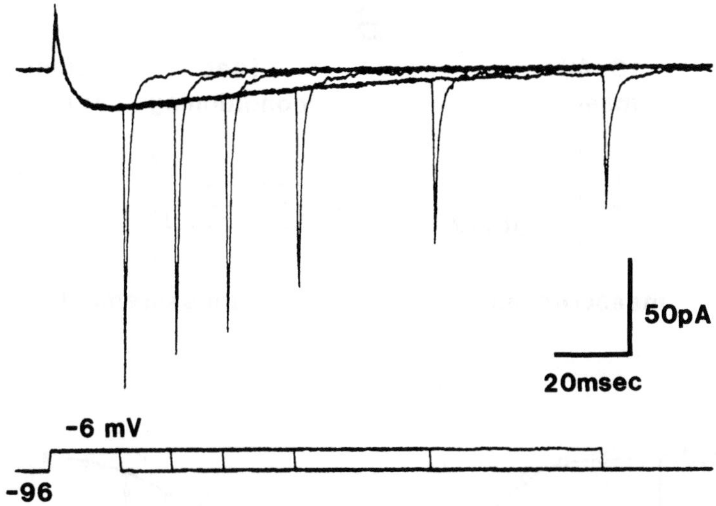

Figure 6: Six superimposed current traces obtained from a mouse bipolar cell evoked by depolarizing commands (to -6 mV from -96 mV V_h) of various duration ranging from 20 ms to 140 ms. Note large tail currents evoked at the end of the command pulses. The amplitude of the tail current is proportional to the amplitude of the inward current immediately before the termination of the commands. Pipette solution contained 120 mM-CsCl.

command voltages, the amplitude of the tail current was directly proportional to gCa. As summarized in Fig. 7, gCa was activated by depolarization of the membrane to potentials more positive than -65 mV. It showed sigmoidal increase with depolarization, showed half maximum value at about -25 mV, and reached saturation at around +30 mV.

Inactivation Inactivation of I_{Ca} was also voltage-dependent. Data illustrated in Fig. 7 were obtained from a series of experiment in which tail currents in response to activating pulses (-6 mV from the V_h of -96 mV) was measured after 1 s conditioning pulses of various voltages. gCa was fully activated when the cell was held at more negative than -80 mV, and was suppressed nearly entirely by keeping the membrane potential at more positive than -30 mV for 1 s. Between these two membrane potentials inactivation of gCa had sigmoidal relation with the membrane potential (half maximum inactivation at about -50 mV).

These observations indicate that I_{Ca} can be activated at membrane potentials between -65 and -10 mV in the superfusate which contains 10 mM $[Ca^{2+}]_o$. Under the physiological condition in which $[Ca^{2+}]_o$ of the extracellular space is approximately 2 mM, the curves must be moved to the hyperpolarizing direction by approximately 20 mV along the voltage axis to compensate the amount of surface charge. Thus, I_{Ca} would be activated at membrane potentials between -80 and -25 mV.

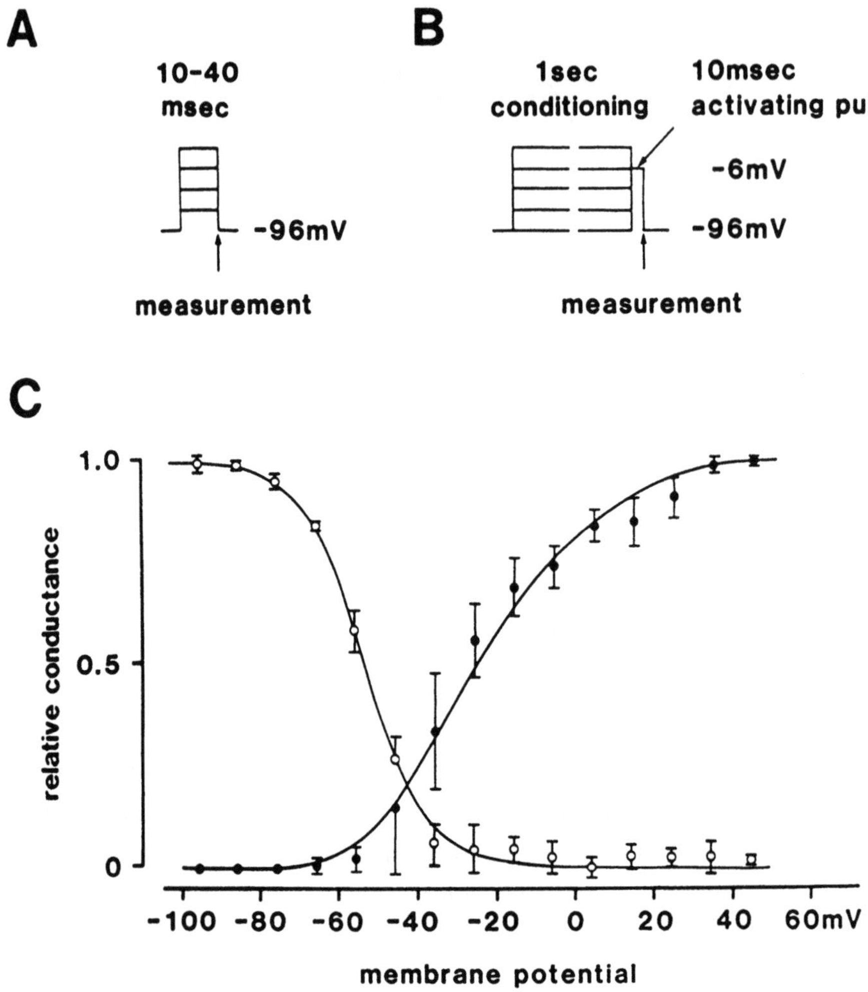

Figure 7: Activation and inactivation curves. •: relative conductance (mean ± S.D., n = 4), measured using tail currents (pulse protocol shown in A), plotted against test pulse voltage. ○: relative conductance (mean ± S.D., n = 4), measured using the tail current to a -6 mV activating pulse that followed a long conditioning pulse of each of many voltages (pulse protocol shown in B), plotted against voltage of the conditioning pulse. V_h = -96 mV. Superfused with solution containing 10 mM $[Ca^{2+}]_o$. Pipette solution contained 120 mM-CsCl. (Reproduced by permission from Kaneko *et al.*, 1989)

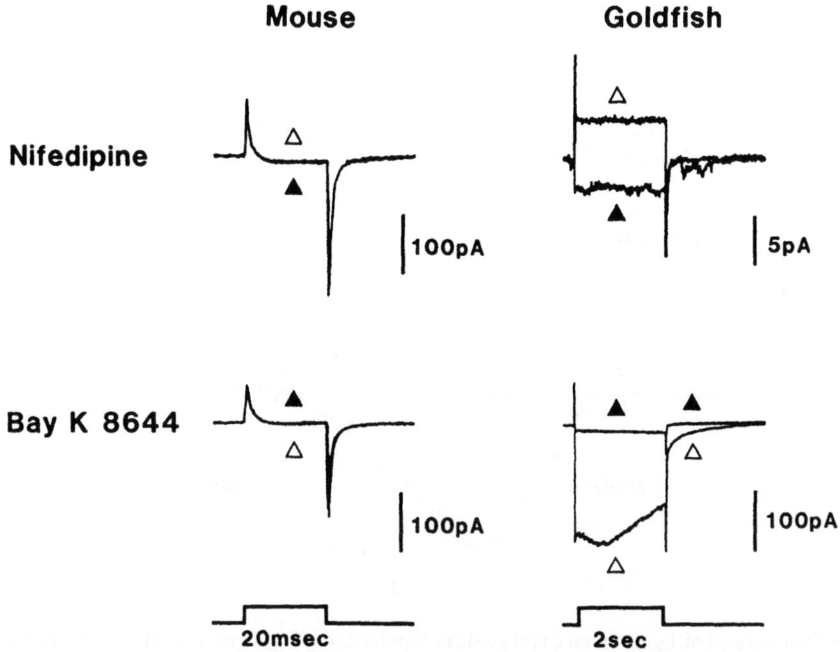

Figure 8: Effects of dihydropyridine compounds on Ca^{2+}-currents of bipolar cells of the mouse and of the goldfish. Each panel consists of two superimposed traces, one obtained under control condition (▲) and another taken during application of test substances (10 μM, each; marked with △). Note identical traces of the mouse records, indicating the absence of any effect by dihydropyridine. Mouse bipolar cells were held at -96 mV and depolarized to -6 mV for 20 ms. Goldfish bipolar cells were held at -46 mV and depolarized to -6 mV for 2 s. Pipette solution contained 120 mM-CsCl. (Reproduced by permission from Kaneko *et al.*, 1989)

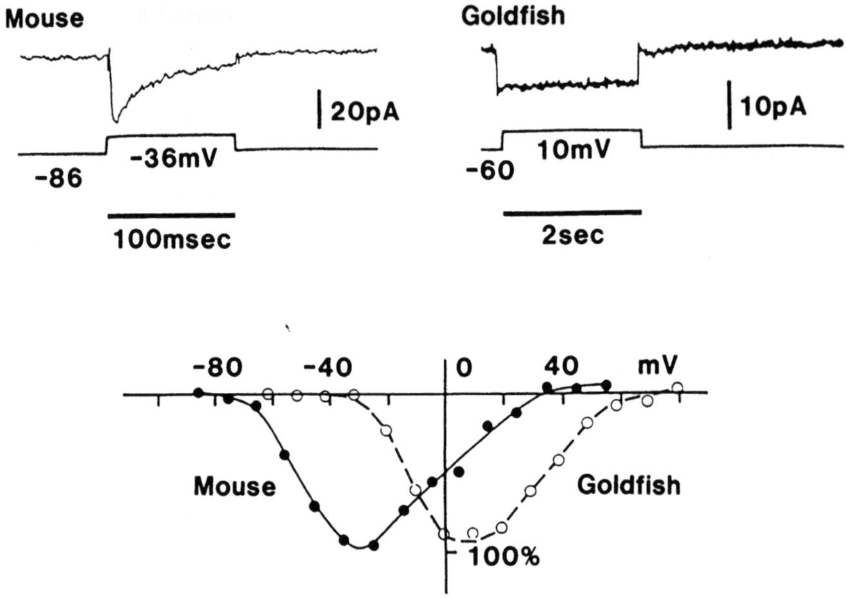

Figure 9: Comparison of I_{Ca} in mouse and goldfish bipolar cells. I_{Ca} in mouse bipolar cell was activated by 100 ms depolarizing command (from V_h of -86 mV to -36 mV), while I_{Ca} in goldfish bipolar cell was activated by 2 s depolarizing command (from V_h of -60 mV to 10 mV). Current-voltage relations of I_{Ca} in mouse (•) and goldfish (○) bipolar cells are shown for comparison after normalization of the peak amplitude.

Effects of dihydropyridines on I_{Ca} of bipolar cells: a comparison of I_{Ca} in mouse and goldfish bipolar cells.

Various kinds of dihydropyridine compounds are known to modify Ca channels, and their effect differs depending on the type of Ca channels (Nilius, Hess, Lansman & Tsien, 1985). I_{Ca} in mouse bipolar cells and in goldfish bipolar cells (either examined by the tail current or by the current during command pulses) showed a remarkable difference in response to dihydropyridine compounds (Fig. 8). I_{Ca} in mouse bipolar cells was unaffected either by nifedipine or Bay K 8644, while I_{Ca} in goldfish bipolar cells was blocked by 10 μM nifedipine and was augmented by nearly 10 fold by 10 μM Bay K 8644. Cd^{2+} also acted differently; 50 μM Cd^{2+} blocked I_{Ca} in goldfish, but had no discernible action on I_{Ca} in mouse bipolar cells.

As has been pointed out in the preceding section, difference in properties of I_{Ca} between mouse and goldfish bipolar cells is also found in their time course (Fig. 9). I_{Ca} in mouse bipolar cells was transient showing a rapid inactivation, while I_{Ca} in goldfish bipolar cells was sustained.

Comparison of I_{Ca} in mouse and goldfish bipolar cells.

The present study has demonstrated that bipolar cells of both mouse and goldfish have Co^{2+}-sensitive, TTX-resistant I_{Ca}, but properties of I_{Ca} in the two species are markedly different. First, the voltage of activation threshold of mouse I_{Ca} is approximately -65 mV, while that of goldfish is about -30 mV. Voltage at which I_{Ca} becomes maximum also differs: approximately -30 mV for the mouse and about +10 mV for the goldfish. Secondly, I_{Ca} in the mouse is transient, reflecting rapid inactivation, but I_{Ca} in goldfish is sustained, keeping almost constant amplitude during the 2 s depolarizing command. Thirdly, I_{Ca} in the two species has different sensitivity to divalent cations and to dihydropyridine; I_{Ca} in the mouse was insensitive to 50 μM Cd^{2+}, 10 μM nifedipine, or 10 μM Bay K 8644. In contrast, I_{Ca} in bipolar cells of the goldfish was suppressed by Cd^{2+} or nifedipine, and was augmented by Bay K 8644. In the light of recent classification of Ca^{2+} channels (Nowycky, Fox & Tsien, 1985; Hagiwara, Irisawa & Kameyama, 1988), I_{Ca} of mouse bipolar cells can be classified as the transient (T) type, while I_{Ca} of goldfish bipolar cells as the long-lasting (L) type.

The finding that I_{Ca} flows both at the perikaryon and at the axon terminal of mouse bipolar cells indicates the possibility that this current may participate in synaptic transmission, and in shaping transient responses of mouse ganglion cells. Activation and inactivation analysis shows that I_{Ca} can be activated at membrane potentials between -65 and -10 mV (in 10 mM $[Ca^{2+}]_o$). Under the physiological condition ($[Ca^{2+}]_o \approx 2$ mM) I_{Ca} activation range would be between -80 and -25 mV. If the light responses of mouse bipolar cells cover a similar potential range as those recorded from the lower vertebrates (Saito, Kondo & Toyoda, 1979), the voltage range at which I_{Ca} of mouse bipolar cells is activated includes the presumed range of membrane potentials accompanying light-evoked responses. Since Ca^{2+} influx is prerequisite to the transmitter release, it may be reasonable to believe that transmitter release from mouse bipolar cells is also transient. It has long been a puzzle how the transient ON- or OFF-responses of amacrine or ganglion cells are formed. The present finding offers a possibility to interpret the transient responses.

Acknowledgement

We thank Michi Hosono for her excellent technical assistance in preparing isolated photoreceptors. This research was supported in part by the Grants in Aid for Scientific Research from the Ministry of Education, Science and Culture (Nos. 61304032,

63638517, 63112007 to AK and 63480111, 63641537 to MT), NEI R01EY01221 (to LHP), NSF INT8613447 and Japan Society for the Promotion of Science.

References

Balkema GW, Mangini NJ, Pinto LH, Vanable JW Jr (1984) Visually evoked eye movements in mouse mutants and inbred strains. Invest Ophthalmol Visual Sci 25:795-800

Baylor DA, Nunn BJ, Schnapf JL (1984) The photocurrent, noise and spectral sensitivity of rods of the monkey *Macaca fascilularis.* J Physiol (Lond) 357:575-607

Hagiwara N, Irisawa H, Kameyama M (1988) Contribution of two types of calcium currents to the pacemaker potentials of rabbit sino-atrial node cells. J Physiol (Lond) 395:233-253

Kaneko A, Pinto LH, Tachibana M (1989) Transient calcium current of retinal bipolar cells of the mouse. J Physiol (Lond) (in press)

Kaneko A, Tachibana M (1985) A voltage-clamp analysis of membrane currents in solitary bipolar cells dissociated from *Carassius auratus.* J Physiol (Lond) 358:131-152

Kolb H, Nelson R (1984) Neural architecture of the cat retina. Prog Retinal Res 3:21-60

Nilius B, Hess P, Lansman JB, Tsien RW (1985) A novel type of cardiac calcium channel in ventricular cells. Nature 291:497-500

Nowycky MC, Fox AP, Tsien RW (1985) Three types of neuronal calcium channels with different calcium agonist sensitivity. Nature 316:440-443

Suzuki H, Pinto LH (1986) Response properties of horizontal cells in the isolated retina of wild-type and pearl mutant mice. J Neurosci 6:1122-1128

IMMUNOCYTOCHEMICAL LOCALIZATION OF PROTEIN KINASE C IN SOME VERTEBRATE RETINAS

K. Negishi, S. Kato and T. Teranishi
Department of Neurophysiology
Neuroinformation Research Institute
University of Kanazawa School of Medicine
13-1 Takara-machi, Kanazawa 920
Japan

Introduction

The enzyme, protein kinase C (PKC), has been found to be destributed throughout mammalian tissues, its highest amounts being present in the CNS (Minakuchi et al, 1981; Kikkawa et al, 1982). Recent immunohistochemistry with poly- and mono-antisera against PKC has shown that its cellular distribution varies with cell type (Girard et al, 1985; Shoji et al, 1986; Wood et al, 1986; Saito et al, 1988). In the present study, an indirect immunohistochemical technique was applied to the retina of several vertebrates to explore the cellular localization of PKC-like immunoreactivity (IR).

Materials and Methods

The retinas examined were from lamprey, fishes (carp, goldfish, dace, loach and channel catfish), frog, turtle, chick, sparrow, guinea-pig, rat, rabbit and cat. The eyes were removed from the animals under deep anesthesia with Nembutal (Abbot, Illinois) for mammals, with MS222 (Sankyo, Tokyo) for fishes, with ether for chick, or immediately after

NATO ASI Series, Vol. H31
Neurobiology of the Inner Retina
Edited by R. Weiler and N.N. Osborne
© Springer-Verlag Berlin Heidelberg 1989

decapitation for frog, turtle and chick. Immediately after eye removal, the mammals were injected into the heart with a large dose of Nembutal to kill them, and the non-mammals were decapitated. The isolated retinas were fixed in ice-cold Zamboni's solution (0.2% picric acid and 2% paraformaldehyde in 130 mM phosphate buffer at pH 7.4) for 2 days and washed in 30% sucrose-containing 0.1 M phosphate buffer saline (PBS at pH 7.4) overnight at 4°C. Frozen sections were made at a thickness of 20 μm. The sections were mounted on chrome-alum-gelatin coated glass slides, air-dried for two hours at room temperature and treated with antisera. All sections were rinsed three times with ice-cold PBS for 15 min prior to the application of antisera according to indirect immunohistochemical procedures (Coons, 1958; Katayama-Kumoi et al, 1985; Kiyama et al, 1985; Negishi et al, 1985).

A monoclonal antibody against PKC (raised in mouse) was purchased from Amersham (clone MC5), and as the second fluorescent antibody Texas Red (TR)-linked anti-mouse IgG (Amersham) or FITC-linked anti-mouse IgG (Cappel, Worthington) was used. According to the instruction attached, the PKC antibody recognizes only the α and β subunits of PKC. To show the coexistence of PKC-like and tyrosine hydroxylase (TH)-like IRs, polyclonal antisera (raised in rabbit) against TH were purchased from Eugene Tech (Allendale, NJ). The second antibody used was FITC-conjugated anti-rabbit IgG (Cappel) for TH and TR-linked anti-mouse IgG for PKC. As references, another monoclonal antibody, PCTH-7 (Hatanaka and Arimatsu, 1984) against TH and other antibodies against some neuropeptides were used. The second antisera were FITC-conjugated anti-rat or anti-rabbit IgG (Cappel) or TR-linked anti-rabbit IgG (Amersham). The dilution of all antisera used was 1:250-1000, depending upon the kinds of antibody and the experimental purposes. For the double labelings, higher concentrations of the 1st and 2nd antibodies were required. The sections were examined under a fluorescence microscope (Nikon EF), provided with a filter system of G (546 nm for excitation) plus 580W (>580 nm for absorption) for TR, or with a system of B (495 nm) plus 515W (>515 nm) for FITC. Photomicrographs were taken on Tri-X pan

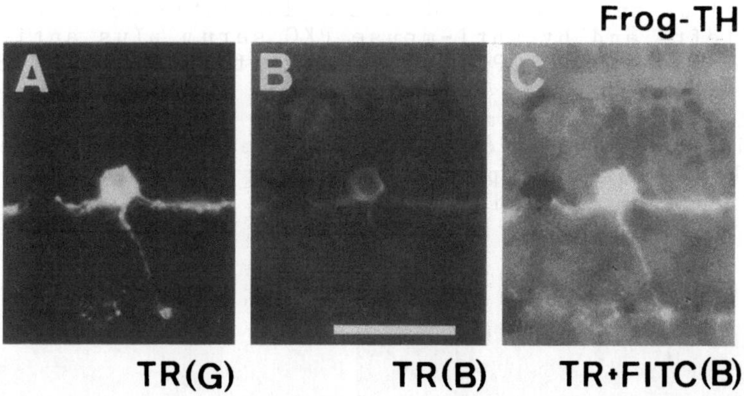

Fig. 1. A TH-like immunorective (ir) cell in the frog retina. A, labeled by anti-rabbit TH serum plus anti-rabbit TR-linked IgG and taken through the G filter system; B, the same cell taken via the B system; C, the same cell was further labeled with anti-rabbit FITC-linked IgG and taken via the B system. Part of TR fluorescence breaks through the B filter (B), but becomes negligible when the cell was doubly stained (C). Scale bar, 50μm.

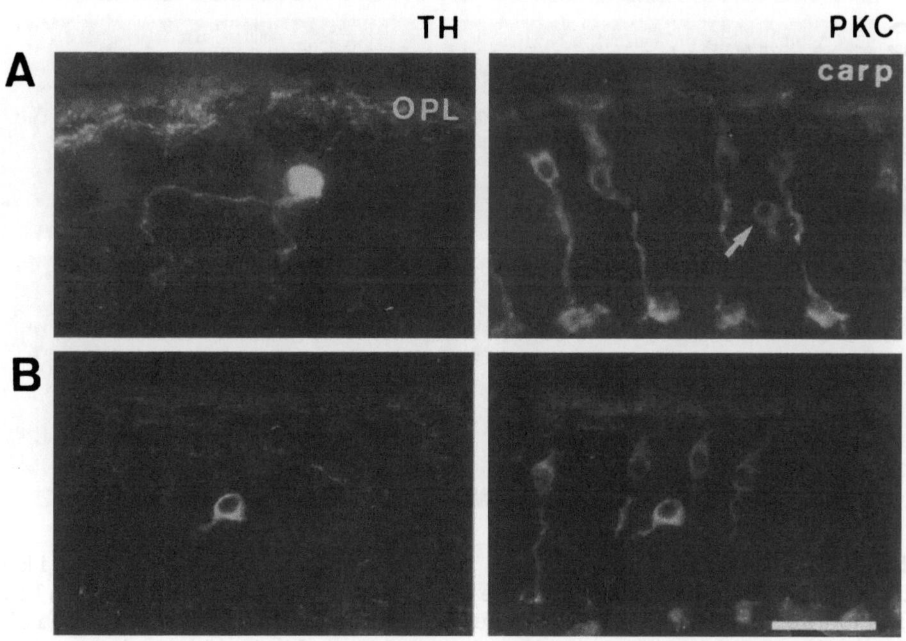

Fig. 2. TH- and PKC-like ir cells in the carp retina. A (left), singly labeled by anti-rat TH serum plus anti-rat FITC-linked IgG; A (right), singly labeled by anti-mouse PKC serum plus anti-mouse FITC-linked IgG; B, doubly

labeled by anti-rabbit TH serum plus anti-rabbit TR-linked IgG (left) and by anti-mouse PKC serum plus anti-mouse FITC-linked IgG; OPL, outer plexiform layer. TH-like interplexiform cells and typical rod bipolar cells contain PKC-like IR. The same results were obtained in goldfish and dace. In the following figures 3 to 5, the same double labeling procedures were applied as in B. Modified from Negishi et al (1988).

TH PKC

A

B

C

D

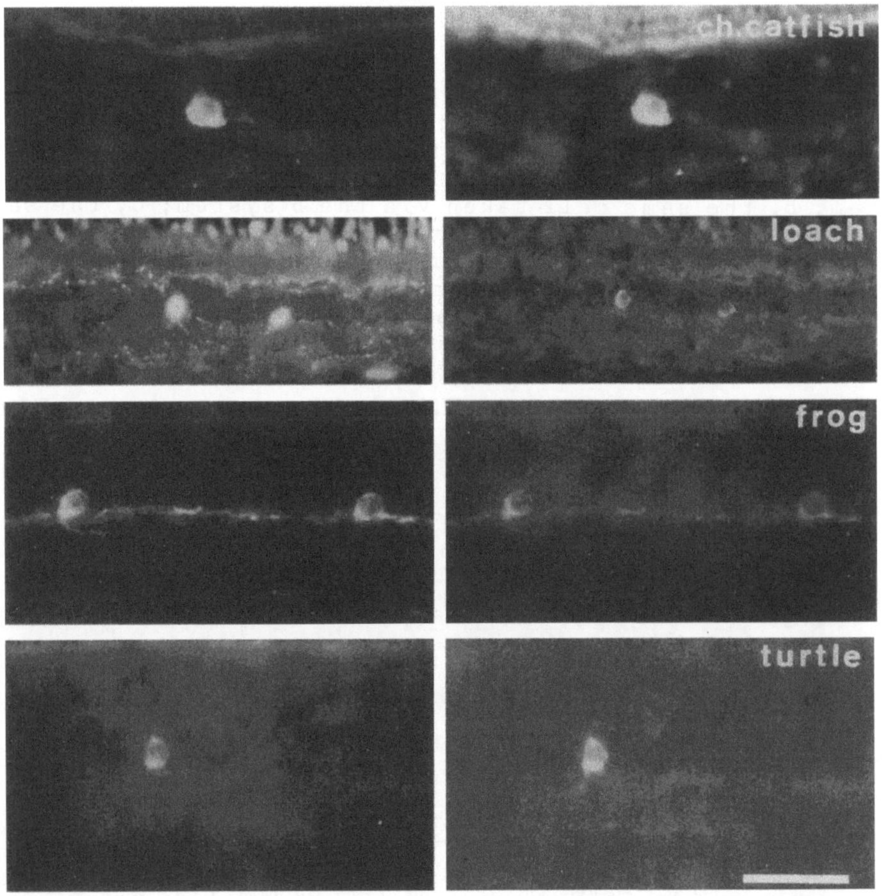

Fig. 3. Double labelings for TH- (left) and PKC-like ir cells (right) in the retina of cold-blooded animals. In these, only TH-like ir cells contain PKC-like IR, which is relatively intense in channel catfish (A) and turtle (D) but weak in loach (B) and frog (C).

or Ektachrome films (Kodak). Part of the present results has been in press elsewhere (Negishi et al, 1988).

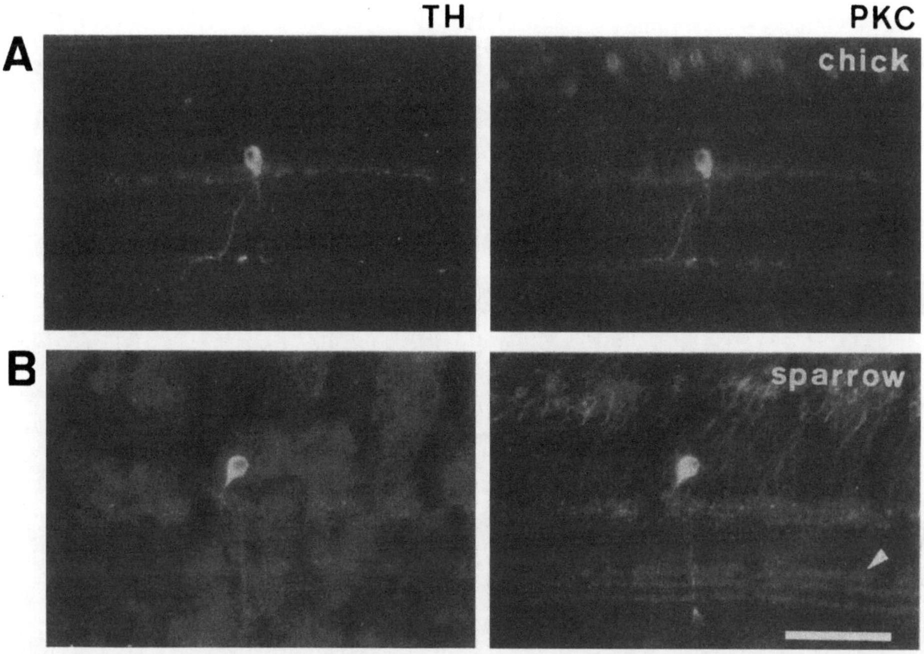

Fig. 4. Double labelings for TH- and PKC-like ir cells in the retina of chick (A) and sparrow (B). In B, the axon terminals ends in forming three layers at the proximal level of the inner plexiform layer (arrowhead).

Results

Part of the TR fluorescence was found to pass through the B filter system. Therefore, reddish TH-like immunoreactive (ir) cells stained with TR might be detected very weakly through the B system, but its intensity was negligible compared with the intense green FITC fluorescence when the cells were doubly stained with TR and FITC (Fig. 1).

PKC-like IR was first found intensely in typical rod bipolar cells and weakly in a class of cells located at the amacrine cell layer of the carp retina (Fig. 2A, arrow) when radial sections were treated with anti-mouse PKC antibody and FITC- or TR-linked anti-mouse IgG. The class of these ir cells had a relatively large soma and nucleus, giving us the

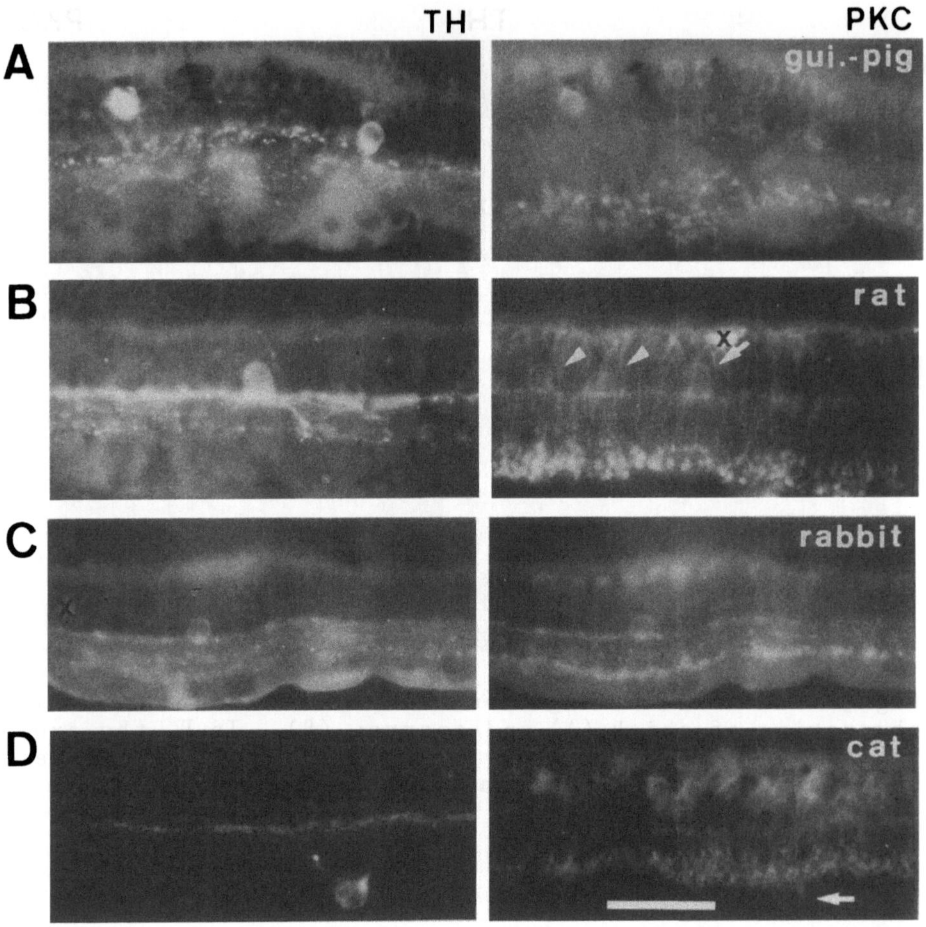

Fig. 5. Double labelings for TH- and PKC-like ir cells in the retina of mammals. PKC-like IR is relatively weak in these animals. In rat (B), some amacrine cells (arrowheads) contain PKC-like IR in addition to a TH-like ir cell (arrow). In cat, a displaced amacrine cell is shown (arrow). Mark "x" in B and C indicates artifact.

immediate impression that they appear to be dopamine (DA) cells, although no PKC-like ir fibers were seen at the outer plexiform layer (OPL) (Fig. 2A, compare TH with PKC).

To ascertain the presence of PKC-like IR in DA interplexiform (IP) cells, double labelings were employed to the carp retina with anti-mouse PKC and anti-rabbit TH antibodies. It was found that PKC- and TH-like IRs coexist in the IP cells (Fig. 2B). Rod bipolar cells were labeled only by PKC-like

IR, which was frequently most intense in the axon terminals. The same results were obtained in the goldfish and dace retinas (not shown).

The same single and double labeling procedures were applied to other vertebrate retinas. We could not stain any cells in lamprey (Lampetra japonica) with anti-TH and anti-PKC sera (not shown), although DA cells have been histochemically shown (Negishi et al, 1982). In channel catfish (Ictalurus punctatus) (Fig. 3A), loach (Misgurnus anguillicaudatus) (Fig. 3B), frog (Rana pipiens) (Fig. 3C) and turtle (Geoclemys reevesii) (Fig. 3D), only DA amacrine cells exhibited both PKC- and TH-like IRs. In all the other animals, chick (2 days after hatching) (Fig. 4A), sparrow (Fig. 4B), albino guinea-pig (Fig. 5A), albino rat (Fig. 5B), albino rabbit (Fig. 5C) and cat (Fig. 5D), bipolar cells and TH-like ir amacrine cells appeared to contain PKC, its reactive fluorescence intensity being relatively low except in the axon terminals of rat bipolar cells. It is of interest that in the sparrow retina the axon terminal endings form three layers at the proximal level of the inner plexiform layer (IPL) (arrowhead in Fig. 4B), and that in the rat retina some other amacrine cells (arrowheads) in addition to TH-like ir cells (arrow in Fig. 5B) were also positive to PKC antibody.

Discussion

We have been exploring for some years what kinds of neuroactive substance coexist with TH in the fish retina, because the argument has been going on as to whether or not fish DA cells include amacrine-type cells (Dowling and Ehinger, 1975; Teranishi and Negishi, 1988). No double labelings were seen using anti-rat or anti-rabbit TH serum and one of other antisera such as those against glucagon, neuropeptide Y, somatostatin, substance P and GABA (Negishi and Teranishi, 1987). The coexistence of PKC- and TH-like IRs in IP cells of the carp retina was our first experience.

PKC is known as an important enzyme which is activated by 1,2-diacylglycerol (1,2-DAG) in the presence of calcium ions and phospholipids. 1,2-DAG can be produced via a rapid receptor-mediated hydrolysis of inositol phospholipids (Takai et al, 1979; Kishimoto et al, 1980). This enzymatic sequence is assumed to be implicated as a common mechanism in the transduction of extracellular signals into cells to control many physiological processes (Nishizuka, 1984, 1986). At present, however, we cannot discuss the functional significance of PKC-like substance in fish DA-IP cells, but can state that DA cells are unique in possessing the α or/and β subunits of PKC among cells, whose somata are located at the amacrine cell layer. It has been shown that aynthetic DAG stimulates release of [^{3}H]-DA but not of [^{3}H]-aspartate, -glycine or -GABA from isolated cell fractions of the carp retina (Ishita and Kato, 1985). The PKC-like ir bipolar cells are certainly morphologically identical to one or more types of rod bipolars in fish retinas (Tachibana and Kaneko, 1987), and they appear to be rod-connected also in the other vertebrates. However, the finding that the axon terminals end, forming three layers, in the proximal level of the IPL in the sparrow retina suggests that PKC-like ir bipolar cells include some other classes in addition to rod bipolars. In the animals (loach, channel catfish, frog and turtle), whose bipolar cells exhibited no PKC-like IR, bipolar cells may contain too low levels of PKC-like substance, which could not be detected by the present technique.

In the cat retina, DA cells have been identified as A18 amacrine cells in morphology (Kolb et al, 1981; Kolb and Nelson, 1985). The postsynaptic cells are often bistratified amacrine cells of AII type, which link rod bipolar cells to off-center ganglion cells via conventional chemical synapses and to on-center ganglion cells via large gap junctions (Pourch, 1982; Famiglietti, 1983; Voigt and Wässle, 1987). The AII amacrine cells have been shown to generate a light-induced depolarization (Nelson, 1982; Dacheux and Raviola, 1986), and to assume to use an inhibitory transmitter, glycine

(Pourcho and Goebel, 1985). In the fish retina, there has been no report concerning the presence of AII type-like amacrine cells and the relationship between DA-IP and glycinergic IP cells (Marc and Liu, 1984). The regional distribution of DA cells has been said to be in parallel to that of rods in the rhesus monkey retina (Mariani et al, 1984), but to be independent from that of ganglion cells in guinea-pig, rat, cat and rabbit (Mitrofanis et al, 1988). In isolated rod bipolar cells of the goldfish retina, the highest sensitivity to exogenous GABA has been shown to localize at their axon terminals (Tachibana and Kaneko, 1987), where the most intense PKC-like IR was seen in the present study. It is thus conceivable that PKC appears to participate in the DAergic system and rod-signal processes, and that DA cells and the axon terminals of rod bipolar cells receive GABAergic inhibition (Negishi et al, 1983; Ishita et al, 1987; Tachibana and Kaneko, 1987) in most vertebrate retinas.

Summary

The immunocytochemical localization of protein kinase C (PKC) and tyrosine hydroxylase (TH) was investigated in the retina of lamprey, fishes (carp, goldfish, dace, loach and channel catfish), frog, turtle, chick, sparrow and some mammals (guinea-pig, rat, rabbit and cat) by means of fluorescence microscopy. PKC-like immunoreactivity was found in TH-like immunoreactive (ir) cells more or less in all the animals examined (except in lamprey) and also in rod bipolar or presumed rod bipolar cells in the animals (except in loach, channel catfish, frog and turtle). In the rat retina, some other amacrine cells in addition to TH-like ir cells were labeled by monoclonal anti-PKC serum.

Acknowledgments

The authors thank Dr. William K. Stell for his suggestion and correction of the text English at the early stage of manuscript, Dr. Hiroshi Hatanaka for his gift of a monoclonal antibody against tyrosine hydroxylase (PCTH-7) and Mrs. Tami Urano for her secretarial assistance. Juvenile channel catfish and dace were given by Dr. Kazunori Tani (the Kanazawa Aquarium) and juvenile carp were supplied by the Ishikawa Prefectural Station for Water Fisheries. Supported in part by the Ministry of Education of Japan (63570054, 62221014, 62105081), the Japan Society for Promotion of Science, and from the Johnan Hospital in Takaoka. Part of traveling expenses was given to K.N. by the Yamada Science Foundation to attend the NATO workshop at Oldenburg.

References

Coons AH (1958) Fluorescent antibody methods. In: Damielli JF (ed) General cytochemical methods. Academic, New York, pp399-422

Dacheux RF, Raviola E (1986) The rod pathway in the rabbit retina; a depolarizing bipolar and amacrine cell. J Neurosci 6:331-345

Dowling JE, Ehinger B (1975) Synaptic organization of the amine-containing interplexiform cells of the goldfish and cebus monkey retinas. Science 188:270-273

Famiglietti EV (1983) 'Starburst' amacrine cells and cholinergic neurons: mirror-symmertic ON and OFF amacrine cells of rabbit retina. Brain Res 261:138-144

Girard PR, Mazzei GJ, Wood JG, Kuo JF (1985) Polyclonal antibodies to phospholipid/Ca^{2+}-dependent protein kinase C and immunocytochemical localization of the enzyme in rat brain. Proc Natl Acad Sci USA 82:3030-3034

Hatanaka H, Arimatsu Y (1984) Monoclonal antibodies to tyrosine hydroxylase from rat pheochromocytoma PC12h cells with special reference to nerve growth factor-mediated increase of the immunoprecipitable enzymes. Neurosci Res 1:253-263

Ishita S, Kato S (1987) Functional significance of protein kinase C in neural retina in particular reference to neurotransmitter release. Jpn J Pharmacol 90:30P

Ishita S, Negishi K, Teranishi T, Shimada Y, Kato S (1988) GABAergic inhibition on dopamine cells of the fish retina: a [^{3}H]-dopamine release study with isolated cell fractions. J Neurochem 50:1-6

Katayama-Kumoi Y, Kiyama H, Emson PC, Kimmel JR, Tohyama M (1985) Coexistence of pancreatic polypeptide and substance P in the chicken retina. Brain Res 361:25-35

Kikkawa U, Takai Y, Minakuchi S, Inohara S, Nishizuka Y (1982) Calcium-activated, phospholipid-dependent protein kinase from rat brain. Subcellular distribution, purification, and properties. J Biol Chem 257:13341-13348

Kiyama H, Katayama Y, Kimmel J, Steinbusch H, Smith D, Tohyama M (1985) Three dimensional analysis of retinal neuropeptides and amines. Brain Res Bull 15:155-166

Kolb H, Nelson R (1985) Functional neurocircuitry of amacrine cells in the cat retina. In: Gallego A, Gouras P (eds) Neurocircuitry of the retina. Elsevier, New York, pp215-232

Kolb H, Nelson R, Mariani A (1981) Amacrine cells, bipolar cells and ganglion cells of the cat retina: a Golgi study. Vision Res 21:1081-1114.

Marc R, Liu LS (1984) Horizontal cell synapses onto glycine-accumulating interplexiform cells. Nature 311:266-269

Mariani AP, Kolb H, Nelson R (1984) Dopamine-containing amacrine cells of rhesus monkey retina parallel rods in spatial distribution. Brain Res 322:1-7

Mitrofanis J, Vigny A, Stone J (1988) Distribution of catecholaminergic cells in the retina of the rat, guinea pig, cat, and rabbit: independence from ganglion cell distribution. J Comp Neurol 267:1-14

Negishi K, Kato S, Teranishi, T (1982) Dopaminergic cells of the river lamprey retina revealed by a histofluorescence study. Acta Histochem Cytochem 15:757-767

Negishi K, Kato S, Teranishi, T (1988) Dopamine cells and rod bipolar cells contain protein kinase C-like immunoreactivity in some vertebrate retinas. Neurosci Lett (in press)

Negishi K, Kato S, Teranishi T, Kiyama H, Katayama Y, Tohyama M (1985) So-called interplexiform cells immunoreactive to tyrosine hydroxylase or somatostatin in rat retina. Brain Res 346:136-140

Negishi K, Teranishi T (1987) A comparative study of immunoreactive cells in cold-blooded vertebrate retinas. J Physiol Soc Jpn 49:458

Negishi K, Teranishi T, Kato S (1983) A GABA antagonist, bicuculline, exerts its uncoupling action on external horizontal cells through dopamine cells in carp retina. Neurosci Lett 37:261-266

Nelson R (1982) AII amacrine cells quicken time course of rod signals in the cat retina. J Neurophysiol 47:928-947

Nishizuka Y (1984) Turnover of inositolphospholipids and signal transduction. Science 255:1365-1370

Nishizuka Y (1986) Studies and perspectives of protein kinase C. Science 233:305-312

Pourcho RG (1982) Dopaminergic amacrine cells in the cat retina. Brain Res 252:101-109

Pourcho RG, Goebel DJ (1985) A combined Golgi and autoradiographic study of (^{3}H)glycine-accumulating amacrine cells in the cat retina. J Comp Neurol 233:473-480

Saito N, Kikkawa U, Nishizuka Y, Tanaka C (1988) Distribution of protein kinase C-like immunoreactive neurons in rat brain. J Neurosci 8:369-382

Shoji M, Girard PR, Mazzei GJ, Vogler WR, Kuo JF (1986) Immunocytochemical evidence for phorbol ester-induced protein kinase C translocation in HL60 cells. Biochem Biophys Res Commun 135:1144-1149

Tachibana M, Kaneko A (1987) γ-Aminobutyric acid exerts a

local inhibitory action on the axon terminal of bipolar cells: evidence for negative feedback from amacrine cells. Proc Natl Acad Sci USA 84:3501-3505

Takai Y, Kishimoto A, Kikkawa U, Mori T, Nishizuka Y. (1979) Unsaturated diacylglycerol as a possible messenger for the activation of calcium-activated, phospholipid-dependent protein kinase system. Biochem Biophys Res Commun 90:1218-1224

Teranishi T, Negishi K (1988) Regional difference in the dendritic morphology of dopaminergic cells in carp retina. Dev Brain Res 39:9-17

Voigt T, Wässle H (1987) Dopaminergic innervation of AII amacrine cells in mammalian retina. J Neurosci 7:4115-4128

Wood JG, Girard PR, Maaei GJ, Kuo JF (1986) Immunocytochemical location of protein kinase C in identified neuronal compartments of rat brain. J Neurosci 6:2571-2577

MUSCARINIC INHIBITION OF ADENYLATE CYCLASE ACTIVITY IN RABBIT RETINAL CELLS.

H. Ghazi and N.N. Osborne
Nuffield Laboratory of Ophthalmology
Oxford University
Walton Str., Oxford
OX2 6AW England

Introduction

Intracellular cyclic AMP levels can be increased or decreased by specific agonists (Gilman, 1987). Levels of cAMP can be reduced by two possible mechanisms, either by inhibition of adenylate cyclase (AC) via the activation of specific guanine nucleotide binding proteins (G_i) (Gilman, 1987) or by increasing the rate of cyclic nucleotide degradation via activation of cAMP phosphodiesterase (Harden et al., 1986). It has been reported that cholinergic muscarinic receptors decrease cAMP production via both inhibitory mechanisms (Harden et al., 1986).

The purpose of the present study is to investigate whether muscarinic receptors negatively linked to cAMP production exist in the rabbit retina and to try to establish the mechanism by which such an effect might take place (i.e. via a G_i protein or by increase in phosphodiesterase activity). Acetylcholine is a retinal transmitter (Masland and Mills, 1979) and muscarinic receptors linked to the second messengers, namely the inositol phosphates/diacylglycerol have previously been described (Cutcliffe and Osborne, 1987; Osborne, 1988; Ghazi and Osborne, 1988a). Furthermore recent experiments on binding studies have suggested that at least two types of muscarinic receptors exist in the retina (Vanderheyden et al., 1988).

NATO ASI Series, Vol. H31
Neurobiology of the Inner Retina
Edited by R. Weiler and N.N. Osborne
© Springer-Verlag Berlin Heidelberg 1989

Results and Discussion

Figure 1 shows that the activity of AC in rabbit retinal homogenates is stimulated in a dose-dependent manner by forskolin (FKL), vasoactive intestinal peptide (VIP) and dopamine (DA) resulting in EC_{50} values of 53.8nM, 1.7nM and 0.16µM, respectively. These agonist-induced stimulations are mediated through a non-receptor mechanism in the case of FKL or via specific VIP and D_1 receptors (unpublished results and Ghazi and Osborne, 1988a). Stimulation of adenylate cyclase by VIP and DA is dependent upon the presence of GTP, but this is not the case for FKL (results not shown). Fig. 2 shows the effect of the cholinergic agonist, carbachol (CCh), on the basal activity of AC in retinal homogenates. CCh causes a reduction in the basal levels of cAMP by 10% maximally. However, when CCh is added 1 min prior to either DA (100µM), VIP (5µM) or FKL (5µM), it decreases their stimulations of AC activity by 44.8%, 77.8% and 80.5%, respectively (Fig. 2 and 3). This carbachol-induced inhibition is significantly greater than the 10% reduction, which it produces on its own.

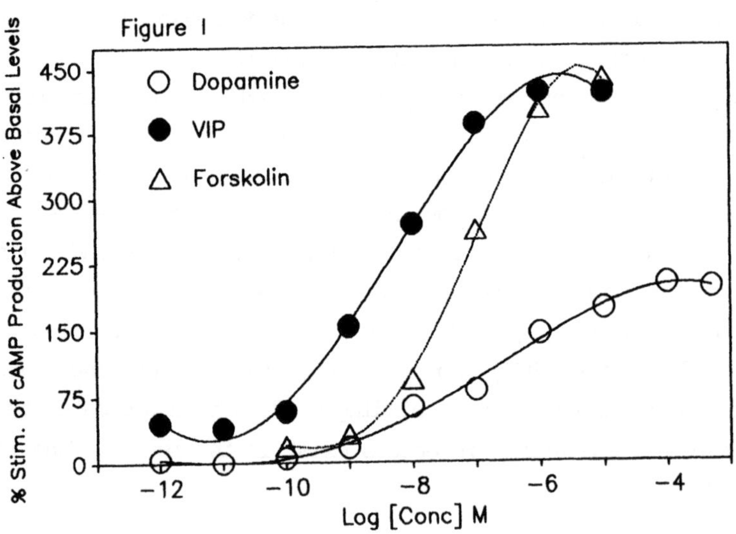

Fig. 1. Dose-response curves for the forskolin-, vasoactive intestinal peptide (VIP)- and dopamine-induced stimulation of cAMP formation in rabbit retinal homogenates. Results are % stimulation above control where n=3.

Figure 2

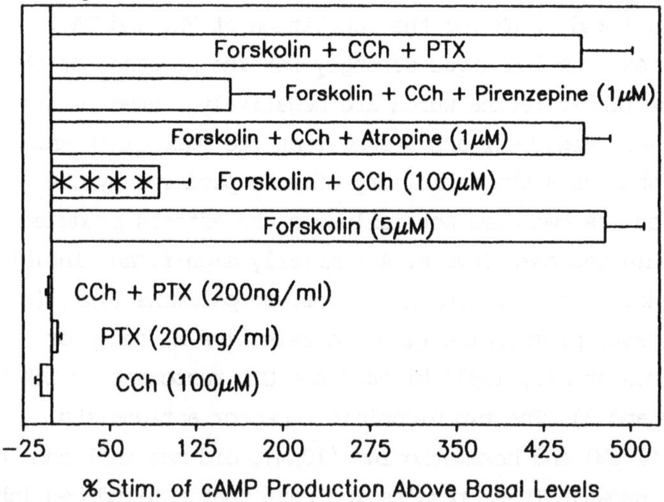

Figure 3

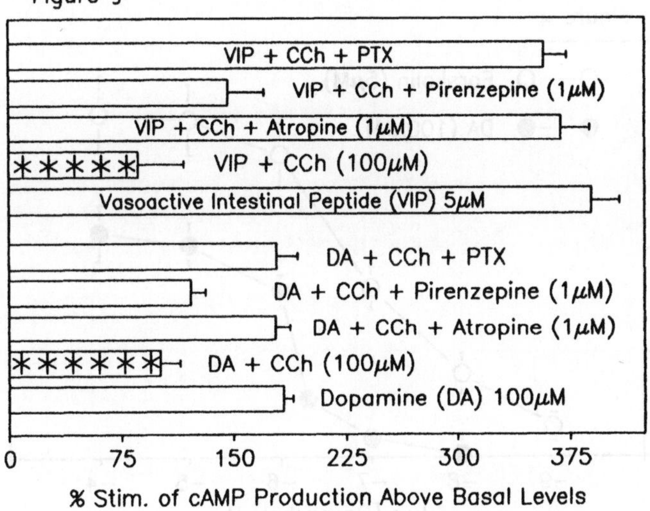

Fig. 2 and 3. Inhibitory effect of Carbachol (CCh) on forskolin (5μM)- and VIP (5μM)-induced stimulation of adenylate cyclase activity in rabbit retinal homogenates. Also shown are the effects of muscarinic antagonists (1μM), atropine, scopolamine and pirenzepine, and of pertussis toxin (PTX) on the CCh-evoked inhibitory responses. Results are expressed as % stimulation of cAMP formation over basal levels and means are ± SEM of 3-6 separate experiments.

Fig. 4 demonstrates that the CCh effect is dose-dependent producing IC_{50} values of 0.1 and 0.7μM for the inhibition of FKL and DA, respectively. These results argue strongly for the presence of cholinergic receptors in the rabbit retina which are negatively linked to the production of cAMP. Figures 2 and 3 show that the muscarinic antagonist, atropine (1μM) abolishes the inhibitory effects produced by CCh. Atropine (1μM) on its own does not influence the effects produced by FKL and VIP in stimulating cAMP levels. A similarly significant inhibition is obtained with another muscarinic antagonist, scopolamine (1μM) (not shown). In contrast, pirenzepine (1μM), a selective M_1 receptor antagonist (Watson et al., 1983) blocked the CCh responses by 16% on the average (Fig. 2 and 3). The two nicotinic receptor antagonists, d-tubocurarine (10μM) and hexamethonium (10μM), did not influence the CCh effect (not shown). These results indicate that CCh-induced inhibition of stimulated AC activity is via specific muscarinic-cholinergic receptors.

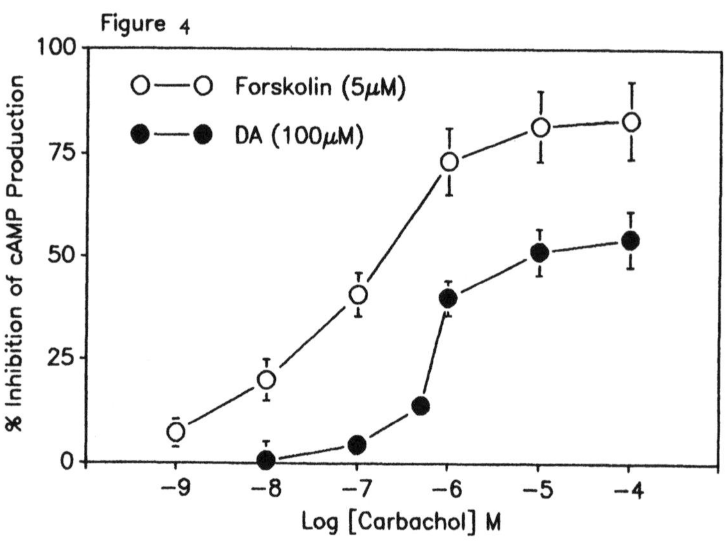

Fig. 4. Inhibitory effects of carbachol on forskolin (5μM)- and dopamine (DA) (100μM)-induced stimulation of cAMP production. The IC_{50} values for the CCh inhibition are 0.1 and 0.7μM for forskolin and DA, respectively. Points are mean values ± SEM of 3 separate experiments.

These receptors clearly have a high affinity for atropine and scopolamine, while pirenzepine has a negligible effect. The muscarinic receptors linked to inositol phosphates formation in rabbit retinal slices also have a strong affinity for atropine and scopolamine (Cutcliffe and Osborne, 1987), where pirenzepine has a similar effect (unpublished results). It may be concluded therefore that CCh induces stimulation of IPs formation through M_1 receptors, and that these receptors have a high affinity for pirenzepine, while the activation of M_2 receptors leads to inhibition of cAMP production with pirenzepine having little effect. This is consistent with studies which show that [^{3}H]pirenzepine binds selectively to muscarinic receptors in rat brain where the density is lowest in regions, such as the cerebullum, where the effect of GTP on agonist binding is greatest. The GTP effect on binding is thought to indicate the presence of receptors linked to AC (Watson et al., 1983; Luthin and Wolfe, 1984). To determine the importance of GTP in experiments on the retina, homogenates were first centrifuged and the pellet then resuspended in buffer. This process produced homogenates free of GTP. In these homogenates, CCh no longer inhibited the FKL-induced stimulation of cAMP. However, addition of 10μM GTP to the homogenates restored the inhibitory effect of CCh. Thus, the CCh-induced inhibition of cAMP production is dependent on the presence of GTP.

In order to establish that the CCh-induced inhibition of AC activity is mediated through a G_i-protein, rabbit retinas were incubated for 4 hours in the presence or absence of pertussis toxin (PTX) (200ng/ml) prior to the addition of CCh. Figures 2 and 3 show that in the PTX-treated tissue, CCh no longer induced a significant inhibition of either DA, VIP or FKL stimulated AC activity. PTX on its own did not significantly affect either basal (Fig. 2) or stimulated levels of AC activity (not shown). This is consistent with the known action of PTX which is thought to catalyze the adenosine diphosphate (ADP)-ribosylation of a guanine nucleotide regulatory protein involved in the receptor-mediated inhibition of AC (G_i) (Hazeki and Ui, 1981; Gilman, 1987). It is worth noting in this context that PTX does not affect the CCh-induced stimulation of inositol phosphates accumulation in rabbit retinal slices (Unpublished results).

The data presented provide additional proof that the muscarinic cholinergic receptors coupled to the inhibition of AC activity are of the M_2 subtype. The inhibition of AC activity which is GTP dependent, is independent of the agonist stimulating AC. However, the FKL- and

VIP-induced stimulations of AC are inhibited to a greater extent than the DA-evoked response. The functional significance of this finding is unclear at present. However, it is possible that a significant component of the dopaminergic response may occur in retinal cells devoid of muscarinic receptors. The present results suggest that cholinergic transmission may be involved in dual regulation of AC activity. One can envisage various scenarios where muscarinic receptor activation modulates a dopaminergic response. DA is known to alter receptive field size in carp horizontal cells (Cohen and Dowling, 1982; Neyton et al., 1982; Piccolino et al., 1982) and reduce conductance at gap junctions between horizontal cells of the turtle, through cAMP production. These effects would decrease if muscarinic (M_2) receptors were activated. GABA released from horizontal cells' neurites is inhibited by DA (Yazulla and Kleinschmidt, 1982). This effect would be counteracted should muscarinic (M_2) receptors be simultaneously activated. Dowling and Walting (1981), reported that a bright flickering red light on the carp retina increased cAMP production, and this was inhibited by the dopaminergic antagonist, haloperidol. In contrast, a non-flickering light caused a decrease in cAMP levels. The latter effect was not blocked by haloperidol. The possibility exists that under steady light conditions, not only DA but also acetlycholine (ACh) are released. The ACh could then act to reduce cAMP formation induced by endogenously released DA.

The fact that inhibition by CCh occurred in the presence of IBMX (1mM), a phosphodiesterase inhibitor, and in the absence of Ca^{2+}, strongly suggests that the decrease in measured cAMP levels is not due to activation of cAMP phosphodiesterase, an enzyme which depends on the presence of Ca^{2+} (Harden et al., 1986). Muscarinic inhibition of cAMP formation was reported in 1321N1 human astrocytoma cells to be due to activation of phosphodiesterase activity by Ca^{2+}-calmodulin protein (Harden et al., 1986). The significant PTX inhibition of the CCh response suggests that the mechanism by which CCh inhibits cAMP production is through the activation of a G_i protein. This mechanism is, however, probably not mediated through deactivation of the dissociated subunit, $G_{\alpha s}$. This is suggested by the fact that the FKL-induced response is not dependent on GTP. Therefore, there would be no $G_{\alpha s}$ in the membrane for binding to $\beta_i \delta_i$ produced from the muscarinic activation of G_i. The suggested mechanism of inhibition where the GTP-$G_{\alpha i}$ complex directly interacts with the catalytic subunit of AC is the most probable (Hildebrandt et al. 1984). The possibility still exists that the inhibition of DA- and

VIP-induced responses occurs through inactivation of $G_{\alpha s}$ by association with $\beta_i\delta_i$. Moreover, it has recently been shown that the dissociated $\beta_i\delta_i$ subunits can directly inhibit Ca^{2+}-calmodulin dependent AC (Katada et al., 1987). This cannot be the case in retinal homogenates since EGTA was included in the assay buffer. Thus this form of AC remains totally inactive. The dissociated $\beta\delta$ subunit has also been shown to inhibit AC directly as a result of direct protein-protein interaction, although the concentration of the $\beta\delta$ subunit required to induce inhibition appears to be excessively high for such an effect under physiological conditions (Katada et al.,1986).

In conclusion, two types of muscarinic receptor appear to exist in the rabbit retina: one coupled to the inositol phosphates/diacylglycerol pathway (M_1) and the other to the AC/cAMP (M_2) second messenger system. The mechanism of inhibition of cAMP production is mediated by G_i, while IPs/DG formation is not linked to a PTX-sensitive G-protein. The results presented here are in contrast to the findings in the rat retina where CCh did not reduce DA-stimulated cAMP production, but rather caused a slight enhancement of that response (Brown and Reitow, 1981). In the present study, activation of the M_2 receptors appears to inhibit, albeit to different degrees, AC activity stimulated by other substances. This is consistent with the proposed mechanism of action whereby the catalytic subunit is directly inhibited by $G_{\alpha i}$.

References

Brown JH, Rietow M (1981) Muscarinic-dopaminergic synergism on retinal cyclic AMP formation. Brain Res 215:388-392

Cohen JL, Dowling JE (1982) Effects of 6-hydroxy-dopamine on the spatial properties of carp horizontal cells. Soc Neurosci Abstr 8:132

Cutcliffe N, Osborne NN (1987) Serotonergic and cholinergic stimulation of inositol phosphates formation in the rabbit reitna. Evidence for the presence of 5-HT_2 and muscarinic receptors. Brain Res 421:95-104

Dowling JE, Watling KJ (1981) Dopaminergic mechanisms in the teleost retina. II. Factors affecting the accumulation of cyclic AMP in pieces of intact carp retina. J Neurochem 36:569-579

Ghazi H, Osborne NN (1988a) Agonist-induced stimulation of inositol phosphates in primary rabbit retinal cultures. J Neurochem 50:1851-1858

Ghazi H, Osborne NN (1988b) Agonist induced glycogenolysis in rabbit retinal slices and cultures. British J Pharmacol (In Press)

Gilman AG (1987) G Proteins: transducers of receptor-generated signals. Ann Rev Biochem 56:615-650

Harden TK, Tanner LI, Martin MW, Nakahata N, Hughes AR, Hepler JR, Evans T, Masters SB, Brown JH (1986) Characteristics of two biochemical responses to stimulation of muscarinic cholinergic receptors. TIPS Suppl:14-18

Hazeki O, Ui M (1981) Modification by islet-activating protein of receptor-mediated regulation of cyclic AMP accumulation in isolated rat heart cells. J Biol Chem 256:2856-2862

Hildebrandt JD, Codina J, Birnbaumer L (1984) Interaction of the stimulatory and inhibitory regulatory proteins of the adenylyl cyclase system with the catalytic components of cyc$^-$ S49 cell membranes. J Biol Chem 259:13178-13185

Katada T, Kusakabe K, Oinuma M, Ui M (1987) A novel mechanism for the inhibition of adenylate cyclase via inhibitory GTP-binding proteins. J Biol Chem 262:11897-11900

Katada T, Oinuma M, Ui M (1986) Mechanisms for inhibition of the catalytic activity of adenylate cyclase by guanine nucleotide-binding proteins serving as the substrate of islet-activating protein, pertussis toxin. J Biol Chem 261: 5215-5221

Luthin GR, Wolfe BB (1984) Comparison of [^{3}H]pirenzepine and [^{3}H]quinuclidinylbenzilate binding to muscarinic cholinergic receptors in rat brain. J Pharmacol Exp Ther 228: 648-655

Masland RH, Mills JW (1979) Autoradiographic identification of acetylcholine in the rabbit retina. J Cell Biol 83:159-178

Neyton J, Piccolino M, Gerschenfeld HM (1982) Dopamine (DA) and drugs that increase intracellular cyclic AMP decrease junctional communication between L-horizontal cells. Soc Neurosci Abstr 8:132

Osborne NN (1988) Muscarinic stimulation of inositol phosphate formation in rat retina: developmental changes. Vision Res 8:875-881

Piccolino M, Neyton J, Witkovsky P, Gerschenfeld HM (1982) gamma-Aminobutyric acid antagonists decrease junctional communication between L-horizontal cells in the retina. PNAS USA 79:3671-3675

Vanderheyden P, Ehinger G, Vanquelin G (1988) Characterization of M1- and M2-muscarinic receptors in calf retina membranes. Vision Res 8:247-250

Watson M, Yamamura HI, Roeske WR (1983) A unique regulatory profile and regional distribution of ^{3}H-pirenzepine binding in the rat provide evidence for distinct M_1 and M_2 muscarinic receptor subtypes. Life Sci 32:3001-3011

Yazulla S, Kleinschmidt J (1982) Dopamine blocks carrier-mediated release of GABA from retinal horizontal cells. Brain Res 233:211-215

SEROTONERGIC CELLS IN THE CHICKEN RETINA

T.J. Millar and I.G. Morgan[1]

Save Sight and Eye Health Institute, Department of Ophthalmology, University of Sydney, New South Wales, Australia.

1. Vision Research Institute, Research School of Biological Sciences, Australian National University, Canberra, Australia.

INTRODUCTION

Although sertonin-accumulating cells have been found in the retinas of most vertebrate species (Osborne '84; Redburn '85), it is only in non-mammals that <u>endogenous</u> serotonin (5HT) has been detected by immunohistochemical techniques (Weiler & Schutte '85a, '85b; Schutte & Weiler '87; Millar et al '87). Consequently there is controversy over the role of 5HT as a retinal neurotransmitter. Commonly, only 5HT- accumulating amacrine cells have been detected but in a few species 5HT-accumulating bipolar cells have also been reported (Osborne '85; Sandell & Masland '86). In the turtle and chicken, serotonin-like immunoreactivity (5HTLI) has been found in subpopulations of both amacrine and bipolar cells without preloading or other pharmacological manipulation of the retina (Weiler & Schutte '85a; Millar et al '87).

The synaptic connections of 5HT-accumulating amacrine cells have been described in several species after labelling of the amacrine cells with 5,6-dihydroxytryptamine (Ehinger & Holmgren '79; Adolph et al '79; Dowling et al '80; Holmgren-Taylor '83 '83), but the synapses of the bipolar cells have not yet been reported. In this study immunohistochemical techniques have been used to examine the 5HT cells of the chicken retina with particular attention being paid to the synaptic connections of the 5HTLI-bipolar cells.

METHODS

One week old chickens (Australorp X White Leghorn) under deep ether anaesthesia were transcardially perfused with 200ml of ice-cold fixative. The fixative consisted of 4% paraformaldehyde in 0.1M sodium phosphate buffer at pH7.4 for light microscopy or 4% paraformaldehyde and 0.05% glutaraldehyde in 0.1M sodium phosphate buffer at pH7.4 for electron microscopy. As the fixative reached the head region, the eyes were pierced several times with a 20 guage

NATO ASI Series, Vol. H31
Neurobiology of the Inner Retina
Edited by R. Weiler and N.N. Osborne
© Springer-Verlag Berlin Heidelberg 1989

needle and 5ml of the same fixative was syringed into the vitreous. Following perfusion the eyes were removed and the cornea, lens and vitreous removed. The eyecups were immersed for 1hr in the same fixative at 4°C. Retinae were separated from the eye-cup washed briefly in 0.1M sodium phosphate buffer at pH 7.4, immersed in 1% sodium borhydride for 30 min, followed by 4hr in several changes of sodium phosphate buffer, and then cryoprotected by immersing in 10%, 20% and 30% buffered sucrose (60min in each). For light microscopy frozen sections were cut on a cryostat and processed for immunohistochemistry using an indirect fluoroscein-labelled double antibody technique as described previously (Millar et al '87). For electron microscopy retinas, vibratome sections were cut (50um), and processed for immunohistochemistry using a peroxidase anti-peroxidase method. The vibratome sections were post-fixed for 1h in 2% buffered osmium tetroxide, dehydrated through alcohols, embedded in araldite. Ultrathin sections were cut using a diamond knife and viewed in a Zeiss 109 or Hitachi 600 electron microscope.

RESULTS

<u>Light microscopy</u>. Immunofluorescence was observed in a large number of bipolar cells located towards the centre of the inner nuclear layer (INL) and a sparse population amacrine cells also situated towards the centre of the INL (Fig 1). The intensity of the fluorescence in the bipolar cells was much weaker than in the amacrine cells. Weakly fluorescent processes emanating from the bipolar cells coursed towards the OPL but their plexuses in the OPL could not be clearly seen. Immunoreactive bipolar cell processes projecting towards the inner plexiform layer (IPL) were rarely seen. In the IPL there were 2 bands of fluorescence - one located at sublamina 1 and the other at sublamina 4, and dendrites from 5HTLI amacrine cells could be traced to both layers. 5HTLI bipolar cell axons were only rarely able to be traced into the IPL, but their lamination could not be discerned.

Figure 1. The two types of 5HTLI cells are shown above. Small bipolar cells (b) and large amacrine cells (a) were detected. Scale = 100μm

<u>Electron microscopy</u>. The diaminobenzidine reaction product in the somas of 5HTLI-bipolar and -amacrine cells was distributed throughout the cytoplasm and was not associated with any particular organelles. The nucleus and mitochondria showed no immunoreactivity.

In the OPL, 5HTLI bipolar cell dendritic terminals were very weakly stained, and received synaptic input from both cone and rod photoreceptors (Figs 2&3). Cone photoreceptor terminals are stratified in three layers within the OPL, and 5HTLI bipolar cell dendrites were postsynaptic at all levels. All contacts observed were at basal or peripheral sites and not associated with synaptic ribbons. Contacts with horizontal cells were mainly either onto the soma of the horizontal cells or onto an initial part of a dendrite and characterized by three or four electron dense vesicles on the horizontal cell side and membrane thickening and vesicle accumulation on the bipolar side (Figs 4&5).

In the IPL, bipolar cell terminals are packed with synaptic vesicles and contain a synaptic ribbon. Amacrine cell terminals lack a synaptic ribbon and are less densely packed with synaptic vesicles than bipolar cells. Ganglion cell dendritic terminals have very pale cytoplasm, very few vesicular-like bodies and commonly contain mitochondria. Because axons of both 5HTLI bipolar and dendrites 5HTLI amacrine cells projected to the IPL, it was necessary to examine serial sections to confirm the identity of many of the synaptic terminals. Identification of 5HTLI amacrine cell profiles is problematic because it relys on the absence of a ribbon synapse and therefore the following strategy was used. Since it was observed from serial sections that after seeing a ribbon in a 5HTLI process, another was always encounted within the next 4 sections (40 examples were studied), profiles were identified as belonging to amacrine cells if no ribbon was seen in 8 consecutive sections.

In the IPL, no obvious differences were noted in the synaptic connections of 5HTLI terminals in the two different sublaminae. A common feature was synapses between 5HTLI processes which took a variety of forms. Non-ribbon synapses between 5HTLI bipolar cell axons were frequently encountered (Figs 6-8). These were characterized by membrane thickening on both sides of the contact and no accumulation of synaptic vesicles. Serial sections of non-ribbon synapses between 5HTLI profiles where one element was identified as a bipolar cell terminal, always showed that the other element was also a 5HTLI bipolar cell axon (15 examples studied). Post-synaptic to 5HTLI ribbon synapses were either a pair of non-immunoreactive amacrine cell dendrites, a ganglion cell dendrite and an amacrine cell dendrite, a 5HTLI profile and a non-immunoreactive amacrine cell dendrite (Fig 9), or rarely, a pair of 5HTLI profiles. Serial sectioning of 5 such 5HTLI showed that in each case the postsynaptic element was the terminal of a 5HTLI amacrine cell. One example was found where the postsynaptic elements were a 5HTLI dendrite and a non-immunoreactive amacrine cell dendrite, and the 5HTLI amacrine cell terminal synapsed onto the non-immuno-reactive amacrine cell dendrite. 5HTLI bipolar axons were postsynaptic to amacrine cells. At these synapses there was accumulation of vesicles on the presynaptic side and the membrane of the 5HTLI axon was thickened in the region of contact.

The other type of contact between 5HTLI profiles was between two amacrine cells where the region of contact was characterized by membrane thickening on both sides and no obvious vesicle accumulation (Fig 10). It was felt that such contacts were common, but only two were followed through 8 serial sections. 5HTLI amacrine cells were both pre- and postsynaptic to non-immunoreactive amacrine cells and presynaptic to ganglion cell dendrites. In addition to the types of synapses already mentioned, a rare example is shown in Fig 11 in which a non-immunoreactive bipolar cell makes a ribbon synapse onto an unidentified 5HTLI profile.

DISCUSSION

This study confirms that the small 5HTLI cell bodies in the middle of the INL are a population of bipolar cells. Photoreceptor input to 5HTLI bipolar cells always occurred at basal or peripheral synapses and there was no evidence of selective input by one particular type of photoreceptor. The response of 5HTLI bipolar cells to kainic acid (Millar et al '88) suggests that they are OFF-centre cells which is consistent with the suggestion that OFF-centre bipolar cells

Figures 2-5. Bipolar dendrites received synaptic input from both cone (3) and rod (4) photoreceptors at basal sites and were not associated with synaptic ribbons. Input from horizontal cell dendrite (5) and cell body (6). Scale 250nm

Figures 6-10. Synapses in the IPL. Figs 6-8. Serial section showing a non-ribbon synapse (6;arrow) between two 5HTLI bipolar cells. Synaptic ribbons in 7&8 arrowed. Fig. 9 shows a 5HTLI bipolar ribbon synapse (arrow) onto 5HTLI amacrine cell dendrite and a non-immunoreactive amacrine cell dendrite. The 5HTLI synapses onto the non-immunoreactive amacrine cell (double arrow). Fig. 10 shows a contact between two 5HTLI amacrine cells with membrane thickening (White arrow). Synaptic input from non-immunoreactive amacrine cell (black arrow). Scale 250nm

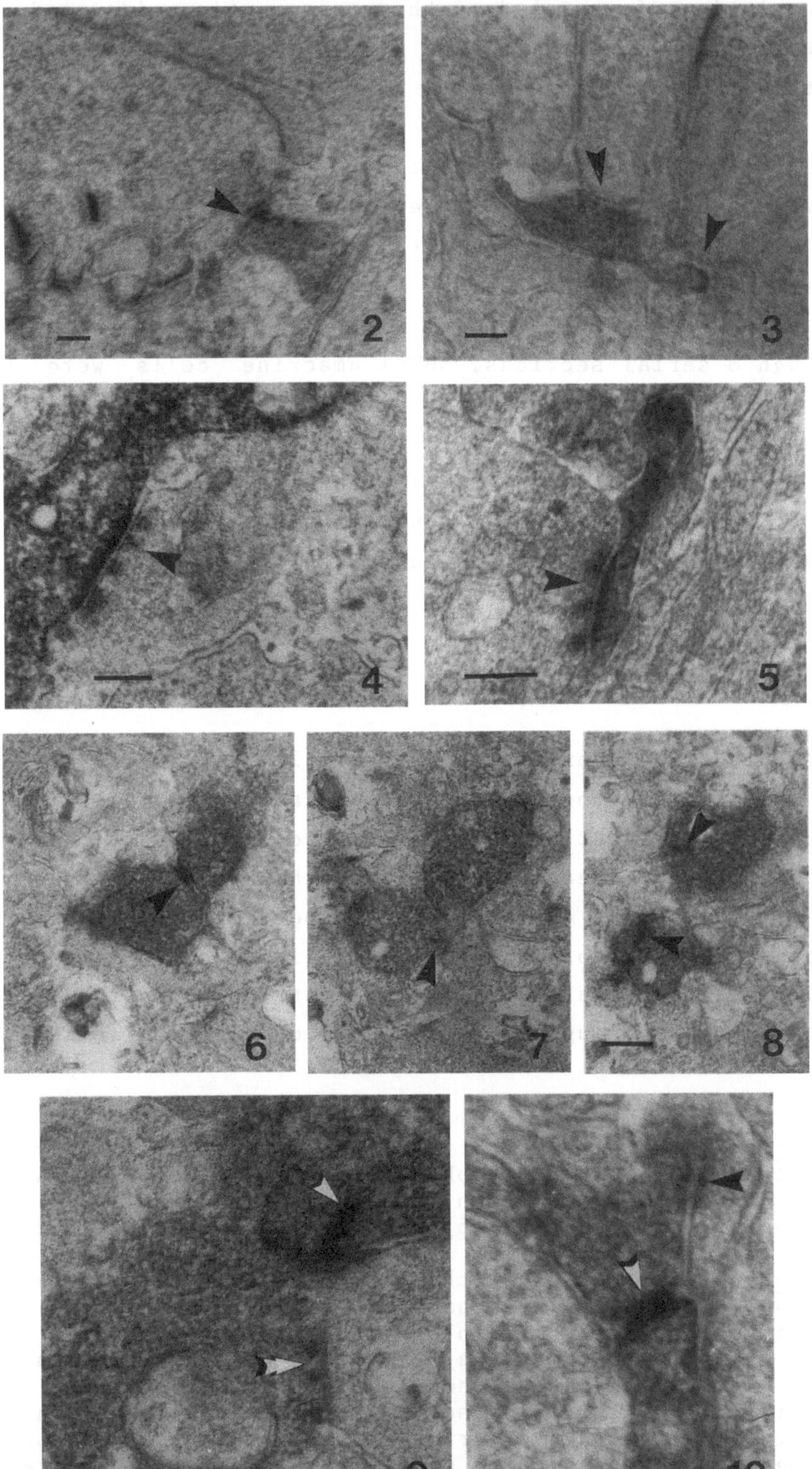
2
3
4
5
6
7
8
9
10

are the postsynaptic elements at basal wide-cleft junctions (Raviola & Gilula '75; Famiglietti & Kolb '76; Stell et al '77; Kolb '79; Nelson et al '81), although this does not appear to be the universal rule (Lasansky '78; Nelson & Kolb '83; Sakai & Naka '83; Saito et al '83 '85). Contacts between 5HTLI bipolar cells and horizontal cells both in the outer part of the OPL and close to the soma of the horizontal cells is indicative of lateral input to 5HTLI bipolar cells. Such input is widely believed to generate antagonistic surrounds of bipolar cells.

In the IPL, 5HTLI bipolar cells were presynaptic to unlabelled amacrine cells at typical synaptic dyads, both in sublaminae 1 and 4 of the IPL. Since the evidence to date suggests that these bipolar cells are a single population, then synapsing in both OFF- (sublamina 1) and ON-sublaminae (sublamina 4) was unexpected. However, a similar finding in the turtle has led to the idea that OFF-bipolar cells may synapse in OFF- or both ON- and OFF-layers of the IPL whereas the ON-centre bipolars synapse only in the ON-sublamina (Weiler '81). 5HTLI bipolar cells were also presynaptic to ganglion cell dendrites and 5HTLI amacrine cell dendrites. A special feature of 5HTLI bipolar cells was their apparent networking with each other through non-ribbon contacts which showed membrane thickening. In the carp retina electrical reciprocal bipolar coupling has been reported (Saito & Kujiraoka '88) and the 5HTLI bipolar contacts described in this paper may be the morphological equivalent.

5HTLI amacrine cells received input from both non-immuno- reactive and 5HTLI-bipolar cells and their output was to non-immunoreactive amacrine cells and ganglion cells. Closely apposed amacrine cell dendrites also appeared to form membrane specializations, but it is not clear whether these were true synapses.

These experiments have shown that an important feature of the organization of 5HTLI-bipolar and amacrine cells is their apparent networking. Although little is known about this possibility in bipolar cells, some amacrine cell populations, such as cholinergic amacrine cells, synapse with each other (Millar & Morgan '87; Millar et al.'87). These networks are not a general feature of amacrine cell organization because other defined amacrine cell populations do not form networks (Ishimoto et al '86). The physiological function of such networking is worthy of further investigation.

REFERENCES

Adolph A, Dowling JE, Ehinger B (1980) Monoaminergic neurons of the mudpuppy retina. Cell Tiss Res 210:269-282.

Ehinger B, Holmgren I (1979) Electron microscopy of the indoleamine-accumulating neurons in the retina of the rabbit. Cell Tiss Res 197:175-194.

Famiglietti EV, Kaneko A, Tachibana M (1977) Neuronal architecture of 'On' and 'Off' pathways to ganglion cells in the carp retina. Science 198:1267-1269.

Famiglietti EV, Kolb H (1976) A structural basis for 'ON'- and 'OFF'-center responses in retinal ganglion cells. Science 194:193-195.

Holmgren-Taylor I (1982) Electron microscopical observations on the indoleamine-accumulating neurons and their synaptic connections in the retina of the cat. J Comp Neurol 208:144-156.

Holmgren-Taylor I (1983) Synaptic organization of the indoleamine-accumulating neurons in the cyprinid retina. Cell Tiss Res 229:317-335.

Ishimoto I, Millar TJ, Chubb IW, Morgan IG (1986) Somatostatin-immunoreactive amacrine cells of the chicken retina: retinal mosaic, ultrastructural features, and light-driven variations in peptide metabolism Neurosci 17:1217-1233.

Lasansky A (1978) Contacts between receptors and electrophysiologically identified neurones in the retina of the larval tiger salamander. J Physiol(Lond) 285:531-542.

Millar TJ, Ishimoto I, Chubb IW, Epstein ML, Johnson CD, Morgan IG. (1987) Cholinergic amacrine cells of the chicken retina: a light and electron microscopic immunocytochemical study. Neurosci 21:725-743.

Millar TJ, Morgan IG (1987) Cholinergic amacrine cells in the rabbit retina synapse onto other cholinergic amacrine cells. Neurosci Lett 74:281-285.

Millar TJ, Winder C, Ishimoto I, Morgan IG (1988) Putative

serotoninergic bipolar and amacrine cells in the chicken retina. Brain Res 439:77-87.
Nelson R, Kolb H (1983) Synaptic patterns and response properties of bipolar and ganglion cells in the cat retina. Vision Res 23:1183-1195.
Osborne NN (1984) indoleamines in the eye with special reference to the serotonergic neurones of the retina. In: Osborne NN, Chader GJ (eds) Progress in Retinal Research. Vol 3. Oxford, Pergammon Press, p61.
Osborne NN (1985) Interplexiform, Horizontal and bipolar- like cells of the rabbit retina take up exogenous serotonin during early developmental stages. Int J Dev Neurosci 3:643-646.
Redburn DA (1985) Serotonin neurotransmitter systems in vertebrate retina. In: Morgan WW (ed) Retinal Transmitters and Modulators: Models for the Brain. Vol.II. Florida: CRC Press, p107.
Raviola E, Gilula NB (1975) Intramembrane organization of specialized contacts in the outer plexiform layer of the retina. A freeze fracture study. J Cell Biol 65:192-222.
Saito T, Kujiraoka T (1988) Characteristics of bipolar-bipolar coupling in the carp retina. J Gen Physiol 91:275-287.
Saito T, Kujiraoka T, Yonaha T (1983) Connections between photoreceptors and horse-radish peroxidase-injected bipolar cells in the carp retina. Vision Res 23:353-363.
Saito T, Kujiraoka T, Yonaha T, Chino Y (1985) Re-examination of photoreceptor-bipolar connectivity patterns in carp retina: HRP-EM and Golgi-EM studies. J Comp Neurol 236:141-160.
Sakai H, Naka K-I (1983) Synaptic organization involving receptor, horizontal and on- and off-center bipolar cells in catfish retina. Vision res. 23:339-351.
Sandell JH, Masland RH (1986) A System of indoleamine-accumulating neurons in the rabbit retina. J Neurosci 6:3331-3347.
Schutte M, Weiler R (1987) Morphometric analysis of serotonergic bipolar cells in the retina and its implications for retinal image processing. J Comp Neurol 260:619-626.
Stell, WK, Ishida, AT, Lightfoot DO (1977) Structural basis for on- and off-center responses in retinal bipolar cells. Science 198:1269-1271.
Weiler R (1981) The distribution of center-depolarizing and center-hyperpolarizing bipolar cell ramifications within the inner plexiform layer of the turtle retina. J Comp Physiol 144:459-464.
Weiler R, Schutte M (1985a) Morphological and pharmacological analysis of putative serotonergic bipolar and amacrine cells in the retina of a turtle, Pseudemys scripta elegans. Cell Tiss Res 241:373-382.
Weiler R, Schutte M (1985b) Kainic acid induced release of serotonin from OFF-bipolar cells in the turtle retina. Brain Res 360:379-383.

PEPTIDE AND INDOLEAMINE MODULATION OF SPATIAL RESPONSE PROPERTIES IN TURTLE RETINAL GANGLION CELLS

Alan R Adolph
Eye Research Institute and
Harvard Medical School
20 Staniford St
Boston MA 02114, USA

Introduction

A number of peptides have been localized in the turtle retina by histochemical techniques. Among these, enkephalin (metENK), somatostatin (SS), and neurotensin (NT) are found in distinct subsets of amacrine cells (AC) (Zucker and Adolph, 1987, 1988; Weiler and Ball, 1984; Eldred and Karten, 1983). The AC processes, apparently confined within the IPL, have trilaminar (metENK,SS) or diffuse (NT) distributions. These include laminae in the outer and inner IPL, a feature which may be functionally significant relative to the segregation of ON- and OFF-pathways. There are histochemical and localization differences in the characteristics of the peptides within the retina. Ten percent of metENK-AC contain colocalized GABA, based on GAD-IR light microscopy and GABA-IR, EM immunohistochemistry (Zucker and Adolph, 1988). NT-AC have a specific uptake mechanism for, and may be colocalized with, glycine (Weiler and Ball, 1984). There is a distinct distribution of SS-AC bodies:they lie only within the visual streak region;their processes are, by contrast, widespread over the retina (Adolph and Zucker,1988). Serotonin (5HT) appears to be localized, in addition to a subset of AC, to certain bipolar cells (BC) (Witkovsky et al, 1984; Weiler and Schutte, 1985). These 5HT-BC have processes always stratifying in the outer IPL with varying degrees of inner IPL stratification, leading to the inference that they are OFF-BC. Weiler finds that the 5HT-BC but not the 5HT-AC have kainate-type receptors (presumably for inputs from photoreceptors and other BC).

The anatomy and histochemistry described above are essential aspects of any attempt to define functional circuitry within the IPL. Correlated physiological information has been sought using a functional pharmacological approach (Adolph 1985, 1988; Ariel and Adolph, 1985). I present here some results on the actions of the aforementioned peptides and indoleamines. Substances were applied to eyecups by superfusion, or focal application using pressure ejection or microiontophoresis from micropipettes, or combinations of these methods. The spatial properties of ganglion cell (GC) receptive fields (RF) were measured by extracellular spike responses to drifting grating and conventional spot/annulus stimulus patterns. The grating paradigm is especially useful for testing the effects of pharmacological manipulations of retinal function. It can give information on the RF center and surround amplitudes and extents, and on trigger features such as orientation sensitivity and directional selectivity, within a time scale appropriate for the changing physiological state of the retina during pharmacological modification.

Results

The action of D-al-met-enkephalinamide (DALA), a hydrolysis-resistant analog of metENK, was tested by focally applying it within the RF of the extracellularly recorded GC using pressure ejection from a micropipette. There is a latency of several seconds before the apparent response to the exogenous opiate, which is an increase in both light-evoked and spontaneous activity. Some part of the latency is a fixed diffusion time, another part decreases as dose increases. The dose-related duration of effects may, to some extent also depend on diffusion time since the agonist is not being chemically inactivated. DALA changes the transient OFF-response (control conditions) to sustained ON- and OFF-responses. These changes in temporal pattern suggest that a variety of pathways influencing the response have been modified by exogenous opiate. Changes seen in the spatial transfer function (STF) confirm this expectation. DALA enhances responses to high spatial

NATO ASI Series, Vol. H31
Neurobiology of the Inner Retina
Edited by R. Weiler and N.N. Osborne
© Springer-Verlag Berlin Heidelberg 1989

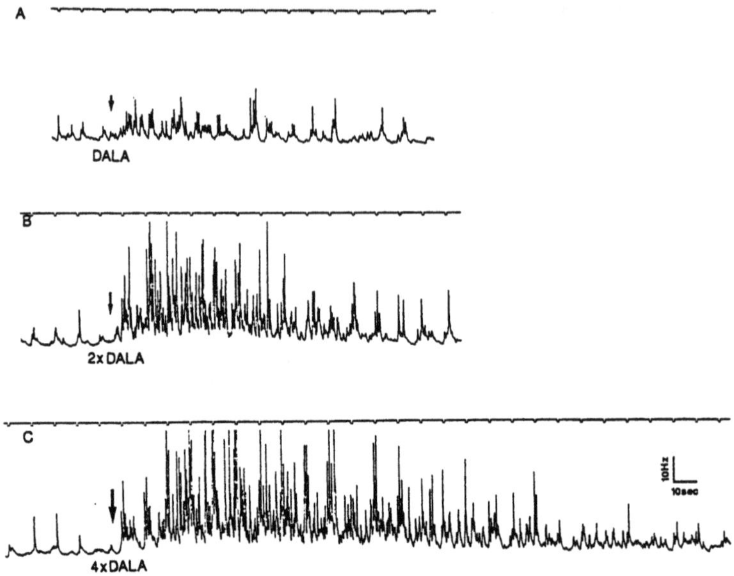

Dose-response series of GC spike response rate to pressure-ejected d-ala-met-enkephalinamide (DALA). Upper trace in all records, shutter signal for 500um, -2log spot stimulus centered on GC receptive field. Lower trace, GC spike rate (output of f/V converter). DALA ca. 400uM in ejection pipette. Eject pulses of 5psi for A) 75msec, B) 150msec, C) 300msec. DALA causes spontaneous activity in addition to increased light-evoked responses. During maximal effect, light responses change from primarily off-transient to on-off transient and ultimately a sustained-like pattern. There are substantial increases in spontaneous activity at the higher doses; the duration of effects is correlated with dose.

frequencies and slightly attenuates low frequencies. One parsimonious explanation of these STF changes suggests primarily a narrowing of RF center. On the other hand, naloxone which is a metENK antagonist, primarily attenuates high spatial frequencies, which may be interpreted as increasing the RF center width. Finally, the response-modulating actions of DALA are greatly enhanced with concurrent application of the GABA-antagonist, picrotoxin. This finding suggests that some part of the opiate action is mediated by GABA effects. It should be recalled that histological results indicate a high degree of synaptic interaction between metENK and GABA: 1) most often, metENK processes are presynaptic to GABA ones, occasionally with the GABA process then being presynaptic to a GC, 2) there is a subset of metENK/GABA colocalized AC synapsing on GC, and 3) there are individual GABA and metENK endings on GC (Zucker and Adolph, 1988). Perhaps the narrowed RF center in response to metENK is due to enhanced, GABA-mediated surround inhibitory effects while the increased spontaneous and light-evoked GC activity results from direct metENK action. Picrotoxin would antagonize only the GABA actions, while naloxone antagonizes both indirect metENK effects via GABA and direct effects. Concerning the receptor for the opiate actions, morphine was tested in a conventional (non-STF) pharmacological paradigm and had analogous actions to exogenous metENK and DALA. This indicates that the receptor may be a mu-opiate type. Naloxone, also an antagonist at delta, kappa, and epsilon-opiate receptors, is most effective for mu-type.

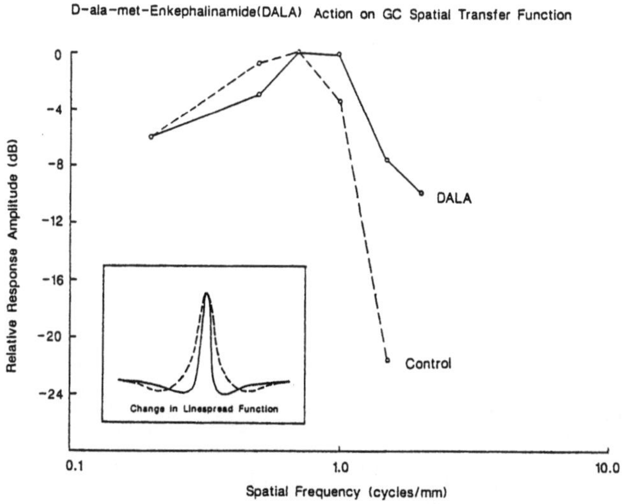

Modification of GC spatial properties by the opiate agonist DALA, an inactivating enzyme-resistant analog of met-enkephalin. Spatial transfer function between drifting square-wave grating stimulus and averaged spike rate response. DALA, applied focally near the recorded GC by pressure ejection, enhances responses to high spatial frequencies and slightly attenuates low frequency responses. These changes, which are the converse of those found with the superfused opiate antagonist, Naloxone, suggest a narrowing of RF center width and inhibitory surround extent (inset sketch).

Neurotensin (NT) produces a slow-onset, sustained excitation of GC activity. This is bimodally mediated via a direct postsynaptic action and an indirect pathway. The latter is blocked by cobalt superfusion. Spontaneous activity stimulated by added [K+] during cobalt blockade, is modulated by NT, which indicates the direct action (Adolph and Bruun,1987). Somatostatin (SS) was also found to have response- and spontaneous activity-enhancing actions on the turtle GC (Adolph and Zucker,1988). Just as with metENK, SS effects are dose-dependent and convert transient light responses to sustained ones as dose is increased. The peculiar spatial localization of SS-containing AC bodies within the visual streak, and the widespread distribution of SS processes, suggests that functional properties dependent on pathways containing SS-AC will be spatially heterogeneous. Witkovsky et al (1984) have also shown a non-uniform distribution of another putative retinal neurotransmitter in turtle. They found that there is an increased density, disproportionate to the general neuronal density increase, of indoleamine within the visual streak.

In contrast to the peptides, the predominant action of the indoleamine, serotonin (5HT), is inhibition of GC activity (Adolph,1986). The inhibition is mediated by direct postsynaptic action on GC (50%) and indirectly via a (cobalt-sensitive) transynaptic pathway (50%). Several indoleamine agonists are effective in modifying GC activity: 5-methoxy-dimethyl-tryptamine (5MeO-DMT), methoxyphenylpiperazine (MOPP), LSD, and cypro-heptadine (CYPRO). 5MEO-DMT and MOPP act as 5HT agonists, with the former somewhat less potent, and the latter somewhat more potent, than the endogenous indoleamine (5HT). LSD can act as an agonist or an antagonist, depending upon receptor type and related functional activity in pathways converging on the subject GC.

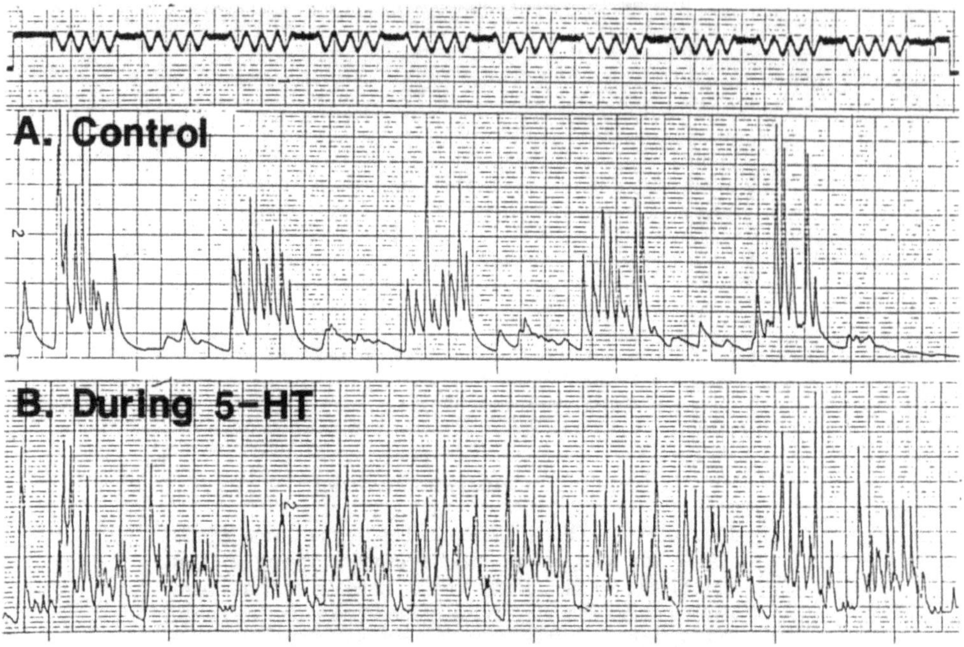

Modulation of Directional Selectivity by 5-HT

A. Directionally selective(DS) GC, whose static response to a spot stimulus was directly inhibited by 5-HT, responding under control conditions to a 1.06 c/mm sinewave grating swept alternately in the preferred then null directions. Upper trace, stimulus marker(photodetector); lower trace, ratemeter record of GC spike response. B. Responses during 5-HT showing some inhibition of preferred response and marked increase in null response, i.e. a loss of DS property. Note that ratemeter sensitivity in B is twice that in A. The direct 5-HT inhibitory effect is reflected by the reduction of the preferred response; loss of DS during 5-HT suggests a more potent action on transynaptic paths involved in this trigger feature.

Although there is no definitive correlation yet available of functional action with (binding-study) defined receptor types, the classes defined in CNS studies seem appropriate, based on our present functional findings. 5HT and the exogenous agonists may be acting via 5HT1 receptors: 5Meo-DMT acts at 5HT1A, and MOPP at 5HT1B, types. The other receptor type which appears functional in this retina is the 5HT2 type, at which LSD and CYPRO act as 5HT antagonists, and neither 5MeO-DMT nor MOPP are agonists. The distinction between 5HT1 and 5HT2 receptors is also interesting in terms of second messengers involved in the neuronal response to the agonists. 5HT1 receptors are associated with a cAMP system, 5HT2 with phosphotidylinositol.

Modification of functional properties of GC by exogenous 5HT is illustrated in two examples. In one, the directional selectivity (DS) exhibited during control conditions is eliminated during 5HT application. A possible basis for this action may be suppression by 5HT inhibition of the GABA-AC source of the time-delayed, asymmetric inhibitory surround which underlies DS (Ariel and Adolph, 1985). In another example, the spatial tuning of a GC is modified by 5HT: STF high spatial frequencies are relatively attenuated compared to control, suggesting a broadened RF center, the result of reduced surround (GABA-mediated) inhibition.

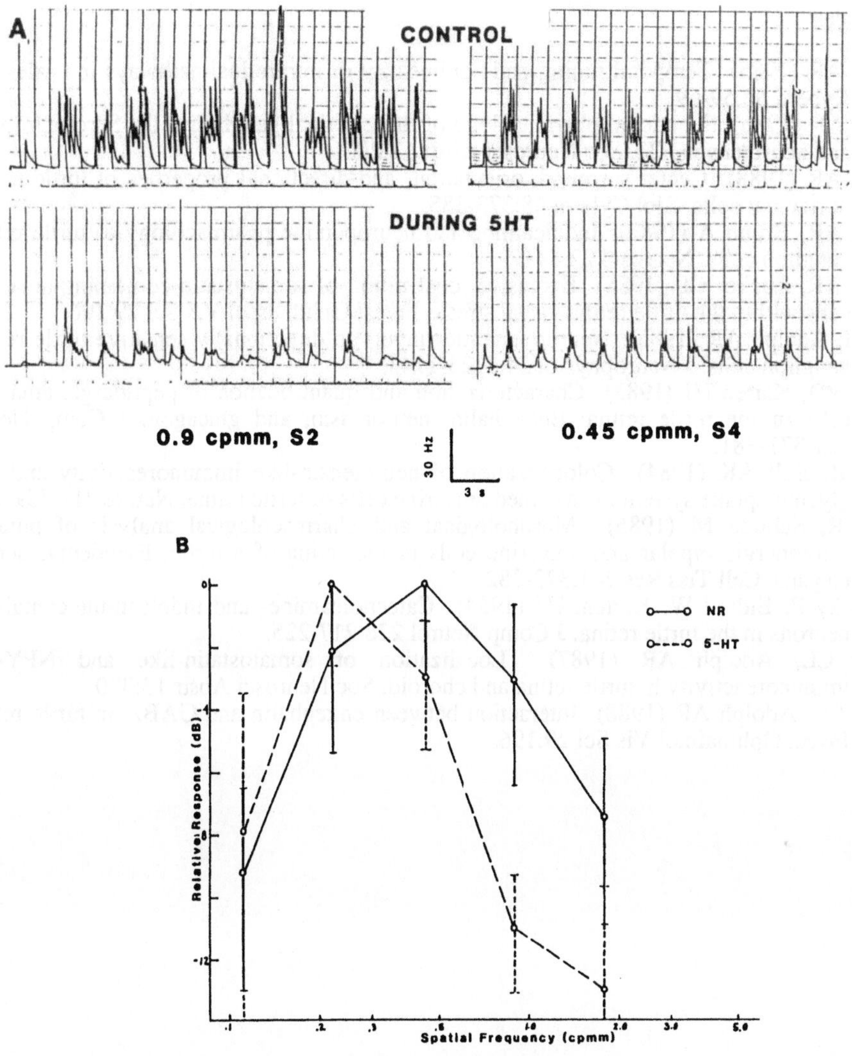

Modulation of GC Spatial Tuning by 5-HT

A. Responses of a non-directionally selective GC to antipodal sweeps of a sinewave grating, at two spatial frequencies(0.9 c/mm and 0.45 c/mm; constant temporal frequency), under control conditions and during retinal superfusion with 5-HT. Upper trace in each panel, stimulus marker indicating movement of stripes past a fixed point in retina, first one direction then other. Lower trace in each panel, ratemeter records of GC firing. B. GC spatial frequency response functions. Mean of peak-to-peak spike rate, normalized to maximum values, as a function of grating spatial frequency. Solid line, control conditions(NR); broken line, during 5-HT. The relative inhibition by 5-HT of the spatial freq. response function at 0.9 c/mm is greater than at 0.45 c/mm. Reduced high-freq. response indicates broadened receptive field center, perhaps as a result of weakened surround inhibition. The RF surround diameter is relatively unaffected as suggested by the insignificant differences in the low-freq. response shape.

References

Adolph AR (1985) Temporal tuning and nonlinearity of intraretinal pathways in turtle. Biol Cybern 52:59-69.

Adolph AR (1986) Functional pharmacology of turtle retinal ganglion cells: Serotonin(5HT) and neurotensin(NT). Soc Neurosci Abstr 12:200.

Adolph AR (1988) Center-surround, orientation, and directional properties of turtle retinal horizontal cells. Biol Cybern 58:373-385.

Adolph AR, Bruun A (1987) Indoleamine and neuropeptide pharmacology of turtle retinal ganglion cells. Neurosci 22:S412.

Adolph AR, Zucker CL (1988) Exclusive localization of somatostatin-containing amacrine cells within the turtle retina visual streak. Invest Ophthalmol Vis Sci 29:196.

Ariel M, Adolph AR (1985) Neurotransmitter inputs to directionally sensitive turtle retinal ganglion cells. J Neurophysiol 54:1123-1143.

Eldred WD, Karten HJ (1983) Characterization and quantification of peptidergic amacrine cells in the turtle retina: Enkephalin, neurotensin, and glucagon. J Comp Neurol 221:371-381.

Weiler R, Ball AK (1984) Colocalization of neurotensin-like immunoreactivity and 3H-glycine uptake system in sustained amacrine cells of turtle retina. Nature 311:759-761.

Weiler R, Schutte M (1985) Morphological and pharmacological analysis of putative serotonergic bipolar and amacrine cells in the retina of a turtle, Pseudemys scripta elegans. Cell Tiss Res 241:373-382.

Witkovsky P, Eldred W, Karten, HJ (1984) Catecholeamine- and indoleamine-containing neurons in the turtle retina. J Comp Neurol 228:217-225.

Zucker CL, Adolph AR (1987) Localization of somatostatin-like and NPY-like immunoreactivity in turtle retina and choroid. Soc Neurosci Abstr 13:320.

Zucker CL, Adolph AR (1988) Interaction between enkephalin and GABA in turtle retina. Invest Ophthalmol Vis Sci 29:196.

MORPHOLOGIES OF SOMATOSTATIN-IMMUNOREACTIVE NEURONS IN THE RABBIT RETINA

*Dennis W. Rickman**
Department of Anatomy and Cell Biology
University of Southern California School of Medicine
Los Angeles, California 90033
U.S.A.

Nicholas C. Brecha
Department of Anatomy and Cell Biology and Department of Medicine
U.C.L.A. School of Medicine and
Veterans Administration Medical Center
Los Angeles, California 90073
U.S.A.

The development of antibodies directed to putative transmitters, or their biosynthetic enzymes, has led to the identification and characterization of several specific cell populations in the vertebrate retina (Brecha, 1983; Ehinger and Dowling, 1987). In addition to defining the distribution of putative transmitter-containing cells within the retina, these studies have identified morphological substrates which might underlie the presumed functional roles of these transmitters in visual processing. These substrates include not only the regional distribution of the identified cell population, but also the morphologies of its individual cells, including both somal position and the laminar distribution of cellular processes within the plexiform layers.

Neurons of the ganglion cell layer (GCL) of the vertebrate retina consist of both ganglion cells, whose axons project upon central visual targets, and a significant number of displaced amacrine cells, intrinsic retinal neurons whose processes are distributed to the inner plexiform layer (IPL) (Ramon y Cajal, 1892). Ganglion cells can be identified experimentally. One commonly used technique exploits the retrograde transport of fluorescent dyes or horseradish peroxidase from the optic nerve, optic tract or central visual nuclei. Nissl counterstaining of the retina reveals a population of neurons which do not contain the retrogradely transported marker. The majority of these cells are of a small size, have a high ratio of nucleoplasm to cytoplasm, and often have a distinctive perinuclear clumping of basophilia (Vaney, 1980). These cells are likely to be displaced amacrine cells. A second approach utilizes optic nerve transection which results, after an appropriate survival period, in the retrograde degeneration of most, if not all, ganglion cells. The surviving cells, which can be visualized with a Nissl counterstain, are probably displaced amacrines. These techniques have demonstrated that displaced amacrine cells account for approximately 50% of all neurons in the GCL of the rat retina (Perry, 1981), while in the cat retina this proportion may be as great as 80% (Wässle, et al., 1987; Wong and Hughes, 1987). In the GCL of the rabbit retina, several specific types of neurons have been defined using both morphological and histochemical approaches

**and Division of Ophthalmology, Childrens Hospital of Los Angeles, Los Angeles, CA 90027, U.S.A.*

NATO ASI Series, Vol. H31
Neurobiology of the Inner Retina
Edited by R. Weiler and N. N. Osborne
© Springer-Verlag Berlin Heidelberg 1989

(Brecha, et al., 1983; Masland, et al., 1984). Using immunohistochemistry, combined with optic nerve transection or retrograde transport of fluorescent dyes from central visual nuclei, Brecha et al. (1987) have identified tachykinin-like immunoreactivity in both ganglion cells and displaced amacrine cells. Although little evidence is available concerning the histochemical identity of other ganglion cells, considerable progress has been achieved in the histochemical characterization of different displaced amacrine cell populations. For instance, in the rabbit retina the majority of displaced amacrine cells appear to contain γ-aminobutyrate (GABA)-like immunoreactivity (Brecha, et al., 1988). Furthermore, a large proportion of these presumed GABAergic cells (about 85%) preferentially accumulate the fluorescent dye, 4,6-diamidino-2-phenylindole, DAPI (Brecha, et al, 1988; Vaney and Young, 1988), which previous studies have shown to be a marker for cholinergic amacrine cells (Masland, et al., 1984). In addition, there are small numbers of nicotinamide adenine dinucleotide phosphate (NADPH) diaphorase-stained (Sagar, 1986) and somatostatin-like (SRIF)-immunoreactive (Rickman, et al., 1986; Sagar, et al., 1986a; Sagar, 1987) cell populations in the GCL.

SRIF immunoreactivity has been localized in a number of vertebrate retinas (Brecha, 1983; Ehinger and Dowling, 1987). Based on electrophysiological studies, this peptide is likely to have a modulatory role in visual information processing in both the inner and outer retina (Zalutsky and Miller, 1987, 1988). The present study describes the distribution of SRIF-immunoreactive cells in rabbit retina, and emphasizes some of their morphological characteristics, including the distinctive laminar arborization of their cellular processes.

Technical methods

These studies used adult New Zealand albino rabbits and a mouse monoclonal antibody (#8) which is directed to the C-terminal portion of $SRIF_{14}$ (Buchan, et al., 1985). In order to increase the level of immunoreactivity in neuronal somata, rabbits, deeply anesthetized with halothane, were given intraocular injections of colchicine just posterior to the *ora serrata*. Following a 12-16 hr survival period, rabbits were anesthetized with a lethal dosage of Nembutal and perfused transcardially with 0.9% saline in 0.1 M sodium phosphate buffer. Following enucleation, retinas were dissected, fixed in 4.0% paraformaldehyde, prepared as whole mounts and processed using standard immunohistochemical techniques as described elsewhere (Brecha, et al., 1988). Specificity of the primary antibody was determined by preabsorption with $SRIF_{14}$ or $SRIF_{28}$ which eliminated immunoreactive staining.

The positions of SRIF-immunoreactive cells in the retina were mapped at a 100x final magnification. Cell densities were determined by counting every immunoreactive cell body along a 1 mm-wide strip from the visual streak to the ventral margin. Using *camera lucida*

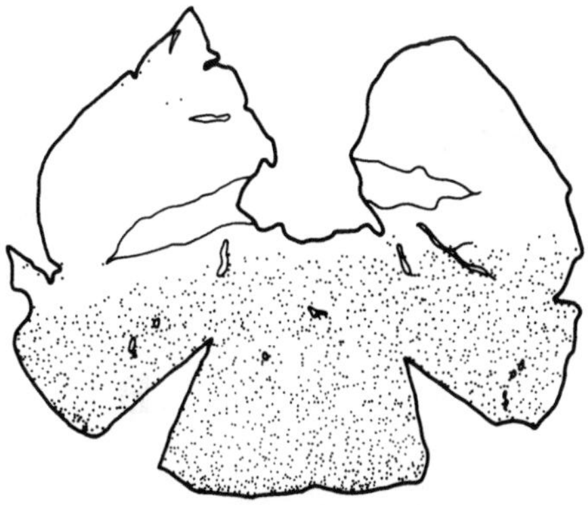

Figure 1. Whole mount preparation of a rabbit retina demonstrating the location of every SRIF-immunoreactive cell body. Labeled somata are confined almost exclusively to the ventral retina where they are distributed sparsely except at the retinal margins where they occur more frequently.

techniques individual cells were drawn at either 400x or 1000x magnification. For measurement of somal areas, drawings were analyzed using conventional morphometric software.

Results and Discussion

SRIF-immunoreactive somata were localized primarily to the GCL of the ventral retina where they were sparsely distributed (Figure 1). From the visual streak to the mid-peripheral retina the density of labeled somata was about 5 cells/mm^2, but increased abruptly in the most peripheral ventral retinal regions to almost 20 cells/mm^2. A few scattered SRIF-immunoreactive somata were also present in dorsal retinal regions. The absolute numbers of SRIF-containing cells in the GCL, determined from two retinas obtained from different animals, were 1018 and 1414. Based on estimates of the total number of displaced amacrine cells (Vaney, 1980; Masland, 1986), SRIF-containing cells comprise about 0.75% of the total number of these cells. Immunoreactive somata were characterized by an ovoid or multipolar shape, medium to large size (~140 µm^2) and 2 to 5 primary processes which ramify in lamina 5 of the IPL (Figure 2). These thick fibers gave rise to a meshwork of fine, beaded processes which arborized primarily in lamina 1. In addition, some fine processes were also observed in lamina 5. Frequently, a long, thin, axon-like process arose from the somal or perisomal region. Often processes with this profile could be followed for several millimeters across the retina in the nerve fiber layer or in lamina 1, but could not always be traced to their somal origin. A moderate number of immunoreactive processes, originating from the plexus

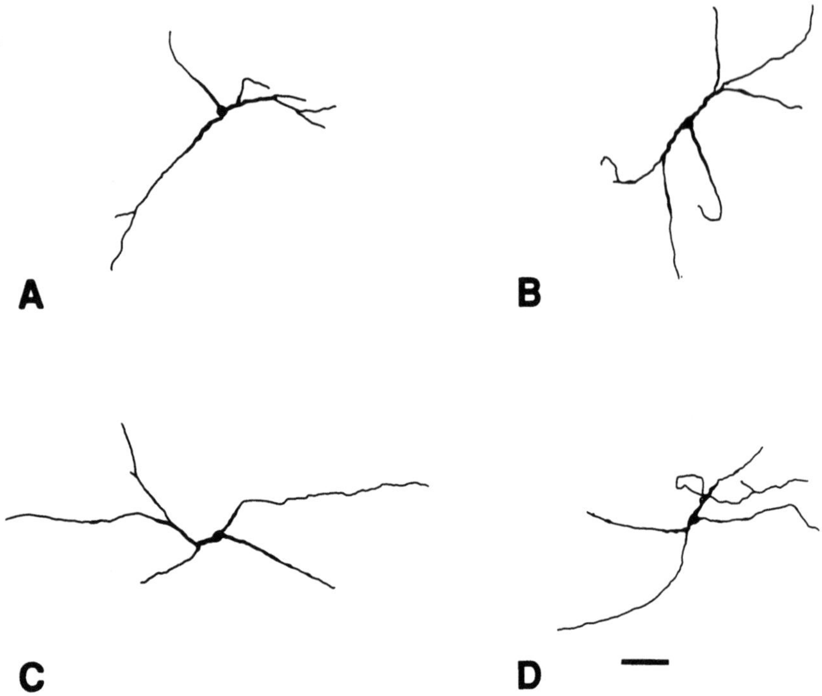

Figure 2. *Camera lucida* drawings of SRIF-immunoreactive cells in the GCL of the rabbit retina. *A. and B.* central retina, *C. and D.* peripheral retina (400x, bar equals 50μm).

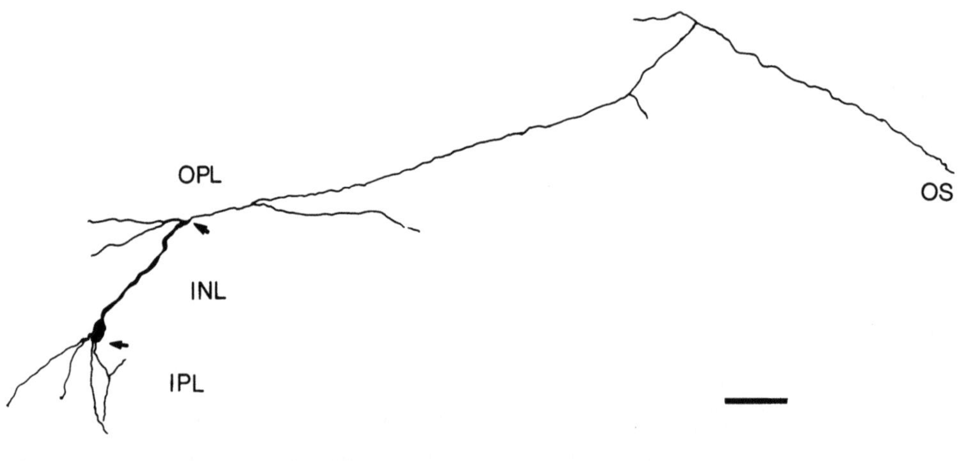

Figure 3. *Camera lucida* of SRIF-immunoreactive interplexiform cell drawn from a whole mount preparation. The cell body is in the proximal INL at the border of the IPL. Processes extend into both the IPL and the OPL. *os, ora serrata*. (1000x, bar equals 50μm).

in lamina 1 of the IPL, were observed in the outer plexiform layer (OPL) of all retinal regions. These processes either coursed directly through the INL perpendicularly to the vitreal plane or traversed the INL obliquely. Some processes ended abruptly upon entering the OPL, while others could be followed for considerable distances in this layer.

A few immunoreactive somata were observed in the proximal INL, giving rise to processes in the IPL. Their somata were medium in size and were similar in appearance to the SRIF-containing cells located in the GCL. On rare occasions an immunoreactive cell body in the proximal INL was seen to give rise to processes to both the IPL and the OPL (Figure 3). These cells resembled one of the interplexiform cells described from Golgi preparations of the rabbit retina by Oyster and Takahashi (1977). However, the infrequent occurrence of these cells in the INL, their lack of retinal coverage and their seemingly random distribution suggest that they are exclusive of the predominant SRIF-immunoreactive cell population. The possibility that some of these cells are developmental errors needs further exploration.

The somal size and location of SRIF-immunoreactive cells suggest that they might be ganglion cells. This possibility was examined in three cases in which the optic nerve was transected and the animal was allowed to survive 110, 120 or 143 days, a sufficient period of time for the elimination of most ganglion cells (Brecha, et al., 1988; Masland et al., 1984). In the optic nerve-transected retinas, there was no apparent change in the number or size of SRIF-immunoreactive somata in the GCL when compared to the normal, contralateral retina (Figure 4). Although a comparison of the number of immunoreactive cells between experimental and control retinas within the same animal is informative, it should be pointed out that small cell losses may be undetecable due to the variability in the absolute number of neurons which may occur between retinas (Rizzo, et al., 1988). Therefore histograms of immunoreactive somal sizes allow for a more sensitive comparison of retinal cell populations, since a shift in somal area would indicate either the loss of, or an atrophic degenerative change in, a subset of retinal neurons. These studies indicate that SRIF-containing cells are not likely to be ganglion cells, although the survival of these cells due to sustaining collateral axons within the retina cannot be discounted (Dacey, 1985).

The classification of SRIF-immunoreactive cells is difficult since the distribution of their processes within the plexiform layers has not yet been fully defined. At present we favor their general classification as displaced amacrine cells. Alternatives are "displaced" interplexiform or bipolar cells, by virtue of their somal location in the GCL and the distribution of processes to both plexiform layers. More information concerning the synaptic organization of these cells is needed, however, to fully resolve this classification. Furthermore, any complete description of these cells must take into account the frequent occurrence of their long intraretinal cellular processes, similar to those observed in other species and defined as "associational" (Gallego and Cruz, 1965) or "proprioretinal" (Catsicas, et al., 1987).

In summary, SRIF-immunoreactive cells have an extremely wide arbor of cellular processes

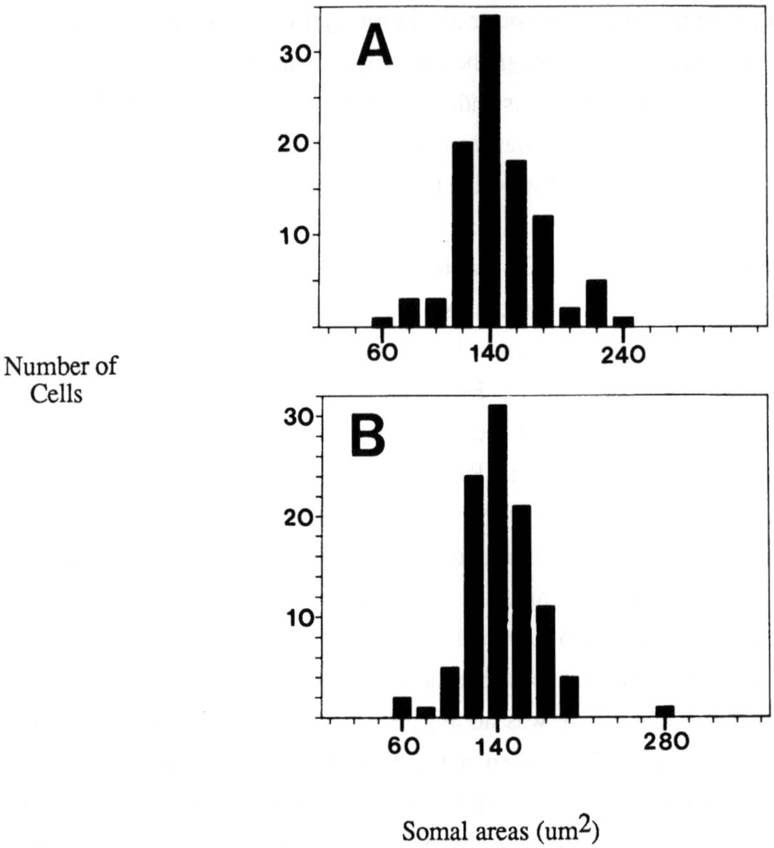

Figure 4. Histograms showing the distribution of somal areas (μm^2) of SRIF-immunoreactive cells in the GCL of the rabbit retina. *A.* Normal distribution of SRIF-immunoreactive somata. *B.* Distribution of SRIF-immunoreactive somata 143 days following optic nerve transection. This figure suggests that there is no shift in the SRIF-immunoreactive population following optic nerve transection.

forming a moderately dense plexus in lamina 1 of the IPL and contributing fibers to the OPL, even in regions of the dorsal retina which may be millimeters removed from the nearest SRIF-containing cell body. This pattern suggests that SRIF-containing cells have a widespread, modulatory influence on retinal processing, a notion which is further supported by the reported pharmacological activities of synthetic $SRIF_{14}$ in the rabbit retina which include long-lasting changes in ganglion cell sensitivity and receptive field properties (Zalutsky and Miller, 1988).

Acknowledgements

The authors wish to thank Dr. Catia Sternini and Dr. Lawrence Kruger for their insightful comments on the manuscript. Supported by Concern Foundation, Knights-Templar Eye Foundation, National Institutes of Health Grant EY04067 and V.A. Merit Review.

References

Brecha, N. (1983) Retinal neurotransmitters: histochemical and biochemical studies. In: P.C. Emson (ed) *Chemical Neuroanatomy*. Raven Press, New York, 85-129.
Brecha, N., Johnson, D., Bolz, J., Sharma, S., Parnavelas, J.G. and Lieberman, A.R. (1987) Substance P-immunoreactive retinal ganglion cells and their central axon terminals in the rabbit. *Nature*, 327:155-158.
Brecha, N., Johnson, D., Peichl, L. and Wässle, H. (1988) Cholinergic amacrine cells of the rabbit retina contain glutamate decarboxylase and γ-aminobutyrate immunoreactivity. *Proc Natl Acad Sci*, 85:6187-6191.
Buchan, A.M.J., Sikora, L.K.J., Levy, J.G., McIntosh, C.H.S., Dyck, I. and Brown, J.C. (1985) An immunocytochemical investigation with monoclonal antibodies to somatostatin. *Histochem*, 83:175-180.
Catsicas, S., Catsicas, M. and Clarke, P.G.H. (1987) Long-distance intraretinal connections in birds. *Nature*, 326(12):186-187.
Dacey, D.M. (1985) Wide-spreading terminal axons in the inner plexiform layer of the cat's retina: evidence for intrinsic axon collaterals of ganglion cells. *J Comp Neurol*, 242:247-262.
Ehinger, B. and Dowling, J.E. (1987) Retinal neurocircuitry and transmission. In: A. Bjorklund, T. Hokfelt and L.W. Swanson (eds) *Handbook of Chemical Neuroanatomy Vol. 5: Integrated Systems of the CNS, Part I.*, Elsevier Science Publishers, 389-446.
Gallego, A. and Cruz, J. (1965) Mammalian retina: association nerve cells in ganglion cell layer. *Science*, 150:1313-1314.
Hughes, A. and Vaney, D.I. (1980) Coronate cells: displaced amacrines of the rabbit retina? *J Comp Neurol*, 189:169-189.
Masland, R.H., Mills, J.W. and Hayden, S.A. (1984) Acetylcholine-synthesizing amacrine cells: identification and selective staining by using radioautography and fluorescent markers. *Proc R Soc Lond*, 223:79-100.
Masland, R.H. (1986) The cholinergic amacrine cell. *Trends Neurosci*, 9:218-223.
Oyster, C.W. and Takahashi, E.S. (1977) Interplexiform cells in rabbit retina. *Proc R Soc Lond*, 197:477-484.
Perry, V.H. (1981) Evidence for an amacrine cell system in the ganglion cell layer of the rat retina. *Neurosci*, 6(5):931-944.
Ramon y Cajal, S. (1892) La retine des vertebres. *La cellule*, 9:119-257.
Rickman, D., Johnson, D., Sharma, S. and Brecha, N. (1986) Distribution of substance P and somatostatin immunoreactive cells in the rabbit retina. *Soc Neurosci Abst*, 12:641.
Rizzo, J.F., Sandell, J.H. and Masland, R.H. (1988) Genetic variation in the numbers of some retinal cells. *Invest Ophthalmol Vis Sci*, 29 (suppl):205.
Sagar, S.M. (1986) NADPH diaphorase histochemistry in the rabbit retina. *Brain Res*, 373:153-158.
Sagar, S.M., Marshall, P.E., Onesti, S.T. and Landis, D.M.D. (1986a) Somatostatin immunocytochemistry in the rabbit retina. *Invest Ophthalmol Vis Sci*, 27:316-322.
Sagar, S.M. (1987) Somatostatin-like immunoreactive material in the rabbit retina: immunohistochemical staining using monoclonal antibodies. *J Comp Neurol*, 266:291-299.
Vaney, D.I. (1980) A quantitative comparison between the ganglion cell populations and axonal outflows of the visual streak and periphery of the rabbit retina. *J Comp Neurol*, 189:215-233.

Vaney, D.I. and Young H.M. (1988) GABA-like immunoreactivity in cholinergic amacrine cells of the rabbit retina. *Brain Res*, 438:369-373.
Wässle, H., Chun, M.H. and Muller, F. (1987) Amacrine cells in the ganglion cell layer of the cat retina. *J Comp Neurol*, 265:391-403.
Wong, R.O.L. and Hughes, A. (1987) The morphology, number, and distribution of a large population of confirmed displaced amacrine cells in the adult cat retina. *J Comp Neurol*, 255:159-177.
Zalutsky, R.A. and Miller, R.F. (1987) The physiology of somatostatin in the rabbit retina. *Soc Neurosci Abst*, 13:380.
Zalutsky, R.A. and Miller, R.F. (1988) Neuropeptide mechanisms in the rabbit retina. *Invest Ophthalmol Vis Sci*, 29 (suppl):271.

ENKEPHALIN RELEASE AND ENKEPHALIN-PRECURSORS IN CHICKEN RETINA

M.K. Boelen, M. Dowton and I.W. Chubb
The Neuroscience Unit
Department of Biology
University of Wollongong
N.S.W. 2500, Australia

INTRODUCTION

The demonstration of enkephalin-like immunoreactive neurons (Brecha, *et al.*, 1979), opiate receptors (Osborne and Hertz, 1981; Slaughter, *et al.*, 1985) and opiate-affected ganglion cell responses (Djamgoz, *et al.*, 1981; Dick and Miller, 1981) favour the acceptance of an opiate-transmitter system in vertebrate retina. However little is known about the participation of such an enkephalinergic transmitter system in visual processing (for reviews see Brecha, *et al.*, 1984; Watt, *et al.*, 1985). In chicken, the retinal levels of Leu^5-enkephalin-like immunoreactivity (LE-LI) vary during the light/dark cycle, and it has been suggested that the decrease in LE-LI during the dark may indicate an increase in the release and subsequent breakdown of this neurotransmitter candidate during the dark (Millar, *et al.*, 1984). To determine whether ambient light modulates the release of LE-LI we have established a technique for superfusing the chicken retina *in vitro*. The endogenous levels and release of LE-LI and LE-precursors (cryptic-LE-LI) were examined. Parts of these results have been published in preliminary form (Boelen, *et al.*, 1988).

MATERIALS AND METHODS

Chickens were kept on a 12 h light/dark cycle. After 12 h light, when LE-LI levels are highest, retinas were isolated, layered onto polysulfone filters (Gellman Sciences), transferred to Whatman Millex filter holders and superfused at 0.6 ml/min with physiological buffer containing (mM): 118.1 NaCl, 4.7 KCl, 1.2 $MgCl_2$, 24.9 $NaHCO_3$, 10 $CaCl_2$, 10 glucose and 100 µg/l bacitracin. The buffer was aerated with carbogen (95 % O_2 and 5 % CO_2) and the pH was adjusted to 7.3 with 1 M HCl. Three ml fractions were collected into ice-cold polypropylene tubes

NATO ASI Series, Vol. H31
Neurobiology of the Inner Retina
Edited by R. Weiler and N. N. Osborne
© Springer-Verlag Berlin Heidelberg 1989

containing 500 µl of 14 M acetic acid. LE-LI levels were determined by RIA (Millar, *et al.*, 1984)

For dark/light modified release, retinas were isolated under dim illumination (Phillips photographic safety light, PF712B). The superfusion chambers were wrapped tightly in aluminium foil. To expose the retinas to the light, the chambers were unwrapped and illuminated by diffuse light (about 800 lux) from an overhead projector.

The presence of LE-precursors in retinal homogenates and superfusate fractions was examined by sequentially incubating lyophilized fractions with trypsin (10 µg, Sigma) in 150 µl of 50 mM Tris-HCl buffer pH 7.8 (containing 5 mM $CaCl_2$) for 6 h, then carboxypeptidase-B (0.05 µg, Boehringer) in 200 µl of the same buffer for 30 min. The cryptic-LE-LI measured in this way probably represent precursors of LE (Lewis, *et al.*, 1980).

The released LE-LI was concentrated using Sep-Pak C_{18}-cartridges (Waters) and characterized by reversed-phase (aquapore RP300 column) high performance liquid chromatography (HPLC). Only one peak of LE-LI was detected in the HPLC-profiles (results not shown). This peak had the same retention time as synthetic Leu^5-enkephalin. The increase in immunoassayable material in fractions from retinas which had been superfused in the dark or with buffer containing high-potassium, was reflected by an increase in LE-LI eluting from the column. The LE-LI measured in this study is thus likely to be Leu^5-enkephalin.

RESULTS AND DISCUSSION

With depolarization by high concentrations of K^+ (60 mM KCl) in the superfusing buffer, the efflux of LE-LI from the retinas approximately doubled (Fig. 1a). On return to normal levels of KCl (4.7 mM), the efflux returned to basal levels. In the presence of 4 mM $CoCl_2$, the basal efflux was reduced and depolarizing concentrations of KCl did not increase the efflux of LE-LI. After the cobalt had been removed, superfusion with buffer containing 60 mM KCl elicited an increase in the efflux of LE-LI from all retinas, indicating that the retinas were able to respond to depolarization.

The effect of changes in ambient lighting on the spontaneous efflux of LE-LI from the isolated retinas was then measured. The total efflux during 25 min in the light was reduced by 59% when compared to the efflux during the previous 25 min in the dark (Fig. 1b). On returning to the dark, the efflux increased to

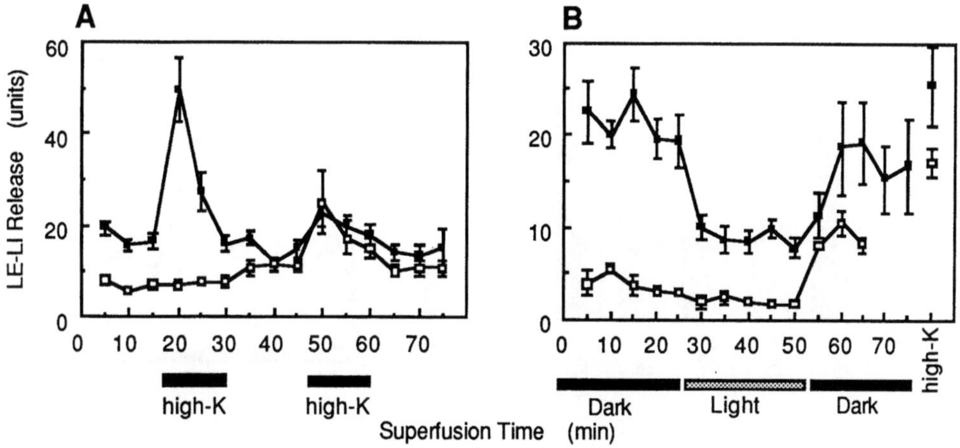

Fig. 1 Depolarization evoked and light/dark modulated release of Leu-enkephalin-like immunoreactivity from chick retina.

A. Retinas were superfused with either normal (4.7 mM KCl) or K^+-rich (60 mM KCl) buffer *in vitro*. (■) Retinas superfused in the absence of Co^{++} (n=12). (□) Retinas superfused in the presence of 4 mM $CoCl_2$ during the first 30 min (n=11).

B. Retinas were superfused in darkness or exposed to diffuse light from an overhead projector as indicated by the bar. (■) In the absence of Co^{++} (n=11). (□) 4mM $CoCl_2$ was present in the buffer during the first 50 min of superfusion (n=4).

The amount of released LE-LI was normalized by expressing the amount in each fraction as a percentage of the total amount of LE-LI recovered from the first 10 superfusate fractions; one unit corresponds to one pg LE-LI. Vertical bars indicate SEMs.

initial rates. Light had no detectable effect on the spontanous release when 4 mM $CoCl_2$ was present (Fig. 1b). However, as the baseline efflux during exposure to cobalt was below the detection limit of the RIA, small effects cannot be excluded. All retinas responded to depolarizing concentrations of K^+ after removal of the $CoCl_2$.

The release of LE-precursors from the superfused retinas was examined by proteolytic digestion of the superfusate fractions (Lewis, *et al.*, 1980). Depolarization by high-K^+ released both LE-LI and cryptic-LE-LI (Fig. 2a). In contrast, no cryptic-LE-LI appeared to be released during the light or dark. Apparently, the two stimuli (high-K^+ and light/dark) do not release identical pools of LE-LI-containing compounds. Thus, synaptic depolarization of the LE-LI amacrine cell evoked by light and dark seemed to be qualitatively distinct from chemical depolarization by high-K^+, and caution seems warranted in interpreting findings of precursor-release triggered by less-physiological stimuli such as high-K^+.

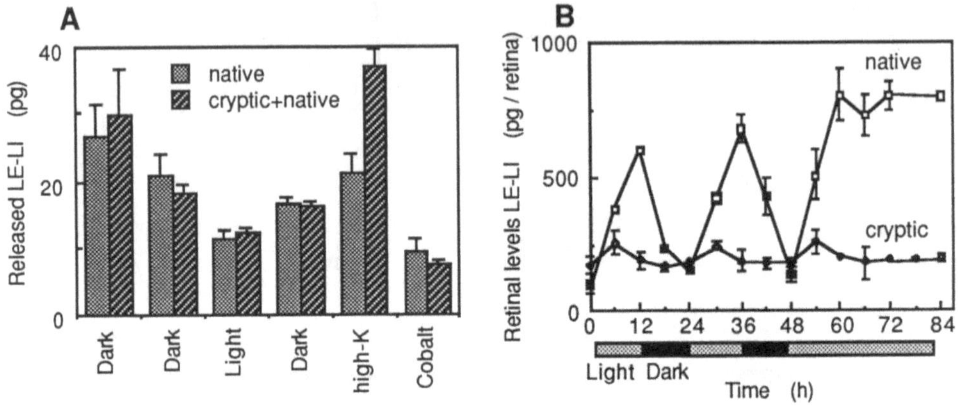

Fig. 2 The release and diurnal variation of enkephalin-precursors in chick retina.

Levels of (native) LE-LI in fractions from superfused retinas or retinal homogenates were measured by RIA. Total (cryptic+native) LE-LI levels were determined after proteolytic digestion of the fractions with trypsin and carboxypeptidase-B. Vertical bars indicate SEMs.

A. Retinas were superfused in the dark or light (n=6-12), with either normal (4.7 mM KCl) or K^+-rich (60 mM KCl) (n=6) buffer, or buffer containing 4mM $CoCl_2$ (n=3).

B. Chickens were placed on a light/dark cycle, as indicated by the bar. Samples were taken and retinal levels of native-LE-LI and cryptic-LE-LI determined (n=12).

Analysis of endogenous levels of enkephalins revealed the presence of cryptic-LE-LI, accounting for 25-50 % of the total LE-LI-containing material in retinal homogenates (Fig. 2b). It is clear that not all the translation products are immediately further processed into LE-LI; too much of the total LE-LI is present as cryptic-LE-LI.

The significant amount of cryptic-LE-LI in retinal homogenates suggests that processing of the precursors is the rate-limiting step in the generation of the pentapeptide. This view is strengthened by the observation that the amount of cryptic-LE-LI shows a small increase in light/dark cycles. The accumulation of cryptic-LE-LI is slight but consistent and peaks after 6 h exposure to light. While this may indicate a small increase in the rate of synthesis towards the end of the dark period, which then takes some time to be down regulated during the light; it also indicates that the rate of processing of the LE-LI-containing peptides is limiting.

In avian retina, LE-LI has only been observed in amacrine cells, including their processes in the inner plexiform layer (Brecha, *et al.*, 1979; Fukuda, 1982; Ishimoto, *et al.*, 1983; Millar, *et al.*, 1984; Watt, *et al.*, 1985b), it is reasonable to conclude that the immunoreactive material measured in the present study originates from these amacrine cells or their processes.

The chicken retina releases more LE-LI during the dark than when the retina is exposed to light, *in vitro*. Retinal LE-LI is released upon depolarization in a calcium-dependent manner, thus, variations in the light/dark conditions may modify the state of polarization of the LE-LI amacrine cell and hence the release of LE-LI. Based on extracellular recordings of ganglion cell activity in goldfish retina, it was noted that light-*off* rather than light-*on* may trigger the release of endogenous opiates (Djamgoz, *et al.*, 1981). Our data is consistent with this notion, but suggests that there is a *sustained* shift in input balance during the dark towards net excitation of the LE-LI amacrine cell. Our results provide a plausible explanation for the observed depletion of retinal levels of LE-LI during the dark (Millar, *et al.*, 1984). Processing of precursor into pentapeptide was rate-limiting for the synthesis of LE-LI, as indicated by the significant amount of cryptic-LE-LI in retinal homogenates, and the rate of processing was at most marginally affected by ambient light. Although a minor metabolic regulation related to light could be present, the major contributor to the diurnal variation in retinal levels of LE-LI seems to be the light modulated release of LE-LI. Retinal levels of LE-LI may therefore be useful as index of neuronal activity.

ACKNOWLEDGEMENTS

We thank Dr. John Oliver for the supply of antibodies to Leu^5-enkephalin, Richard Smith, Irene Backen and Mardi Silburn for technical assistance, and Dr Ian Morgan for valuable discussions. The work was supported by a grant from the National Health and Medical Research Foundation of Australia.

REFERENCES

Boelen MK, Dowton M, Smith R, Chubb IW (1988) Comparison of the effects of light on the release of leu-enkephalin-like immunoreactivity and 3H-dopamine from chicken retina *in vitro*. Neurosci Lett (Suppl.) 30: S49

Brecha NC, Eldred W, Kuljis RO, Karten HJ (1984) Identification and localization of biologically active peptides in the vertebrate retina. In: Osborne NN, Chader GJ (eds) Progress in Retinal Research 3. Pergamon Press, New York, p 185-226

Brecha N, Karten HJ, Laverack C (1979) Enkephalin-containing amacrine cells in the avian retina: Immunohistochemical localization. Proc Natl Acad Sci USA 76: 3010-3014

Dick E, Miller RF (1981) Peptides influence retinal ganglion cells. Neurosci Lett 26: 131-135

Djamgoz MBA, Stell WK, Chin C-A, Lam DMK (1981) An opiate system in the goldfish retina. Nature (London) 292: 620-623

Fukuda M (1982) Localization of neuropeptides in the avian retina: an immunohistochemical analysis. Cell Mol Biol 28: 275-283

Ishimoto I, Shiosaka S, Shimizu Y, Kuwayama Y, Fukuda M, Inagaki S, Takagi H, Sakanaka M, Sasaoka A, Senba E, Sakiyama T, Tohyama M (1983) Leucine-enkephalin-like immunoreactivity in the chicken retina with a special reference to its fine structures. Invest Ophthalmol Vis Sci 24: 879-885

Lewis RV, Stern AS, Kimura S, Rossier S, Stein S, Udenfriend S (1980) An about 50,000-Dalton protein in adrenal medulla: a common precursor of [Met]- and [Leu]enkephalin. Science 208: 1459-1461

Millar TJ, Salipan N, Oliver JO, Morgan IG, Chubb IW (1984) The concentration of enkephalin-like material in the chick retina is light dependent. Neuroscience 13: 221-226

Osborne H, Herz A (1981) Opioid binding properties in bovine retina. Neurochem Int 3: 143-148

Slaughter MM, Mattler JA, Gottlieb DI (1985) Opiate binding sites in the chick, rabbit and goldfish retina. Brain Res 339: 39-47

Watt CB, Su Y-YT, Lam DM-K (1985a) Enkephalins in the vertebrate retina. In: Osborne NN, Chader GJ (eds) Progress in Retinal Research 4. Pergamon Press, New York, p 221-246

Watt CB, Su Y-YT, Lam DM-K (1985b) Opioid pathways in an avian retina II Synaptic organization of enkephalin-immunoreactive amacrine cells. J Neurosci 5: 857-865

PUTATIVE GABAERGIC GANGLION CELLS IN THE RETINA OF SALAMANDRIDS: EVIDENCE BY IMMUNOCYTOCHEMISTRY AND ELECTROPHYSIOLOGY

G. Gläsener, G. Manteuffel*
Institute of Zoology
TH Darmstadt
Schnittspahnstr. 3
6100 Darmstadt
Federal Republic of Germany

It has since long been known that the ganglion cells of the vertebrate retina are the neurons that project to the central nervous system and thus transmit the visual information from the retina to the brain (Ramõn y Cajal 1893). Although it is widely accepted that the majority of these cells uses an excitatory neurotransmitter, there is a growing body of evidence that a small subpopulation of GABAergic ganglion cells might exist in various vertebrates (Mosinger et al. 1986, Caruso et al. 1988, Yu et al. 1988). Since GABA is known to be a major inhibitory transmitter in the central nervous system these ganglion cells probably inhibit central neurons. We have investigated the retinal ganglion cells in salamandrids (Amphibia) by GABA-immunocytochemistry and the responses of their target cells by electrophysiology to identify putative inhibitory retinal ganglion cells.

* University of Bremen, Fachbereich Biologie, 2800 Bremen 33, FRG

NATO ASI Series, Vol. H31
Neurobiology of the Inner Retina
Edited by R. Weiler and N.N. Osborne
© Springer-Verlag Berlin Heidelberg 1989

GABA-Immunocytochemistry

GABA-like-immunohistochemical stainings in the retina of Triturus alpestris reveal labelings in about 40 % of the cells of the ganglion cell layer (GCL) (Gläsener et al. 1988). Ganglion cells in the GCL are judged to make up only 50-60 % of the total cell population of the GCL. Thus this quantitative relationship does not provide unambiguous evidence for the existence of putative GABAergic ganglion cells.

In the optic nerve GABA-like-immunoreactive (IR) fibers are clearly visible. In contrast to the number of labeled neurons in the GCL, the number of labeled fibers in the optic nerve is relatively low, indicating that only a small subpopulation of putative GABAergic ganglion cells might exist. Another explanation for the GABA-like-IR optic fibers could be, that they might belong to an efferent system. Although a number of anatomical and physiological reports support the existence of a centrifugal projection to the retina in almost all vertebrate classes, in amphibians, however, this issue remains controversial. In frogs and toads, a retinopetal projection is not detectable with HRP-tracing (Scalia and Teitelbaum 1978), although there is some physiological evidence that suggests the existence of retinal efferents (Ewert and Borchers 1974). To determine whether the GABA-like-IR fibers in the optic nerve belong to ganglion cells or to an efferent projection, a double label method was used.

The optic nerve was cut shortly behind the bulbus of the eye and HRP was injected, resulting in a selective filling of all ganglion cells. HRP was not visualized by peroxidase activity, but by FITC coupled immunocytochemistry. GABA was shown in one and the same section using TRITC coupled immunohistochemistry. Double labeled neurons were clearly detectable among the cells of the GCL. According to the number of labeled fibers in the optic nerve, putative GABAergic ganglion cells are not very numerous. This indicates, that a projection system of putative GABAergic neurons from the retina to the brain exists.

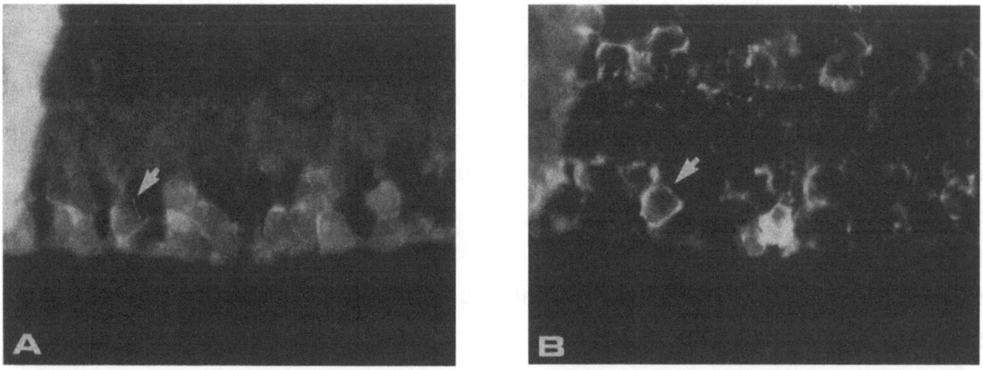

Fig. 1. Double labeling of HRP-traced ganglion cells (A) und GABA-like-IR neurons (B). The double labeled GABA-like-IR ganglion cell in the GCL is marked (arrow).

In the brain, a lot of fine GABA-like-IR processes are present in the optic tract which we suppose to be of glial origin. These processes make it impossible to localize the detailed projection targets of the GABA-like-IR fibers. For this reason, electrophysiological recordings were used in a retino-recipient nucleus to provide evidence for the central projections of the GABA-like-IR fibers.

Electrophysiological Data

Intracellular recordings of pretecal and adjacent tegmental neurons of Salamandra salamandra were combined with electrical stimulation of both optic nerves to investigate the latency distribution of excitatory and inhibitory visual inputs. Since HRP-filled glass fiber capillaries were used, which allow HRP-injection in the electrophysiologically characterized neurons, additionally the morphology of the cells could be investigated.

The latency distributions of excitatory (depolarizing) and inhibitory (hyperpolarizing) inputs to postsynaptic pretectal neurons show that inhibitory effects are found among the fastest responses (2-3 ms latency) providing evidence for a direct monosynaptic input. This was confirmed by the HRP injections which showed that neurones with short latencies always had parts of their dendrites within the ocular projection field, a feature that was restricted to cells with short latencies (Manteuffel 1985). Longer latencies are probably due to indirect inputs, involving one or several central interneurons.

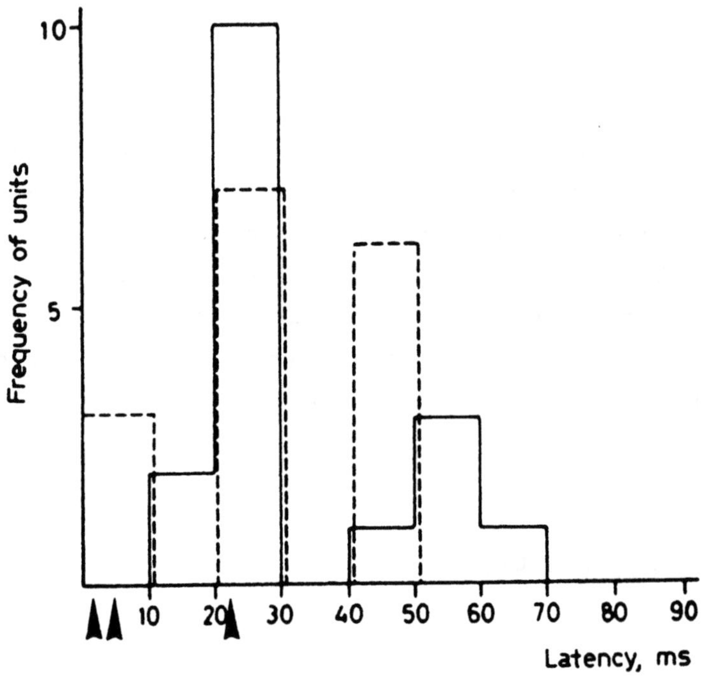

Fig. 2. Latency histogram of electrophysiologically recorded units in the pretectal area. Contralateral (solid line) and ipsilateral (dashed line) inputs show similar peaks with the exception of the very short latency group (< 10 ms) of the ipsilateral inputs. Inhibitory effects appear among the shortest latencies recorded (arrowheads).

Conclusions

The data presented here provide strong evidence for the existence of GABAergic ganglion cells in the retina of salamandrid amphibians. Although some immunohistochemical investigations support this for other vertebrate species (Mosinger et al. 1986, Caruso et al. 1988, Yu et al. 1988), it had never been shown by a double labeling involving retrograde tracing of ganglion cells and GABA-immunocytochemistry.
The electrophysiological data presented here indicate that a monosynaptic inhibitory input from the retina to the central nervous system is likely to exist and that the pretectal area might be one of the projection sites of the GABA-like-IR ganglion cells. We can not exclude that other primary visual projection centers also receive input from putative GABAergic ganglion cells. For example it has been shown in the frog Rana pipiens that fibers of class-5-ganglion cells which terminate in the optic tectum possess collateralisations in the pretectal area (Stirling and Merrill 1987). In the pigeon, there is electrophysiological evidence that retinal ganglion cells inhibit tectal cells, too (Leresche et al. 1986). The detailled role of these neurons in visual processing remains to be determined.

Acknowledgements. This work was supported by a research grant of the Deutsche Forschungsgemeinschaft (Hi 306/1-1 and SFB 45/D3). The authors want to thank Drs. I. Wulle and J. Schnitzer for supplying them with the antibody against GABA, and C. Schmidt for her excellent technical assistance.

References

Caruso DM, Owczarak T, Hazlett JC, Pourcho RG (1988) GABA-immunoreactive ganglion cells in rat retina project to the superior colliculus. Invest Ophthalmol Vis Sci (Suppl) 29:197

Ewert JP and Borchers HW (1974) Inhibition of toad (Bufo bufo L.) retinal on-off and off-ganglion cells via active eye clsoing. Vis Res 14:1275-1276

Gläsener G, Himstedt W, Weiler R, Matute C (1988) Putative neurotransmitters in the retinae of three urodele species (Triturus alpestris, Salamandra salamandra, Pleurodeles waltli). Cell Tissue Res 252:317-328

Leresche N, Hardy D, Audinat E, Jassik-Gerschenfeld D (1986) Synaptic organization of inhbitory circuits in the pigeon's optic tectum. Brain Res 365:383-387

Manteuffel G (1985) Monocular and binocular optic inputs to salamander pretectal neurons: intracellular recording and HRP labelling study. Brain Behav Evol 27:1-10

Mosinger JL, Yazulla S, Studholme KM (1986) GABA-like immunoreactivity in the vertebrate retina: a species comparison. Exp Eye Res 42:631-644

Ramõn y Cajal SR (1893) La rētine des vertēbrēs. La Cellule 9:17-257

Scalia F and Teitelbaum I (1978) Absence of efferents to the retina in the frog and toad. Brain Res 153:340-344

Stirling RV and Merrill EG (1987) Functional morphology of frog retinal ganglion cells and their central projections: the dimming detectors. J Comp Neurol 258:477-495

Yu BCY, Watt CB, Lam DMK, Fry KR (1988) GABAergic ganglion cells in the rabbit retina. Brain Res 439:376-382.

HOST AND GRAFT GLIAL CELL ACTIVITIES FOLLOWING RETINAL TRANSPLANTATION TO THE ADULT RAT EYE

M. Seiler and J. E. Turner*
Eye Research Institute of Retina Foundation
20 Staniford Str.
Boston MA 02114
U.S.A.

Transplantation of immature retina into adult retina is a model which has been developed rather recently (Turner & Blair 1986, Del Cerro et al. 1987, Aramant et al. 1988). The donor embryonic or neonatal (rat) retinal tissue is placed as an aggregate into a small retinal lesion site in the dorsal retina. The transplanted tissue continues to differentiate, developing most retinal layers. Neonatal, as well as embryonic retina, can be successfully grafted. However, embryonic day 15 (E 15) grafts tend to develop more distinctive laminae and to fuse better with the host retina in a fresh, but not a conditioned (previously made), lesion site (Aramant et al. 1988). In E 15 grafts, photoreceptor cells are frequently arranged around an external limiting membrane to form a "rosette" within the graft.

The rat retina contains two types of glial cells, astrocytes and Mueller cells. Astrocytes are confined to the ganglion cell layer and the inner limiting membrane, whereas the radial processes of Mueller cells reach through the entire thickness of the retina, forming the inner limiting membrane with their endfeet (see Newman 1986 for review).

The present study is part of a project about the time course of the development of rat E 15 grafts. We were interested in the reaction of host glial cells to the presence of the graft as well as in the intrinsic development of graft glial cells. The expression of glial-specific markers in host and graft was investigated by immunohistochemistry for GFAP, S-100, Vimentin and Laminin.

The methods and the staining results with antibodies against GFAP and S-100 are published in detail elsewhere (Seiler & Turner 1988, in press). To summarize these results briefly : up to 8 d after transplantation, there was little intrinsic S-100 and GFAP staining in the graft. Host glial cells (S-100+, GFAP+) appeared to migrate into the graft (see Fig. 1 A,B). Graft glial cells developed relatively normally, but became partially reactive (GFAP+) at 3 weeks and completely reactive at 5 weeks after transplantation (see Fig. 1 C,D).

In normal, unlesioned retina, Mueller cells contain only Vimentin as their intermediate filament protein (Shaw & Weber 1983, Draeger et al. 1984), but they express GFAP after injury (Bignami & Dahl 1979). Vimentin is contained in all retinal cells at early stages of development; in adult retina, it is restricted to processes of Mueller cells, astrocytes and horizontal cells (Bignami et al. 1982, Shaw & Weber 1983, Draeger et al. 1984).

*Bowman Gray School of Medicine, Wake Forest University, 300 S. Hawthorne Rd., Winston-Salem, NC 27103

NATO ASI Series, Vol. H31
Neurobiology of the Inner Retina
Edited by R. Weiler and N. N. Osborne
© Springer-Verlag Berlin Heidelberg 1989

Our results obtained with a commercially available goat antiserum against Vimentin (ICN) are presented in Fig. 2 . Radial fibers in almost all cells were faintly Vimentin-positive at early stages of development in the graft(Fig.2D)as well as in age-matched controls of E 15 and postnatal day 1 (PN 1) retina (Fig. 2 A,B). Astrocytes were clearly recognizable in PN 1 retina (Fig. 2 B); Mueller cell fibers extending from the external limimting membrane could be identified at PN 8 (not shown). The adult staining pattern was seen from 3 weeks after transplantation (Fig. 2 E,F) corresponding to PN 15 retina (Fig. 2 C).

Laminin is an extracellular matrix protein which promotes neuronal outgrowth from retinal neurons in culture (Ford-Holevinski et al. 1986, Smalheiser et al. 1984). It is thought to be present at the time of axonal outgrowth and neuronal migration during development (Liesi 1985a). After injury, brain astrocytes transiently express Laminin (Liesi et al. 1984), but astrocytes in the optic nerve do not (Ford-Holevinski et al. 1986). Laminin is continuously present in CNS glia of lower vertebrates capable of regeneration (Liesi 1985b). Therefore, we were interested to know whether Laminin was expressed at the host/graft interface thus promoting neuronal outgrowth from the graft.

Our (preliminary) results obtained with a rat antiserum against Laminin (Chemicon) are shown in Fig. 3 . (Laminin is optimally demonstrated in unfixed tissue; however, our tissue was fixed with 4% Paraformaldehyde.) In E 15 retina, the antiserum stained the inner limiting membrane (Fig. 3 A). Radial processes (Mueller cell fibers ?) extending from the external limiting membrane were faintly stained at PN 1 (not shown) and PN 8 (Fig. 3 B). At PN 15 (Fig. 3 C) and in adult retina, the staining of Mueller cell processes was even fainter; the antiserum stained faintly blood vessels, the inner limiting membrane and (diffusely) the inner and outer plexiform layers. 2 d after transplantation (Fig. 3 D,E), numerous new Laminin+ blood vessels were formed in and near the lesion site, invading the graft along the open host/graft interface. The inner limiting membrane near the lesion site showed increased Laminin-immunoreactivity. The Laminin-immunoreactivity of blood vessels decreased gradually, but was still higher than in the rest of the host retina after 5 weeks. Host Mueller cell fibers near the lesion site showed increased Laminin-staining, which was persistent for at least up to 5 weeks (not shown). So far, no Laminin+ Mueller cell fibers have been found in the graft . Most likely host-derived astrocyte-like cells migrating on the graft vitreal surface expressed strong Laminin-immunoreactivity (Fig. 3 F).

In summary : After transplantation of embryonic retina into an adult retinal lesion site, host glial cells become reactive, express Laminin (near the lesion site) and migrate into the graft. Numerous Laminin+ blood vessels are formed in the first days after transplantation; the graft seems to be vascularized by the host. Intrinsic graft glial cells develop approximately normally, but fail to develop an inner limiting membrane.

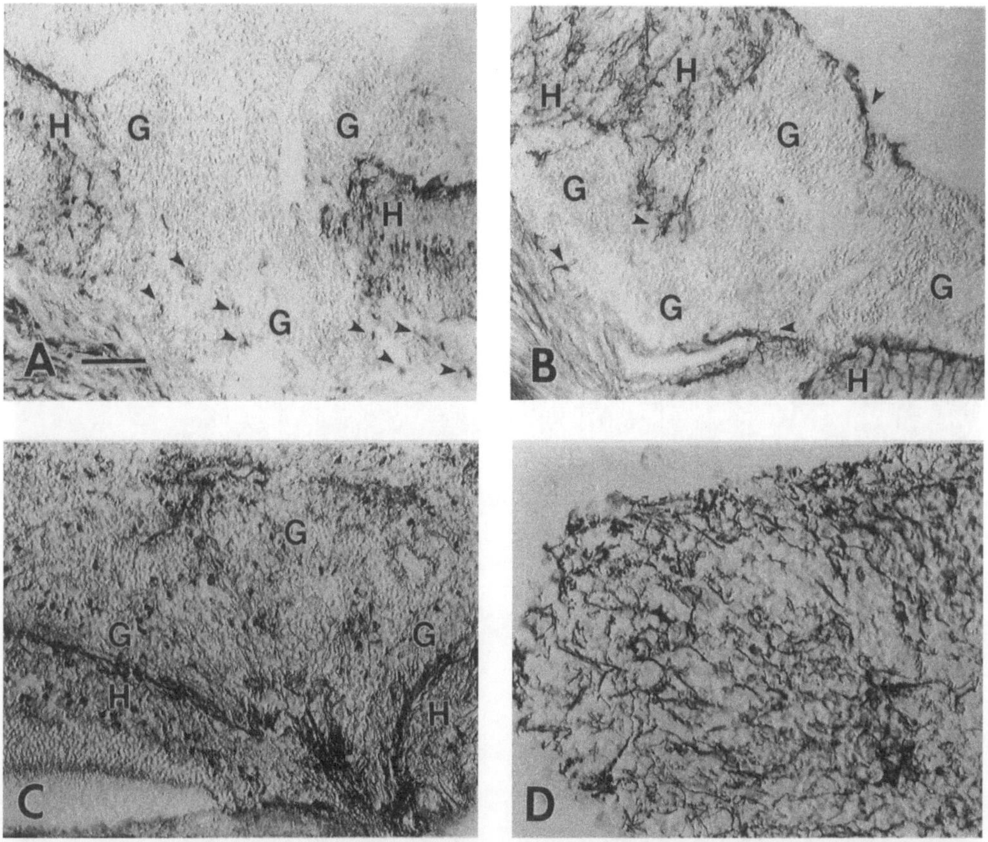

Fig. 1 : S-100 and GFAP staining of E 15 retinal grafts (Bar = 50 um). Nomarski Optics, Peroxidase staining. A+B) 8 d after transplantation : A) S-100 staining, B) GFAP staining. The graft (G) extends from the subretinal space into the vitreous. Host(H)-derived glial cells (arrowheads) apparently have migrated into the graft along the open host/graft interface, blood vessels and surfaces. There is no intrinsic staining of the graft at that stage.
C+D) 5 weeks after transplantation : C) S-100 staining. The graft glial cells stain with the same intensity as the host. Note fiberbundles extending from the host to the graft base. D) GFAP staining. A lot of reactive Mueller cells (parallel fibers) in the graft; no continuous inner limiting membrane on the graft surface.

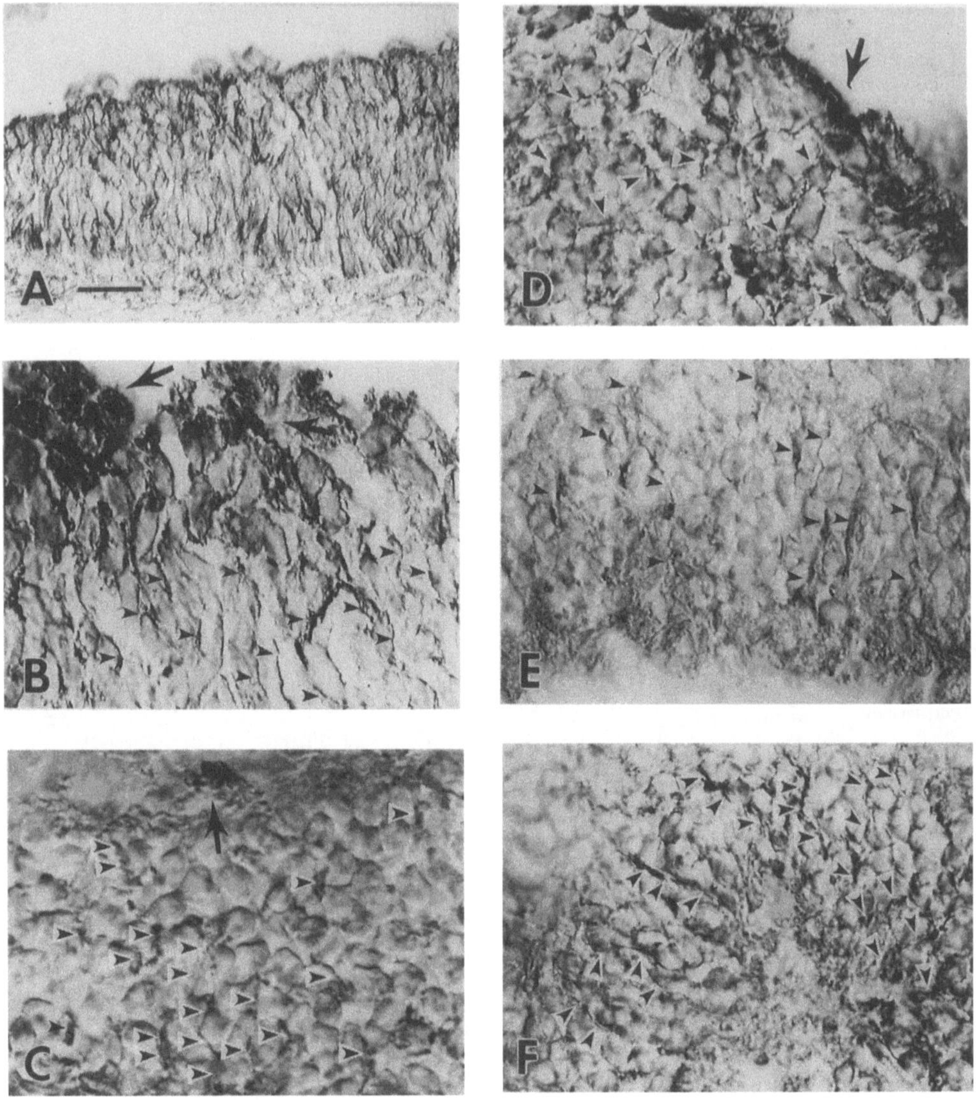

Fig. 2 : Vimentin-staining (goat antiserum, ICN) (Bar = 25 um(A), 10 um (B-F)). All photographs oriented with inner retina up and outer retina down. A) E 15 retina: faint staining of radial fibers all over retina. B) PN 1 retina : Astrocytes (arrow) at inner limiting membrane and radial fibers (arrowheads) extending through retina. C) PN 15 retina : Staining restricted to Mueller cell processes (arrowheads) in outer nuclear layer and horizontal cell (arrow). D) 4 d old graft: darkly staining host-derived astrocyte (arrow) on vitreal surface; fainter stained fibers (arrowheads) in graft. E) 3 weeks and F) 5 weeks after transplantation : stained Mueller cell processes (arrowheads) in graft rosette.

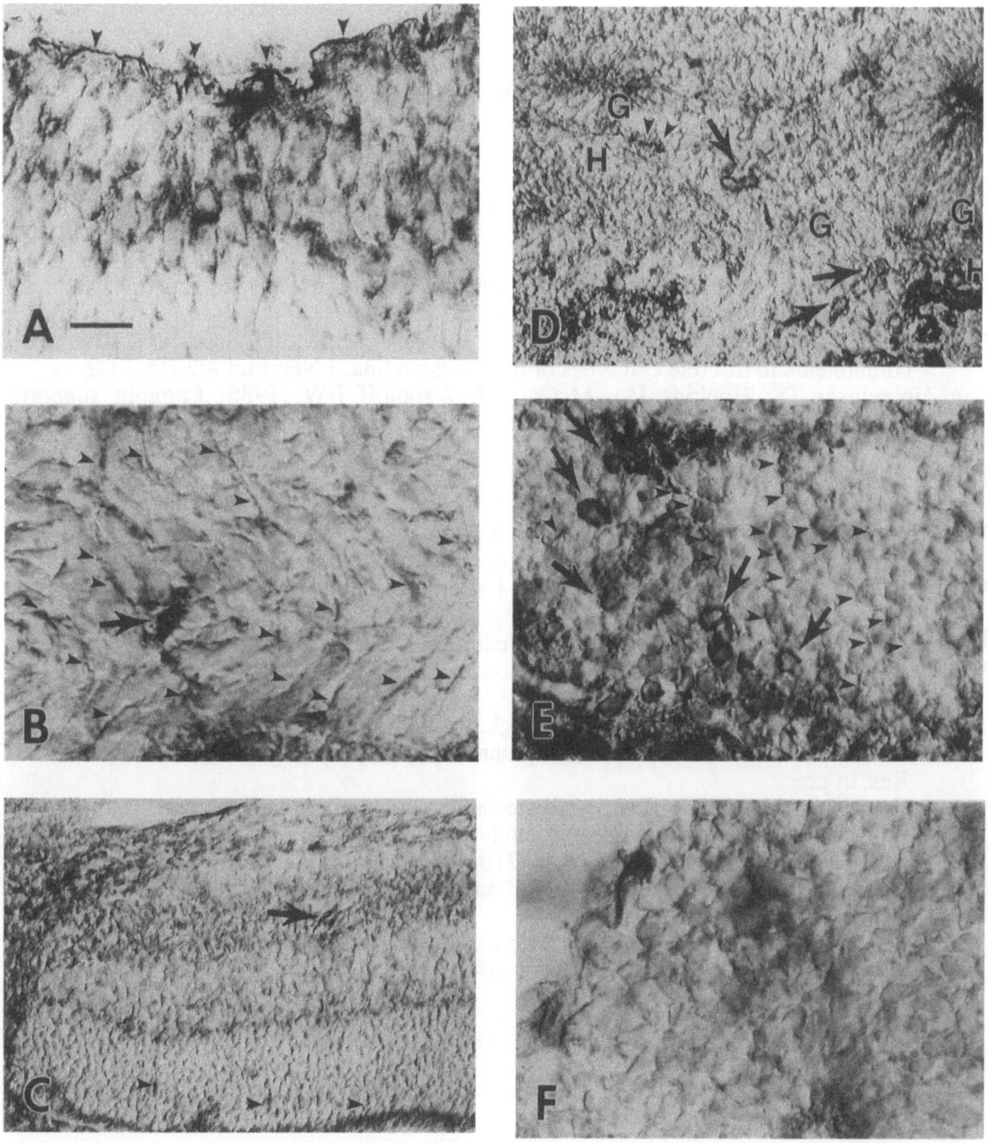

Fig. 3 : Laminin-staining (rat antiserum, Chemicon) (Bar = 25 um (C,D), 10 um (A,B,E,F)). A) E 15 retina : distinct staining of the inner limiting membrane (arrowheads) and diffuse staining of extracellular spaces in the inner retina. B) PN 8 retina : Stained radial fibers (arrowheads). Staining of one cell (arrow) in outer retina. C) PN 15 retina at the optic nerve head: staining pattern similar to adult retina. Diffuse staining of inner limiting membrane, inner and outer nuclear layer. Very faint staining of Mueller cell processes (arrowheads) in outer nuclear layer and of blood vessels (arrow). Unspecific staining of external limiting membrane and inner photoreceptor segments. D,E) 2 d after transplantation : D) Graft (G) in lesion site, open interface to host (H). Newly formed blood vessels (arrows) extending into the graft. The host inner limiting membrane is darkly stained near the lesion site (arrowhead). E) Host retina near the lesion site : Laminin+ newly formed blood vessels (arrows) and Mueller cell fibers (arrowheads). F) 2 weeks after transplantation : Laminin+ host-derived glial cell on the vitreal surface of the graft.

References :

Aramant R, Seiler M, Turner JE (1988) Donor age influences on the success of retinal grafts to adult rat retina. Invest Ophthalmol Vis Sci 29:498-503

Bignami A, Raju T, Dahl D (1982) Localization of Vimentin, the non-specific intermediate filament protein, in embryonal glia and in early differentiating neurons. Dev Biol 31:286-295

Bignami A, Dahl D (1979) The radial glia of Mueller in the rat retina and their response to injury - an immunofluorescence study with antibodies for the glial fibrillary acidic protein. Exp Eye Res 28:63-69

Del Cerro M, Notter MF, Jiang LQ, Grover DA, del Cerro C (1987) Retinal transplants in phototoxic retinopathy. Soc Neurosci Abstr 13:568 (160.1)

Draeger UC, Edwards DL, Barnstable CJ (1984) Antibodies against filamentous components in discrete cell types of the mouse retina. J Neurosci 4:2025-2042

Ford-Holevinski TS, Hopkins JM, McCoy JP, Agranoff BW (1986) Laminin supports neurite outgrowth from explants of axotomized adult rat retinal neurons. Dev Brain Res 28:121-126

Liesi P (1985a) Do neurons in the vertebrate CNS migrate on Laminin ? EMBO J 4:1163-1170

Liesi P (1985b) Laminin-immunoreactive glia distinguish regenerative adult CNS systems from non-regenerative ones. EMBO J 4:2505-2511

Liesi P, Kaakola S, Dahl D, Vaheri A (1984) Laminin is induced in astrocytes of adult brain by injury. EMBO J 3:683-686

Newman EA (1986) The Mueller cell. In: Fedoroff S, Vernandakis A (eds) Development, morphology and regional specialization of astrocytes. Academic Press, New York, pp 149-171

Seiler M, Turner JE (1988) The activities of glial cells following retinal transplantation into the lesioned adult rat eye : developmental expression of glial markers. Dev Brain Res, in press

Shaw G, Weber K (1983) The structure and development of the rat retina : an immunofluorescence microscopic study using antibodies specific for intermediate filament proteins. Eur J Cell Biol 30:219-232

Smalheiser NR, Crain SM, Reid LM (1984) Laminin as a substrate for retinal axons in vitro. Dev Brain Res 12:136-140

Turner JE, Blair JR (1986) Newborn rat retinal cells transplanted into a retinal lesion site in adult host eyes. Dev Brain Res 26:91-104

ACETYLCHOLINESTERASE ACTIVITY AND ALPHA-BUNGAROTOXIN BINDING IN THE INNER RETINA OF A MARINE TELEOST

S. Deplano
Institute of Comparative Anatomy
University of Genoa
Viale Benedetto XV, 5
16132 Genova
Italy

Acetylcholine (ACh) is accepted as a retinal neurotransmitter, and a specific class of amacrine cells in mammals have been elucidated as cholinergic neurons (Masland et al., 1984, Voigt, 1986). Specific identification of such neurons provides new impetus for an investigation both of the enzymes involved in the cholinergic system (e.g., the ACh-hydrolyzing enzyme, and acetylcholinesterase, AChE) and of the locations of acetylcholine receptors.

The distribution of AChE in the retina has been the subject of many histochemical studies which have shown the enzyme to be concentrated in cell somas of amacrine and ganglion cells and in their processes in the IPL, arranged in a laminar distribution (Hayashi, 1980, Neal, 1983).

The binding of alpha-bungarotoxin, a specific nicotinic ligand present at the vertebrate neuromuscular junction (Chang, 1963) and in the CNS (Oswald & Freeman, 1981) was located in plexiform layers in the goldfish retina, and amacrine, ganglion, and bipolar cells were identified as postsynaptic elements (James & Klein, 1985, Zucker & Yazulla, 1982).

The present study is an attempt to correlate the morphology of AChE--containing neurons and elements with alpha-bungarotoxin binding in the inner retina of a marine teleost.

Two series of experimental procedures were applied to dark-adapted retinas of marine teleost, the bogue (<u>Boops boops</u>, Sparidae family):

1- histochemical staining with fluorescein-conjugated-alpha-bungarotoxin (α-BTX/FITC) (Sigma). Freshly dissected tissue was incubated for 1

NATO ASI Series, Vol. H31
Neurobiology of the Inner Retina
Edited by R. Weiler and N. N. Osborne
© Springer-Verlag Berlin Heidelberg 1989

hour at 4°C with 10^{-7}M αBTX/FITC in phosphate buffer (PB) (1:1), followed by several rinses in PB to remove the unbound toxin derivative. Then the retina was flat-mounted on a glass slide, ganglion side up, with a mixture of pyrogallate-glycerol (1:1). Control tissues were exposed for 1 hour either to native αBTX (10^{-8}M) or to d-tubocurarine chloride (10^{-3}M); αBTX/FITC was added to each control for 1 hour at a final concentration of 10^{-7}M. Fluorescence was observed with a Zeiss epifluorescence photomicroscope.

2- histochemical procedure for the detection of acetylcholinesterase activity (AChE): unfixed retinas were processed according to the method of Karnowski & Roots (1964). An incubation time of 24 hours, at room temperature, allowed visualization of the retinal neurons. BW284c51 was used as an inhibitor for acetylcholinesterase, and iso-OMPA was used as an inhibitor for non-specific cholinesterase. The tissue was washed in phosphate buffer, fixed in glutaraldehyde 2% for 24 hours, and embedded in JB4 preparation (Polyscience). Ten micrometer sections of embedded material were used for light microscopy (Zeiss photomicroscope).

Acetylcholinesterase activity was concentrated in the IPL of the bogue retina, forming a wide band near the border of the INL from stratum S_1 to stratum S_2 and a narrow band corresponding to stratum S_4 (Fig. 1a). The cell bodies showing AChE activity included amacrine cells and ganglion cells (Fig. 1). Reaction product was not seen in retinas incubated in the presence of inhibitors, suggesting that the observed activity was not artifactual. The AChE-containing amacrine cells belong to the "coarse" and "tapered" classes previously described by Vallerga & Deplano (1984) from silver-impregnated bogue retinas. The coarse AChE-positive amacrine cells are monostratified on stratum S_1 (Fig. 1b), which extends close to the amacrine cell bodies. The tapered AChE-containing amacrine cells are stratified on stratum S_4 (Fig. 1e), or are bistratified on strata S_2 and S_4 (Fig. 1d). In the ganglion cell layer, AChE activity was present in oval, large somata from which there departed three or four stout dendrites (Fig. 1c) or two primary

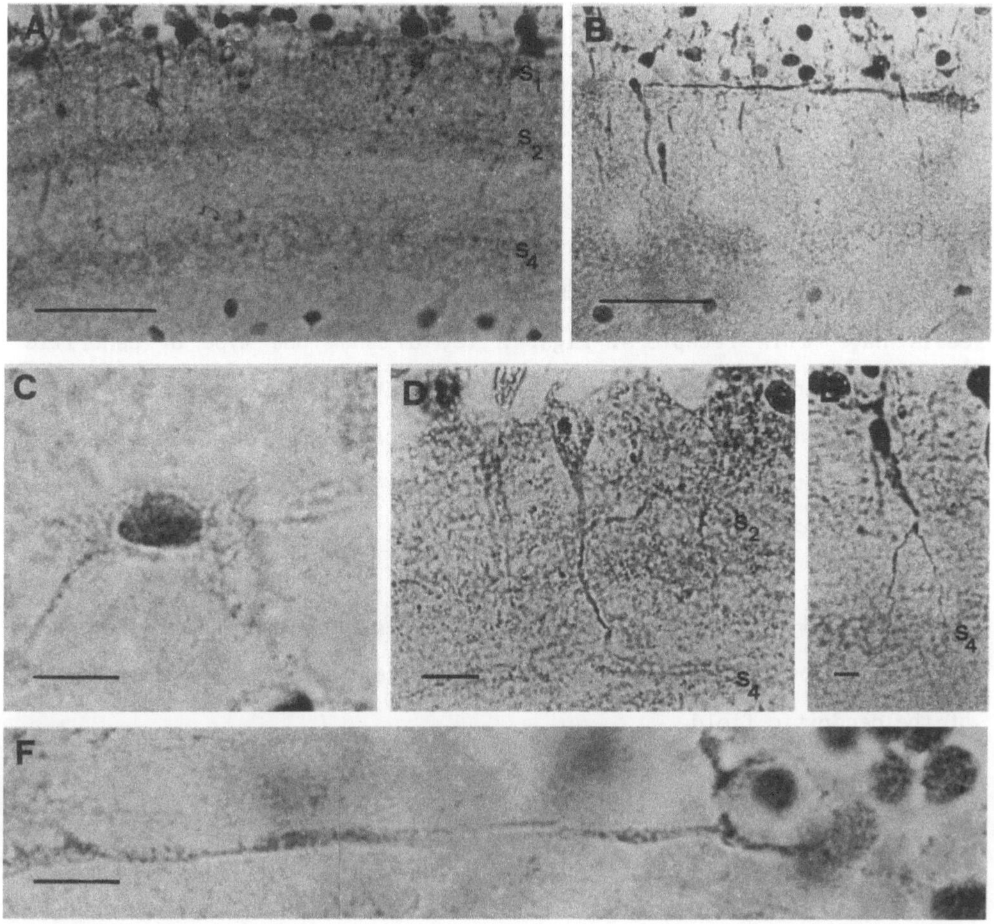

Fig. 1

Acetylcholinesterase activity in the inner retina.

A: Localization on strata, S_1, S_2, and S_4, of the IPL. Bar 50 µm. B: "coarse" amacrine cell stratified on stratum S_1. Bar 50 µm. C: tangential section of a ganglion cell. Bar 10 µm. D: "tapered" amacrine cell bistratified on strata S_2 and S_4. Bar 10 µm. E: "tapered" amacrine cell whose dendrites stratify on stratum S_4. Bar 10 µm. F: tangential section of a ganglion cell. The long dendrite located in the GCL, ramifies for several micrometers and then it spread up to stratum S_4. Bar 10µm.

dendrites running for several micrometers along the nuclear layer; from such dendrites, secondary dendrites spread up to stratum S_4. Labelling in transverse sections reveals αBTX cell bodies at two levels in the INL, corresponding to bipolar and amacrine cells, and three bands of dendritic labelling in the IPL from stratum S_1 to stratum S_4 (Fig. 1a). Flat-mounted preparations show the density of labelled somata in the inner board of the INL (Fig. 2b). Two classes of amacrine cells have αBTX receptors. Figs. 2 (c, d) show one type of dendritic morphology in two cells with different dendritic branching, whose dendrites exhibit the main feature of coarse-class amacrine cells. The second type characterized by thinner and beaded processes, is shown in Fig. 2e.

Few elements are labelled in the GCL. Figs. 2 (f, g, h) show the main types of ganglion cells with toxin receptors. The distinctive features of these labelled neurons are a large-sized soma and the extension of the dendritic field.

There exist no descriptions of the morphology and stratification of retinal neurons involved in the postsynaptic cholinergic circuitry. Famiglietti (1987) suggested by a Golgi study that the on-off directionally selective ganglion cells of cat retina make selective connections with starburst cholinergic amacrine cells. However, the neurons involved in the postsynaptic cholinergic pathway are amacrine, bipolar and ganglion cells, as demonstrated by ultrastructural studies on goldfish (Zucker & Yazulla, 1982) and on mouse retina (Pourcho, 1979).

Fig. 2
Fluorescence micrographs of BTX/FITC binding in the IPL of the bogue retina.
A: laminar distribution in the IPL. Labelled somata of amacrine (arrow), bipolar (arrowhead), and ganglion (open arrow) cells are present. Bar 50 µm. B: flat-mount preparation focused on the inner board of the INL, where amacrine cell somata are present. C: amacrine cell with thick dendrites layered just below the amacrine cell somata. D: amacrine cell with thick and ramified processes. E: amacrine cell with fine, beaded dendrites. F, G, H: ganglion cells. Bars 10µm.

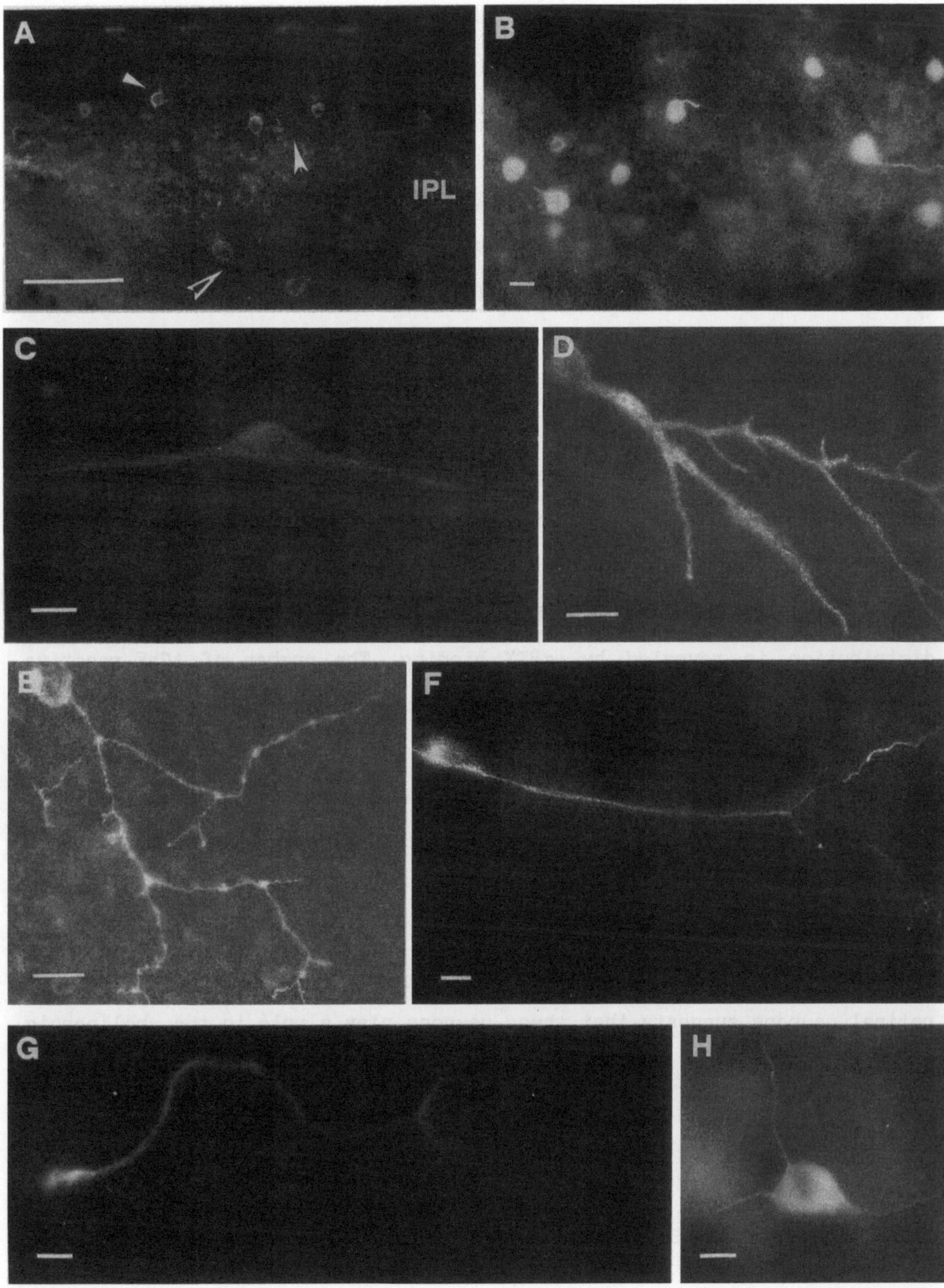
A
IPL
B
C
D
E
F
G
H

The aim of the present study is to find out which neurons have nicotinic receptors and to define the morphology of postsynaptic neurons in the cholinergic pathway. The significance of a combined study using AChE detection and αBTX binding can be useful in determining the distribution and the morphology of retinal neurons linked to such cholinergic circuitry. The involvement of AChE in ACh metabolism enables one to predict that this enzyme will be present in both cholinergic and cholinoceptive cells, even though the distribution of AChE occurs beyond the sites of direct cholinergic input (Pourcho & Osman, 1986). However, the morphology of "coarse" and "tapered" AChE-containing amacrine cells closely resembles the morphology of αBTX labelled amacrine cells with thick dendrites. Furthemore, at least two strata with AChE activity in the IPL (S_2 and S_4) correspond to those with ChAT immunoreactivity in goldfish retina (Tumosa et al., 1984), and a laminar distribution is revealed by αBTX binding. The number of AChE-reactive ganglion cells in the bogue retina exceeds the number of ganglion cells with αBTX receptors, but at least one type of AChE-containing neurons in the GCL resembles the large-sized dendritic field of αBTX labelled ganglion cells. The function of αBTX receptors is controversial, but it seems that αBTX binding sites in poikilothermic vertebrate optic tectum (Freeman et al., 1980, Henley et al., 1986) are acetylcholine receptors. The location of this binding in a specific population of retinal neurons suggests that these neurons play a role in the cholinergic circuitry, and that nicotinic receptors are likely to be present in amacrine and ganglion cells with definite dendritic morphology.

REFERENCES

Chang CC, Lee Y (1963) Isolation of neurotoxin from the venom of <u>Bungarus multicinctus</u> and their modes of neuromuscular blocking action. Archs Int Pharmacody 144:241-257

Famiglietti EV (1987) Starburst amacrine cells in cat retina are associated with bistratified presumed directionally selective, ganglion cells. Brain Res 413:404-408

Freeman JA, Schmidt JT, Oswald RE (1980) Effect of alpha-bungarotoxin in retino-tectal synaptic transmission in goldfish and toad. Neuroscience 5:929-942

Hayashi T (1980) Histochemical localization of dopamine and acetyl cholinesterase activity in the carp retina. Acta Histochem Cytochem 13:330-342

Henley JM, Mynlieff M, Lindstrom JM, Oswald RE (1986) Interaction of monoclonal antibodies to electroplaque acetylcholine receptors with the α-bungarotoxin binding site of goldfish brain. Brain Res 364:405-412

James WM, Klein WL (1985) α-bungarotoxin receptors on neurons isolated from turtle retina: molecular heterogeneity of bipolar cells. J Neurosci 5:352-361

Karnowsky MJ, Roots L (1964) A "direct coloring" thiocholine method for cholinesterase. J Histochem Cytochem 12:219-221

Neal MJ (1983) Cholinergic mechanisms in the vertebrate retina. In: Osborne N, Chader G (eds) Progress in Retinal Research 2, Pergamon Press, Oxford, 191-212

Oswald RE, Freeman JA (1981) α-bungarotoxin binding and central nervous system nicotinic receptors. Neuroscience 6:1-14

Pourcho RG (1979) Localization of cholinergic synapses in mammalian retina with peroxidase conjugated α-bungarotoxin. Vision Res 19:287-293

Pourcho RG, Osman K (1986) Acetylcholinesterase localization in cat retina: a comparison with choline acetyltransferase. Exp Eye Res 43:585-594

Tumosa N, Eckenstein F, Stell WK (1984) Immunocytochemical localization of putative cholinergic neurons in the goldfish retina. Neurosci Lett 48:255-259

Vallerga S, Deplano S (1984) Differentiation, extent and layering of amacrine cells in the retina of a sparid fish. Proc R Soc Lond B 221:465-477

Voigt T (1986) Cholinergic amacrine cells in the rat retina. J Comp Neurol 248:19-35

Zucker C, Yazulla S (1982) Localization of synaptic and nonsynaptic nicotinic-acetylcholine receptors in the goldfish retina. J Comp Neurol 204:188-195

VOLTAGE CLAMP STUDY OF AMACRINE CELLS IN CARP RETINA

J.C. Low, M. Yamada & M.B.A. Djamgoz
Neurobiology Group
Department of Pure & Applied Biology
Imperial College
London, SW7 2BB. U.K.

Light-evoked amacrine cell (AC) responses in the vertebrate retina have traditionally been classified into three types; transient ON-OFF, and sustained which subdivide into ON (depolarizing) or OFF (hyperpolarizing) (e.g. Kaneko 1973; Famiglietti et al., 1977; Djamgoz 1986). Within this scheme some variations in photoresponse waveform have been noted (Djamgoz, 1986; Teranishi et al. 1987). It has been suggested however, that all ACs have similar membrane properties, and that differences in their synaptic circuitry lead to the variety of photoresponse waveforms seen (Djamgoz & Ruddock, 1978; Toyoda & Fujimoto, 1984; Djamgoz, 1986). Little information has been published on 'resting' membrane characteristics of ACs or their light-evoked synaptic currents. We have addressed these issues by determining current-voltage relationships of ACs and studying their light-evoked synaptic currents under 'voltage-clamp' conditions. This article reports our initial findings.

Materials and Methods

The fish, mirror carp (*Cyprinus carpio*) was dark adapted for 20 minutes then pithed and the eyeball enucleated and hemisected. The retina was isolated and mounted receptor-side up on a piece of filter paper in a perfusion chamber. The retina was perfused by saline with the following compostion (in mM): NaCl 102, $NaHCO_3$ 28, $CaCl_2$ 1, $MgCl_2$ 1, KCl 2.6 and glucose 5. The saline was aerated with a gas mixture (95% O_2 and 5% CO_2) giving a pH of 7.6. Double-barrelled glass microelectrodes filled with 3M KCl were used for intracellular recording and voltage clamp, using two pre-amplifiers (WPI 767 and Nihon

NATO ASI Series, Vol. H31
Neurobiology of the Inner Retina
Edited by R. Weiler and N.N. Osborne
© Springer-Verlag Berlin Heidelberg 1989

Kohden MEZ 8201). The resistance of each barrel was between 100 & 150 MOhm and only electrodes that had coupling resistances of less than 1MOhm (measured by passing $\pm$ 1nA) were used. The retina was stimulated by a 2-3 mm diameter spot of 640nm light delivered by a single beam photostimulator. Cells were penetrated from the receptor side of the retina, and amacrine cells identified on the basis of (a) the depth of the electrode tip in the retina and (b) the characteristic light-evoked responses. In many ACs the resting potential we observed was close to -30mV and this value was used as the holding potential in the voltage clamp. A microcomputer was used to generate two types of command voltages, first a positive going (depolarizing) ramp from -80 to +80mV at 20mV/sec, and then steps (4.25 sec duration, 50% duty cycle) from holding to the following voltages 0,-20,+20,-40,+40,-60,+60,-80mV. We were thus able to obtain two types of current-voltage curve, a ramp I-V and a step I-V curve. At the end of a clamp sequence, the response of the cell to light was checked, and unless the cell was still clearly responding to light the data were not used for analysis. No estimates of leakage currents have been made in this study.

Results & Discussion

We have so far investigated 36 ACs and have observed a variety of forms of both ramp and step I-V curves. The membrane resistance in the dark was determined by measuring the slope of the I-V curves at 0nA. The measurements made so far gave a value for membrane resistance of 26 $\pm$ 13 MOhm (mean $\pm$ s.d., n=36), and this was not markedly different for different AC types. This value is close to that found in isolated retina by Watanabe & Murakami (1985), but much less than in slices of salamander retina (Eliasof, Barnes & Werblin, 1987).

A region of negative conductance at membrane voltage levels slightly below 0mV, typically -8mV, was seen in many of the transient type (ON-OFF) amacrine cells (T-AC) investigated

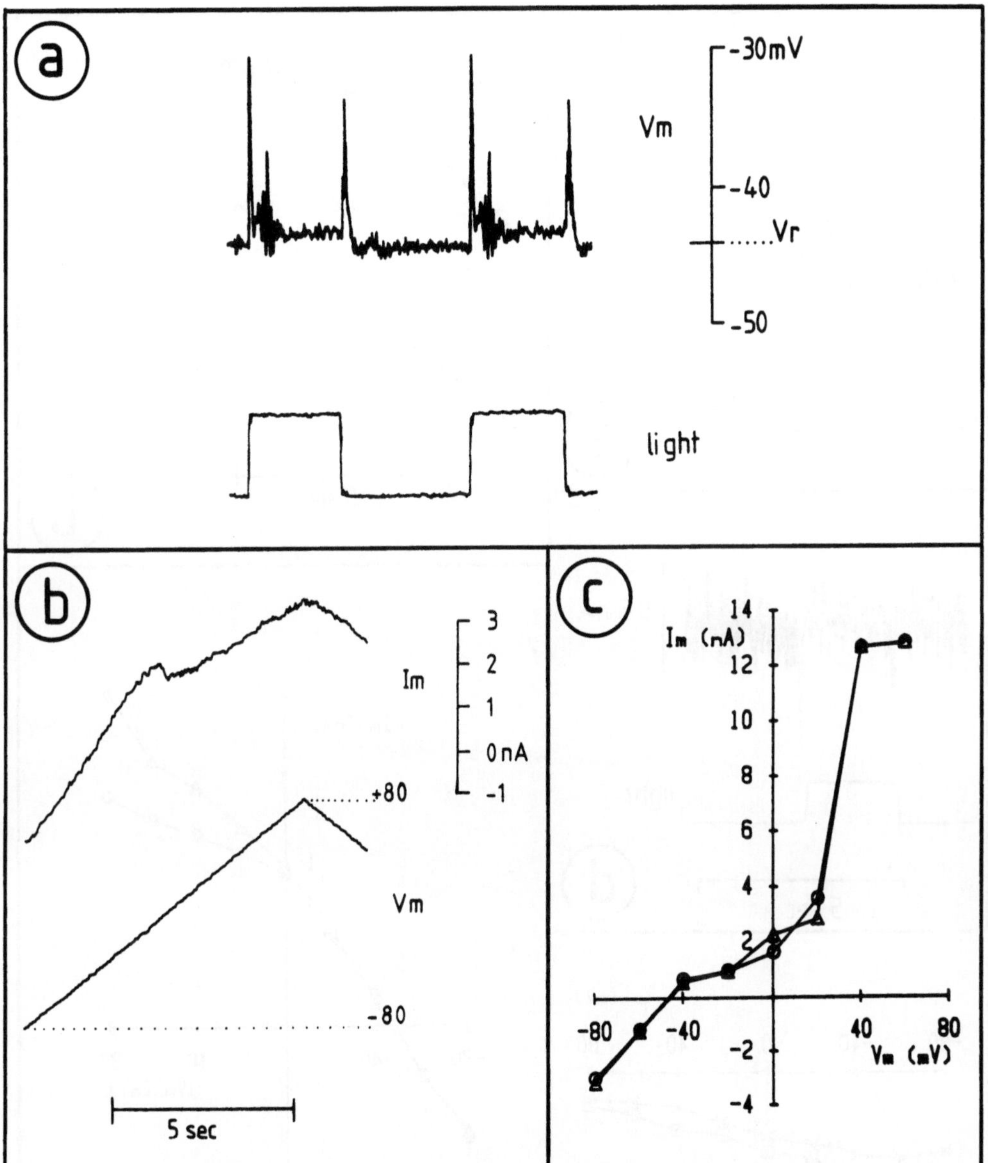

Fig. 1. a) The voltage response to red light of a T-AC; b) membrane current (upper trace, ordinate shows current with outward current displayed upwards) as a function of the command voltage ramp (the clamped membrane voltage is shown beneath); c) I-V curve determined by voltage clamp to steps of command voltage for this cell measured 100ms (triangle), and 2s (circle) after the onset of the clamping voltage from a holding potential of -30mV.

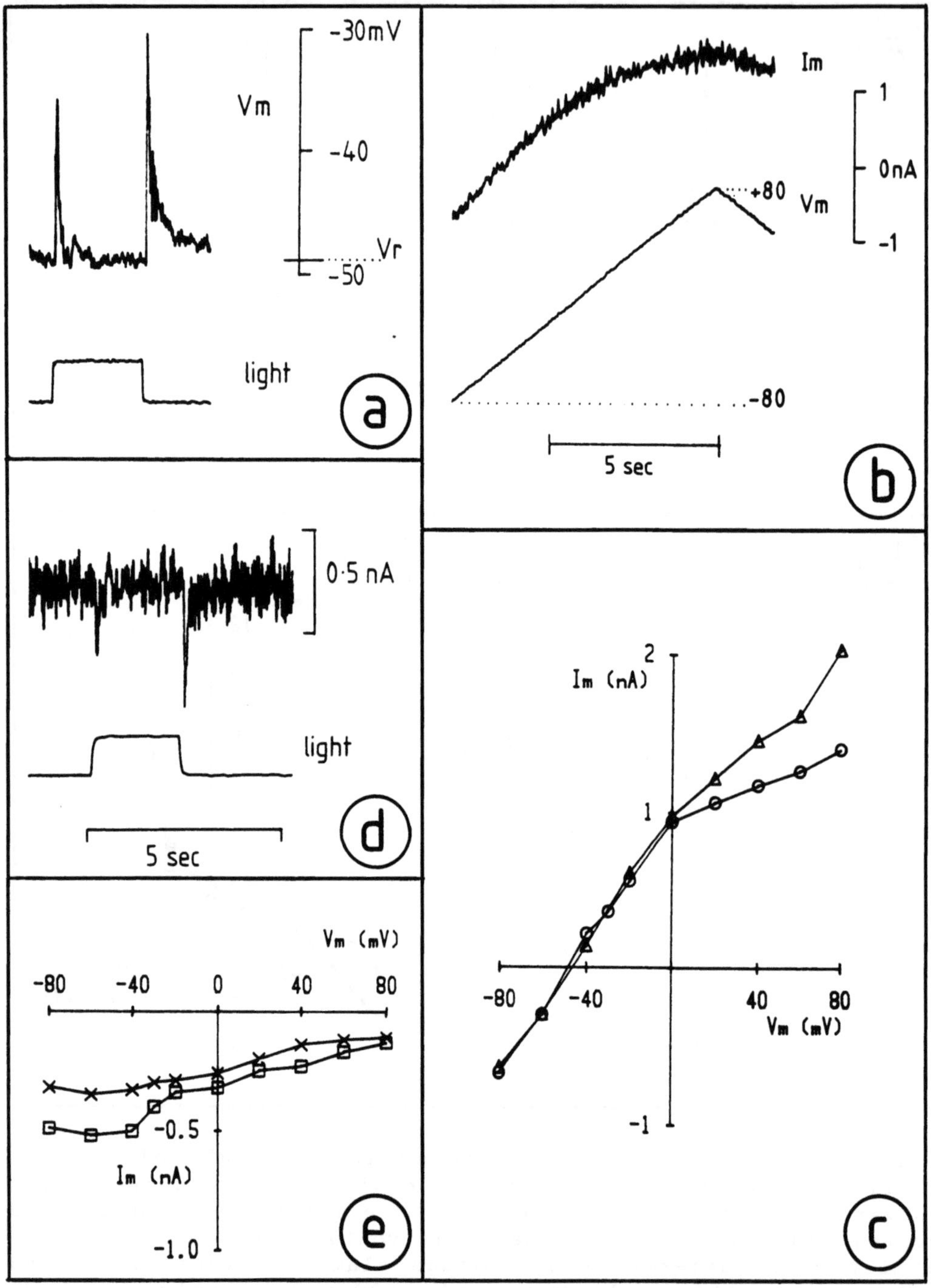

Fig. 2 a) The voltage response to red light of a T-AC; b) I-V curve determined by voltage clamp to a command voltage ramp; c) I-V curves for this cell measured 100ms (triangle), and 2s (circle) after the onset of the clamping voltage from a holding potential of -30mV; d) synaptic current by photo-stimulus under voltage clamp at -40mV; e) I-V curves of the ON (cross) and OFF (square) synaptic currents as in (d).

(Fig. 1b). Although the photoresponse of the cell in Fig. 2a was similar in many respects to that of Fig. 1a no negative conductance region was seen in the ramp I-V curve (Fig. 2b). An example of an I-V curve obtained using voltage step intervals of 20mV is shown in Fig. 1c. Typically for T-ACs, step I-V curves had steep slopes at membrane voltages below -40mV and above +20mV, the slope conductances being least in the mid-range between these two values.

Regions of negative conductance in the ramp I-V curve have been noted for T-ACs in the mudpuppy (Werblin, 1977) but at more negative voltages than we have observed, typically spanning the range -35 to -20mV. One possible explanation for the difference may be that negative slope conductance regions occur at different membrane potentials in different species. Such a phenomenon occurs in cat spinal motoneurons (Schwindt & Crill, 1981), where two types of negative slope conductance have been reported; one at depolarizations of 10-20mV relative to a resting potential of -60mV and another at larger depolarizations of 60 to 120mV. Another possibility is that the differing rate of voltage ramp used in the two studies contributed an additional factor. More recent studies of ACs in the tiger salamander retina (Barnes & Werblin, 1986; Eliasof, Barnes & Werblin, 1987) have shown that four types of ionic current are involved in the generation and control of the T-AC spike. One of the two major ionic contributions to the spike response, a calcium -activated potassium conductance, has kinetics which vary depending on the direction of voltage polarization, i.e. is dependent upon the recent voltage history of the membrane. In molluscan neurones also a negative slope conductance due to a calcium-mediated potassium current has been reported and this progressively deactivates as the membrane potential approaches the calcium equilibrium potential (Meech & Standen, 1975).

Synaptic currents of amacrine cells, elicited by photostimuli, were obtained at each membrane potential under the voltage clamp condition. These currents represent the change in membrane conductances which are affected by synaptic inputs, such as ON and OFF bipolar cells and other amacrine cells. The photoresponse (e.g Fig. 1a, 2a) was observed in conditions in which rod horizontal cell responses were not detected, thus simplifying the input to the amacrine cells. Synaptic currents of T-ACs corresponding to transitions in light intensity were monophasic inward, and mediated by a conductance increase to ion(s) (Fig. 2d). In this cell the synaptic currents increased by about 0.05nA for each 20mV polarization of the membrane (Fig. 2e), corresponding to a peak conductance increase of about 2.5nS. The inward synaptic currents became larger at increasing levels of hyperpolarization and reduced with depolarization. Slender processes and / or electrical coupling between cells might have made it difficult to completely space clamp the sub-synaptic membrane. In most ACs we were not able to get reversal potentials of synaptic currents and they could only be estimated by extrapolation to be above 80mV. This means that most of the inputs to the T-ACs were conductance increasing synapses with positive reversal potentials and were, therefore, excitatory. This finding is in agreement with earlier observations that amacrine cells receive mainly excitatory input(s) (Miller & Dacheux,1976; Toyoda, Hashimoto & Ohtsu, 1973 ; Werblin, 1977; Toyoda & Fujimoto, 1984; Kujiraoka, Saito & Toyoda, 1986). A sustained component of synaptic current was often observed even in T-ACs, it corresponded with a small sustained component of the voltage response to light between ON & OFF transient components (Djamgoz & Ruddock, 1983; Watanabe & Murakami, 1985). Inward sustained currents corresponded to membrane depolarization and outward ones to membrane hyperpolarization.

In conclusion, the form of the I-V curves and other indices of membrane behaviour for the cells so far investigated do not seem to correlate with particular physiological amacrine cell types. Our initial experiments have shown apparent differences in AC membrane input characteristics. It is not clear however, whether these differences reflect intrinsic or imposed properties, derived from interaction with other cells.

Acknowledgement

This research was funded by The Wellcome Trust

References

Barnes S & Werblin F (1986) Gated currents generate single spike activity in amacrine cells of the tiger salamander. Proc Natl Acad Sci USA 83: 1509-1512

Djamgoz MBA (1986) Common features of light-evoked amacrine cell responses in vertebrate retina. Neurosci Lett 71: 187-191

Djamgoz MBA & Ruddock KH (1978) Properties of amacrine cell responses recorded from isolated fish retinae. Neurosci Lett 7: 89-93

Eliasof S Barnes S & Werblin F (1987) The interaction of ionic currents mediating single spike activity in retinal amacrine cells of the tiger salamander. J Neurosci 7: 3512-3524

Famiglietti EV, Kaneko A & Tachibana M (1977) Neuronal architecture of ON and OFF pathways to ganglion cells in carp retina. Science 198: 1267-1268

Kaneko A (1973) Receptive field organization of bipolar and amacrine cells in the goldfish retina. J Physiol (Lond) 235: 133-153

Kujiraoka T Saito T & Toyoda J (1986) Bipolar-amacrine synaptic transmission: effect of polarization of bipolar cells on amacrine cells in the carp retina. Neurosci Res suppl 4 S111-S119

Meech RW & Standen NB (1975) Potassium activation in Helix aspersa neurones under voltage clamp: a component mediated by calcium influx. J Physiol (Lond) 249: 211-239

Miller RF & Dacheux RF (1976) Synaptic organization and ionic basis of On and Off channels in mudpuppy retina. III. A model of ganglion cell receptive field organization based on chloride free experiments. J gen Physiol 67: 679-690

Schwindt PC & Crill WE (1981) Negative slope conductance at large depolarizations in cat spinal motoneurons. Brain Res 207: 471-475

Teranishi T, Negishi K & Kato S (1987) Functional and morphological correlates of amacrine cells in carp retina. Neuroscience 20:935-950

Toyoda J Hashimoto H & Ohtsu K (1973) Bipolar-amacrine transmission in the carp retina. Vis Res 13: 295-307

Toyoda J & Fujimoto (1984) Application of transretinal current stimulation for the study of bipolar-amacrine transmission. J gen Physiol 84: 915-925

Werblin FS (1977) Regenerative amacrine cell depolarization and formation of on-off ganglion cell response. J Physiol (Lond) 264: 767-785

Watanabe S & Murakami M (1985) Electrical properties of ON-OFF transient amacrine cells in the carp retina. Neurosci Res suppl 2. S201-S210.

CARP RESPONSES OF COLOR CODED GANGLION CELLS EVOKED BY CURRENT INJECTION INTO HORIZONTAL CELLS

M. Kamermans, B.W. van Dijk and H. Spekreijse
Laboratory of Medical Physics, Academic Medical Center, Meibergdreef 9, Amsterdam-Zuidoost; The Netherlands Ophthalmic Research Institute, P.O.Box 12141, 1100 AC Amsterdam-Zuidoost, the Netherlands; and the National Institute for Physiological Sciences, Dept. of Information Physiology, Okazaki, Japan

Correspondence should be addressed to B.W. van Dijk, at the Netherlands Ophthalmic Research Institute

In catfish and dogfish retina where a single receptor type determines the light response, the projections from horizontal cells (h.c.s) to ganglion cells (g.c.s) are all sign inverting. Injection of a hyperpolarizing current into the h.c. results in a g.c. response of equal sign as the g.c. surround response, whereas a depolarizing current results in a response of the center sign (Naka & Nye, 1971; Naka & Witkovsky, 1972; Naka, 1976). In pike retina and in turtle retina also On/Off g.c. responses to extrinsic current injected into h.c.s have been described (Maksimova, 1969; Schwartz, 1973); while in carp retina an even more complex response behavior is observed.

It may be argued that this complex response behavior is due to the existence of color coding in carp retina. To study this possibility we performed experiments in Carp retina in which we related the response behavior of g.c.s to extrinsic current injection into h.c.s to the type of h.c. into which the current was injected and to the color coding of the g.c.

METHODS

Setup

Retinas of dark adapted common carp (Cyprinus Carpio) were isolated and superfused using standard procedures (Yang et al., 1982).

The optical stimulator used is described in detail elsewhere (Yang et al.,1982); in short it allowed the projection of spots and annuli of different sizes, wavelengths and intensities from the receptor side on the

NATO ASI Series, Vol. H31
Neurobiology of the Inner Retina
Edited by R. Weiler and N.N. Osborne
© Springer-Verlag Berlin Heidelberg 1989

retina. The maximum available intensity was 5.10^{16} photons $s^{-1}m^{-2}$; the intensity of light stimuli will be specified in relative log units with respect to this intensity level.

Responses from g.c.s were recorded extra-cellularly using PtIr electrodes (g.c.-electrode). Recorded spikes were shaped and fed into a PDP 11/34 computer which stored the event times and calculated peri-stimulus time histograms (p.s.t.h.s) by counting all spikes in successive 1 ms time bins.

Responses from h.c.s were recorded intra-cellularly using micro-pipettes filled with 3M KCl, which had a resistance of 50-100 MOhm when measured in ringer's. The headstage amplifier allowed steady current injection through the electrode tip as long as the potential drop over the tip was smaller than 3.5 V, hence the maximum current was in the range of 30-60 nA.

The h.c.- and g.c.-electrodes were mounted on two independent micromanipulators. The electrodes were carefully aligned and positioned under a preparation microscope immediately before the experiment. The tip of the g.c.-electrode was approx. 100 um lower than the tip of the h.c.-electrode. The manipulator settings were marked and the electrodes withdrawn before placing the retina into the chamber. The h.c.-electrode could be moved by means of a micrometer in small steps in one horizontal direction. Figure 1 gives a schematic overview of the electrode configuration.

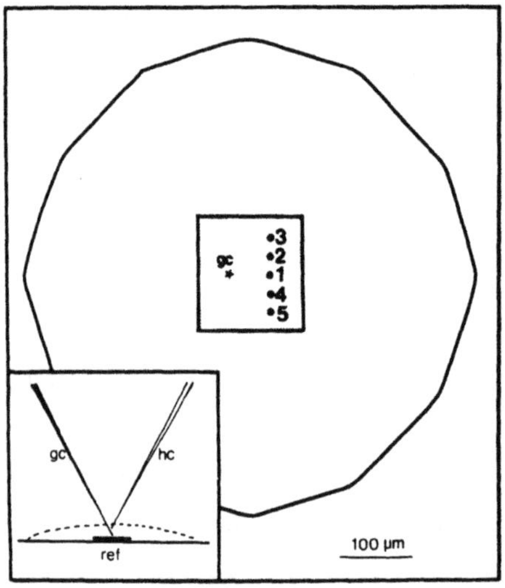

Figure 1. The electrode configuration. The two electrodes penetrate the retina under an angle of approx. 60 deg. With both electrode tips centered in the stimulus the tip of the g.c.-electrode is approx. 100 um lower than that of the h.c. electrode. Five successive penetrations were made with the h.c. electrode for each g.c. studied. All recorded cells were located within the shaded area.

This recording method made it possible to record from h.c.-g.c. pairs with close receptive field centers. A disadvantage was that a new retina had to be used when the h.c.-electrode had to be replaced after approx. 10 penetrations. The reasons for this complicated procedure were that it proved to be very difficult to evoke h.c. polarizations of the size of light responses with current injection, and more importantly that it would be unlikely to find other projections than to the g.c. surround when the two electrodes would be widely separated. With the procedure used the distance between the receptive field centers of the h.c.-g.c. pair were never larger than 170 um.

Classification

Light responses from g.c.s were classified according to the cone mechanisms feeding the center and surround. The classification scheme used is identical to that reported in Van Dijk and Spekreijse, 1984.

Light responses from h.c.s were characterized according to their spectral behavior into monophasic, biphasic or triphasic responses (Norton et al. 1968). For monophasic h.c.s the site of penetration of the cell by the electrode, i.e. soma or axon terminal, could be distinguished on the basis of the size of the receptive field, which was measured using spots of different diameters (Yagi & Kaneko, 1986).

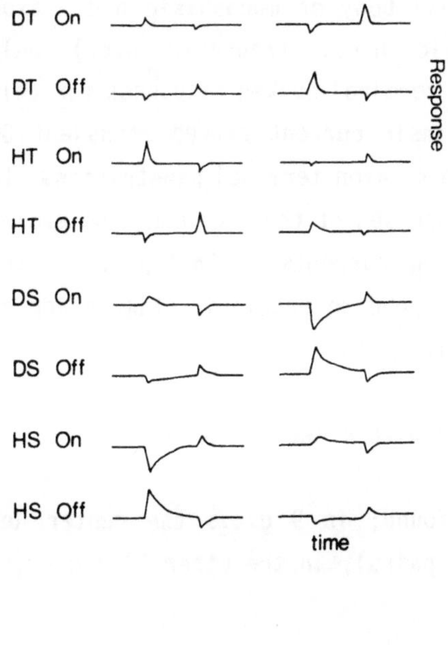

Figure 2. Schematic overview of the classification of g.c. responses to extrinsic current injection in a h.c. The first column depicts the response type, the second column the hypothetical response to hyperpolarizing current, the third column to depolarizing current.

The g.c.-responses to extrinsic current were classified according to the scheme depicted in fig. 2. Responses to extrinsic current were classified as **On** or **Off**. They could be **T**ransient or **S**ustained (i.e. lasting shorter than or throughout the applied 0.5s current pulse); and they could be evoked best by **D**epolarizing or by **H**yperpolarizing current or **B**oth polarities could yield equally strong responses. Of the 12 possible response types to extrinsic current the DTOn, BSOn and BSOff types were never observed.

RESULTS

In total we studied 92 g.c.s and 321 g.c.-h.c. pairs. The results are presented per g.c. type. All results are summarized in table I. It should be noted that the various types of color coded g.c.s were selected in the experiments in such a way that about equal numbers for each class were studied.

Red On/Red Off ganglion cells

Red On/Red Off ganglion cells receive input only from Red cones; these g.c.s have a low mean spontaneous activity; their responses are transient. The ratio of the sensitivities of the On and Off mechanisms is often little sensitive for the spatial properties of the stimulus (Van Dijk and Spekreijse, 1984). Four Red On/Red Off g.c.s have been studied (13 cell pairs). Of the four types of h.c.s (cell body of monophasic h.c., axon terminal of monophasic h.c., biphasic h.c., triphasic h.c.) only polarization of the monophasic h.c. axon terminal evoked responses for this type of g.c. Hyperpolarization by extrinsic current evoked transient ON g.c. responses in 4 out of 7 monophasic h.c. axon terminal penetrations. In a fifth monophasic h.c. axon terminal responses of the g.c. could be evoked by both hyperpolarizing and depolarizing currents (BTOn-type). In the remaining two cases, the g.c. did not respond to extrinsic stimulation of the monophasic h.c. axon terminal at all.

Center Red/Green opponent ganglion cells with a Red surround

Nineteen cells of this category were found; in 9 g.c.s the center Red response component was On (43 g.c.-h.c. pairs), in the other 10 g.c.s (51

Ganglion cell type	Response (columns)	Axon terminal, monophasic	Cell body, monophasic	Biphasic	Triphasic	
$R^{on} R^{off}$, 4 cells	TOn (D B H)	– 1 4	– – –	– – –	– – –	
	TOff (D B H)	– – –	– – –	– – –	– – –	
	SOn (D H)	– –	– –	– –	– –	
	SOff (D H)	– –	– –	– –	– –	
	Tot.	7	4	2	0	13
	N.R.	2	4	2	0	8
$R^{on} G^{off}/R$, 9 cells	TOn (D B H)	– 4 –	– – –	– – –	– – –	
	TOff (D B H)	– – –	– – –	– 7 –	– – –	
	SOn (D H)	6 2	2 4	– –	– –	
	SOff (D H)	– 1	– 7	– –	– –	
	Tot.	15	20	7	1	43
	N.R.	2	7	0	1	8
$R^{off} G^{on}/R$, 10 cells	TOn (D B H)	– – –	– – 1	– – 8	– – –	
	TOff (D B H)	– – –	– – –	– 1 1	– – –	
	SOn (D H)	– –	2 5	2 –	– –	
	SOff (D H)	4 11	1	– –	– –	
	Tot.	17	17	14	3	51
	N.R.	2	8	2	3	15
$R^{on} G^{off}/RG$, 22 cells	TOn (D B H)	– – –	1 1 –	– – 5	– – –	
	TOff (D B H)	– 8 –	– 4 –	– – 2	– – –	
	SOn (D H)	4 –	– –	– –	– –	
	SOff (D H)	– 12	1 2	– –	– –	
	Tot.	28	26	12	2	68
	N.R.	4	17	5	2	28
$R^{off} G^{on}/RG$, 21 cells	TOn (D B H)	– – –	– 4 2	– – –	– – –	
	TOff (D B H)	– – –	– – –	– 7 –	– – –	
	SOn (D H)	– 12	– 8	– –	– –	
	SOff (D H)	7 –	– –	– 3	– –	
	Tot.	21	31	13	1	66
	N.R.	2	17	3	1	23
$R^{on} G^{off} B^{on}/RG$, 8 cells	TOn (D B H)	– 1 –	– – –	– – –	– – –	
	TOff (D B H)	– 4 –	– 3 –	– – 3	– – –	
	SOn (D H)	– 2	– 1	– 4	– 1	
	SOff (D H)	– 2	– –	1 –	– –	
	Tot.	12	7	10	2	31
	N.R.	3	3	2	1	9
$R^{off} G^{on} B^{off}/RG$, 8 cells	TOn (D B H)	– 5 –	– 5 –	– – –	– – –	
	TOff (D B H)	– – –	– – –	– – –	– – –	
	SOn (D H)	– 4	– 1	2 2	– –	
	SOff (D H)	5 –	– –	– 2	– 2	
	Tot.	20	19	6	3	48
	N.R.	6	13	1	1	21
		120 (21)	124 (69)	65 (15)	12 (9) [5]	

Table I. Overview of the collected data. Each row represents a ganglion cell type, each column a horizontal cell type. In each field the occurrence of projection types (fig.2) for each g.c. - h.c. combination is depicted. "Tot." gives the total number of pairs, "N.R." the number of non-projecting pairs.

pairs) the center Red response component was Off.

Of the various types of h.c.s, polarization of the biphasic h.c. was most effective. Hyperpolarization by extrinsic current yielded responses in 19 out of the 21 biphasic h.c.s. These responses were predominantly of the same sign as the light responses due to stimulation of the Green cones, i.e. hyperpolarizing current decreased the spike rate for the center Green Off cells and increased the rate for the center Green On cells[1]. These responses were more transient than the Green light responses. Extrinsic stimulation of monophasic h.c. cell bodies was less effective. In only 22 out of 37 cases responses of the g.c.s could be evoked by extrinsic current injection. If the g.c.s did respond, the response pattern was sustained and of the surround Red sign, i.e. HSOff (or DSOn) for the center Red On cells and HSOn for center Red Off g.c.s. Extrinsic polarization of monophasic h.c. axon terminals was more effective than polarization of the cell bodies (in 28 out of 32 cell pairs), and these responses were always opposite to the responses (if any) evoked by polarization of the cell bodies.

Double Opponent R/G ganglion cells

The most common receptive field coding of carp retinal ganglion cells is double opponent R/G, i.e. opponent input from Red and Green cones in both center and surround of the receptive field. We studied 43 g.c.s of this type, 22 g.c.s with a center Red On component (68 g.c.-h.c. pairs) and 21 with a center Red Off component (66 g.c.-h.c. pairs).

Figure 3 shows typical responses. Light responses to center stimulation with 689 nm light (showing the center Red On response) and to 536 nm light (showing the center Red On and Green Off responses) are shown in fig. 3a. Light responses to surround stimulation are shown in fig. 3b.

Monophasic h.c.s projected to this type of g.c. Again polarization of the cell bodies was less effective (23 out of 57 pairs yielded responses) than polarization of the axon terminals (43 out of 49 pairs). Responses to stimulation of cell body and axon terminal were similar. The responses caused by hyperpolarization of monophasic h.c.s had the same sign as the surround responses to deep red light (with 2 exceptions in 66 projecting cells). The response shapes to light and current were rather similar.

---

[1]We assume that monophasic h.c.s project to Red mechanisms and biphasic h.c.s to Green.

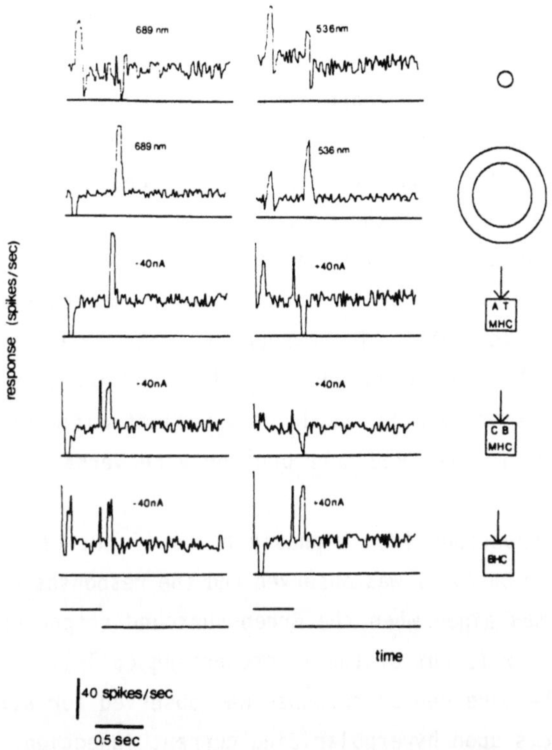

Figure 3. Responses from a center Red On, Green Off double opponent g.c. to:
a. center stimulation; intensity 0 log unit; spot size 1 mm;
b. surround stimulation; intensity 0 log unit; annulus size i.d. 2 mm, o.d. 2.5 mm;
c. current injection into the axon terminal of a monophasic h.c.;
d. current injection into the soma of a monophasic h.c.;
e. current injected into a biphasic h.c.

Examples of responses from extrinsic polarization of monophasic h.c.s are shown in fig. 3c (for current injection into an axon terminal) and in fig. 3d (for injection into a cell body). Biphasic h.c.s also projected. The responses due to polarization of biphasic h.c.s were of the same sign as the surround Green responses, i.e. HTOn for g.c.s with a surround Green ON and BTOff or HSOff for g.c.s with a surround Green Off component (fig. 3e). In only 2 out of 25 cases the projections were of center Green sign. None of the three triphasic h.c.s found simultaneously with double opponent g.c.s projected.

Red/Green/Blue ganglion cells

In carp retina a (small) number of g.c.s receive input from Blue cones. These g.c.s have a receptive field coding that is Red/Green double opponent and frequently a Blue center input of equal sign as the Red center. Blue surround receptive fields are never found (Van Dijk and Spekreijse, 1984). We studied 16 such g.c.s, 8 with center Red Off input (49 cell pairs) and

8 center Red On input (31 cell pairs).

Light responses of these cells are commonly transient, except for the center Blue and surround Red responses that are often sustained.

All h.c. types projected to this g.c. type. In most cases monophasic h.c.s projected to the Red surround, biphasic h.c.s to the Green surround and triphasic cells to the Blue center of the g.c. receptive fields, i.e. for a center Red Off g.c. the projections from monophasic h.c.s (both from cell body and from axon terminal) were mostly BTOn or HSOn, from biphasic h.c.s HSOff, and from triphasic h.c.s HSOff. For monophasic h.c.s a reverse mode behavior was observed for 7 out of 33 projecting units, and often the responses from extrinsic current stimulation could be transient while the light responses were sustained (for the Red surround) or vice versa (for the Green surround).

In a small number of projections from biphasic h.c.s (2 out of 14 projecting cells) also a reverse behavior was observed and the responses to extrinsic current were often sustained when the Green surround responses were transient (this occurred for 11 out of the 14 projecting cells).

A response similar to the Blue center response was observed for all three projecting triphasic h.c.s upon hyperpolarizing current injection.

DISCUSSION

Since we quite frequently found non-projecting cell pairs it is obvious that the responses to extrinsic current are not due to passive field effects.

Our data in general support the view that h.c.s feed to the surround of g.c.s:

1. Extrinsic current injected into monophasic h.c.s commonly evokes g.c. responses similar to responses evoked by stimulating the surround with red light. This is particularly true for the projections to double opponent g.c.s.
2. Extrinsic current injected in biphasic h.c.s generally evokes g.c. responses of equal sign as the g.c. responses evoked by green light stimulation of the surround.
3. Only a limited number of triphasic h.c.s was found (12 units). Almost all of these did not project to g.c.s. Current injected in the three units that did project always resulted in responses similar to the light

responses from the Blue center mechanism of R/G/B g.c.s.

However, it should be noted that exceptions from the above rules were numerous. In particular:

1. The projection from the monophasic h.c. axon terminal to R/G g.c.s with only a Red surround was in all cases of the center sign.
2. The time course of the light responses very frequently did not correspond to the time course of the responses to extrinsic current injection. Often the light responses were transient whereas the current responses were sustained. Again this occurred most consistently in the projection from monophasic h.c.axon terminal to R/G g.c.s with only a Red surround. Some caution seems indicated since the dynamics of the light response depends strongly on the intensity and spatial properties of the stimulus (Schellart and Spekreijse, 1973). The same spot yields generally a more sluggish response for the color coded center mechanism without an antagonistic surround.

Another unexpected result of this study is that we consistently observed that current injected into the axon terminals of monophasic h.c.s was more effective in evoking g.c. responses than current injected into the soma. This might be related to the high specific resistance of the axon terminal membrane (Yagi and Kaneko, 1987).

REFERENCES

Maksimova, Ye.M. (1969) Effect of intracellular polarization of horizontal cells on the activity of ganglionic cells of the retina of fish. Biofizika, 14, 537 - 544.

Naka, K.I. (1976) Neuronal circuitry in the catfish retina. Invest. Ophthalmol. Vis. Sci., 15, 926-936.

Naka, K.I. and Nye, P.W. (1971) Role of the horizontal cells in organization of the catfish retinal receptive field. J. Neurophysiol., 34, 785-801.

Naka, K.I. and Witkovsky, P. (1972) Dogfish ganglion cell discharge resulting from extrinsic polarization of the horizontal cells. J. Physiol., 223, 449-460.

Negishi, K., Kato, S., Teranishi, T. and Hayashi, T. (1978) Effects of extrinsic horizontal cell polarization on spike discharges in the carp retina. Brain Res., 148, 95-104.

Norton, A.L., Spekreijse,H. Wolbarsht, M.L. and Wagner, H.G. (1968) Receptive field organization of the S-potential. Science, 160, 1021-1022.

Schellart, N.A.M. and Spekreijse, H. (1972) Dynamic characteristics of retinal ganglion cell responses in goldfish. J.gen. Physiol., 59, 1-21.

Schwartz, E.A. (1973) Organization of on-off cells in the retina of the turtle. J. Physiol., 230, 1-14.

Van Dijk, B.W. and Spekreijse H., (1984) color fundamentals deduced from carp ganglion cell responses. Vision Res., 24, 211-220.

Yagi, T. and Kaneko, A. (1988) The axon terminal of goldfish retinal horizontal cells: a low membrane conductance measured in solitary preparations and its implication to the signal conduction from the soma. J. Neurophysiol., 59, 482-494.

Yang, X.L., Tauchi, M. and Kaneko, A. (1982) Quantitative analysis of photoreceptor inputs to external horizontal cells in the goldfish retina. Jpn. J. Physiol., 32, 399-420.

Acknowledgement:

The authors thank Drs A. Kaneko & M. Tachibana for the facilities given at the Institute for Physiological Sciences and support throughout the experiments.

M.K. is supported through the Foundation for Biophysics of NWO, and B.W.v.D is a recipient of the C&C Huygens Stipend of NWO.

BIPOLAR CELLS, AMACRINE CELLS AND GANGLION CELLS IN THE RETINA OF THE CALIFORNIAN SLENDER SALAMANDER, BATRACHOSEPS ATTENUATUS (AMPHIBIA, URODELA, PLETHODONTIDAE)

Rüdiger Linke
University of Bremen
Department of Biology
P.O. Box 330 440
2800 Bremen
Fed. Rep. Germany

The miniaturized californian slender salamander Batrachoseps attenuatus has excellent visual abilities despite a visual system that is largely reduced both centrally and peripherally with regard to the numbers of neurons (Linke et al., 1986; Roth et al., 1988). In contrast to the extended knowledge about physiologigcal properties of single retinal neurons, limited knowledge exists about the morphology of retinal neurons in salamanders. A morphological characterization of the retinal neurons that contribute to the inner plexiform layer (ipl), i.e. bipolar cells (bc), amacrine cells (ac) and ganglion cells (rgc) will enhance our knowledge of the morphological variations of cell types in vertebrate retinas.

Golgi-Kopsch (Collonier, 1964) and rapid Golgi (Valverde, 1970) methods were used to study retinal neurons in transverse sections. The application of HRP into the tectum opticum, the thalamus and the optic chiasm and subsequent processing with DAB (Adams, 1977; Stuermer, pers. com.) resulted in retrograde labeling of rgc, which were studied in flatmounts and in transverse sections.

Three types of bc, seven types of ac, and four classes of rgc could be distinguished.

Bipolar cells were differentiated by their axonal arborization. Type 1 arborizes near the ganglion cell layer in the supposed ON lamina whereas type 2 arborizes near the inner nuclear layer in the supposed OFF lamina. Also bc were found which have their axonal arborization in the middle of the ipl. In the tiger salamander the latter type is shown to be either ON or OFF type (Hare et al., 1986). Most of these bc possess a

NATO ASI Series, Vol. H31
Neurobiology of the Inner Retina
Edited by R. Weiler and N. N. Osborne
© Springer-Verlag Berlin Heidelberg 1989

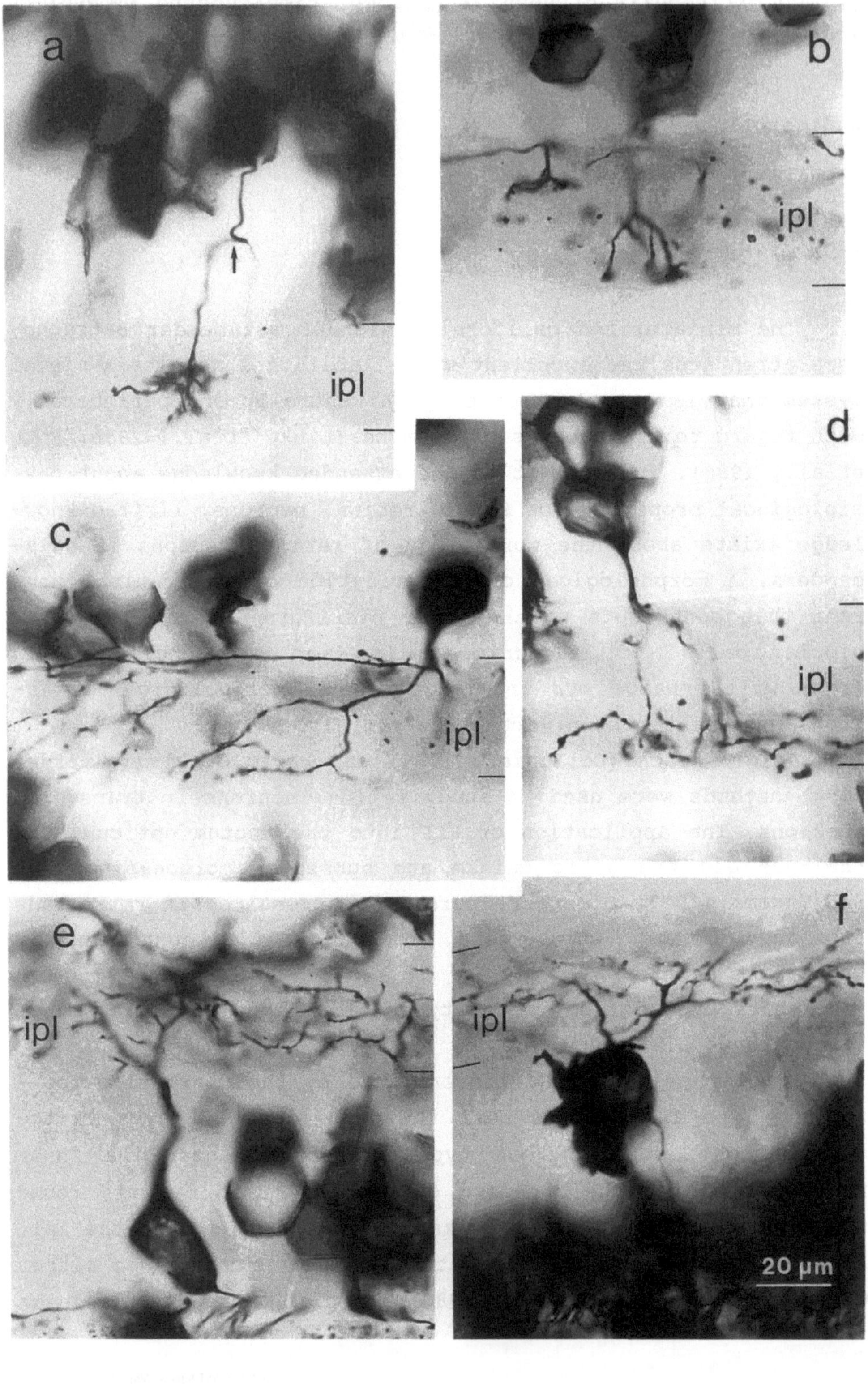
a
ipl
b
ipl
c
ipl
d
ipl
e
ipl
f
ipl
20 µm

Fig. 1: Golgi stained retinal neurons which have processes in the ipl. The photographs are orientated in a manner that the top is towards the outer retina and the bottom towards the inner retina. The inner and outer border of the ipl are marked by small bars. a: Axon of a bc arborizing mostly in the proximal ipl. The arrow points to a bifurcation of the axon. b: Two axons of bc, one arborizing in the proximal and the other in the distal ipl. Note the different types of arborization of the bc axons in the proximal ipl in a and b. c: Broad-field ac with dendrites mostly parallel to the inner and outer margin of the ipl. The dendrites inbetween are orientated nearly perpendicular to the retinal layers. d: Small-field ac with dendrites mostly in the proximal lamina of the ipl. Small appendages are seen near the margin of the inner nuclear layer. e: Small-field rgc with diffuse dendritic arborization and the soma placed at the inner margin of the rgc-layer. d: Broad-field rgc with the soma deep in the rgc-layer and dendritic branching confined to the distal ipl.

Landolt club and show rather large dendritic arborizations suggesting that they receive input from several photoreceptors. A small number of all observed bipolar cells have small dendritic arborizations and seem to receive input only from a single receptor. A Landolt club could not be identified in this type.

Amacrine cells were classified according to their stratification pattern in the ipl. The dendrites of the first type are widely unistratified in the most scleral layer of the ipl. The second type has a very small arborization in layer 2 and 3 of the ipl. The somata of both types are situated close to the border of the ipl in the inner nuclear layer (inl). Type 3 has its cell body somewhat deeper in the inl and its dendrites arborize mostly in layer 5 of the ipl. Additionally, several appendages are seen in layers 1 and 2. Type 4 has a small dendritic arborization with its dendrites mostly in the presumptive ON lamina of the ipl and, too, a small appendage in layer 1 of the ipl. Type 5 has a larger dendritic arborization than type 4. The dendrites are confined to layers 1 and 5. The dendrites of type 6 and 7 arborize in the whole ipl. Whereas in type 6 the dendrites run without any bifurcations in a perpendicular manner through the ipl, the dendrites of type 7 show numerous bifurcations of secondary and tertiary order which run in a tortuous fashion through the whole depth of the ipl thus giving the type 7 ac a bushy appearence.

Retina ganglion cells. From both Golgi and HRP material at least 4 classes of rgc could be distinguished by both size and pattern of dendritic arborization. Class-1 rgc have small dendritic fields with a dense pattern of dendritic arborization. Numerous small swellings are seen on the dendrite, which diffusely arborizes in the ipl. Class-2 rgc mostly have smooth dendrites with small swellings only at the finest tips of the dendrite. The dendritic fields are larger than those of class-1 cells. The cells of this class arborize in a bistratified, tristratified, or diffuse manner in the ipl. Class-3 consists of cells which have medium to large dendritic fields with smooth, thin dendrites which sparsely branch in the proximal ipl only. Class-4 rgc have large dendritic fields which dendrites arborize only in layer 1 of the ipl. The dendrites are mostly smooth. Sometimes small spiny protrusions can be seen. Small swellings occur only at the finest tips of the dendrites.

This study indicates that there are more different morphological types of ac and rgc (and perhaps bc) in the salamander retina than have hitherto been described. From studies of retinal neurons in the tiger salamander and mudpuppy it is known that different types of bc, ac, and rgc exist in the salamander retina and that these types arborize according to the proposed ON/OFF dichotomy of the ipl (Vallerga, 1981; Hare et al., 1986; Yang and Yazulla, 1986; Arkin and Miller, 1988), i.e. ON cells arborize in the proximal lamina of the ipl, OFF cells arborize in the distal lamina and ON-OFF cells arborize in both (Vallerga and Usai, 1986). No morphological details are available for the physiologically defined "hybrid" rgc type of Wunk and Werblin (1979).

ON and OFF rgc are mainly driven by excitatory input from ON and OFF bc, respectively, whereas in ON-OFF and "hybrid" rgc the excitatory input cannot simply be correlated with the excitatory activity of any presynaptic cell. The separate ON and OFF pathways are reflected by the distribution of the bc axons in the ipl. The differing appearence of ac and rgc dendrites in the ipl of these salamanders is comparable with the situation found in frogs (Ramon y Cajal, 1933; Frank and Hollyfield, 1987) and suggests an important functional role of single ac and rgc cell types in this salamander retina similar to the situation found in higher vertebrates.

Acknowledgement: The results are part of the Ph.D. thesis of the author. I want to thank Prof. Roth for constant encouragement and support during the course of this study.

Literature

Adams JC (1977) Technical considerations on the use of horseradish peroxidase as a neuronal marker. Neuroscience 2:141.
Arkin MS, Miller RF (1988) Mudpuppy retinal ganglion cell morphology revealed by an HRP impregnation technique which provides golgi-like staining. J Comp Neurol 270:185-208.
Collonier, M (1964) The tangential organization of the visual cortex. J Anat 98:327-344.
Dowling JE (1968) Synaptic organization of the frog retina: An electron microscopic analysis comparing the retinas of frogs and primates. Proc R Soc Lond Ser B 170:205-227.
Dowling JE, Werblin FS (1969) Organization of the retina of the mudpuppy, Necturus maculosus. I. Synaptic structure. J Neurophysiol 32:315-338.
Frank BD, Hollyfield JG (1987) Retinal ganglion cell morphology in the frog Rana pipiens. J Comp Neurol 266: 413-434.
Hare WA, Lowe JS, Owen G (1986) Morphology of physiologically identified bipolar cells in the retina of the tiger salamander Ambystoma tigrinum. J Comp Neurol 252:130-138.
Linke R, Roth G, Rottluff B (1986) Comparative studies on the eye morphology of lungless salamanders, Family Plethodontidae, and the effect of miniaturization. J Morphol 189:131-143.
Ramon y Cajal, S (1933) The structure of the retina. Transl. by SA Thorpe, M Glickstein (1972), Springfield, Ill., Thomas.
Roth G, Rottluff B, Linke R (1988) Miniaturization, genome size and the origin of functional constraints in the visual system of salamanders. Naturwiss. 75: in press.
Vallerga S, Usai C (1986) Relation between light responses and dendritic branching in the salamander ganglion cells. Exp Biol 45:81-91.
Valverde F (1970) The Golgi method. A tool for comparative structural analyses. In: WJH Nauta, SOE Ebbesson (eds) Contemporary Research Methods in Neuroanatomy. Berlin Heidelberg New York, Springer, pp 12-28.
Wong-Riley MTT (1974) Synaptic organization of the inner plexiform layer in the retina of the larval tiger salamander. J Neurocytol 3:1-33.
Wunk DF, Werblin FS (1979) Synaptic inputs to the ganglion cells in the tiger salamander retina. J Gen Physiol 79:265-286.
Yang C-Y, Yazulla S (1986) Neuropeptide-like immunoreactive cells in the retina of the larval tiger salamander: Attention to the symmetry of dendritic projections. J Comp Neurol 248:105-118.

ARE ORTHOGONAL ARRAYS OF PARTICLES IN RETINAL MÜLLER (GLIAL) CELLS RELATED TO K⁺ CHANNELS?

Karin Berg and Hartwig Wolburg
Institute of Pathology, University of Tübingen
Liebermeisterstraße 8
D-7400 Tübingen, FRG

Introduction

The Müller cells (MC) of the vertebrate retina are radial glial elements extending from the inner to the outer limiting membrane. MC are believed to be related to astroglial cells; however, occurrence and distribution of both glial cell types in the retina differ considerably. Astrocytes are strictly confined to the nerve fiber layer and to the retinal areas supplied by blood vessels. Outside the vascularized areas MC fulfill the perivascular functions of astrocytes.

According to the proposal of Newman astrocytes and MC take up K^+ ions where the neuronal or synaptic activity is increased and K^+ is released out of the neurons. In the retina, this is predominantly the case in the two plexiform layers. K^+ is postulated to be released out of the glial cell into large extracellular spaces which serve as K^+ sinks (Newman et al 1984). Such spaces are capillaries, the subarachnoidal space or the vitreous body. In fact, the membranes of glial endfeet contacting capillaries or the vitreous body are characterized by an increased K^+ conductance (Newman 1986, 1987).

From freeze-fracture observations on astroglial cells it is well-known that the most conspicuous compounds of the glial membrane, the orthogonal arrays of particles (OAP), are distributed non-uniformly over the cell surface. Opposite perivascular or subarachnoidal spaces OAP are accumulated in the glial membrane; within the neuropil where astrocytes and neurons or astrocytes and myelin sheaths contact each other the OAP expression is diminished (Massa and Mugnaini 1982). Since

NATO ASI Series, Vol. H31
Neurobiology of the Inner Retina
Edited by R. Weiler and N. N. Osborne
© Springer-Verlag Berlin Heidelberg 1989

the distribution of the OAP in astrocytes resembles that of the K^+ conductance the question arises whether this situation is confirmed in MC. From the review of Raviola (1977) it was known that OAP occur in MC of mammals. Wolburg and Berg (1987, 1988) described the morphology of MC in the retina of the rabbit and the mouse using freeze-fracturing. In this paper, we give a short comparative overview over the ultrastructure of the vertebrate MC membrane and the most recent status of the discussion about the putative functional role of the OAP.

Results

Because of the geometrical constraints in the retina the convenient exposition of the inner membranes of MC on the one and their paravitreous endfeet on the other hand requires different orientations of the fracture plane. Transversal fracture through the frozen retina results in the visualization of the MC membranes in the OPL, INL and IPL. The paravitreous endfeet are mostly not exposed, because the fracture plane runs through the cytoplasm of the endfoot and not through its membrane. These membranes can be inspected only by fracturing the retina parallel or slightly tilted to the layers. The fracture plane dips from the vitreous body gently into the inner retinal regions, exposing a variously extended mosaic-like pattern of paravitreous MC endfeet the size and OAP-density of which are easily to determine.

Rabbit

The retina of the rabbit differs from that of most other mammals in possessing an intraretinal portion of the optic nerve. This results in a considerable increase in retinal thickness where the bundles of myelinated nerve fibers (meduallary rays) run over the ganglion cell layer. The medullary rays are the only site where astrocytes surround blood vessels (Schnitzer 1988). Outside the medullary rays MC are the only glial elements. In freeze-fracture replicas, only the paravitreous endfeet reveal OAP, the longer the MC (within the medullary rays) the higher their OAP density. Most important is the observation that the OAP expression requires the real contact with the vitreous body: already a slight deflection of a given endfoot membrane away from the vitreous body results in an abrupt disappearance of the OAP. Consequently, in inner regions of the retina, MC membranes which can be identified because of their high density of intramembranous particles (IMP) lack OAP completely.

Mouse

The vascularized mouse retina is, in contrast to the rabbit retina, equal in thickness throughout the eye cup. Therefore, the length of the MC is equal everywhere. In the nerve fiber layer, astrocytes, and, in deeper layers, MC surround the capillaries.

The OAP density of MC membranes is very high in paravitreous as well as in pericapillary endfeet. There is no difference between MC endfeet from peripheral and central areas. In contrast to the rabbit, also the non-endfoot membranes reveal a relatively high, although decreased, OAP density (see Fig.1).

Turtle

The turtle (*Pseudemys scripta elegans*) possesses MC with a high density of OAP in their paravitreous endfeet membranes which decreases considerably in inner MC-membranes.

Axolotl

In the MC membranes of the Axolotl (albino form of *Ambystoma mexicanum*) we found a relatively high density of OAP where endfeet contacted the vitreous body. In the deeper regions of the retina, the OAP-density in MC membranes dropped to low values, but not to zero. Unfortunately, we had no opportunity so far to look for the astrocytes outside the retina; however, all amphibia investigated have OAP-negative astrocytes (Korte and Rosenbluth 1981; Wujek and Reier 1984; Kästner 1987).

Goldfish

In the goldfish (*Carassius auratus*) retina, we found in the membranes of the MC paravitreous endfeet a high density of OAP.

Again, within deeper layers only few OAP were expressed. As in the Axolotl, this finding is unexpected, because astrocytes in goldfish lack OAP completely (Wolburg et al 1983; Kästner 1987).

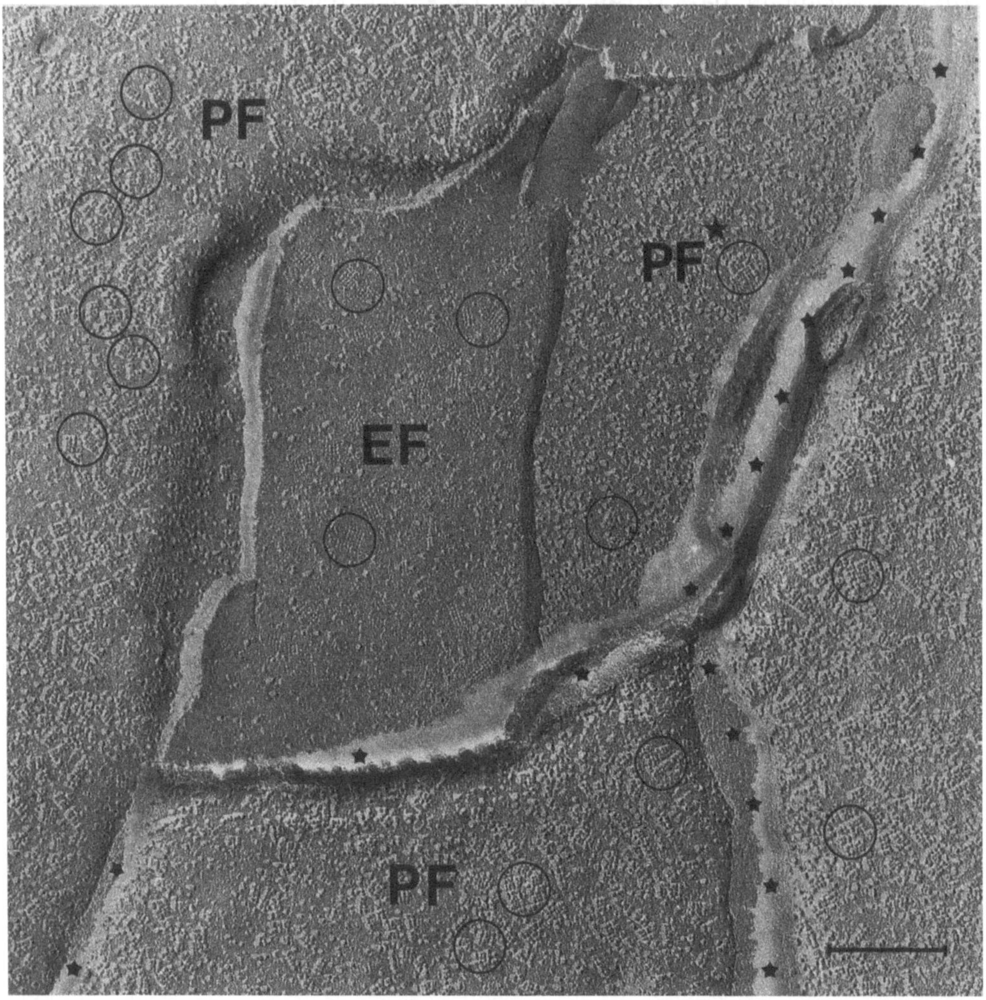

Fig.1: Freeze-fracture replica of several MC endfeet membranes of the mouse retina. PF P-face of the endfeet contacting directly the vitreous body. PF* labels that part of the left endfoot which is indented and deflected away from the direct contact with the vitreous by another endfoot the E-face (EF) of which is shown. Note the difference in OAP densities between areas with and without direct MC-vitreous contact. Some OAP are circled. The lines of asterisks label the border between adjacent endfeet. Bar represents 0.2 µm.

Discussion

With respect to the expression of OAP in retinal MC, one general principle is realized in all species investigated so far: where the MC contact mesenchymal spaces outlined by a basal lamina, OAP expression is stimulated (Fig.1); where the membranes contact the non-mesenchymal compartment, OAP expression is more or less inhibited. Further, the OAP-density on the MC endfeet in a given retina seems to correlate positively with the MC length. In retinae of equal thickness the OAP-densities at the MC endfeet do not differ in different areas.

Unfortunately, nothing is known about the biochemical identity of the OAP or their physiological functions. However, it is interesting that the K^+ conductance of the MC membrane (Newman 1987) is codistributed with OAP: in the rabbit, both the high K^+ conductance and the OAP are confined to the paravitreous endfeet. In the MC of the vascularized mouse retina, the K^+ conductance is distributed all over the membrane; the OAP are found in non-endfoot membranes, too. This provides indirect evidence for the hypothesis that OAP may be related to sites of elevated K^+ conductance. Nilius and Reichenbach (1988) described in the endfoot membrane of the rabbit a non-rectifying very high conductance (VHC) K^+ channel. Unfortunately, the authors were not able to form patches at the "true" endfoot membrane contacting the vitreous body. Since the non-paravitreous endfoot membrane occupied by the VHC channel do not express OAP, it seems unlikely that OAP are related to the VHC channel. Brew et al (1986) described in the MC endfoot membrane of the Axolotl a rectifying low conductance K^+ channel. If this channel would be related to OAP (and the OAP of different species would be functionally identical!), it should be present also in the "true" endfoot membrane of rabbit MC. However, a low-conductance rectifier channel in the rabbit MC does not codistribute with OAP (Wolburg and Berg, 1987; Nilius and Reichenbach, 1988).

It is generally accepted that the efficiency of spatial buffering in the CNS is more advanced in mammals than in lower vertebrates. However, it is completely unknown, how the channels involved in spatial buffering cooperate; the quality of channel cooperation may determine the efficiency of spatial buffering. What we need, is the comparative investigation of the physiology and biochemistry of both the K^+ channels and the OAP of astrocytes and MC in different species, and the explanation of the enigmatic observation that in fish and amphibia membranes of astrocytes and MC differ morphologically, although exerting equivalent functions.

References

Brew H, Gray PTA, Mobbs P, Attwell D (1986) Endfeet of retinal glial cells have higher densities of ion channels that mediate K^+ buffering. Nature 324: 466-468

Kästner R (1987) Comparative studies on the astrocytic reaction in the lesioned central nervous system of different vertebrates. J Hirnforsch 28: 221-232

Korte GE, Rosenbluth J (1981) Ependymal astrocytes in the frog cerebellum. Anat Rec 199: 267-279

Newman EA (1986) High potassium conductance in astrocyte endfeet. Science 233: 453-455

Newman EA (1987) Distribution of potassium conductance in mammalian Müller (glial) cells: a comparative study. J Neurosci 7: 2423-2432

Newman EA, Frambach DA, Odette LL (1984) Control of extracellular potassium levels by retinal glial cell K^+ siphoning. Science 225: 1174-1175

Nilius B, Reichenbach A (1988) Efficient K^+ buffering by mammalian retinal glial cells is due to cooperation of specialized ion channels. Pflügers Arch 411: 654-660

Massa PT, Mugnaini E (1982) Cell junctions and intramembrane particles of astrocytes and oligodendrocytes. A freeze-fracture study. Neurosci 7: 523-538

Raviola G (1977) The structural basis of the blood-ocular barriers. Exp Eye Res Suppl: 27-63

Schnitzer J (1988) Astrocytes in mammalian retina. Progr in Ret Res 7: 209-231

Wolburg H, Kästner R, Kurz-Isler G (1983) Lack of orthogonal particle assemblies and presence of tight junctions in astrocytes of goldfish. A freeze-fracture study. Cell Tissue Res 234: 389-402

Wolburg H, Berg K (1987) Heterogeneity of Müller cell endfeet in the rabbit retina as revealed by freeze-fracturing. Neurosci Letters 82: 273-277

Wolburg H, Berg K (1988) The distribution of orthogonal arrays of particles in the Müller cell membrane of the mouse retina. Glia (in press)

Wujek JR, Reier PJ (1984) Astrocytic membrane morphology: differences between mammalian and amphibian astrocytes after axotomy. J Comp Neurol 222: 607-620

SUBJECT INDEX

NATO ASI Series H

Vol. 1: **Biology and Molecular Biology of Plant-Pathogen Interactions.** Edited by J.A. Bailey. 415 pages. 1986.

Vol. 2: **Glial-Neuronal Communication in Development and Regeneration.** Edited by H.H. Althaus and W. Seifert. 865 pages. 1987.

Vol. 3: **Nicotinic Acetylcholine Receptor: Structure and Function.** Edited by A. Maelicke. 489 pages. 1986.

Vol. 4: **Recognition in Microbe-Plant Symbiotic and Pathogenic Interactions.** Edited by B. Lugtenberg. 449 pages. 1986.

Vol. 5: **Mesenchymal-Epithelial Interactions in Neural Development.** Edited by J.R. Wolff, J. Sievers, and M. Berry. 428 pages. 1987.

Vol. 6: **Molecular Mechanisms of Desensitization to Signal Molecules.** Edited by T.M. Konijn, P.J.M. Van Haastert, H. Van der Starre, H. Van der Wel, and M.D. Houslay. 336 pages. 1987.

Vol. 7: **Gangliosides and Modulation of Neuronal Functions.** Edited by H. Rahmann. 647 pages. 1987.

Vol. 8: **Molecular and Cellular Aspects of Erythropoietin and Erythropoiesis.** Edited by I.N. Rich. 460 pages. 1987.

Vol. 9: **Modification of Cell to Cell Signals During Normal and Pathological Aging.** Edited by S. Govoni and F. Battaini. 297 pages. 1987.

Vol. 10: **Plant Hormone Receptors.** Edited by D. Klämbt. 319 pages. 1987.

Vol. 11: **Host-Parasite Cellular and Molecular Interactions in Protozoal Infections.** Edited by K.-P. Chang and D. Snary. 425 pages. 1987.

Vol. 12: **The Cell Surface in Signal Transduction.** Edited by E. Wagner, H. Greppin, and B. Millet. 243 pages. 1987.

Vol. 13: **Toxicology of Pesticides: Experimental, Clinical and Regulatory Perspectives.** Edited by L.G. Costa, C.L. Galli, and S.D. Murphy. 320 pages. 1987.

Vol. 14: **Genetics of Translation. New Approaches.** Edited by M.F. Tuite, M. Picard, and M. Bolotin-Fukuhara. 524 pages. 1988.

Vol. 15: **Photosensitisation. Molecular, Cellular and Medical Aspects.** Edited by G. Moreno, R.H. Pottier, and T.G. Truscott. 521 pages. 1988.

Vol. 16: **Membrane Biogenesis.** Edited by J.A.F. Op den Kamp. 477 pages. 1988.

Vol. 17: **Cell to Cell Signals in Plant, Animal and Microbial Symbiosis.** Edited by S. Scannerini, D. Smith, P. Bonfante-Fasolo, and V. Gianinazzi-Pearson. 414 pages. 1988.

Vol. 18: **Plant Cell Biotechnology.** Edited by M.S.S. Pais, F. Mavituna, and J.M. Novais. 500 pages. 1988.

Vol. 19: **Modulation of Synaptic Transmission and Plasticity in Nervous Systems.** Edited by G. Hertting and H.-C. Spatz. 457 pages. 1988.

Vol. 20: **Amino Acid Availability and Brain Function in Health and Disease.** Edited by G. Huether. 487 pages. 1988.

NATO ASI Series H

Vol. 21: **Cellular and Molecular Basis of Synaptic Transmission.**
Edited by H. Zimmermann. 547 pages. 1988.

Vol. 22: **Neural Development and Regeneration. Cellular and Molecular Aspects.**
Edited by A. Gorio, J.R. Perez-Polo, J. de Vellis, and B. Haber. 711 pages. 1988.

Vol. 23: **The Semiotics of Cellular Communication in the Immune System.**
Edited by E.E. Sercarz, F. Celada, N.A. Mitchison, and T. Tada. 326 pages. 1988.

Vol. 24: **Bacteria, Complement and the Phagocytic Cell.**
Edited by F.C. Cabello und C. Pruzzo. 372 pages. 1988.

Vol. 25: **Nicotinic Acetylcholine Receptors in the Nervous System.**
Edited by F. Clementi, C. Gotti, and E. Sher. 424 pages. 1988.

Vol. 26: **Cell to Cell Signals in Mammalian Development.**
Edited by S.W. de Laat, J.G. Bluemink, and C.L. Mummery. 322 pages. 1989.

Vol. 27: **Phytotoxins and Plant Pathogenesis.**
Edited by A. Graniti, R.D. Durbin, and A. Ballio. 508 pages. 1989.

Vol. 28: **Vascular Wilt Diseases of Plants. Basic Studies and Control.**
Edited by E.C. Tjamos and C.H. Beckman. 590 pages. 1989.

Vol. 29: **Receptors, Membrane Transport and Signal Transduction.**
Edited by A.E. Evangelopoulos, J.P. Changeux, L. Packer, T.G. Sotiroudis, and K.W.A. Wirtz. 387 pages. 1989.

Vol. 30: **Effects of Mineral Dusts on Cells.**
Edited by B.T. Mossman and R.O. Bégin. 470 pages. 1989.

Vol. 31: **Neurobiology of the Inner Retina.**
Edited by R. Weiler and N.N. Osborne. 529 pages. 1989.

Vol. 32: **Molecular Biology of Neuroreceptors and Ion Channels.**
Edited by A. Maelicke. 675 pages. 1989.